AF745373

Advances in Environmental Microbiology

Volume 3

Series Editor
Christon J. Hurst
Cincinnati, Ohio
USA

and

Universidad del Valle
Santiago de Cali, Valle
Colombia

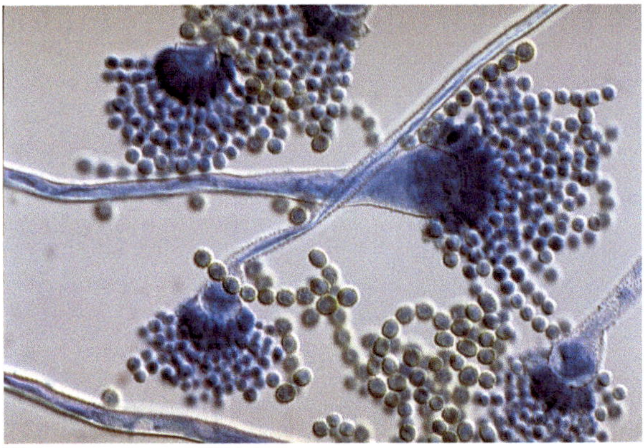

Aspergillus flavus. Courtesy of Hossein Mirhendi

More information about this series at http://www.springer.com/series/11961

Christon J. Hurst
Editor

The Rasputin Effect: When Commensals and Symbionts Become Parasitic

Springer

Editor
Christon J. Hurst
Cincinnati, Ohio
USA

Universidad del
Valle Santiago de Cali
Valle, Colombia

ISSN 2366-3324 ISSN 2366-3332 (electronic)
Advances in Environmental Microbiology
ISBN 978-3-319-28168-1 ISBN 978-3-319-28170-4 (eBook)
DOI 10.1007/978-3-319-28170-4

Library of Congress Control Number: 2016935618

© Springer International Publishing Switzerland 2016
This work is subject to copyright. All rights are reserved by the Publisher, whether the whole or part of the material is concerned, specifically the rights of translation, reprinting, reuse of illustrations, recitation, broadcasting, reproduction on microfilms or in any other physical way, and transmission or information storage and retrieval, electronic adaptation, computer software, or by similar or dissimilar methodology now known or hereafter developed.
The use of general descriptive names, registered names, trademarks, service marks, etc. in this publication does not imply, even in the absence of a specific statement, that such names are exempt from the relevant protective laws and regulations and therefore free for general use.
The publisher, the authors and the editors are safe to assume that the advice and information in this book are believed to be true and accurate at the date of publication. Neither the publisher nor the authors or the editors give a warranty, express or implied, with respect to the material contained herein or for any errors or omissions that may have been made.

Cover illustration: "Volvox reimagined" is a montage image created in 2015 and is being used with permission of the artist, Christon J. Hurst. Those images incorporated into this montage were: center image "Aspergillus flavus" (author: Hossein Mirhendi, image supplied by and used with author's permission); clockwise from upper right "Micrasterias" (author: Rogelio Moreno Gill, image supplied by and used with author's permission), "BrownGiantKelp3600ppx" (author: FASTILY, Creative Commons Attribution-Share Alike 3.0 Unported license), "Stumpfungus" (author: Ecornerdropshop, public domain image), "Letharia vulpina JHollinger crop" (author: Jason Hollinger, Creative Commons Attribution-Share Alike 3.0 Unported license), "Phytoplankton - the foundation of the oceanic food chain", (author: NOAA MESA Project, public domain image), "Spirogyra cell" (author: Jasper Nance, Creative Commons Attribution-Share Alike 3.0 Unported license), and "Morelasci" (author: Peter G. Werner, Creative Commons Attribution 3.0 Unported license).

Printed on acid-free paper

This Springer imprint is published by Springer Nature
The registered company is Springer International Publishing AG Switzerland

Dedication

I met Donald Berman through one of his daughters when both she and I were undergraduate students at the University of Cincinnati. She knew that I was interested in studying viruses and told me her father did that kind of work. I met her father soon afterwards and he undertook the task of encouraging my pursuit of virology. One summer, I took some time away from my undergraduate job of making flavorings and fragrances to instead learn from Don about propagating polioviruses and performing viral plaque assays. My first day spent working with him was interesting in two ways, one of which held true for the rest of my career in science. That afternoon we had birthday cake because it happened to be the birthday of someone in the virology group. Don and I stayed and worked far beyond normal quitting time on that day, not because we had been eating cake but simply because laboratory research always seems to take longer than you optimistically anticipate. The concept of having cake in the afternoon turned out not to be a normal part of the work days in science. My understanding that laboratory research always took more time than anticipated did continue to hold true for all of the years that were to follow.

I spent many happy hours of my undergraduate years working with Don, and I appreciate that his family always welcomed me very kindly into their home as if I naturally belonged there. Don also helped by guiding me to graduate school for studying virology. After I finished my formal education, Don helped me to find a job where he was employed, and I then happily anticipated seeing him each workday for perhaps an additional 15 years until he retired. I wish that I could relive the summer when I learned to do plaque assays by working alongside Don. Instead, since reliving that summer is not possible, I will derive pleasure from my remembrances and in gratitude I dedicate this book to him.

Donald Berman (1925–2011)

Series Preface

The light of natural philosophy illuminates many subject areas including an understanding that microorganisms represent the foundation stone of our biosphere by having been the origin of life on Earth. Microbes therefore comprise the basis of our biological legacy. Comprehending the role of microbes in this world which together all species must share, studying not only the survival of microorganisms but as well their involvement in environmental processes, and defining their role in the ecology of other species, does represent for many of us the Mount Everest of science. Research in this area of biology dates to the original discovery of microorganisms by Antonie van Leeuwenhoek, when in 1675 and 1676 he used a microscopc of his own creation to view what he termed "animalcula," or the "little animals" which lived and replicated in environmental samples of rainwater, well water, seawater, and water from snow melt. van Leeuwenhoek maintained those environmental samples in his house and observed that the types and relative concentrations of organisms present in his samples changed and fluctuated with respect to time. During the intervening centuries we have expanded our collective knowledge of these subjects which we now term to be environmental microbiology, but easily still recognize that many of the individual topics we have come to better understand and characterize initially were described by van Leeuwenhoek. van Leeuwenhoek was a draper by profession and fortunately for us his academic interests as a hobbyist went far beyond his professional challenges.

It is the goal of this series to present a broadly encompassing perspective regarding the principles of environmental microbiology and general microbial ecology. I am not sure whether Antonie van Leeuwenhoek could have foreseen where his discoveries have led, to the diversity of environmental microbiology subjects that we now study and the wealth of knowledge that we have accumulated. However, just as I always have enjoyed reading his account of environmental microorganisms, I feel that he would enjoy our efforts through this series to summarize what we have learned. I wonder, too, what the microbiologists of still future centuries would think of our efforts in comparison with those now unimaginable discoveries which they will have achieved. While we study the many

Christon J. Hurst in Heidelberg

wonders of microbiology, we also further our recognition that the microbes are our biological critics, and in the end they undoubtedly will have the final word regarding life on this planet.

Indebted with gratitude, I wish to thank the numerous scientists whose collaborative efforts will be creating this series and those giants in microbiology upon whose shoulders we have stood, for we could not accomplish this goal without the advantage that those giants have afforded us. The confidence and very positive encouragement of the editorial staff at Springer DE has been appreciated tremendously and it is through their help that my colleagues and I are able to present this book series to you, our audience.

Cincinnati, OH Christon J. Hurst

Volume Preface

Our goal as the authors of this book is to share a collective understanding that normally benign interspecies relationships do sometimes undergo changes whereby those relationships become detrimental.

Biologists have assigned a variety of definitions to each of the terms commensalism, symbiosis, and parasitism, with those definitions seeming to cross paths and disagree equally as do the biologists. We generally tend to view commensal relationships as being associations without inherent obligation, and for which there is no definable cost to any participant although some beneficial enticements can be involved. Symbionts are partners, by strict definition, with the organisms living together in a joint existence which seems more tightly involved and perhaps more mutually beneficial as compared to a commensal relationship. Symbiotic relationships often are so involved as to seem nearly obligatory for the participant species. But still, the principle assumption remains that both participants in a symbiotic association are deriving benefit from the relationship rather than receiving harm. In those cases where enticements are offered to favor the interspecies relationship, and often those enticements are nutritional, the term host may be used to help describe the major provider. Any energetic cost paid by the host to support presence of either commensal or symbiotic species presumably is outweighed by the beneficial and often protective nature of such relationships, with commensals and symbionts sometimes serving either to prevent or restrict the presence of other organisms that may be less favorably described as parasitic. Not all guests are welcome, and some initially may be considered benign but subsequently lose their welcome. The ecological definition of parasitism includes those less favorable situations that occur when a guest species obviously becomes deleterious. Thus, the dividing distinction between parasitism and these other types of interspecies relationships becomes a matter of detriment to the host.

Those microorganisms which normally might be considered either benign or even beneficial, but opportunistically become far more dangerous, very often are represented under the broad term 'opportunistic pathogens'. However, rather than simply relying upon that term as a general cliché, the purpose of this book is helping to explain the current state of knowledge regarding conditions and mechanisms which either allow or facilitate opportunistic pathogenicity. The trigger which allows that change can come in many ways. Sometimes, the effect results from a change in the host's capacity for mounting an effective immune response due to factors such as nutritional deprivation and coinfections. At other times, virus species either may have changed the opportunist or attacked the host's protective natural microflora. Even seemingly subtle environmental changes such as the amount of available sunlight, temperature, water and air quality parameters, can be enough to trigger dramatic shifts in delicately balanced interspecies relationships. The result of those shifts can be perceived as either a temporary bonanza for the pathogen or a disaster for the host. Knowledge regarding the nature of interactions which represent opportunistic pathogenicity in any single host–guest relationship valuably may then assist us towards unlocking the mystery of opportunistic pathogenicity for yet other systems.

We hope that you, our audience, will continue to carry forward the goal and purpose of this knowledge and of these efforts.

I am tremendously grateful to Hanna Hensler-Fritton, Andrea Schlitzberger, and Isabel Ullmann at Springer DE, for their help and constant encouragement which has enabled myself and the other authors to achieve publication of this collaborative project.

Cincinnati, OH Christon J. Hurst

Contents

Part I
Understanding Interspecies Relationships

Chapter 1
How Well Do Surrogate Hosts Serve as Model Systems for Understanding Pathogenicity

Christine Fink and Thomas Roeder

Abstract Experimental infection studies are of crucial importance to find and characterize virulence factors of pathogens or to identify novel compounds that can be used to treat the corresponding infections. The use of mammalian infection models including mice, rats, and guinea pigs is restricted due to several reasons including high costs, low statistical power, and ethical reservations. Simple, invertebrate models have been introduced as surrogate hosts as they are inexpensive, they can be used in great numbers, and doing experiments with them is not accompanied by ethical reservations. The soil nematode *Caenorhabditis elegans* and the fruit fly *Drosophila melanogaster* have served as the most important surrogate hosts. Both organisms have served as workhorses in various biomedical disciplines. They combine simple and cheap handling and housing with an enormous armamentarium of genetic tools available to the scientific community. As their innate immune systems share substantial similarities with our own one, human bacterial and fungal pathogens often also infect these surrogate hosts. Nevertheless, it has to be kept in mind that both hosts share some drawbacks such as the apparent lack of adaptive immunity or the inability to survive at 37 °C. The latter point is relevant for especially those pathogens that require higher temperatures to become pathogenically triggered, and thus it would be helpful to seek the introduction of alternative models that can be used under these conditions. The greater wax moth *Galleria mellonella* exactly fits into this gap although it lacks most of the benefits supplied by the "classical" model organisms.

C. Fink • T. Roeder (✉)
Department of Zoology, Molecular Physiology, Christian-Albrechts University Kiel, Kiel, Germany
e-mail: troeder@zoologie.uni-kiel.de

© Springer International Publishing Switzerland 2016

C.J. Hurst (ed.), *The Rasputin Effect: When Commensals and Symbionts Become Parasitic*, Advances in Environmental Microbiology 3,
DOI 10.1007/978-3-319-28170-4_1

1.1 Introduction

Experimental infection biology completely relies on having appropriate model systems to elucidate important factors relevant for the pathogen's capacity to colonize and damage the human or animal host. Ideally, these models should mirror the situation found in the customary host (in most cases the human host) as nearly as possible. Thus, mammalian models including mice, rats, rabbits, or guinea pigs that share much of our biology have served as model hosts for a huge variety of different bacterial and fungal pathogens. Although these models are characterized by a common set of advantages, their use in infection experiments is limited due to their huge biophysical complexity, exorbitant housing costs, and, most importantly, ethical reasons aiming to limit the use of mammals for animal experiments (Steinert et al. 2003). Thus, alternative models are required that might serve as valuable hosts, while excluding the drawbacks mentioned above.

One alternative approach that has gained some interest is the use of cell culture systems, ideally based on immortalized human cell lines that open the possibility to study at least some aspects of the pathogen's virulence and mechanisms of infection. Unfortunately, cell culture-based systems cannot represent the different levels of the complex interaction between host and pathogen. Thus, whole-animal-based infection systems are required to cover all major aspects relevant for this complex interaction of host and pathogen. Simple, nonvertebrate model organisms have come into the focus of experimental infection biology as they combine the advantages of cell culture systems (low costs, high ethical acceptance) with a whole-animal setting incorporating all major aspects of the infection process. Using invertebrate surrogate hosts became reasonable because the important findings of the last decades have revealed that the great majority of signaling pathways relevant for system development, including tissue homeostasis and innate immune responses, apparently evolved before vertebrates and invertebrates evolutionarily split. Thus, these important systems are conserved throughout the majority of the animal kingdom (Hemmrich et al. 2007; Salzet 2001). Moreover, those organisms that are genetically tractable open the opportunity to use yet that additional feature for mechanistic studies. This latter idea emanated almost 40 years ago, when the slime mold *Dictyostelium discoideum* was proposed to serve as a valuable model host (Depraitere and Darmon 1978). This simple, mostly unicellular organism has since been used to study the infection process and the virulence of intracellular bacteria such as *Legionella pneumophila* and *Listeria monocytogenes* (Steinert et al. 2003).

More recently, the use of more complex invertebrate models became popular, with the soil nematode *Caenorhabditis elegans* and the fruit fly *Drosophila melanogaster* as their most important representatives. The major advantage relative to working with these two species is the unmatched wealth of information available for them combined with the easiness of their genetic manipulation, which is a prerequisite for understanding molecular infection mechanisms. Both organisms share a huge list of advantages including low-cost culturing, adaptability to high-throughput systems, immune systems with similarities to our own innate immune system, fully sequenced genomes, and, most importantly, a plethora of available mutants and genetic tools allowing us to manipulate almost every gene in these model hosts. Moreover, both model organisms share a common lifestyle characterized by feeding on microorganisms such as bacteria and yeast. This natural way of confrontation with microorganisms predestines these models for infection experiments via the oral route. Although both organisms share this large list of pros, some cons have to be kept in mind. Neither worms nor flies are humans; they have a significantly different physiology and biology. Relevant for infection biology is that both seemingly lack any signs of adaptive immunity. Although this appears to represent an important drawback, most infections have to be managed, at least in the initial time of infection, by our innate immune system alone, and that result has its counterpart in these simple organisms. Moreover, the virulence of pathogens mainly relies on their ability to specifically interact with mucosal surfaces to reside there or to enter the body. The underlying processes involved with initiating infection appear to be well conserved from invertebrates to mammals (Steinert et al. 2003). Although using invertebrate surrogate hosts in experimental infection studies has been a success story, the limitations of the systems have always to be kept in mind. For instance, temperature-sensitive virulence factors that are operative only at the human body temperature of 37 °C could not be studied in the two model organisms *Drosophila melanogaster* and *Caenorhabditis elegans* as this would represent an abnormally elevated body temperature for those two invertebrate model organisms, and prolonged exposure of them to this elevated temperature is lethal. Nevertheless, pathogens adapted to this temperature can be studied in alternative models such as the greater wax moth *Galleria mellonella* that has been introduced into the field to exactly fill this gap (Lionakis 2011). Using one or more of these invertebrate hosts that tolerate higher temperature for infection studies offers not only the advantages mentioned above, but moreover it allows us to identify if a certain virulence factor is species specific or if it is relevant for a broad host range. In the following text, we want to focus on these three model hosts with respect to their recent and future contributions for the field of experimental infection biology.

1.2 *Caenorhabditis elegans*

Caenorhabditis elegans is a small free-living soil nematode that lives in temperate environments all over the world. Most individuals are hermaphrodites that self-fertilize. Only a small percentage of males can be found in natural populations. *Caenorhabditis elegans* had been introduced in 1974 by Sydney Brenner as a versatile model for developmental biology and neurobiology (Brenner 1974), an endeavor that was honored with the Nobel Prize for Physiology or Medicine in 2002 (Brenner 2003). *Caenorhabditis elegans* is small (1 mm) and translucent and can easily be cultivated on petri dishes using *Escherichia coli* or other bacteria as a food source. These very simple growth conditions combined with the rapid generation time (2–3 days) and the incredibly high number of available mutants make this model ideally suited for a great number of study fields in biomedical research (Riddle et al. 1997). In addition, the ease of producing transgenic mutants or applying of RNAi targeted against endogenous target genes makes *C. elegans* ideally suited for experimental infection studies. Moreover, the extremely rich and versatile supportive resources available to all researchers make it easy to start working with this model (http://www.wormbook.org). Among the novel technical advantages that have become available in the most recent years is knowledge of the whole set of this species' genes as a functional study tool (Ashrafi et al. 2003; Poole et al. 2011; Yanos et al. 2012). Coming along in parallel with these technical breakthroughs, high-throughput approaches that can be used for infection studies or pharmacological studies have become popular (Burns et al. 2006; Okoli et al. 2009). In order to understand *C. elegans* as something more than a black box that is used in high-throughput screens, a basic understanding of its anatomy and immune system is expedient. Although *C. elegans* is a complex metazoan organism, it shows some peculiarities not shared by other invertebrate organisms. Nematodes are eutelic, meaning that adult worms have a constant number of cells that are slightly below 1000 for hermaphrodites and slightly greater than 1000 for males. Consequently, the organ composition of *C. elegans* is very simple. Most important for all aspects of the immune response is the gastrointestinal tract that encompasses a complex pharynx and a very simple intestine. As the lifestyle of *C. elegans* depends upon grazing on microorganisms, it is at constant risk of becoming infected via the oral route. Thus, it is especially well suited for all pathogens that usually infect humans via the oral route (Hilbi et al. 2007). To fight these potential pathogens that may be ingested by the nematode, a sophisticated, innate immune system is active (Irazoqui et al. 2010; Marsh and May 2012). It comprises signaling systems similar to the mammalian TLR-p38 MAPK and TGF-β pathways and an array of antimicrobial peptide compounds such as nlp-29 and cnc-2 (Pujol et al. 2008), as well as the large peptide family of so-called caenopores (Roeder et al. 2010; Irazoqui et al. 2010). Although substantial effort has been invested to elucidate the immune system of the nematode, some very important parts are still unknown, including the proteins that recognize bacterial or fungal components to trigger the above mentioned signaling cascades. For working with *C. elegans* as a surrogate host, one very big advantage is the possibility of using death as the important readout endpoint of an infection. In the following text, we will highlight only a very few examples of the numerous which have been published using *C. elegans* as a surrogate host of human bacterial and fungal pathogens (Table 1.1).

Table 1.1 Selected surrogate host models established in *Caenorhabditis elegans*

Microorganism	Inoculation method	References
Gram-negative bacteria		
Aeromonas hydrophila	Oral uptake	Couillault and Ewbank (2002)
Agrobacterium tumefaciens	Oral uptake	Couillault and Ewbank (2002)
Burkholderia cepacia	Oral uptake	Kothe et al. (2003)
Burkholderia pseudomallei	Oral uptake	O'Quinn et al. (2001)
Pseudomonas aeruginosa	Oral uptake, confrontation	Mahajan-Miklos et al. (1999); Tan et al. (1999b)
Salmonella typhimurium	Oral uptake	Aballay et al. (2000)
Serratia marcescens	Oral uptake	Mallo et al. (2002)
Yersinia spp.	Oral uptake	Darby et al. (2002)
Gram-positive bacteria		
Enterococcus faecalis	Oral uptake	Sifri et al. (2002)
Staphylococcus aureus	Oral uptake	Begun et al. (2005)
Streptococcus pyogenes	Confrontation	Jansen et al. (2002)
Fungi		
Histoplasma spp.	Oral uptake	Muhammed et al. (2012)
Candida albicans	Oral uptake	Okoli et al. (2009); Pukkila-Worley et al. (2011)
Cryptococcus spp.	Oral uptake	Mylonakis et al. (2002)

1.2.1 Infection of C. elegans *with Bacteria*

Caenorhabditis elegans was used as a surrogate host for a number of different human pathogens prior to the eventual identification of natural infection models for this species. Among the very few, naturally occurring bacterial pathogens that can infect the nematode in a natural environment, *Microbacterium nematophilum* is the best-studied example (Hodgkin et al. 2000; Gravato-Nobre et al. 2005). In contrast to the great variety of other pathogens that have been studied in *C. elegans*, *M. nematophilum* infects the anal region rather than attacking the intestine via the oral route. In the anal region, this microorganism induces a protective swelling response. Other naturally occurring pathogens including *Leucobacter* strains also have been identified that possess the potential to infect and kill the nematode in a natural setting (Hodgkin et al. 2013).

A huge list of potential bacterial pathogens encompassing both gram-positive and gram-negative species among which are intracellular bacteria has been studied in different *C. elegans* systems with the aim of understanding both virulence factors of the pathogen and mechanisms used by the host to fight these pathogens (Gravato-Nobre and Hodgkin 2005; Darby 2005). In the following, we want to focus on some pathogens in more detail including "major" human pathogens such as *Staphylococcus aureus* and *Pseudomonas aeruginosa*. The latter of these was among the first bacteria for which virulence mechanisms were studied utilizing *C. elegans*. *Pseudomonas aeruginosa* is a common bacterium found in water and soil. Although it is almost completely innocuous for healthy persons, immunocompromised patients are highly endangered. It can kill *C. elegans* by two completely different mechanisms, a fast killing caused by the secretion of toxins and a slow killing induced by conventional infection processes (Tan et al. 1999a, b). Studies using *P. aeruginosa* may have served a blueprint to utilize *C. elegans* as surrogate host. Researchers have in total tested more than 2000 bacterial mutants regarding their ability to kill *C. elegans*. Based on these studies, transcriptional activators such as LasR have been determined to be one of the quorum sensing systems that represent a major virulence factor (Tan et al. 1999a, b).

Staphylococcus aureus is frequently found on the human skin and within the respiratory tract. In most cases, this colonization is symptomless. The great number of antibiotic-resistant *S. aureus* strains causes huge problems in clinical settings. In *C. elegans*, *S. aureus* can cause a classical intestinal infection characterized by colonization of the intestinal lumen (Sifri et al. 2002). Screening efforts similar to those already described for *P. aeruginosa*, encompassing large mutant libraries, have enabled the identification of a large set of potential bacterial virulence factors (Bae et al. 2004; Begun et al. 2005).

A highly interesting example using infection caused by *Yersinia pestis*, the causative agent of bubonic plague, revealed microbial biofilm production as being a highly relevant virulence factor (Darby et al. 2002). The biofilm is formed around the mouth and subsequently that prevents feeding, which eventually leads to host death. Very similar to this biofilm formation is the situation found in the natural vector of *Yersinia*, which is the flea, where this biofilm formation is required for transmission (Darby et al. 2002).

1.2.2 Infection of C. elegans *by Fungi*

Nematodes have evolved a highly effective armamentarium of antifungal compounds. Upon fungal infection with *Drechmeria coniospora*, a diversified family of potential antifungal peptides is activated, which presumably offers the ability to fight a greater diversity of fungal pathogens (Pujol et al. 2008). Not only naturally occurring pathogens of *C. elegans* are able to trigger an immune response in the nematode following an infection, but also human pathogens such as *Candida albicans* hold this potential. Infection with *C. albicans* has been observed to induce characteristic sets of host genes that appeared to be yeast or fungus specific as very similar responses could be induced by confrontation with heat-killed yeast but not with bacterial pathogens (Pukkila-Worley et al. 2011). Based on comparable studies, a set of different assays has been developed that allow using *C. elegans* as a surrogate host for fungal infections. These include not only the conventional killing assay but also more sophisticated approaches such as the progeny permissive assay and the antifungal compound assay (Muhammed et al. 2012).

Thus, three major lines of assays have been used: those that focus on the fungal site utilizing panels of mutants, those that aim to identify novel antifungal components using infection models, and those that focus on the host site that aim to identify novel antifungal compounds (Anastassopoulou et al. 2011).

The human pathogens *Candida albicans* and *Candida neoformans* have been extensively used to infect *C. elegans* in different experimental settings, aiming to identify different virulence factors. Both fungi are able to establish a lethal infection in the intestine of *C. elegans*. Successful infection with these pathogens starts by breaking up yeast-infected hosts by means of a grinder to end up with a preparation of hyphal filaments, thus enabling the fungi to escape the body and proceed with infection of other animals. The subsequent appearance of filaments that have broken through the new host nematode's cuticle represents the endpoint of the infection process as this result always is accompanied by death of the host. In contrast to conventional life span assays, scoring these dead worms via the occurrence of fungal filaments is much easier, making it simple to adapt these screening systems to high throughput. Moreover, using temperature-sensitive mutants that are not able to produce any progeny above, e.g., 25 °C has the advantage that even longer infection periods can be scored quantitatively without being compromised by differing numbers of progeny.

A broad variety of assay techniques ranging in complexity from the very sophisticated to the very simple have been developed for screening antifungal compounds using a *C. albicans* infection. Green fluorescent protein-tagged pathogens are characterized by increasing green fluorescence in the intestine if an infection was successful. Compounds that reduce the fungal burden and therewith the green fluorescence in living nematodes can easily be identified using high-throughput screens. One of the various screens utilizing the *C. albicans*/*C. elegans* infection system led to the identification of novel antifungal compounds by Breger et al. (2007). In this comprehensive screen, two interesting compounds were revealed, namely, caffeic acid phenethyl ester, which is a component of the honey bee product propolis, and enoxacin, both of which also exhibit antifungal activities in murine infection models of candidiasis (Breger et al. 2007). This type of assay has been improved further to cope with greater numbers of compounds to be tested, which is a prerequisite to become a valuable model in screening assays for pharmacological companies. Both reduction of GFP signals within the animals in those assays that utilize GFP-tagged *C. albicans* and the occurrence of filamentous fungi indicative for killed worms in conventional assays are techniques ideally suited for high-throughput, quantitative approaches (Tampakakis et al. 2008).

The easiness of using these assays has enabled the study of more complex interactions, e.g., the interaction of *C. albicans* and *Acinetobacter baumannii* during a *C. elegans* infection. Surprisingly, coinfection with both microbes revealed a reduced virulence of *C. albicans* indicative for an interaction of *C. albicans* and *Acinetobacter baumannii* that inhibits important virulence factors of the fungus (Peleg et al. 2008). This is one showcase study that points to a complex interplay between *C. albicans* and *A. baumannii* that apparently is the consequence of reciprocal adaptation processes to suppress growth of the other microbe. Thus, virulence factors of, e.g., *A. baumannii* that were identified in this screen hold the potential to represent interesting target molecules that interfere with the virulence of *C. albicans* for metazoan hosts (Peleg et al. 2008).

On the other hand, some *Candida* strains used for infection experiments are highly attenuated. This attenuation of virulence opens the possibility of identifying the underlying virulence factors. Those *C. albicans* strains that are defective in hyphenation including efgl deficiency and flo8 deficiency are much less virulent as compared to normal strains (Pukkila-Worley et al. 2009). Moreover, other sets of deficiencies have been identified, including some of which represent defectiveness in biofilm formation, and there are yet others whose deficient nature remains to be understood indicating that virulence of *C. albicans* for metazoan hosts is more complex than previously anticipated.

Besides *C. albicans*, *Cryptococcus neoformans* is the second fungal pathogen that has been studied in great detail utilizing *C. elegans* as a surrogate host infection model. Similarly as with *C. albicans*, the *C. neoformans* model is especially relevant for immunocompromised patients. *Cryptococcus neoformans* similarly is taken up orally by the host and establishes an intestinal infection. As already shown for a number of different human pathogens, yeast with higher virulence in murine models also shows a higher killing ability in the *C. elegans* assay. Especially noted is the finding that a component of the *C. neoformans* capsule is toxic to nematodes, as identified by the fact that heat-inactivated pathogens were still able to kill the nematodes (Mylonakis et al. 2002). In a larger screen utilizing several hundred *C. neoformans* insertion mutants, only very few were identified as having attenuated virulence. Of special interest is a mutation in the kin-1 gene that has shown reduced killing although no effect on colonization could be detected. This bipolar phenotype could be recapitulated in murine infection models (Mylonakis et al. 2004).

1.3 *Drosophila melanogaster*

The fruit fly *Drosophila melanogaster* has served as a workhorse for genetic studies since more than a century ago. It shares most of the same advantages for serving as a surrogate host in infection studies as noted above with the soil nematode *C. elegans*. The genome of *D. melanogaster* was sequenced more than a decade ago (Adams et al. 2000), and a plethora of mutants covering almost every gene is available (Ryder et al. 2007). Moreover, modern, *Drosophila*-specific, or *Drosophila*-adopted methods allow complex ways of genetic manipulation opening the opportunity to produce tailored fly models (Pfeiffer et al. 2008; Pfeiffer et al. 2010; Pandey and Nichols 2011). Most important in this context is the availability of bi- or tripartite expression control systems that allow for a tight spatial or even spatiotemporal expression control (Brand and Perrimon 1993; McGuire et al. 2003; Manning et al. 2012). Compared with *C. elegans*, *Drosophila* is characterized by an organ composition that more closely resembles the one seen in mammals. This is comprised of a structured intestine, normal storage organs, and a body fluid system that is equipped with patrolling hemocytes. The fruit fly has served as the model for studying innate immunity since the rediscovery of this important aspect of its immune response repertoire almost two decades ago (Lemaitre et al. 1996). This "rediscovery" was initiated by the identification of Toll receptors serving as pattern recognition receptors of the innate immune system, which was honored by awarding of the Nobel Prize in Physiology or Medicine to Jules Hoffmann in 2011. Based on these early studies, it became apparent that *Drosophila* immunity not only serves as a blueprint for the mammalian innate immune system but that certain pathogens used the same strategies in flies and men to establish an infection in the host (Lemaitre and Hoffmann 2007; Ganesan et al. 2011; Hultmark 2003). The fly can react with a number of different responses toward the encounter with a pathogen. Exactly as in mammals, flies have potent humoral and local immune responses (Lemaitre and Hoffmann 2007; Hultmark 2003). While the major aim of the humoral immunity is to fight those pathogens that managed to invade into the body cavity, the local immune response aims to control the animals' surfaces, especially those in the intestine and the airways as they represent the most relevant entry points for pathogens. Protecting these mucosal surfaces is achieved by a multifaceted local immune system that is independent of Toll signaling. Instead, conventional innate immune responses that culminate in the production and release of antimicrobial peptides from these mucosal surfaces solely rely on IMD-signaling, a system that is homologous to our own TNF-a signaling system (Wagner et al. 2008, 2009; Tzou et al. 2000). This effective arm of the innate immune system is complemented by others controlling ROS production through the dual oxidase enzyme (Duox) (Ryu et al. 2010; Ha et al. 2005) and by danger signal-induced responses that are mediated through the transcription factor FoxO (Becker et al. 2010). These local, epithelial immune systems have not only the task to fight pathogens directly at the mucosal surfaces but also to shape the microbial community especially in the intestine to maintain a homeostatic situation between the host epithelia and the indigenous microbiota (Buchon et al. 2009) (Table 1.2).

For those pathogens that managed to penetrate these mucosal surfaces and that get access to the body cavity, the systemic immune system comes into play. It is composed of two major arms, the humoral and the cellular immune systems. The humoral arm of the systemic immune response reacts with release of antimicrobial compounds, namely, of antimicrobial peptides, into the hemolymph. Of central importance for this reaction is the fat body, the main immune-relevant organ in insects. This response can be triggered via two different signaling systems, the Toll and the IMD pathways. Whereas the Drosophila Toll pathway is homologous in all major components to the mammalian Toll-like signaling pathways, the IMD pathway is the insect counterpart of the mammalian TNF-a signaling system. Both signaling systems converge onto activation of NF-kB factors inducing transcription of relevant target genes. Besides these two pathways, others known from mammalian immune systems to be relevant, such as the JNK or the JAK/STAT pathways, are also involved in the fly's immune responses. The humoral immune system is

Table 1.2 Selected surrogate host models established in *Drosophila melanogaster*

Microorganism	Inoculation method	References
Gram-negative bacteria		
Burkholderia cepacia	Injection	Castonguay-Vanier et al. (2010)
Burkholderia thailandensis	Injection, oral uptake	Pilatova and Dionne (2012)
Pseudomonas aeruginosa	Injection	D'Argenio et al. (2001); Fauvarque et al. (2002)
Salmonella typhimurium	Injection	Shinzawa et al. (2009)
Serratia marcescens	Oral uptake	Cronin et al. (2009)
Yersinia spp.	S2 cells	Walker et al. (2013)
Gram-positive bacteria		
Enterococcus faecalis	Oral uptake	Teixeira et al. (2013)
Listeria monocytogenes	Injection, S2 cells	Cheng and Portnoy (2003); Ayres et al. (2008)
Staphylococcus aureus	Oral uptake, injection	Shiratsuchi et al. (2012)
Streptococcus pneumoniae	Injection	Chambers et al. (2012)
Fungi		
Aspergillus spp.	Injection, oral uptake, skin assay	Lionakis and Kontoyiannis (2010)
Fusarium spp.	Injection	Lamaris et al. (2007)
Scedosporium spp.	Injection	Lamaris et al. (2007)
Candida albicans	Injection, oral uptake	Glittenberg et al. (2011a, b)
Cryptococcus spp.	Oral uptake	Apidianakis et al. (2004)

supplemented by a highly effective cellular immune system. Three different types of hemocytes have been described in *Drosophila* that take different roles in the cellular immune response. Macrophage-like cells (plasmatocytes) ingest bacterial and fungal spores, while lamellocytes have the capacity to encapsulate and kill larger intruders. Crystal cells are a category of hemocytes that can release cytotoxic compounds and are involved in melanization (Lemaitre and Hoffmann 2007). Moreover, a potent, crystal cell-independent melanization cascade supplements the immune system at different levels. Taken together, the multiple layers of the fly's innate immune system are very similar to the defense systems of mammals that aim to inhibit colonization by potential pathogens.

Consequently and taking advantage of the various tools available, *Drosophila* has been used as a surrogate host for a number of different human pathogens with the aim to learn more about the infection mechanisms and the role of virulence factors. A number of different studies initially built the framework for later studies utilizing not just human pathogens but also insect- and invertebrate-specific pathogens including *Pseudomonas entomophila* and *Erwinia carotovora* (Liehl et al. 2006; Vodovar et al. 2005; Basset et al. 2000). Both of these pathogens are able to infect flies via the oral route and have extensively been used to study the basic aspects characteristic for infections introduced via the oral route.

Based on these studies and the finding that the human and the fly's intestine shares a surprisingly high degree of similarities, those infection models utilizing oral infection procedures have become especially popular.

1.3.1 **Drosophila** *as a Surrogate Host for Bacterial Pathogens*

Principally, two types of infection are used. Although the "natural" oral infection that is achieved by mixing living bacteria to be tested with the fly food has several advantages, septic injury by pricking with needles that carry the bacteria into the body cavity often is also used. Infection with *Pseudomonas aeruginosa* was among the first attempts to use the fruit fly as a surrogate host, and this pathogen shows a strikingly broad host range covering not only mammals and invertebrates but also plants, making this potential pathogen ideally suited to be studied using surrogate hosts. In one of the pioneering studies in this field (D'Argenio et al. 2001), a panel of different *P. aeruginosa* mutants was used to assess their ability to kill the host within a certain time. They identified mutants defective in the gene switching ability involved in virulence factor regulation as being less effective in killing. In another study, type III secretion systems which inject effector proteins into host cells were found to be highly relevant for the severity of the infection (Fauvarque et al. 2002). Some clinical isolates (*P. aeruginosa* CHA as an example) of this bacterium are defective in the type III secretion system and as a consequence are less pathogenically effective in the *Drosophila* system. Moreover, the functionality of the quorum sensing system has been shown to be of great importance for enabling of a highly effective infection leading to quick host death (Chugani et al. 2001). More recent studies utilizing this infection model have revealed that biofilm formation during the infection process is highly relevant for virulence. During infection, a biofilm is formed in the crop, which is part of the digestive tract. Those strains defective in biofilm formation show not only a changed biofilm formation, triggering an immune response by the host, but they are also attenuated in the *Drosophila* infection model (Mulcahy et al. 2011).

Elucidating the infection mechanisms of *Serratia marcescens* has been tackled using different approaches (Nehme et al. 2007). Both injection by septic injury as well as oral infection with different *S. marcescens* isolates and clones have been performed and used to quantify the microbe's ability to kill the host. Septic injury with the *S. marcescens* isolate DB11 killed the host within the first day of infection. In contrast, lethality induced by the oral exposure route occurred after a few days of incubation within the host. Moreover, *S. marcescens* strains either deficient in O-antigen biosynthesis or characterized by reduced protease release show a reduced killing capacity, indicative for reduced virulence. On the host side, it became apparent that two different arms of the fly's immune system are required for fighting the infection, local immunity at the site of gut epithelium, and the activity of hemocytes targeted against invaded bacteria (Nehme et al. 2007).

1.3.2 Use of Drosophila *to Study Infections Caused by Fungi*

Although the majority of infection studies performed with the fly have focused on bacteria as pathogens, fungi have recently been introduced in this field of research (Hamilos et al. 2012). As some fungi have been shown to cause mortality especially in immunocompromised patients, a better understanding of fungal-related infection mechanisms became mandatory. Experimental infection with fungi is mainly achieved via direct injection into the hemolymph. More natural physiological routes of infection, either via the oral route or the skin, have also been employed. Matching the situation found in humans, wild-type flies show a relatively low susceptibility to fungal infections. Thus, immunocompromised flies, namely, those deficient in the Toll pathway, which is specifically tailored to protect against fungal pathogens, have been introduced into these screening systems (Lionakis and Kontoyiannis 2010). Although it takes some time to produce Toll-deficient animals, they can be used in high-throughput assays and are amenable to different routes of infection including injection, rolling the animals in fungal spores, or oral uptake. The fly system thus allows testing for the effectiveness of antifungal compounds as it has been demonstrated by the authors (Lionakis and Kontoyiannis 2010). *Drosophila* appears to be well protected against the great majority of fungal pathogens. Nevertheless, some naturally occurring fungal pathogens such as *Beauveria bassiana* managed to induce an effective infection (Kraaijeveld and Godfray 2008; Kirsanova et al. 1975). Taking into account the advantages of *Drosophila* serving as a surrogate host, it should be ideally suited for studying complete libraries of pathogenic fungi such as those that have been produced to identify virulence factors from *Aspergillus* species. Indeed, a number of different *Aspergillus* virulence factors have been identified based upon such studies including DeltasidA and DeltasidD that are defective in siderophore biosynthesis, DgliP that are impaired in secondary metabolite mechanism, and H515 with deficiencies in PABA metabolism (Lamaris et al. 2007; Chamilos et al. 2008). In contrast, a very important *Aspergillus fumigatus* virulence factor, CgrA, could not be identified in flies simply because its expression requires a higher body temperature (37 °C) to become effective. This latter problem may be experienced for all temperature-sensitive virulence factors if *Drosophila* is used as the sole infection model (Hamilos et al. 2012).

1.3.3 Candida albicans *Infections*

Presumably, *Drosophila* has gained its greatest importance as a surrogate host for studying fungal virulence factors of different *Candida* species (Junqueira 2012). Among the various *Candida* species that are usually associated with mammals, *Candida albicans* appears to be most relevant (Edmond et al. 1999). As already pointed out in the *C. elegans* section, *C. albicans* is part of our normal gut flora and is found on the skin of about 80 % of all persons in a population, usually without causing health problems. It is an opportunistic pathogen causing oral and genital infection especially in immunocompromised patients.

The suitability of *Drosophila* to act as a surrogate host for *Candida albicans* has been established in a proof-of-principle study. For this, the virulence of a set of *Candida* strains that previously had been scored in mice was analyzed (Alarco et al. 2004). Although Toll-deficient flies had previously been used for infection studies with fungi, they gave no satisfactory congruence with the mouse data regarding their virulence. In contrast, albeit wild-type flies are generally less amenable to fungal infections, [comma] virulence parameters quantified using wild-type *Drosophila* showed a very good correlation with those results obtained from the corresponding mouse studies, implying that these flies are well suited to act as surrogate hosts in *Candida albicans* infection experiments (Glittenberg et al. 2011a, b).

1.4 *Galleria mellonella*

As already pointed out above, the two major invertebrate model organisms *C. elegans* and *D. melanogaster* have one very important limitation, they cannot be reared at elevated temperatures. This is especially relevant for all those pathogens that have temperature-sensitive virulence factors. Thus, the need for an invertebrate model that can be reared at 37 °C was urgent. Although the greater wax moth *Galleria mellonella* is no classical model organisms, it has been established as a surrogate host for some important human pathogens. Two major advantages predestine *Galleria* for this purpose. It can be reared at elevated temperatures representing the human body temperature (37 °C), and it is large enough to easily inject well-defined amounts of the pathogen into the hemolymph. *Galleria* has an immune system that shares numerous similarities with that of other insects including *Drosophila*, although it appears to show a slightly higher complexity (Vogel et al. 2011; Jiang et al. 2010) than that of the fruit fly. The immune system of the greater wax moth can be primed by incubation with nonpathogenic yeast to protect from an otherwise lethal infection by *Candida albicans* (Bergin et al. 2006). Thus it serves as a very valuable additional surrogate host especially for those pathogens that cannot be studied in the two major invertebrate models *Drosophila melanogaster* and *Caenorhabditis elegans* (Table 1.3).

Table 1.3 Selected surrogate host models established in *Galleria mellonella*

Microorganism	Inoculation method	References
Gram-negative bacteria		
Burkholderia cepacia	Injection	Seed and Dennis (2008)
Burkholderia thailandensis	Injection	Wand et al. (2011)
Burkholderia pseudomallei	Injection	Wand et al. (2011)
Pseudomonas aeruginosa	Injection	Koch et al. (2014)
Salmonella typhimurium	Injection	Bender et al. (2013)
Yersinia spp.	Injection	Erickson et al. (2011)
Gram-positive bacteria		
Enterococcus faecalis	Injection	Luther et al. (2014)
Listeria monocytogenes	Injection	Mukherjee et al. (2013)
Staphylococcus aureus	Injection	Gibreel and Upton (2013)
Streptococcus spp.	Injection	Loh et al. (2013); Olsen et al. (2011)
Fungi		
Aspergillus spp.	Injection	Jackson et al. (2009)
Fusarium spp.	Injection	Coleman et al. (2011)
Histoplasma spp.	Injection	Thomaz et al. (2013)
Candida albicans	Injection	Brennan et al. (2002); Fuchs et al. (2010a)
Cryptococcus spp.	Injection	Mylonakis et al. (2005)

Despite the lack of sophisticated genetic tools available for *Galleria*, it has some advantages as compared with the fruit fly, principle of these being that the large size of the animal makes it easy to reproducibly infect this surrogate host, a task that usually requires sophisticated apparatus and much experience if performed in the fly.

1.4.1 Studying Fungal Pathogens in Galleria

Wild-type *Galleria* can successfully be infected by fungal pathogens, whereas *Drosophila* wild types are almost insensitive to a great variety of different fungal pathogens (Fuchs et al. 2010b; Lionakis 2011). As outlined above, the two major advantages of *Galleria*, namely, incubation at human body temperature and the reliability of consistently inoculating the host with exactly the same doses of pathogens, have enabled its use for informative experimental infection studies with fungal pathogens. Thus, even subtle variations in parameters of the larval immune system can be useful to quantify the effectiveness of an infection (Bergin et al. 2003).

Candida albicans, as one of the most important fungal pathogens, can infect a huge number of different hosts with varying efficiencies. A recent study employing a set of *C. albicans* strains tested in murine models regarding their major infection characteristics revealed a surprising overlap regarding the infection parameters between both models (mice and *Galleria*), further supporting the hypothesis that the Galleria system is well suited for experimental infection studies using *C. albicans* mutants (Brennan et al. 2002). Other fungal pathogens, such as *Aspergillus fumigatus, Cryptococcus neoformans*, and *Fusarium oxysporum*, are also highly effective in infecting *Galleria* larvae. Of particular noteworthiness are highly informative studies of *A. fumigatus* mutants in this surrogate host (Jackson et al. 2009; Mylonakis et al. 2005; Navarro-Velasco et al. 2011).

1.4.2 Bacterial Pathogens and Galleria

A number of different bacterial pathogens have been used to infect the greater wax moth to learn more about both the mechanisms of infection and bacterial virulence factors.

One very impressive example of using this system as a surrogate host for bacteria has been presented recently. *Listeria monocytogenes*, the causative agent of food-borne listeriosis in humans, can infect parts of the brain, leading to meningitis and meningoencephalitis (Disson and Lecuit 2012). This type of complication caused by infection of brain tissue with *Listeria* could be recapitulated almost completely in *Galleria* (Mukherjee et al. 2013). Moreover, it has been shown possible to interfere with the infection in Galleria by treatment with a number of pharmacological agents known to act similarly in humans. Clinically relevant *Streptococcus* strains also have been tested in *Galleria*. For them, ability of the moth to be reared at elevated temperatures is of prime importance. Most important for the validation of an invertebrate model organism is the use of a panel of pathogen mutants that already have been scored in mammalian infection models. In the case of the greater wax moth, the correlation of data between mammalian and invertebrate hosts is very good, meaning that virulent *Streptococcus* strains (as scored in murine models) are also more effective in killing *Galleria* (Olsen et al. 2011). However, it has to be kept in mind that while the experimental conditions can be well controlled in the greater wax moth model of *Streptococcus* infection, not all virulence factors and in particular the virulence of clinical strains are exactly replicated (Loh et al. 2013).

1.5 Conclusions

The interest for using invertebrate surrogate hosts to study human pathogens has developed in a succession of waves. Currently, a steep increase in the use of several corresponding systems can be observed, which is shown by the huge number of publications dealing with this topic in the last years. This expansive recent development is the result of a “maturation” of the corresponding disease models combined with a greater knowledge not only about their strengths but, even more importantly, about their weaknesses. The two classical model organisms *Caenorhabditis elegans* and *Drosophila melanogaster* have to be mentioned in a special context as they supply unrivaled genetic toolboxes. This research field has diversified substantially in the last years by using these simple invertebrate surrogate hosts and following a variety of different research strategies. Testing sets of human pathogens that are expected to differ in their virulence remains among the most important research questions to be addressed by using these surrogate host systems. However, this primary field of work has recently been supplemented by utilizing these whole-animal infection models for screening studies aimed at finding new antibacterial or antifungal compounds. Last but not least, these models are used to decipher host factors relevant for virulence.

Taken together, invertebrate surrogate host models for human pathogens are not only a good alternative to mammal-based infection studies due to their lower costs and their much higher ethical acceptance; they open the potential to be part of larger screening pipelines as a first whole-animal screen that can be adapted to high-throughput scenarios.

Acknowledgments We would like to thank all members of our research lab and the DFG and the BMBF for financial support.

References

Aballay A, Yorgey P, Ausubel FM (2000) Salmonella typhimurium proliferates and establishes a persistent infection in the intestine of Caenorhabditis elegans. Curr Biol 10(23):1539–1542

Adams MD, Celniker SE, Holt RA, Evans CA, Gocayne JD, Amanatides PG, Scherer SE, Li PW, Hoskins RA, Galle RF, George RA, Lewis SE, Richards S, Ashburner M, Henderson SN, Sutton GG, Wortman JR, Yandell MD, Zhang Q, Chen LX, Brandon RC, Rogers YH, Blazej RG, Champe M, Pfeiffer BD, Wan KH, Doyle C, Baxter EG, Helt G, Nelson CR, Gabor GL, Abril JF, Agbayani A, An HJ, Andrews-Pfannkoch C, Baldwin D, Ballew RM, Basu A, Baxendale J, Bayraktaroglu L, Beasley EM, Beeson KY, Benos PV, Berman BP, Bhandari D, Bolshakov S, Borkova D, Botchan MR, Bouck J, Brokstein P, Brottier P, Burtis KC, Busam DA, Butler H, Cadieu E, Center A, Chandra I, Cherry JM, Cawley S, Dahlke C, Davenport LB, Davies P, de Pablos B, Delcher A, Deng Z, Mays AD, Dew I, Dietz SM, Dodson K, Doup LE, Downes M, Dugan-Rocha S, Dunkov BC, Dunn P, Durbin KJ, Evangelista CC, Ferraz C, Ferriera S, Fleischmann W, Fosler C, Gabrielian AE, Garg NS, Gelbart WM, Glasser K, Glodek A, Gong F, Gorrell JH, Gu Z, Guan P, Harris M, Harris NL, Harvey D, Heiman TJ, Hernandez JR, Houck J, Hostin D, Houston KA, Howland TJ, Wei MH, Ibegwam C, Jalali M, Kalush F, Karpen GH, Ke Z, Kennison JA, Ketchum KA, Kimmel BE, Kodira CD, Kraft C, Kravitz S, Kulp D, Lai Z, Lasko P, Lei Y, Levitsky AA, Li J, Li Z, Liang Y, Lin X, Liu X, Mattei B, McIntosh TC, McLeod MP, McPherson D, Merkulov G, Milshina NV, Mobarry C, Morris J, Moshrefi A, Mount SM, Moy M, Murphy B, Murphy L, Muzny DM, Nelson DL, Nelson DR, Nelson KA, Nixon K, Nusskern DR, Pacleb JM, Palazzolo M, Pittman GS, Pan S, Pollard J, Puri V, Reese MG, Reinert K, Remington K, Saunders RD, Scheeler F, Shen H, Shue BC, Siden-Kiamos I, Simpson M, Skupski MP, Smith T, Spier E, Spradling AC, Stapleton M, Strong R, Sun E, Svirskas R, Tector C, Turner R, Venter E, Wang AH, Wang X, Wang ZY, Wassarman DA, Weinstock GM, Weissenbach J, Williams SM, Woodage T, Worley KC, Wu D, Yang S, Yao QA, Ye J,

Yeh RF, Zaveri JS, Zhan M, Zhang G, Zhao Q, Zheng L, Zheng XH, Zhong FN, Zhong W, Zhou X, Zhu S, Zhu X, Smith HO, Gibbs RA, Myers EW, Rubin GM, Venter JC (2000) The genome sequence of Drosophila melanogaster. Science 287(5461):2185–2195

Alarco AM, Marcil A, Chen J, Suter B, Thomas D, Whiteway M (2004) Immune-deficient Drosophila melanogaster: a model for the innate immune response to human fungal pathogens. J Immunol 172(9):5622–5628

Anastassopoulou CG, Fuchs BB, Mylonakis E (2011) Caenorhabditis elegans-based model systems for antifungal drug discovery. Curr Pharm Des 17(13):1225–1233

Apidianakis Y, Rahme LG, Heitman J, Ausubel FM, Calderwood SB, Mylonakis E (2004) Challenge of Drosophila melanogaster with Cryptococcus neoformans and role of the innate immune response. Eukaryot Cell 3(2):413–419

Ashrafi K, Chang FY, Watts JL, Fraser AG, Kamath RS, Ahringer J, Ruvkun G (2003) Genome-wide RNAi analysis of Caenorhabditis elegans fat regulatory genes. Nature 421(6920): 268–272. doi:10.1038/nature01279

Ayres JS, Freitag N, Schneider DS (2008) Identification of Drosophila mutants altering defense of and endurance to Listeria monocytogenes infection. Genetics 178(3):1807–1815. doi:10.1534/genetics.107.083782

Bae T, Banger AK, Wallace A, Glass EM, Aslund F, Schneewind O, Missiakas DM (2004) Staphylococcus aureus virulence genes identified by bursa aurealis mutagenesis and nematode killing. Proc Natl Acad Sci U S A 101(33):12312–12317. doi:10.1073/pnas.0404728101

Basset A, Khush RS, Braun A, Gardan L, Boccard F, Hoffmann JA, Lemaitre B (2000) The phytopathogenic bacteria Erwinia carotovora infects Drosophila and activates an immune response. Proc Natl Acad Sci U S A 97(7):3376–3381. doi:10.1073/pnas.070357597

Becker T, Loch G, Beyer M, Zinke I, Aschenbrenner AC, Carrera P, Inhester T, Schultze JL, Hoch M (2010) FOXO-dependent regulation of innate immune homeostasis. Nature 463(7279):369–373. doi:10.1038/nature08698

Begun J, Sifri CD, Goldman S, Calderwood SB, Ausubel FM (2005) Staphylococcus aureus virulence factors identified by using a high-throughput Caenorhabditis elegans-killing model. Infect Immun 73(2):872–877. doi:10.1128/IAI.73.2.872-877.2005

Bender JK, Wille T, Blank K, Lange A, Gerlach RG (2013) LPS structure and PhoQ activity are important for Salmonella Typhimurium virulence in the Galleria mellonella infection model [corrected]. PLoS One 8(8):e73287. doi:10.1371/journal.pone.0073287

Bergin D, Brennan M, Kavanagh K (2003) Fluctuations in haemocyte density and microbial load may be used as indicators of fungal pathogenicity in larvae of Galleria mellonella. Microbes Infect 5(15):1389–1395

Bergin D, Murphy L, Keenan J, Clynes M, Kavanagh K (2006) Pre-exposure to yeast protects larvae of Galleria mellonella from a subsequent lethal infection by Candida albicans and is mediated by the increased expression of antimicrobial peptides. Microbes Infect 8(8): 2105–2112. doi:10.1016/j.micinf.2006.03.005

Brand AH, Perrimon N (1993) Targeted gene expression as a means of altering cell fates and generating dominant phenotypes. Development 118(2):401–415

Breger J, Fuchs BB, Aperis G, Moy TI, Ausubel FM, Mylonakis E (2007) Antifungal chemical compounds identified using a C. elegans pathogenicity assay. PLoS Pathog 3(2):18. doi:10.1371/journal.ppat.0030018

Brennan M, Thomas DY, Whiteway M, Kavanagh K (2002) Correlation between virulence of Candida albicans mutants in mice and Galleria mellonella larvae. FEMS Immunol Med Microbiol 34(2):153–157

Brenner S (1974) The genetics of Caenorhabditis elegans. Genetics 77(1):71–94

Brenner S (2003) Nature's gift to science (Nobel lecture). Chembiochem 4(8):683–687. doi:10.1002/cbic.200300625

Buchon N, Broderick NA, Chakrabarti S, Lemaitre B (2009) Invasive and indigenous microbiota impact intestinal stem cell activity through multiple pathways in Drosophila. Genes Dev 23(19):2333–2344. doi:10.1101/gad.1827009

Burns AR, Kwok TC, Howard A, Houston E, Johanson K, Chan A, Cutler SR, McCourt P, Roy PJ (2006) High-throughput screening of small molecules for bioactivity and target identification in Caenorhabditis elegans. Nat Protoc 1(4):1906–1914. doi:10.1038/nprot.2006.283

Castonguay-Vanier J, Vial L, Tremblay J, Deziel E (2010) Drosophila melanogaster as a model host for the Burkholderia cepacia complex. PLoS One 5(7):e11467. doi:10.1371/journal.pone.0011467

Chambers MC, Lightfield KL, Schneider DS (2012) How the fly balances its ability to combat different pathogens. PLoS Pathog 8(12):e1002970. doi:10.1371/journal.ppat.1002970

Chamilos G, Lewis RE, Hu J, Xiao L, Zal T, Gilliet M, Halder G, Kontoyiannis DP (2008) Drosophila melanogaster as a model host to dissect the immunopathogenesis of zygomycosis. Proc Natl Acad Sci U S A 105(27):9367–9372. doi:10.1073/pnas.0709578105

Cheng LW, Portnoy DA (2003) Drosophila S2 cells: an alternative infection model for Listeria monocytogenes. Cell Microbiol 5(12):875–885

Chugani SA, Whiteley M, Lee KM, D'Argenio D, Manoil C, Greenberg EP (2001) QscR, a modulator of quorum-sensing signal synthesis and virulence in Pseudomonas aeruginosa. Proc Natl Acad Sci U S A 98(5):2752–2757. doi:10.1073/pnas.051624298

Coleman JJ, Muhammed M, Kasperkovitz PV, Vyas JM, Mylonakis E (2011) Fusarium pathogenesis investigated using Galleria mellonella as a heterologous host. Fungal Biol 115(12): 1279–1289. doi:10.1016/j.funbio.2011.09.005

Couillault C, Ewbank JJ (2002) Diverse bacteria are pathogens of Caenorhabditis elegans. Infect Immun 70(8):4705–4707

Cronin SJ, Nehme NT, Limmer S, Liegeois S, Pospisilik JA, Schramek D, Leibbrandt A, Simoes Rde M, Gruber S, Puc U, Ebersberger I, Zoranovic T, Neely GG, von Haeseler A, Ferrandon D, Penninger JM (2009) Genome-wide RNAi screen identifies genes involved in intestinal pathogenic bacterial infection. Science 325(5938):340–343. doi:10.1126/science.1173164

D'Argenio DA, Gallagher LA, Berg CA, Manoil C (2001) Drosophila as a model host for Pseudomonas aeruginosa infection. J Bacteriol 183(4):1466–1471. doi:10.1128/JB.183.4.1466-1471.2001

Darby C (2005) Interactions with microbial pathogens. WormBook: the online review of C. elegans biology, pp 1–15. doi: 10.1895/wormbook.1.21.1

Darby C, Hsu JW, Ghori N, Falkow S (2002) Caenorhabditis elegans: plague bacteria biofilm blocks food intake. Nature 417(6886):243–244. doi:10.1038/417243a

Depraitere C, Darmon M (1978) [Growth of "Dictyostelium discoideum" on different species of bacteria (author's transl)]. Ann Microbiol 129B(3):451–461

Disson O, Lecuit M (2012) Targeting of the central nervous system by Listeria monocytogenes. Virulence 3(2):213–221. doi:10.4161/viru.19586

Edmond MB, Wallace SE, McClish DK, Pfaller MA, Jones RN, Wenzel RP (1999) Nosocomial bloodstream infections in United States hospitals: a three-year analysis. Clin Infect Dis 29(2): 239–244. doi:10.1086/520192

Erickson DL, Russell CW, Johnson KL, Hileman T, Stewart RM (2011) PhoP and OxyR transcriptional regulators contribute to Yersinia pestis virulence and survival within Galleria mellonella. Microb Pathog 51(6):389–395. doi:10.1016/j.micpath.2011.08.008

Fauvarque MO, Bergeret E, Chabert J, Dacheux D, Satre M, Attree I (2002) Role and activation of type III secretion system genes in Pseudomonas aeruginosa-induced Drosophila killing. Microb Pathog 32(6):287–295

Fuchs BB, Eby J, Nobile CJ, El Khoury JB, Mitchell AP, Mylonakis E (2010a) Role of filamentation in Galleria mellonella killing by Candida albicans. Microbes Infect 12(6): 488–496. doi:10.1016/j.micinf.2010.03.001

Fuchs BB, O'Brien E, Khoury JB, Mylonakis E (2010b) Methods for using Galleria mellonella as a model host to study fungal pathogenesis. Virulence 1(6):475–482

Ganesan S, Aggarwal K, Paquette N, Silverman N (2011) NF-kappaB/Rel proteins and the humoral immune responses of Drosophila melanogaster. Curr Top Microbiol Immunol 349: 25–60. doi:10.1007/82_2010_107

Gibreel TM, Upton M (2013) Synthetic epidermicin NI01 can protect Galleria mellonella larvae from infection with Staphylococcus aureus. J Antimicrob Chemother 68(10):2269–2273. doi:10.1093/jac/dkt195

Glittenberg MT, Kounatidis I, Christensen D, Kostov M, Kimber S, Roberts I, Ligoxygakis P (2011a) Pathogen and host factors are needed to provoke a systemic host response to gastrointestinal infection of Drosophila larvae by Candida albicans. Dis Model Mech 4(4):515–525. doi:10.1242/dmm.006627

Glittenberg MT, Silas S, MacCallum DM, Gow NA, Ligoxygakis P (2011b) Wild-type Drosophila melanogaster as an alternative model system for investigating the pathogenicity of Candida albicans. Dis Model Mech 4(4):504–514. doi:10.1242/dmm.006619

Gravato-Nobre MJ, Hodgkin J (2005) Caenorhabditis elegans as a model for innate immunity to pathogens. Cell Microbiol 7(6):741–751. doi:10.1111/j.1462-5822.2005.00523.x

Gravato-Nobre MJ, Nicholas HR, Nijland R, O'Rourke D, Whittington DE, Yook KJ, Hodgkin J (2005) Multiple genes affect sensitivity of Caenorhabditis elegans to the bacterial pathogen Microbacterium nematophilum. Genetics 171(3):1033–1045. doi:10.1534/genetics.105.045716

Ha EM, Oh CT, Bae YS, Lee WJ (2005) A direct role for dual oxidase in Drosophila gut immunity. Science 310(5749):847–850. doi:10.1126/science.1117311

Hamilos G, Samonis G, Kontoyiannis DP (2012) Recent advances in the use of Drosophila melanogaster as a model to study immunopathogenesis of medically important filamentous fungi. Int J Microbiol 2012:583792. doi:10.1155/2012/583792

Hemmrich G, Miller DJ, Bosch TC (2007) The evolution of immunity: a low-life perspective. Trends Immunol 28(10):449–454. doi:10.1016/j.it.2007.08.003

Hilbi H, Weber SS, Ragaz C, Nyfeler Y, Urwyler S (2007) Environmental predators as models for bacterial pathogenesis. Environ Microbiol 9(3):563–575. doi:10.1111/j.1462-2920.2007.01238.x

Hodgkin J, Kuwabara PE, Corneliussen B (2000) A novel bacterial pathogen, Microbacterium nematophilum, induces morphological change in the nematode C. elegans. Curr Biol 10(24):1615–1618

Hodgkin J, Felix MA, Clark LC, Stroud D, Gravato-Nobre MJ (2013) Two Leucobacter strains exert complementary virulence on Caenorhabditis including death by worm-star formation. Curr Biol 23(21):2157–2161. doi:10.1016/j.cub.2013.08.060

Hultmark D (2003) Drosophila immunity: paths and patterns. Curr Opin Immunol 15(1):12–19

Irazoqui JE, Urbach JM, Ausubel FM (2010) Evolution of host innate defence: insights from Caenorhabditis elegans and primitive invertebrates. Nat Rev Immunol 10(1):47–58. doi:10.1038/nri2689

Jackson JC, Higgins LA, Lin X (2009) Conidiation color mutants of Aspergillus fumigatus are highly pathogenic to the heterologous insect host Galleria mellonella. PLoS One 4(1):e4224. doi:10.1371/journal.pone.0004224

Jansen WT, Bolm M, Balling R, Chhatwal GS, Schnabel R (2002) Hydrogen peroxide-mediated killing of Caenorhabditis elegans by Streptococcus pyogenes. Infect Immun 70(9):5202–5207

Jiang H, Vilcinskas A, Kanost MR (2010) Immunity in lepidopteran insects. Adv Exp Med Biol 708:181–204

Junqueira JC (2012) Models hosts for the study of oral candidiasis. Adv Exp Med Biol 710:95–105. doi:10.1007/978-1-4419-5638-5_10

Kirsanova RV, Levitin MM, Lekarkina LP, Usenko LI, Sharygin VI (1975) Drosophila and the entomopathogenic fungus Beauveria bassiana as a model for the study of host and parasite interrelationships. Zh Obshch Biol 36(2):251–258

Koch G, Nadal-Jimenez P, Cool RH, Quax WJ (2014) Assessing Pseudomonas virulence with nonmammalian host: Galleria mellonella. Methods Mol Biol 1149:681–688. doi:10.1007/978-1-4939-0473-0_52

Kothe M, Antl M, Huber B, Stoecker K, Ebrecht D, Steinmetz I, Eberl L (2003) Killing of Caenorhabditis elegans by Burkholderia cepacia is controlled by the cep quorum-sensing system. Cell Microbiol 5(5):343–351

Kraaijeveld AR, Godfray HC (2008) Selection for resistance to a fungal pathogen in Drosophila melanogaster. Heredity 100(4):400–406. doi:10.1038/sj.hdy.6801092

Lamaris GA, Chamilos G, Lewis RE, Kontoyiannis DP (2007) Virulence studies of Scedosporium and Fusarium species in Drosophila melanogaster. J Infect Dis 196(12):1860–1864. doi:10.1086/523765

Lemaitre B, Hoffmann J (2007) The host defense of Drosophila melanogaster. Annu Rev Immunol 25:697–743. doi:10.1146/annurev.immunol.25.022106.141615

Lemaitre B, Nicolas E, Michaut L, Reichhart JM, Hoffmann JA (1996) The dorsoventral regulatory gene cassette spatzle/Toll/cactus controls the potent antifungal response in Drosophila adults. Cell 86(6):973–983

Liehl P, Blight M, Vodovar N, Boccard F, Lemaitre B (2006) Prevalence of local immune response against oral infection in a Drosophila/Pseudomonas infection model. PLoS Pathog 2(6):e56. doi:10.1371/journal.ppat.0020056

Lionakis MS (2011) Drosophila and Galleria insect model hosts: new tools for the study of fungal virulence, pharmacology and immunology. Virulence 2(6):521–527. doi:10.4161/viru.2.6.18520

Lionakis MS, Kontoyiannis DP (2010) The growing promise of Toll-deficient Drosophila melanogaster as a model for studying Aspergillus pathogenesis and treatment. Virulence 1(6):488–499

Loh JM, Adenwalla N, Wiles S, Proft T (2013) Galleria mellonella larvae as an infection model for group A streptococcus. Virulence 4(5):419–428. doi:10.4161/viru.24930

Luther MK, Arvanitis M, Mylonakis E, LaPlante KL (2014) Activity of daptomycin or linezolid in combination with rifampin or gentamicin against biofilm-forming Enterococcus faecalis or E. faecium in an in vitro pharmacodynamic model using simulated endocardial vegetations and an in vivo survival assay using Galleria mellonella larvae. Antimicrob Agents Chemother 58(8):4612–4620. doi:10.1128/AAC.02790-13

Mahajan-Miklos S, Tan MW, Rahme LG, Ausubel FM (1999) Molecular mechanisms of bacterial virulence elucidated using a Pseudomonas aeruginosa-Caenorhabditis elegans pathogenesis model. Cell 96(1):47–56

Mallo GV, Kurz CL, Couillault C, Pujol N, Granjeaud S, Kohara Y, Ewbank JJ (2002) Inducible antibacterial defense system in C. elegans. Curr Biol 12(14):1209–1214

Manning L, Heckscher ES, Purice MD, Roberts J, Bennett AL, Kroll JR, Pollard JL, Strader ME, Lupton JR, Dyukareva AV, Doan PN, Bauer DM, Wilbur AN, Tanner S, Kelly JJ, Lai SL, Tran KD, Kohwi M, Laverty TR, Pearson JC, Crews ST, Rubin GM, Doe CQ (2012) A resource for manipulating gene expression and analyzing cis-regulatory modules in the Drosophila CNS. Cell Rep 2(4):1002–1013. doi:10.1016/j.celrep.2012.09.009

Marsh EK, May RC (2012) Caenorhabditis elegans, a model organism for investigating immunity. Appl Environ Microbiol 78(7):2075–2081. doi:10.1128/AEM.07486-11

McGuire SE, Le PT, Osborn AJ, Matsumoto K, Davis RL (2003) Spatiotemporal rescue of memory dysfunction in Drosophila. Science 302(5651):1765–1768. doi:10.1126/science.1089035

Muhammed M, Coleman JJ, Mylonakis E (2012) Caenorhabditis elegans: a nematode infection model for pathogenic fungi. Methods Mol Biol 845:447–454. doi:10.1007/978-1-61779-539-8_31

Mukherjee K, Hain T, Fischer R, Chakraborty T, Vilcinskas A (2013) Brain infection and activation of neuronal repair mechanisms by the human pathogen Listeria monocytogenes in the lepidopteran model host Galleria mellonella. Virulence 4(4):324–332. doi:10.4161/viru.23629

Mulcahy H, Sibley CD, Surette MG, Lewenza S (2011) Drosophila melanogaster as an animal model for the study of Pseudomonas aeruginosa biofilm infections in vivo. PLoS Pathog 7(10):e1002299. doi:10.1371/journal.ppat.1002299

Mylonakis E, Ausubel FM, Perfect JR, Heitman J, Calderwood SB (2002) Killing of Caenorhabditis elegans by Cryptococcus neoformans as a model of yeast pathogenesis. Proc Natl Acad Sci U S A 99(24):15675–15680. doi:10.1073/pnas.232568599

Mylonakis E, Idnurm A, Moreno R, El Khoury J, Rottman JB, Ausubel FM, Heitman J, Calderwood SB (2004) Cryptococcus neoformans Kin1 protein kinase homologue, identified through a Caenorhabditis elegans screen, promotes virulence in mammals. Mol Microbiol 54(2):407–419. doi:10.1111/j.1365-2958.2004.04310.x

Mylonakis E, Moreno R, El Khoury JB, Idnurm A, Heitman J, Calderwood SB, Ausubel FM, Diener A (2005) Galleria mellonella as a model system to study Cryptococcus neoformans pathogenesis. Infect Immun 73(7):3842–3850. doi:10.1128/IAI.73.7.3842-3850.2005

Navarro-Velasco GY, Prados-Rosales RC, Ortiz-Urquiza A, Quesada-Moraga E, Di Pietro A (2011) Galleria mellonella as model host for the trans-kingdom pathogen Fusarium oxysporum. Fungal Genet Biol 48(12):1124–1129. doi:10.1016/j.fgb.2011.08.004

Nehme NT, Liegeois S, Kele B, Giammarinaro P, Pradel E, Hoffmann JA, Ewbank JJ, Ferrandon D (2007) A model of bacterial intestinal infections in Drosophila melanogaster. PLoS Pathog 3(11):e173. doi:10.1371/journal.ppat.0030173

O'Quinn AL, Wiegand EM, Jeddeloh JA (2001) Burkholderia pseudomallei kills the nematode Caenorhabditis elegans using an endotoxin-mediated paralysis. Cell Microbiol 3(6):381–393

Okoli I, Coleman JJ, Tampakakis E, An WF, Holson E, Wagner F, Conery AL, Larkins-Ford J, Wu G, Stern A, Ausubel FM, Mylonakis E (2009) Identification of antifungal compounds active against Candida albicans using an improved high-throughput Caenorhabditis elegans assay. PLoS One 4(9):e7025. doi:10.1371/journal.pone.0007025

Olsen RJ, Watkins ME, Cantu CC, Beres SB, Musser JM (2011) Virulence of serotype M3 Group A Streptococcus strains in wax worms (Galleria mellonella larvae). Virulence 2(2):111–119

Pandey UB, Nichols CD (2011) Human disease models in Drosophila melanogaster and the role of the fly in therapeutic drug discovery. Pharmacol Rev 63(2):411–436. doi:10.1124/pr.110.003293

Peleg AY, Tampakakis E, Fuchs BB, Eliopoulos GM, Moellering RC Jr, Mylonakis E (2008) Prokaryote-eukaryote interactions identified by using Caenorhabditis elegans. Proc Natl Acad Sci U S A 105(38):14585–14590. doi:10.1073/pnas.0805048105

Pfeiffer BD, Jenett A, Hammonds AS, Ngo TT, Misra S, Murphy C, Scully A, Carlson JW, Wan KH, Laverty TR, Mungall C, Svirskas R, Kadonaga JT, Doe CQ, Eisen MB, Celniker SE, Rubin GM (2008) Tools for neuroanatomy and neurogenetics in Drosophila. Proc Natl Acad Sci U S A 105(28):9715–9720. doi:10.1073/pnas.0803697105

Pfeiffer BD, Ngo TT, Hibbard KL, Murphy C, Jenett A, Truman JW, Rubin GM (2010) Refinement of tools for targeted gene expression in Drosophila. Genetics 186(2):735–755. doi:10.1534/genetics.110.119917

Pilatova M, Dionne MS (2012) Burkholderia thailandensis is virulent in Drosophila melanogaster. PLoS One 7(11):e49745. doi:10.1371/journal.pone.0049745

Poole RJ, Bashllari E, Cochella L, Flowers EB, Hobert O (2011) A genome-wide RNAi screen for factors involved in neuronal specification in Caenorhabditis elegans. PLoS Genet 7(6):e1002109. doi:10.1371/journal.pgen.1002109

Pujol N, Zugasti O, Wong D, Couillault C, Kurz CL, Schulenburg H, Ewbank JJ (2008) Antifungal innate immunity in C elegans is enhanced by evolutionary diversification of antimicrobial peptides. PLoS Pathog 4(7):e1000105. doi:10.1371/journal.ppat.1000105

Pukkila-Worley R, Peleg AY, Tampakakis E, Mylonakis E (2009) Candida albicans hyphal formation and virulence assessed using a Caenorhabditis elegans infection model. Eukaryot Cell 8(11):1750–1758. doi:10.1128/EC.00163-09

Pukkila-Worley R, Ausubel FM, Mylonakis E (2011) Candida albicans infection of Caenorhabditis elegans induces antifungal immune defenses. PLoS Pathog 7(6):e1002074. doi:10.1371/journal.ppat.1002074

Riddle DL, Blumenthal T, Meyer BJ, Priess JR (1997) Introduction to C. elegans. In: Riddle DL, Blumenthal T, Meyer BJ, Priess JR (eds) C. elegans II, 2nd edn. Cold Spring Harbor Laboratory Press, New York

Roeder T, Stanisak M, Gelhaus C, Bruchhaus I, Grotzinger J, Leippe M (2010) Caenopores are antimicrobial peptides in the nematode Caenorhabditis elegans instrumental in nutrition and immunity. Dev Comp Immunol 34(2):203–209. doi:10.1016/j.dci.2009.09.010

Ryder E, Ashburner M, Bautista-Llacer R, Drummond J, Webster J, Johnson G, Morley T, Chan YS, Blows F, Coulson D, Reuter G, Baisch H, Apelt C, Kauk A, Rudolph T, Kube M, Klimm M, Nickel C, Szidonya J, Maroy P, Pal M, Rasmuson-Lestander A, Ekstrom K, Stocker H, Hugentobler C, Hafen E, Gubb D, Pflugfelder G, Dorner C, Mechler B, Schenkel H, Marhold J, Serras F, Corominas M, Punset A, Roote J, Russell S (2007) The DrosDel deletion collection: a Drosophila genomewide chromosomal deficiency resource. Genetics 177(1): 615–629. doi:10.1534/genetics.107.076216

Ryu JH, Ha EM, Lee WJ (2010) Innate immunity and gut-microbe mutualism in Drosophila. Dev Comp Immunol 34(4):369–376. doi:10.1016/j.dci.2009.11.010

Salzet M (2001) Vertebrate innate immunity resembles a mosaic of invertebrate immune responses. Trends Immunol 22(6):285–288

Seed KD, Dennis JJ (2008) Development of Galleria mellonella as an alternative infection model for the Burkholderia cepacia complex. Infect Immun 76(3):1267–1275. doi:10.1128/IAI.01249-07

Shinzawa N, Nelson B, Aonuma H, Okado K, Fukumoto S, Miura M, Kanuka H (2009) p38 MAPK-dependent phagocytic encapsulation confers infection tolerance in Drosophila. Cell Host Microbe 6(3):244–252. doi:10.1016/j.chom.2009.07.010

Shiratsuchi A, Mori T, Sakurai K, Nagaosa K, Sekimizu K, Lee BL, Nakanishi Y (2012) Independent recognition of Staphylococcus aureus by two receptors for phagocytosis in Drosophila. J Biol Chem 287(26):21663–21672. doi:10.1074/jbc.M111.333807

Sifri CD, Mylonakis E, Singh KV, Qin X, Garsin DA, Murray BE, Ausubel FM, Calderwood SB (2002) Virulence effect of Enterococcus faecalis protease genes and the quorum-sensing locus fsr in Caenorhabditis elegans and mice. Infect Immun 70(10):5647–5650

Steinert M, Leippe M, Roeder T (2003) Surrogate hosts: protozoa and invertebrates as models for studying pathogen-host interactions. Int J Med Microbiol 293(5):321–332. doi:10.1078/1438-4221-00275

Tampakakis E, Okoli I, Mylonakis E (2008) A C. elegans-based, whole animal, in vivo screen for the identification of antifungal compounds. Nat Protoc 3(12):1925–1931. doi:10.1038/nprot.2008.193

Tan MW, Mahajan-Miklos S, Ausubel FM (1999a) Killing of Caenorhabditis elegans by Pseudomonas aeruginosa used to model mammalian bacterial pathogenesis. Proc Natl Acad Sci U S A 96(2):715–720

Tan MW, Rahme LG, Sternberg JA, Tompkins RG, Ausubel FM (1999b) Pseudomonas aeruginosa killing of Caenorhabditis elegans used to identify P. aeruginosa virulence factors. Proc Natl Acad Sci U S A 96(5):2408–2413

Teixeira N, Varahan S, Gorman MJ, Palmer KL, Zaidman-Remy A, Yokohata R, Nakayama J, Hancock LE, Jacinto A, Gilmore MS, de Fatima Silva Lopes M (2013) Drosophila host model reveals new enterococcus faecalis quorum-sensing associated virulence factors. PLoS One 8(5):e64740. doi:10.1371/journal.pone.0064740

Thomaz L, Garcia-Rodas R, Guimaraes AJ, Taborda CP, Zaragoza O, Nosanchuk JD (2013) Galleria mellonella as a model host to study Paracoccidioides lutzii and Histoplasma capsulatum. Virulence 4(2):139–146. doi:10.4161/viru.23047

Tzou P, Ohresser S, Ferrandon D, Capovilla M, Reichhart JM, Lemaitre B, Hoffmann JA, Imler JL (2000) Tissue-specific inducible expression of antimicrobial peptide genes in Drosophila surface epithelia. Immunity 13(5):737–748

Vodovar N, Vinals M, Liehl P, Basset A, Degrouard J, Spellman P, Boccard F, Lemaitre B (2005) Drosophila host defense after oral infection by an entomopathogenic Pseudomonas species. Proc Natl Acad Sci U S A 102(32):11414–11419. doi:10.1073/pnas.0502240102
Vogel H, Altincicek B, Glockner G, Vilcinskas A (2011) A comprehensive transcriptome and immune-gene repertoire of the lepidopteran model host Galleria mellonella. BMC Genomics 12:308. doi:10.1186/1471-2164-12-308
Wagner C, Isermann K, Fehrenbach H, Roeder T (2008) Molecular architecture of the fruit fly's airway epithelial immune system. BMC Genomics 9:446. doi:10.1186/1471-2164-9-446
Wagner C, Isermann K, Roeder T (2009) Infection induces a survival program and local remodeling in the airway epithelium of the fly. FASEB J 23(7):2045–2054. doi:10.1096/fj.08-114223
Walker KA, Maltez VI, Hall JD, Vitko NP, Miller VL (2013) A phenotype at last: essential role for the Yersinia enterocolitica Ysa type III secretion system in a Drosophila melanogaster S2 cell model. Infect Immun 81(7):2478–2487. doi:10.1128/IAI.01454-12
Wand ME, Muller CM, Titball RW, Michell SL (2011) Macrophage and Galleria mellonella infection models reflect the virulence of naturally occurring isolates of B. pseudomallei, B. thailandensis and B. oklahomensis. BMC Microbiol 11(1):11. doi:10.1186/1471-2180-11-11
Yanos ME, Bennett CF, Kaeberlein M (2012) Genome-wide RNAi longevity screens in Caenorhabditis elegans. Curr Genomics 13(7):508–518. doi:10.2174/138920212803251391

Chapter 2
Host–Symbiont Relationships: Understanding the Change from Guest to Pest

Robin M. Overstreet and Jeffrey M. Lotz

Abstract The several meanings for the term "symbiosis" create confusion, which can be avoided when the author provides details of the interrelationships between the symbiotic organism and the "host" so that a reader can clearly understand what definition is implied in each case. For example, we, as opposed to many other mentioned readers, consider a symbiont as an organism living in an association with another regardless of whether it causes a pathologic response or not, but from our title, the reader may incorrectly infer that we consider a parasite to be different from a symbiont. A symbiont is an organism that uses another organism as a habitat. This chapter discusses the primary associations and associated conflicts involving the terminology. It also provides both differentiation between and conflicting views regarding the interpretation of the terms "infect" and "infest," "infection" and "disease," and other terms. Many seemingly harmless symbionts of a wide array of taxonomic groups are triggered to become pathogenic or virulent, and we provide several examples of the provoking (stimulating) triggers, with the understanding that in most cases, the conditions for the triggered activities are much more complex and complicated than presented. Examples of triggers follow: environmental ones like temperature, toxic chemicals (dose), chemotherapeutics, dietary changes, and geographic habits; internal ones like host site, host resistance or susceptibility, and host modifications; and combinations of these and other conditions. We provide examples involving multiple triggers for organisms associated with termites, for an endemic virus being affected by multiple factors and having multiple effects on its commercial penaeid shrimp hosts, and for contrasting variables associated with two exotic viruses in wild and cultured commercial penaeid shrimps with an emphasis on hypothesizing how the pathogenicity developed in these two viruses. The chapter ends by trying to answer the question of why would a symbiont become pathogenic in some hosts and not in others from an evolutionary perspective. It uses two hypotheses to explain the increased virulence.

R.M. Overstreet (✉) • J.M. Lotz
The University of Southern Mississippi, 703 East Beach Drive, Ocean Springs, MS 39564, USA
e-mail: Robin.Overstreet@usm.edu

© Springer International Publishing Switzerland 2016
C.J. Hurst (ed.), *The Rasputin Effect: When Commensals and Symbionts Become Parasitic*, Advances in Environmental Microbiology 3,
DOI 10.1007/978-3-319-28170-4_2

2.1 Introduction

When reading an article on symbiosis, most readers assume they know the definition of all the associated words in the document. This is not the case; even the term "symbiosis" is defined differently by different authors in different fields, by those in different geographic areas, and by those taught by different mentors. The term "symbiosis" was originally used by the German de Bary (1879) to mean "living together." His meaning referred to all situations where either similar or unlike organisms or species live together in an intimate association. This is typically thought to include "commensalism," "mutualism," and "parasitism." We prefer to include all such associations under the general term "symbiosis" but realize that our title does not comply, and many readers prefer to use a more specific term for symbiosis and other associations. In the preface of his book (1970), the eminent parasitologist William Trager quoted "the conflict in nature between dissimilar kinds of organisms has been popularly expressed in phrases like 'struggle for existence' and 'survival of the fittest.' Yet few people realize that mutual cooperation between different kinds of organisms—symbiosis—is just as important, and that the 'fittest' may be the one that most helps another to survive."

For purposes of this series, we restrict "symbiont" to organisms that use other organisms as habitat. Therefore, symbiont is a division of nature similar to "terrestrial" or "marine." The symbiont is characterized by where it lives rather than the quality of the relationship. For our purposes, we divide up the world into two kinds of organisms, free living and not free living. This accords us the advantage of comparing the quality of the relationship between free-living organisms and their habitat with that between host and symbiont. The symbiont is dependent upon the host as any inhabitant is on its habitat. We consider symbionts to include bacteriophages (viruses that infect bacteria), bacteria, viruses, "protozoans," and metazoans and symbiosis to include commensalism, mutualism, parasitism, and other relationships. Moreover, all symbionts, even metazoans, observed best with a microscope will be considered microbes. More important, we will examine in this series a shift in the interaction among a host and some stage in the life cycle of the symbiont, whether it be free living, facultative, or obligate. We will emphasize that some symbiotic relationships can change from harmless to something that becomes far less favorable for the host.

2.2 Definitions

Symbiosis This term as defined above is usually thought of as an association (mutualism, commensalism, or parasitism) between organisms of different species involving a unilateral or bilateral exchange of material or energy. The symbiont, or symbiote, is any member of a pair of organisms involved in this symbiotic relationship, with the larger member usually designated as the host. Barrows (2011) provided a classification of symbiosis separating nonsocial symbiosis from social symbiosis, and others classify various terms differently. Most of these terms and classifications will not be treated in this chapter.

Commensalism A symbiotic relationship in which one of two partner species benefits and the other shows no apparent beneficial or harmful effect.

Mutualism A symbiotic relationship in which two or more partners gain reciprocal benefits, usually mutual ones.

Parasitism A symbiotic relationship in which a symbiont lives all or part of its life in or on a living host, usually benefiting while harming the host in some way and usually having a higher reproductive potential than the host. Noble et al. (1989) define it as an association between two different species of organisms in which the dependence of the parasite on its host is metabolic and involves mutual exchange of substances; this dependence is the result of a loss of genetic information by the parasite. There are several atypical kinds of parasites. An accidental one infects an unusual or unnatural host; a commensal one derives its substances from the food of its host; and an erratic, or aberrant, one infects an unusual site. A facultative parasite is usually parasitic but is capable of an independent existence, while an obligate one cannot lead an independent, nonparasitic existence. The late Gerald Schmidt's definitions are listed by Roberts and Janovy (2009). He considered a parasite the "raison d'être" for parasitologists, a parasitologist as a quaint person who seeks truth in strange places and one who sits on one stool while staring at another, and parasitology as a study of the most common mode of life on earth. Many parasites have complicated life cycles with multiple hosts and with a variety of stages, not all of which harm or depend on the host. One can glean that students find it difficult to define a parasite. This series will treat a variety of facultative and other relationships.

Predatory association A relationship in which a predator obtains its living by preying on other animals, usually consuming all or part of it, usually killing it. This relationship can be symbiotic when a predator feeds on one or a few species or when it is relatively small and not harming its host. For example, a trematode (all trematodes are considered parasites) may live in the lumen of a host without causing any harm; that trematode is occasionally called a micropredator.

Other terms that will be causing confusion in this series are *infection* and *infestation*. American parasitologists usually define an internal association as an infection, whether harm is caused or not or the organism is small or large. If the association is external, it is referred to as an infestation, and the symbiont is called an ectosymbiont or an ectoparasite. For an internal association to be called an infection, microbiologists usually restrict the agents to viruses, bacteria, and fungi. Consequently, some authors restrict an infestation to a metazoan parasite (e.g., Maggenti et al. 2005). Barrows (2011) provided three meanings for an infestation: (1) a parasite's colonization, utilization, or both of the host; (2) a host being colonized, utilized, or both by parasites; and (3) and environment being colonized, utilized, or both by pests. Other authors consider an infestation to refer to a population rather than to individuals. Still, others use the word "infestation" to suggest an action and the term "infection" to suggest a condition or a state. Microorganisms such as oral bacteria that live naturally in the mouth or elsewhere in a body are not considered infections or infectious agents by many microbiologists, but the organisms are symbionts.

Another important term for this series is *disease*. In medical cases, this often refers to dose. A well-nourished person infected with the American hookworm (*Necator americanus*) with a normal hemoglobin and five eggs per milligram of feces, perhaps resulting from up to 50 adult worms, has an "infection" but without a corresponding disease caused by the loss of blood or other detrimental signs. An egg count of 20 per mg in a healthy person or 5 in one anemic with iron or protein deficiency usually indicates the threshold of disease. Massive infections with disease can produce as many as 50 eggs per mg of feces (Beaver et al. 1984). Lightner and Redman (1998) discussed this term when dealing with shrimp infections. They said "In veterinary and human pathology, the terms 'disease' and 'syndrome' have a range of definitions. Dorland's Medical Dictionary (1968) defines disease as 'a definite morbid process, often with a characteristic train of symptoms' and syndrome as 'a combination of symptoms [signs] resulting from a single cause or so commonly occurring together as to constitute a distinct clinical entity.' A simpler definition of disease is 'any alteration from the normal state of health.' The latter definition permits the inclusion of alterations in health that may result in subtle conditions in which poor health or reduced resistance to stress are the only signs of disease, as well as those disease syndromes on the other extreme, that may be accompanied by catastrophic losses."

An important, prudent course of action for an author is to carefully define a term or description that is being treated in any manuscript. A reader, on the other hand, may have serious problems understanding what is being said without that definition. Terms being used in symbiosis are especially difficult because research conducted to define an entire relationship seldom exists. Several glossaries are careful not to include definitions of either infections or infestations, and others are careful not to define a parasite! Many terms used for defining ecology of parasites such as "incidence," "prevalence," "intensity," "mean intensity," "component population," and many others are used differently by different people, and we strongly recommend the article by Bush et al. (1997) as a standard or at least a source of discussion.

2.3 Factors Triggering a Harmful Relationship

Usually, a combination of factors triggers a harmful condition, but only one or two of these factors are known in any detail for most cases. Furthermore, most of the examples below treat only one or two of these triggering factors. Differentiating the triggering factors into those from the external environment and those originating from the host is, for the most part, impossible to characterize because the two are typically entwined. Moreover, genetic triggers that involve expression such as upregulation, downregulation, and gene knockdown may each necessitate a variety of triggers for expression. Triggers causing a harmless symbiont to transform into a harmful one can also involve the initiation, interruption, or inhibition of biochemical pathways, complicated actions that sometimes incorporate a cascade of reactions. This chapter will not treat the details of mechanisms but rather present examples of what appear like results of cause-and-effect. As indicated above, the headings indicate only some of the triggers in each case, with an emphasis on the primary one.

2.3.1 *Environmental*

2.3.1.1 Temperature

Temperature often regulates disease in marine animals to benefit stocks of all parties. For example, the English sole, *Parophrys vetulus*, arrives into its estuarine nursery area in the northern Oregon Coast from January to April, grows dramatically in the upper estuary during the summer, and then migrates offshore in the autumn, never to return inshore. When inshore, the sole gets infected by the microsporidian *Glugea stephani* that develops in the intestine at temperatures >15 °C. This agent appears to cause mortalities from September through November, but this period is when most of the fish routinely migrate offshore where the temperature remains less than 10 °C, a temperature that was shown experimentally to inhibit the microsporidian agent's development, which ultimately results in uninfective spores within a few months. Consequently, the parasite seldom harms the host stock. However, approximately 10 % of the sole stock remained in the upper estuary where an estimated 80 % of those fish had a rapidly developing infection and most probably died and decomposed or were eaten, dispersing the spores. Those fish plus the overwintering infected fish and few reservoir starry flounder juveniles (*Platichthys stellatus*) (a reservoir host is a primary host that harbors the pathogen but typically exhibits no ill effect and serves as a source of infection) provided infective spores for next year-class of sole (Olson 1981; Overstreet 1982).

Temperature also has a great bearing on many other pathogenic triggers. For example, the coccidian *Calyptospora funduli* commonly infects the liver, pancreas, and occasionally other tissues in the Gulf killifish, *Fundulus grandis*, in the coastal Gulf of Mexico. The fish becomes infected by feeding on either the infected intermediate host *Palaemonetes pugio* (the common daggerblade grass shrimp) or related species. Infections are synchronous, meaning that all the developing stages are the same. Early developing stages and mature oocyst stages are rarely seen in the same fish. In some cases, over 95 % of the liver can be infected without causing any notable severe harm to the host. When heavily infected and lightly or noninfected killifish were maintained in outdoor raceways and freezing occurred, the heavily infected fish all selectively died. Presumably, that was because the liver serves as a storage reservoir for necessary glycogen, vitamins, minerals, and other necessary nutritional resources. Those resources found in 5–10 % of the liver were presumably not enough to satisfy the needs of the fish during stressful low-temperature conditions. A series of experimental infections conducted at about 22, 10, and 7 °C for periods between 5 and 20 days in length (Solangi et al. 1982) showed that both low-temperature treatments inhibited all developmental stages. When the fish were returned from the low temperatures to 22 °C, development of all the stages resumed except in fish exposed to 7 °C for 20 days. In those infections, many of the coccidian organisms were atrophied or disintegrated within their parasitophorous vacuoles, and necrosis also occurred in some of the pancreatic tissue. In any event, decreased parasitism during inhibition was not linked with a leucocytic inflammatory response. When fish remained at a constant temperature of about 24 °C, an inflammatory response to the organism commenced at day 18, intensified at day 20, and diminished by about day 30 (Hawkins et al. 1981, Solangi and Overstreet 1980). In other words, inflammation associated with infections in warm water begins as gamonts developed and ends after formation of the oocyst wall. Consequently, infections that occur during cold winters can be either helpful or harmful to the host, depending on the stage of the parasite, the extent of low-temperature exposure, the age of the fish, and other variables.

Atypical temperatures, such as warm water associated with power plants, can cause infections of a specific parasite during periods when the hosts are more likely to be consumed by predators, more susceptible to disease, or more susceptible to interactions among parasites that can occur and result in unusual pathogenic conditions.

As also suggested above, temperature can have an effect on development of a parasite and can occasionally result in a harmful effect. The ascaridoid nematode *Contracaecum multipapillatum* in a coastal lagoon at Celestún, State of Yucatán, Mexico, infects as definitive hosts the olivaceous cormorant and the great egret. The intermediate host for this avian nematode is the Mayan cichlid (*Cichlasoma urophthalmus*). The juvenile in the fish host for all members of the genus *Contracaecum* is usually a third stage; however, in the warm area of Mexico, many of the juveniles had developed into the fourth stage. When these fourth-stage juveniles were fed to a kitten, they developed into adults in the intestine, often associated with hemorrhaging small ulcers (Vidal-Martínez et al. 1994); they did not develop or cause a pathogenic response in rats, ducks, or chickens. People commonly eat the cichlid in Mexico and, consequently, are potential hosts. When Deardorff and Overstreet (1980b) fed third-stage juveniles, the typical stage occurring in fish, to day-old chicks and ducklings and to mammals, they neither developed nor survived; however, when the third-stage juvenile was surgically inserted within tied off semipermeable dialysis tubing into the abdominal cavity of the animals, the worms did mature.

The gnathostomatid nematode *Echinocephalus sinensis* matures in the intestine of the eagle ray, which acquires the third-stage juveniles by eating the infected oyster *Crassostrea gigas*. Infections in this commercial oyster were most prevalent and had the highest intensity in Hong Kong between July and September. Kittens, monkeys, and puppies were fed the infective stage from oysters during every month, but those worms collected only from the warm months of August to October infected the mammals (Ko 1976). The worms penetrated the wall of the stomach and intestine, migrated to and lodged in various tissues, and often killed the hosts. To test this temperature-triggering condition, Ko (1977) obtained oysters during the infective season and acclimated them to 5, 15, about 24, 28, and 33 °C. Worms from these different temperature groups were fed to kittens, and infections were most abundant in the 33 °C group and were less so in the 28 °C group; only one kitten became infected from the 24 °C group, and no kitten from the 5 or 15 °C group became infected. Since people eat raw oysters, the nematode constitutes a potential public health risk.

2.3.1.2 Environmental Habitat

Under normal environmental conditions, parasites seldom harm their hosts. When the habitat is a series of aquaculture ponds, several conditions are modified. The pond design usually accommodates a heavy density of snail intermediate hosts near the shallow shore where fish fry occur, and the fish stock inhabiting the middle of the ponds creates an abundance of prey for fish-eating birds. For example, primarily in Mississippi but also less so in adjacent states, the channel catfish, *Ictalurus punctatus*, is reared, comprising a multimillion dollar industry. In the natural environment, catfishes, including their madtom host relatives, do not occur as juveniles in habitats where the host marsh ramshorn snail, *Planorbella trivolvis*, and American white pelican, *Pelecanus erythrorhynchos*, occur and defecate in any abundance. Consequently, infections of the metacercaria (larval stage acquired from a cercaria shed by the ramshorn snail) of the pathogenic diplostomoid trematode *Bolbophorus damnificus* are relatively rare in those natural waters. A total of five different diplostomoid trematode species infecting the catfish were able to kill it in aquaculture environments. Two of these developed from cercariae shed from *P. trivolvis* that ultimately mature in the American white pelican; two were from worms shed from the ash gyro, *Gyraulus parvus*, that mature in the double-crested cormorant; and the fifth species is one that matures in seagulls. The most harmful species, *B. damnificus*, can be present in the catfish with infections of less than 40 metacercariae without harming it. Heavier infections seem to have an effect on the kidney of the fish aiding in their mortalities. In the shallow water where large numbers of infected snails occur, thousands of cercariae can penetrate the young catfish and kill it quickly, even before developing into metacercariae. This is in contrast to two other trematodes, *Bursacetabulus pelecanus* and *Austrodiplostomum compactum*, which both infected the nerve cord, brain, optic nerve, and eye and at least in experimental infections can occur as several hundred individuals without killing the host (Overstreet and Curran 2004). The trigger for mortality regarding *B. damnificus* can be more complicated than pure numbers. In experimental infections with fingerling channel catfish exposed to a sublethal infection of trematode cercariae per fish, they died when additionally exposed to the bacterium *Edwardsiella ictaluri*, the agent of enteric septicemia of catfish. Controls given either the cercariae or nothing did not die. Those given 7.5×10^5 colony-forming units per mL of bacteria for 30 min singularly without cercarial exposure started dying at day 7, and, by day 21, the percent cumulative mortality leveled at about 46 compared with 84 % for the group given both the cercariae and bacteria. Then at day 28 when the metacercariae fully developed, groups consisting of the remaining fish given only the trematode and of the negative controls were exposed to the bacteria. Both groups starting dying at day 7, but, by day 21, there was no significant difference in the percent cumulative mortality of about 18 % (Labrie et al. 2004). When contaminating forage fish that occur in the ponds in addition to the commercial catfish became exposed to the cercariae of *B. damnificus*, they did not become infected. On the other hand, there is an unnamed species of *Bolbophorus* which also infects both the same snail and pelican, concurrently with *B. damnificus*. However, it does not use the catfish as the second intermediate host, but it uses mosquitofish and sunfish, which it can kill when in high numbers (Overstreet et al. 2002). Also of note is the finding that channel catfish exposed to the *E. ictaluri* bacterium 1 day prior to being exposed to the pathogenic ciliate *Ichthyophthirius multifiliis* had 71 % mortality compared with 27, 29, and 0 % for those respectively given only the bacterium or the ciliate or given neither. The bacterium could be detected by PCR in the gills, brain, liver, and kidney of the fish whether infected with the ciliate or not, but, by day 8, the bacterium no longer persisted in the bacteria-only group except in the kidney (Shoemaker et al. 2012).

A few free-living amebas are well known because they can infect humans and occasionally cause fatalities (see Overstreet 2013). Best known is *Naegleria fowleri* because it causes fatal "primary amebic meningoencephalitis (PAM)," but it is restricted to freshwater, and victims are typically those who swim for extended periods underwater in open bodies with a silty-muddy substratum. The swimmers' nasal mucosa becomes weakened to the point that the ameba can penetrate the membrane and enter the central nervous system (CNS) along the olfactory nerve. Because that site contains little cellular inflammatory response but provides a good medium for amebic growth, the organism replicates rapidly, usually before the disease can be diagnosed and treated. Marine free-living species of the genus *Acanthamoeba* do not grow as rapidly and, depending on the species, invade different sites. They enter the skin, lower respiratory tract, or nasopharynx, and the vegetative trophozoite can reach the CNS through the circulatory system. These acanthamebic infections may take weeks or months to cause death, and often infections occur in patients that are immunocompromised (unlike the otherwise healthy hosts that become ill due to PAM) but are still difficult to diagnose. Six different ameba species have also been associated with painful amebic keratitis, a difficult to treat corneal disease. Even though recognized in 1973, amebic keratitis was not common until 1985 when contact lenses became popular; these lenses typically are maintained overnight in a saline solution that can become a source of contamination.

2.3.1.3 Chemical

Toxic waste products can trigger a reduction in host resistance, resulting in susceptibility to rapidly reproducing agents that would normally be held in check or not able to produce infections. The myxosporidian *Henneguya gambusi* infects the western mosquitofish, *Gambusia affinis*, and is rare, not previously considered pathogenic, and not known from any other fish (Parker et al. 1971). We have also seen it in mosquitofish but only from a location in Mississippi receiving wastes from a timber treatment facility, involving the wood preservative chromated copper arsenate, a mixture of chromium, copper, and arsenic (as copper(II) arsenate) (Overstreet 1997). Presumably, the mosquitofish becomes infected from being in close contact with actinospores shed from a tubificid oligochaete. A small percentage of the few fish in the contaminated creek exhibited mass infection (Figs. 2.1 and 2.2), and it was typically histozoic with plasmodia throughout the skeletal muscle mixed with tissue debris not associated with an obvious inflammatory response rather than in pseudocysts located in the epidermis, corium, and subdermal connective tissue as originally described. A few of those fish from the wild had a severely pathogenic infection, with mature spores liberated from the plasmodia and spread into adjacent muscles, replacing most tissues throughout the body (e.g., Dyková and Lom 2007). However, when we collected 50 fish from this station and 50 from another station containing tubificids and then maintained them in aquaria in our laboratory and fed them commercial flakes daily, over half the fish from the contaminated location became moribund or died within 2 months. No fish from the other location died or appeared unhealthy. The moribund fish were sectioned or examined fresh and demonstrated the myxosporidian with its 10 by 6 μm spores

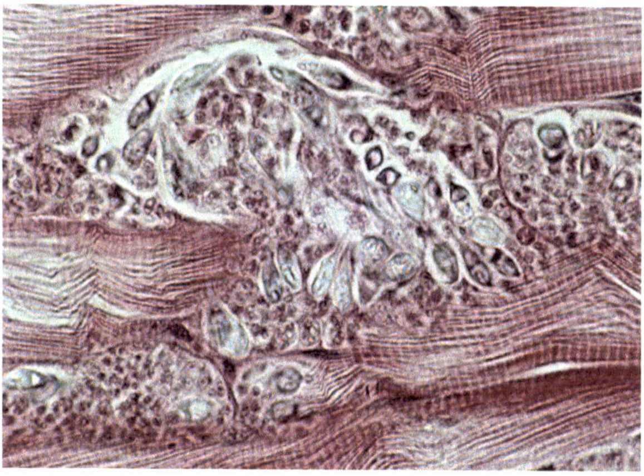

Fig. 2.1 Mass infection of the myxosporidian *Henneguya gambusi* triggered within the skeletal muscle of the western mosquitofish, *Gambusia affinis*, from a location in coastal Mississippi that contains heavy metal contamination. Vegetative and mature stages in plasmodia located between and within muscle fibers, with mature triggered spores being liberated and destroying muscle and showing no plasmalemma separating spores from fibers

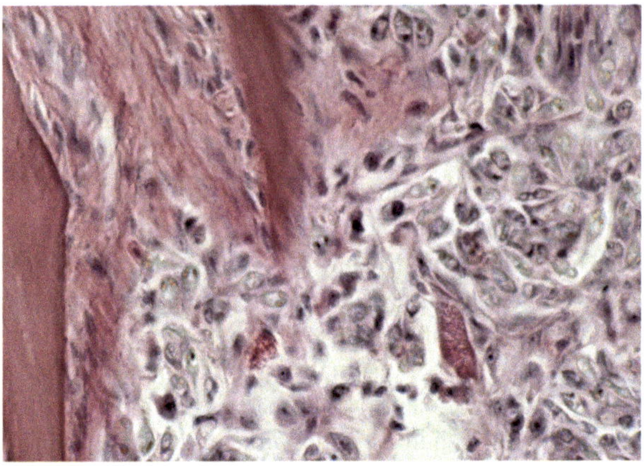

Fig. 2.2 *Henneguya gambusi* in *Gambusia affinis*. Mature spores with some liberated ones from plasmodia leaving only remnants of muscle fibers. Note lack of host inflammatory cells

invading most tissues throughout the samples. Representatives of live fish from both groups did not exhibit infection in section, and none of the remaining fish exhibited infection when critically examined as fresh. The contamination apparently reduced the host's protective immune response which would normally keep a low infection in check. Light infections had to be present in about half the fish from the contaminated location when brought into the laboratory, even though not evident in fresh or sectioned material. Under normal conditions, spores of all histozoic species occur in small to large pseudocysts, the latter containing millions of spores. When the plasmodium is extremely small or when few vegetative or spore individuals are present, an infection is difficult to see, and polymerase chain reaction (PCR) probes are necessary to detect infections. In fact, the only person we have seen skilled enough to routinely detect minimal infections is Iva Dyková of the Academy of Sciences of the Czech Republic.

Another example of a pathogenic agent in the western mosquitofish deals with a free-living organism rather than a known parasite. The free-living organism, the ciliate *Tetrahymena corlissi*, has been reported from fish in aquaria and hatcheries, causing pathological alterations and mortality by Hoffman et al. (1975). They were unable to experimentally infect fish with the ciliate and suggested that infection resulted from a wound or stress. We (Overstreet et al. 1995) sampled the mosquitofish from an outlet canal from an integrated pulp and paper mill, 2.5 km upriver from the canal where water entering from a dam kept mill effluent from flowing upstream, and three downstream locations. Two of 99 fish from upstream, but none from the other locations, had a proliferation of the ciliate in the head and in the musculature, brachial chamber, and pericardial sac; that location also had fish with a significantly higher prevalence of macrophage aggregates in the spleen, a good indicator of stress. After publication, we learned of a long submerged pipe emptying toxic wastes upstream from our locations and originating from a nonrelated facility located several km from the river and apparently promoting the ciliate infection. Some ciliate specimens measured larger than those reported by Hoffman et al. (1975), but sequencing of presumed free-living ciliates infecting various aquatic hosts should allow identifications and experimentation to determine details of the mechanisms that shift a free-living organism to become a pathogenic one. For parasites or diseases of a host organism to be used as monitors of environmental health or biological activities, both the host and the symbionts or pathological responses need to fit several criteria (Overstreet 1997).

Chemotherapeutics can also serve as toxins. A CDC report (U.S. Department of Health and Human Services, Centers for Disease Control and Prevention 2013) estimated that about 100 million people worldwide possess a chronic infection with the nematode *Strongyloides stercoralis*, and prevalence of the infection in refugee populations ranged from 11 to 69 % by serosurveys. This species has the unusual ability to replicate and auto-infect its human host, persisting for decades. *Strongyloides* hyperinfection syndrome may be triggered many years after migration of a prior refugee to a nonendemic locality, with large numbers of the parasite infiltrating internal organs, resulting in fatality rates exceeding 50 %. The syndrome is generally induced when an individual is placed on corticosteroids, although other immunosuppressive conditions such as cancer and transplant chemotherapeutic immunosuppression may also trigger the hyperinfection.

A change in diet as well as some chemotherapeutic compounds can serve as a trigger and induce a parasite located in one internal habitat such as the intestine, blood, visceral organ, or muscle to migrate to more sensitive tissue or be released from a cyst or an encapsulation and migrate or replicate.

2.3.2 *Site Within Host*

Free-living or symbiotic metazoans, protozoans, algae, fungi, and bacteria can get into abnormal sites (locations in host) or habitats and develop into pathological agents. Some cases are rare and others are common. A rare case consisted of finding the diatom *Amphora* sp., which normally is a free-living alga, in the white shrimp, *Litopenaeus setiferus*, presumably after the shrimp's carapace had been punctured or abraded. The diatom occurred in abundance as clusters in the hemolymph, plus some individuals were present in the gills associated with a melanistic response. The shrimp died. To test the effect of the related *Amphora coffeaeformis* on shrimp, Overstreet and Safford (1980) injected cultures into shrimp resulting in melanistic responses, but no extensive replication occurred as had been observed in the shrimp killed with *Amphora* sp. A more common opportunist is the fungus *Fusarium solani*, which also developed in penaeid shrimp. *Farfantepenaeus californiensis*, a species of penaeid initially being considered for aquaculture, is a highly susceptible species to this fungus. When experimentally infected with a cultured conidial suspension of this free-living species (Hose et al. 1984), all shrimp became infected within 14 days, tissue destruction and a strong but often unsuccessful hemocytic inflammatory response occurred, and over half of the shrimp died within 24 days. Biochemical and hemocytic parameters of the hemolymph changed significantly as the infection developed when compared with noninfected shrimp. When *Farfantepenaeus aztecus* or *Litopenaeus setiferus* was held in seawater with macroconidia spores cultured from an infection from *Farfantepenaeus californiensis* or injected with those spores, resistance to infection occurred. A strong hemocytic encapsulation and melanization of the macroconidia and lysis of the conidiospores occurred in the gills by 28 h. If spore dosage was excessive, 3.2×10^6 or greater, all brown shrimp died within 24 h from the macroconidia and hyphae blocking the distal portion of the gill lamellae (Solangi and Lightner 1976). In *Litopenaeus vannamei*, the primary cultured penaeid today, naturally infected individuals in culture are moderately susceptible to infection and the spore aggregate, often grossly apparent in the distal portion of the eyestocks; *Penaeus monodon*, which used to be the primary commercially cultured species, is relatively resistant (Lightner 1996). For pathogenicity by free-living algae and fungi, the triggering factor seems to be a wound or being compromised by other infectious agents or toxicants, but, as pointed out, the host species is also critical.

One or more ciliates identified as or presumed to be the scuticociliate *Orchitophrya stellarum* has an extensive host and geographic range in Europe, Australia, and North America as determined by morphological and sometimes molecular techniques. It is considered a facultative parasite that can live indefinitely outside the host when cultured with bacteria and tissue detritus or yeast but with different size and morphology when compared with material in male sea stars. Cultured specimens enter male but not female sea stars, probably through the gonopores, where it enters the testes and feeds on sperm of fully mature individuals only (Stickle et al. 2007). It has been reported from several sea star species and may have been inadvertently introduced into the Northeast Pacific region in the late 1980s (Boom, in Bates et al. 2010). Recently, Small et al. (2013) determined using ITS (internal transcribed spacer) sequences and PCR (polymerase chain reaction) analyses that the histophagous ciliate infecting the blue crab, *Callinectes sapidus*, in research facilities in Virginia and previously thought to be *Mesanophrys chesapeakensis* was actually *O. stellarum*. The same or similar infectious ciliate was probably responsible for histophagous disease in wild and captive blue crabs in Mississippi (Shields and Overstreet 2007), penaeid shrimps in Mississippi, and the wild lined shore crab, *Pachygrapsus crassipes*, from Carpinteria Salt Marsh, California, that we examined for Ryan F. Hechinger, University of California, Santa Barbara. This infection spreads in the partially closed circulatory system of the decapod (McGaw 2005) from the hemolymph to muscle, visceral organs, and other tissues (Figs. 2.3, 2.4, 2.5, and 2.6), usually killing the host. It has been assumed to enter into wounds of its hosts and now shown by Miller et al. (2013) to cause rapidly developing fatal infections in the blue crab inoculated with the ciliate or exposed to ciliates after experimental autonomy. When exposed to ciliates and not wounded, the crab seldom died. For comparisons, the fiddler crab *Uca minax* was inoculated with doses of either 10 or over 500 ciliates per crab, and crabs with 10 sometimes established infections, but those with the higher doses developed them rapidly. The infections developed at 10–15 °C, and ciliates were attracted to blue crab serum over other nutrient sources, suggesting the facultative nature of a blue crab parasite.

Viridans streptococci typically occur harmlessly in the mouth. These bacteria can be differentiated from *Streptococcus pneumoniae* using several tests.

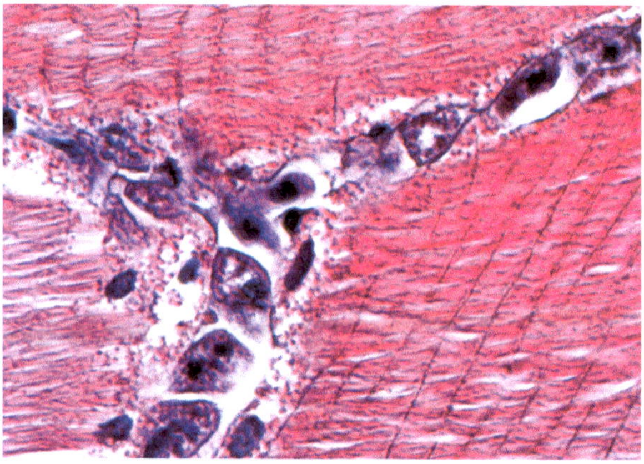

Fig. 2.3 The scuticociliate *Orchitophrya stellarum* or related ciliate in wild *Pachygrapsus crassipes* from Carpinteria Salt Marsh, California. Ciliates in hemolymph vessels beginning to erode adjacent skeletal myofibrils

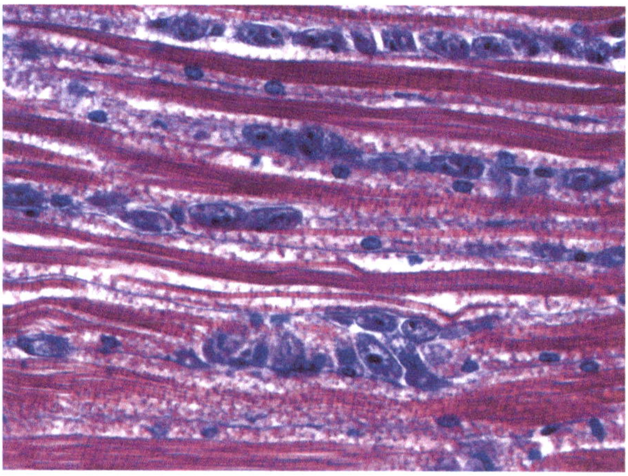

Fig. 2.4 *Orchitophrya stellarum* or related ciliate in *Pachygrapsus crassipes*. Heavily eroded muscle fibers

Moreover, they lack either the polysaccharide-based capsule typical of *S. pneumoniae* or the Lancefield antigens, based on the carbohydrate composition of bacterial antigens found on the cell walls in beta-hemolytic bacteria of the pyogenic, or pus-producing, members of the genus. Some may be involved in other mouth or gingival infections as pericoronitis or inflammation of the gums around molar teeth. However, if they are introduced into the bloodstream from surgical or other lesions, they have the potential of causing endocarditis, especially in individuals with damaged heart valves. Viridans streptococci have the unique ability to synthesize dextrans from glucose, allowing them to adhere to fibrin-platelet aggregates at damaged heart valves.

Often there occurs an indirect site-associated triggering mechanism for parasite establishment, with a corresponding associated ability to enhance or decrease pathogenic effects. The bothriocephalid *Anantrum tortum*, a long, up to over 15 cm, cestode, twists either singly or in groups of up to eight, within the intestine of its relatively small fish host, *Synodus foetens* (inshore lizardfish). It can occur

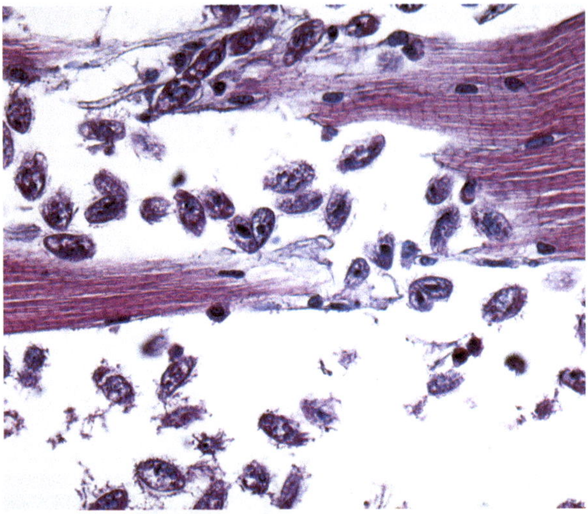

Fig. 2.5 *Orchitophrya stellarum* or related ciliate in *Pachygrapsus crassipes*. Severed muscle fiber showing clear view of ciliates

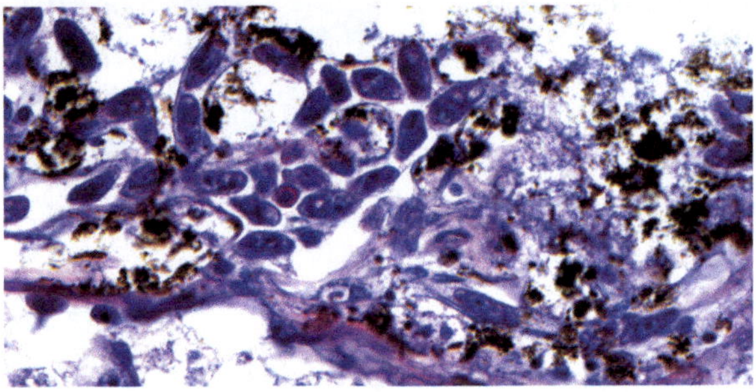

Fig. 2.6 *Orchitophrya stellarum* or related ciliate in *Pachygrapsus crassipes*. Infection showing associated melanistic response

near the pyloric ceca, along the intestine, or near the anus of the samples, measuring up to 300 mm long. Overstreet (1968) detected a 99 % level of significance over a 2-year period between the monthly prevalence of worms located near the anus and water temperature. Salinity was held constant, and an inverse sine transformation was applied to the percentage data. This anal location favors loss of the worm during warm weather, possibly assisting probability of best completing life cycle, but to be expelled, the lizardfish appears to have needed a several-mm-long packet of prey fish scales with a diameter much greater than that of the intestine. The carnivorous lizardfish can eat fishes and shrimps larger than itself. Consequently, fish prey with large scales tends to be advantageous for the lizardfish to rid itself of the parasite.

Sometimes, a fortuitous site creates a trigger for pathogenicity. For example, a diver accidently punctured his hand while reloading his spear gun. Later, unconcerned about his wound, he was filleting a jackfish, and a female of the nematode *Philometra* sp. invaded the wound and attached deeply. After several unsuccessful attempts to remove the worm, the pain-inflicting infection required surgery (Deardorff et al. 1986).

Not to be overlooked is the obvious matter of dose. This obvious matter is treated in discussions and is important for all agents from viruses to metazoans. In the case of metazoans, a host can come into contact with an especially high, unnatural dose of infective agents such as thousands of cercariae killing an intermediate host well before the corresponding metacercariae develop (e.g., Overstreet and Curran 2004). Intermediate hosts under specific conditions can be heavily infected simultaneously with several parasitic species. Some parasites such as microphallid trematodes infect a variety of birds and mammals. Therefore, a bird could acquire a harmful infection by feeding on individual heavily infected intermediate hosts, or source communities (Bush et al. 1993).

2.3.3 Host Species

The host species is also critical for many protozoans that are acquired by feeding on an intermediate host in which replication or maturation of a stage is necessary. For example, whereas a few killifish species are natural hosts of *Calyptospora funduli*, several related atheriniform fishes can be experimentally infected, but no nonatheriniform could be infected (Fournie and Overstreet 1993). Those experimentally infected demonstrated a variety of abnormalities, including asynchronous development, degeneration of early developmental stages, formation of macrophage aggregates, and a granulomatous inflammatory response, especially one exhibiting liver destruction in *Fundulus olivaceus* and *Rivulus marmoratus*. Of course, feeding behavior, environmental conditions, and geographic isolation can also serve as barriers to infection in fish in addition to innate immune barriers.

When introduced species become established in a system, they can be resistant to harmful factors in the environment, or they can be susceptible to extensive predation, diseases, and parasites. Species of the ascaridoid genus *Goezia* typically embed in an encapsulated ulcer in the wall as well as free in the lumen of the stomach of their natural fish hosts. These associations are described by Deardorff and Overstreet (1980a) for the nematode species *Goezia pelagia* in the cobia, *Rachycentron canadum*, and the Atlantic spadefish, *Chaetodipterus faber*, as well as for *Goezia minuta* in the gafftopsail catfish, *Bagre marinus*; the hardhead catfish, *Ariopsis felis*; and the inshore lizardfish, *Synodus foetens*. In none of these hosts does there appear to be any significant harm associated with the presence of the nematode (Deardorff and Overstreet 1980a). However, when the introduced host species such as the blue tilapia, *Oreochromis aureus*; striped bass, *Morone saxatilis*; and hybrid striped bass become infected with *Goezia sinamora*, the nematode produces massive fibrotic nodules in the fish stomach, and the nematode can also cause mortality. *Goezia sinamora* was implicated in mortality of hatchery-reared striped bass and tilapia introduced into a series of lakes in Florida, including Lake Parker [Gaines and Rogers 1972, see correction by Deardorff and Overstreet (1980a)]. In the tilapia, this roundworm has been observed to migrate through the intestinal wall, causing extensive lesions, in addition to forming nodules in the stomach. When the largemouth bass, *Micropterus salmoides*, and other native species of fish were exposed to this nematode, the resulting infections contained relatively few of the roundworms and demonstrated no conspicuous harmful effect (Deardorff and Overstreet 1980a).

2.3.4 Host Modifications

Several books and chapters [such as those by Overstreet (1983), Barnard and Behnke (1990), Lewis et al. (2002), Moore (2002), Lefèvre et al. (2009), Adamo (2012), Adamo and Webster (2013)] have reported on parasite infections resulting in virulence, evolutionary, and behavioral changes in intermediate hosts such that the intermediate host has a better opportunity to be preyed upon by the appropriate definitive host than by chance alone. Whereas these changes are not necessarily triggers that make an agent more pathogenic, they often are changes that reflect the point at which the undeveloped, noninfective agent becomes infective to the definitive host and pathogenic to the intermediate host. A good example is the microphallid trematode *Levinseniella byrdi*, which infects the intestinal ceca of a few bird species such as the seaside sparrow, clapper rail, willet, and semipalmated sandpiper. These birds acquire the infection from one of a few talitrid amphipods such as *Uhlorchestia uhleri* from salt marshes of Texas to North Carolina and *Uhlorchestia spartinophilia* in a similar habitat from Cape Canaveral, Florida, to central Maine. In the Gulf of Mexico, the metacercaria also occurs in the amphipod species *Orchestia grillus* and *Chelorchestia forceps* (Bousfield and Heard 1986). The dramatic modification in the amphipod usually occurs after about a month, once the metacercaria becomes infective. At this point, the host amphipod turns from greenish or grayish to a translucent or bright orange (Fig. 2.7). The infected amphipod also unusually slows down its movements and, in contrast to its uninfected, negatively phototactic cohorts, does not always hide under wracks of dead, dissociated leaves and stems of spartina grass or other of their dietary debris shelters. Apparently, the amphipod carotenoids become unbound from protein of the infected host, allowing the bird host to be preferentially attracted to the now brightly colored amphipods containing infective metacercariae (Bousfield and Heard 1986). Johnson et al. (2009) experimentally manipulated a few tidal salt marsh creeks in Plum Island Estuary, Massachusetts, by nutrient fertilizer enrichment and exclusion of the killifish *Fundulus heteroclitus* (mummichog), a primary predator in the system. Interaction of the two treatments reduced abundance of the common *Uhlorchestia spartinophilia*, and apparently infected amphipods moved from the marsh edge to the adjoining creek-wall habitats during 1 year, resulting in 97 % of the amphipods clinging on the creek walls and exhibiting the bright orange. A subsample of the colored amphipods was confirmed to be infected, and they were seen being fed upon by the semipalmated sandpiper and seaside sparrow.

In regard to the above host modification, other microphallid species in those amphipods did not turn their hosts orange or noticeably modify their behavior. A similar situation has been noted to occur where *Levinseniella tasmaniae* but not

Fig. 2.7 Specimens of the amphipod *Orchestia grillus* in coastal Mississippi showing on the *top* an uninfected one and on the *bottom* a transformed orangish one that is infected with the microphallid *Levinseniella byrdi*, a trematode that infects the intestinal ceca of a few bird species. The transformation triggers phenotypic and behavioral changes specifically attracting infective specimens to predatory birds in which the trematode matures

other microphallids induced an orangish color in the amphipod *Austrochiltonia australis* in Tasmania (Smith 1981). That is not to say all species of *Levinseniella* induce coloration and other modifications in their amphipod hosts since *Levinseniella tridigitata* does not modify *Gammarus aequicauda* (see Thomas et al. 1996) and *Levinseniella carteretensis* (or *Levinseniella hunteri*) does not induce a color change in any of the five talitrids, including *U. uhleri*, when experimentally infected (Bousfield and Heard 1986, Heard personal communication). Another trigger involves photoreception. The snail host typically sheds large quantities of cercariae of *Levinseniella tasmaniae* infective for the amphipod host when under light conditions, whether natural or experimentally reversed (Smith 1981). And yet, cercariae of other species are shed during dark or other physical conditions.

2.3.5 Combination of Many Factors

Almost all symbiotic relationships, including those discussed above, involve a combination of triggering factors. Those factors and host–symbiont relationships involving termites are particularly well studied. We, ourselves, have examined in considerable detail the host–symbiont relationships affecting the outcome of pathogenic viruses in populations of commercial penaeid shrimp. We will address in this section some of the symbiotic relationships and interrelationships that can involve a shift from harmless to harmful relative to three host groups.

2.3.5.1 Termites

The relationships between termites or related insects and their symbionts are numerous and provide examples of an abundance of triggers, including diet, to control the various relationships. Some of these involving termites will be discussed elsewhere by David Bignell (Volume 2, Chapter 6 of this series). Investigations by L. R. Cleveland (e.g., 1926) were designed to better understand symbiosis using termites and their intestinal flagellates. He was well aware that bacteria and yeasts were involved in the relationships and that there was a fine line separating where one symbiotic association ended and the other began. For example, when one agent in a so-called mutualistic relationship could survive without the host, it nevertheless also can become a parasite living off the host. This situation, however, is complicated and difficult to assess. He tried to remove one or more flagellate microorganisms from a host without harm to the host so that the association could be manipulated. He also thought it best to consider all components, which he defined differently than we do (he considered a commensal association to be one when neither party was benefited nor injured), of an association as symbionts. Depending on a specific termite host, he could void the flagellates with starvation, a high temperature not lethal to the host, a specific level of moisture, and oxygen under pressure. For example, by oxygenating the large Pacific Coast termite (*Zootermopsis nevadensis*) for 7 h at 1.5 atm, two (*Leidyopsis sphaerica* and *Trichonympha campanula*) of the four flagellates survived, and the host lived "indefinitely." Additionally, starving the termite at the above conditions for 6 days left only one flagellate (*L. sphaerica*), showing that both *L. sphaerica* and the termite do together constitute a necessary and mutualistically beneficial symbiotic relationship. Without any flagellates but with a diet of wood, the termite lived for about 3 weeks. Reintroducing either *L. sphaerica* or *T. campanula* allowed the termite to experience its normal much longer longevity (usually 60–70 days). Cleveland realized that the presence of the other two flagellates (*Trichomonas termopsidis* and *Streblomastix strix*, both having a poorly understood symbiosis with epibiotic bacteria) in the termite starved for 8 days would help the termite survive for about 10 weeks but were not necessary symbionts; he realized that bacteria played a role in digesting the diet. The "primitive" termite *Mastotermes darwiniensis* from Australia was studied by Li et al. (2003), and it had in its hindgut six flagellates, none of which can yet be cultured. Historically, these flagellates were considered to use their digestive enzymes to digest cellulose for the benefit of the termite. However, they determined using PCR technology that the main endoglucanase activity in the flagellates appears to originate from termite cellulases produced in the salivary glands. At least two of the flagellates possess their own endoglucanase genes, which are expressed but without significant enzyme activity in their nutritive vacuole. After millions of years of evolution, these flagellates, suggested by Li et al. (2003), are heading for a secondary loss of their own endoglucanases to exclusive use of the termite cellulases. Feeding on the symbionts still seems to be an important nutritional component of the termite diet.

Some termites, such as those which are soil-feeders, depend entirely on bacterial symbionts (Bignell et al. 1980). This association, often involved with coprophagy, feeding on fecal pellets containing termite bacteria that digest the cellulose, will be treated by D. E. Bignell elsewhere in this series (Volume 2, Chapter 6). Other termites, such as members of the *Macrotermitinae*, depend on a relationship with mutualistic fungal symbionts of the genus *Termitomyces*, which form fungal combs in the nests. These and attine ants can produce nests that often are thousands of liters in volume, able to persist for decades, and contain millions of sterile helper individuals usually resulting as offspring from a single queen. Those termites are major decomposers of the Old World tropics, and the ants are dominant herbivores of the New World tropics. The fungal symbionts of the termites can produce sexual fruiting bodies allowing their horizontal acquisition, but those of the attine ants rarely fruit and are typically propagated clonally and vertically by the dispersing queens. The life cycles of the fungus are again shown to help trigger different symbiotic relationships. Aanen et al. (2002) present phylogenies of the termites (about 330 species in 11 genera) and their fungal symbionts which number about 40 and are shared by different termite species. They show significant congruence between the termite and fungal phylogenies because the interactions at higher taxonomic levels show considerable specificity. They also considered the trait of biparental colony founding to constrain evolution of vertical symbiont transmission in termites, where the male survives and mates repeatedly with the female for life but not in ants where males die after a single mating. The Formosan subterranean termite pest (*Coptotermes formosanus*) builds a large nest with spongelike networks of intricate feces-lined tunnels (carton material). Fungal pathogenic agents are used, usually unsuccessfully, to kill the termite. Studies by Chouvenc et al. (2013) have shown that environmental conditions within termite nests promote the growth of Actinobacteria, whose presence in turn seems to protect the termite colony against fungal entomopathogens, including *Metarhizium anisopliae*. In other words, the Actinobacteria, which represented a nonnutritional exosymbiosis in the termite, was a defensive mutualism that increased survival of the termite and was additive to the termite's individual immunity and social defensive capacity, which in turn increased survival of the termite.

2.3.5.2 An Endemic Virus

Baculovirus penaei, commonly called BP or BP (PvSNPV), has a widespread distribution in cultured and wild penaeids, and it can cause severe epizootics in larval, post-larval, and juvenile stages. It is an enveloped, polyhedrosis, rod-shaped, double-stranded, intranuclear, DNA virus infecting epithelial cells of either the

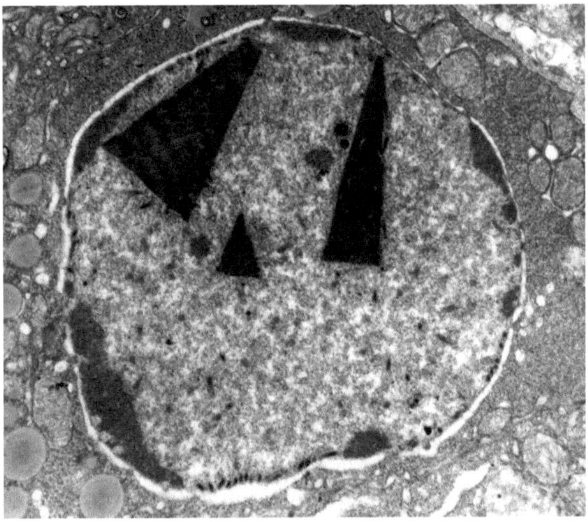

Fig. 2.8 About 2–3 day infection of *Baculovirus penaei* (BP) infecting hepatopancreatic tubules (HP) of larvae of *Litopenaeus vannamei* demonstrating triggering condition for viral replication and resulting pathogenic features in larval and young post-larval individuals that are rare or absent in juvenile and adult shrimp. Ultrastructure of hypertrophied nucleus of infected cell showing emarginated chromatin, developed virions lining internal surface of the nuclear membrane, replicating developing nucleocapsids in nucleoplasm, and polyhedra incorporating virions. Note abnormal mitochondria in cytoplasm surrounding nucleus

anterior midgut and midgut gland, or the hepatopancreas (HP; R-cells, F-cells, and B-cells, most commonly found near the base where the HP tubules join the anterior midgut). The E-cell populations (stem cells) at the distal portion of the hepatopancreatic tubules produce cells so quickly as to assure noninfected new cells (Figs. 2.8, 2.9, and 2.10) (Overstreet et al. 1988). Baculoviral infections of BP can involve nearly 90 % of the shrimp HP during mysis and early postlarvae stages without necessarily causing death. A number of triggering factors can tip these acutely infected shrimp into mortality. The virus is one out of 25 or so that infects penaeid shrimp and one of a few thousand known from invertebrates. Because it has a proteinaceous tetrahedral occlusion body, infections can be detected and followed with a light microscope; because of these bodies, it was the first species reported and characterized from a shrimp (Summers 1977).

The virus has a simple direct life cycle, although unknown means allow the agent to remain dormant or in a reservoir host. It can infect an acceptable penaeid either from free or occluded virions (Fig. 2.8) in the surrounding water or through a carrier host. Experimental infections are best accomplished by feeding the virus concentrated in a rotifer to protozoeal or early-stage mysis larvae or concentrated in a brine shrimp to infect late-stage mysis or early postlarvae (Overstreet et al. 1988). Depending on a number of factors, the virus replicates in the host alimentary cells and is most prevalent and infective at about day 3 after being fed. The nucleus of those cells enlarges, and the polyhedra and associated virions rupture into the lumen

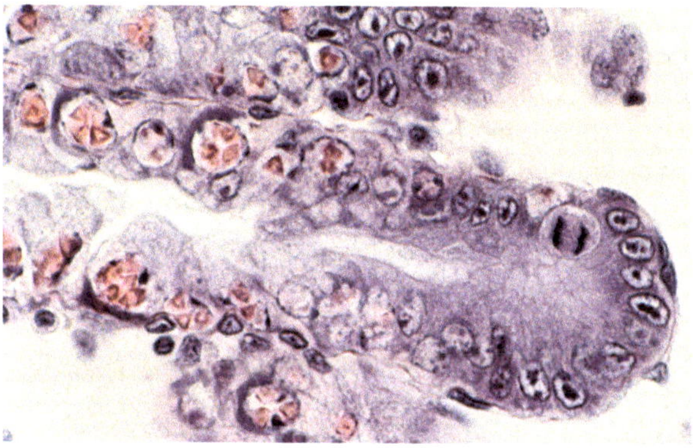

Fig. 2.9 *Baculovirus penaei* in HP of 2–3 day old larva of *Litopenaeus vannamei*. Histological section (stained with hematoxylin and eosin) showing tip of HP tubule with few actively dividing embryonic cells (E-cell, note large mitotic nucleus) with distal to proximal cells showing a progression of enlarged nuclei and large polyhedra

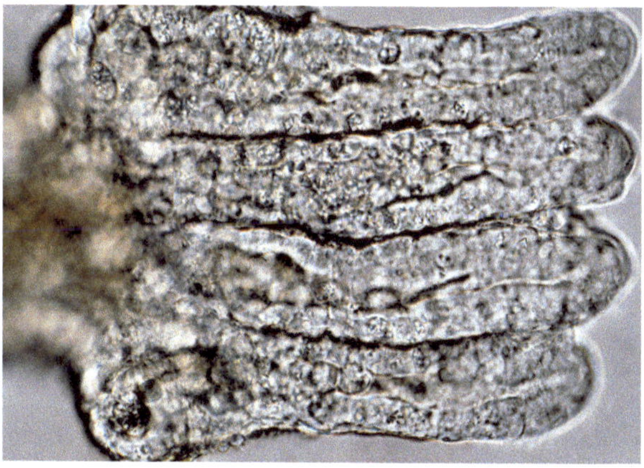

Fig. 2.10 *Baculovirus penaei* in HP of 2–3 day old larva of *Litopenaeus vannamei*. Fresh HP tubules showing small- to medium-sized polyhedra infecting most cells other than those at the tips

of the host midgut. We have observed as many as 24 relatively large tetrahedral bodies (about 13 μm on a side) or 100 small distinct bodies in experimental infections. The infective agent is then passed out through the intestine or it can be released when the infected larva is eaten by an appropriate predator. Unlike some insect baculoviruses that are used to kill agricultural pests, the BP virion remains active in the polyhedron for less than 2 weeks rather than for 80 or so years. On the other hand, BP maintained in an ultralow freezer remained active for at least three and a half years.

Natural infections of BP in the brown shrimp, *Farfantepenaeus aztecus*, were monitored monthly between January 1989 and November 2000 with seasonal samples conducted until May 2004. The monitoring took place in the northern Gulf of Mexico in or off Mississippi waters, but shrimp from Florida were also examined. During the mid-1970s, infections occurred in wild populations of the pink shrimp, *Farfantepenaeus duorarum*, but when we monitored all commercial penaeid shrimps for those infections, the virus primarily infected the brown shrimp and only occasionally infected the pink or white shrimp, *Litopenaeus setiferus*, which exhibited the otherwise rare infection. During the monitoring, we tried to use 4-cm long shrimp because they exhibited the highest prevalence and intensity of infection. Infection usually occurred only about 3 or 4 months of the year between about April and June, with the highest prevalence occurring in 1998 when we detected values of 56, 10, and 4 %. Values reaching 40 and 36 % occurred in May 1993 and 1997, respectively. Typically, the prevalence reached about 10–25 % during the remaining years, and infections were almost always absent in fall and winter. Rarely was the intensity of infection great, as usually observed in experimental infections with larvae and early postlarvae. The outbreak dynamics probably resulted from a variety of factors, including genetics (primarily brown shrimp), the interaction between the life cycle of the shrimp and the presence of the virus, the age of shrimp when initially infected, and resistance to infection, which occurred primarily in larger shrimp. These factors will be discussed below in more detail such as with infections and problems in aquaculture hatcheries.

Viral Factors

Strains There are at least four strains of BP as differentiated with molecular probes. Each seems to be restricted to specific penaeid shrimps. One strain occurs in some commercially important shrimp in the northern Gulf of Mexico; another occurs in noncommercially important shrimp (*Trachypenaeus* spp.) in the northern Gulf of Mexico; another occurs in *Melicertus marginatus* in Hawaii; and the last occurs in *Litopenaeus vannamei* in Ecuador. We have used for our experimental research the latter strain originally obtained and frozen from Ecuadorian aquaculture facilities by biologist James Brock. There is also a series of strains of MBV (*monodon* baculovirus), a baculovirus that produces spherical polyhedral bodies but which have many similarities to BP. We attempted cross infections with BP for all but the Hawaiian strain, using presumably susceptible penaeid postlarvae without success.

Free and occluded virions In addition to the virions occurring in the tetrahedron, which occasionally does not contain any virions, free ones lined up along the abnormally-distorted internal surface of the nuclear membrane. The free viral particles were collected from a freeze-thaw preparation either by filtration through a 0.45 μm filter or from processing of the preparation by relatively low-speed centrifugation to prepare a cell-free supernatant, and when next exposed to shrimp larvae, they produced infections. These subsequent infections did not appear to be more virulent than those initiated from virions incorporated in the tetrahedron.

However, the free virions of some insect baculoviruses produce secondary infections in the hemocoel, rather than gut cells, and those are considerably more virulent (Kelly 1982).

Dose Dose of BP is difficult to determine because there is no appropriate crustacean cell line available to culture the virus. However, we conducted a replicated relative study using the same standard viral stock suspension and determined a clear dose response involving both prevalence of BP and mortality of shrimp (Overstreet 1994).

Environmental factors Temperature and probably other factors affect both the production of free virus and the relationship between the virus and its shrimp host. When a homogenate of BP-infected cells is placed in seawater (32 ppt), the virus becomes completely inactivated between 7 and 14 days when maintained at 22 °C. However, when maintained at 5 °C, some virions remained active after 14 days. A low temperature also delayed or inhibited infections in larvae and presumably adults. Virions were inactivated by a 10-min exposure to temperatures of 60–90 °C.

A variety, but not all, of toxicants can probably affect the relationship between the viral agent and shrimp host. Couch (1976) reported an increase in prevalence of infection in pink shrimp exposed to low levels of Aroclor 1254®. We tried to duplicate his experiments using brown shrimp and 3 ppb Aroclor 1254 and also 2 ppm nickel but with unsuccessful outcomes. We later discovered from Couch (personal communication) that the Aroclor 1254 he used was probably unknowingly contaminated with another toxin (Overstreet 1994).

Studies were conducted by treating free BP in a variety of ways and then feeding it to larvae of *L. vannamei* as a bioassay (Overstreet et al. 1988). Unlike insect baculoviruses, all of the BP virions that we tested were inactivated when desiccated for 48 h. Tests conducted for possible treatments in aquaculture also showed that the virus was completely inactivated by a 30-min exposure to pH 3; pH 11 extended the pre-patency period of infection, but it did not inactivate the virus. Ultraviolet irradiation for 40 min at a wavelength of 254 nm when the virus was 5 cm from the light source also inactivated the virus (LeBlanc and Overstreet 1991a). The virus was also completely inactivated by a chlorine (in the form of calcium hypochlorite) concentration of 200 mg/L when treated for 1 h and by 1600 mg/L when treated for a time period as short as 20 s (LeBlanc and Overstreet 1991b). Other than for sterilizing aquariums, these methods other than desiccation and steam cleaning seemed impractical for an aquaculture facility.

Host Factors

Research host shrimp To conduct experimental infections, we used mostly *Litopenaeus vannamei* cultured at the Gulf Coast Research Laboratory Consortium either at Oceanic Institute (OI) in Hawaii or at GCRL in Ocean Springs. Our early studies with larvae required antibiotics to counter bacterial infections, which otherwise often reduced a larval stock by 50 % and would have been considered an acceptable loss in commercial hatcheries at that time. Even with antibiotics, we obtained BP infections with associated catastrophic mortalities when the shrimp larvae were exposed to the virus. After hybridization crosses of shrimp and introducing special wild brood stock, the result was creation of shrimp stocks with improved resistance to BP, and we noted as a bonus our having created crosses exhibiting resistance to other microbial agents. The BP prevalence could be variable, but mortality was reduced or eliminated. In a few cases, we obtained shrimp from commercial hatcheries or other research facilities and had to use antibiotics.

Host resistance The genetic families of *Litopenaeus vannamei* produced at OI gradually became more resistant to BP. In some cases, it was easy to obtain 100 % prevalence but without the associated mortalities observed using other stocks of shrimp. Middlebrooks et al. (1989) examined lectin levels in the hemolymph of monthly samples of wild *Farfantepenaeus aztecus*, and the levels were found to vary by season and among individuals, suggesting that levels could serve as indicators of health or immune status. High titers of activity for lectin agglutination occurred in September, and low levels occurred in April. Those observations of positive BP infection results when lectin levels were low corresponded with the time when BP first became apparent in the wild shrimp population, and polyhedra subsequently were no longer apparent when lectin levels were high.

Age and growth The age of *Litopenaeus vannamei* influences BP infections in a variety of ways. Stuck and Overstreet (1994) exposed pathogen-free shrimp ages mysis 2–3 through 25-day-old postlarvae (PL) from different sources, each with a single viral exposure, and cultured them for 15–21 days. All groups PL9 or younger became heavily infected within 2–5 days, some experiencing high mortalities compared with controls. Surviving postlarvae had reduced growth as determined by dry weight. One of these groups was cultured for an additional 49 days, and the smaller postlarvae that survived the infection appeared similar to the controls after 4 weeks by which time viral prevalence had decreased. Exposure of the virus to older postlarvae produced a high prevalence of infection but with little effect on either survival or growth. When 13- to 14-day-old postlarvae in similar size groups of previously infected and noninfected individuals were starved for 10 days, less than 2 % of the infected postlarvae survived the 10-day period compared with 52 % of the noninfected ones. Leblanc and Overstreet (1990) exposed infected BP to groups of *L. vannamei* 3, 39, 63, 120, 157, 325, and 454 days after reaching postlarvae. The postlarvae exposed when 3-days-old (PL3) exhibited infections in 77 % of those examined at 5 days postinfection (PI) and in 100 % of the same group when examined at day 9 compared with 0 % prevalence in the controls; 14 % of the shrimp that were exposed when postlarvae for 39 days (PL39) had developed an infection at day 9 PI when 20 % of those exposed as PL63s exhibited infections; and the prevalence in the latter group increased to 42 % at day 14. With the exception of one shrimp exposed when a PL120, none developed an observable infection at 19 days, and only 2 of 20 of the exposed PL157 group examined at day 16 exhibited an infection. No control became infected. Other experiments also showed this decrease in prevalence as well as intensity of infection and mortality with age. In general, most postlarvae 12 days old or older did not become infected, and those that became infected did not die. This age corresponds with the age that most postlarvae leave the offshore waters and settle on the substratum in the estuarine area where the infective agents occur.

Energy The nutritional condition of *Litopenaeus vannamei* can be assessed biochemically in terms of available energy reserve. The principal energy storage materials in the penaeid shrimp are lipids and proteins, with carbohydrates considered to be a minor energy reserve. Triacylglycerol (TAG), an ester derived from glycerol, and three fatty acids serve as the primary classes of lipid used for energy storage. In patent infections of BP conducted by Stuck et al. (1996), the polyhedra first appeared in the HP 18–24-h postexposure with a maximal infection usually occurring at 72 h (also see Hammer et al. 1998); however, progression of the infection after the initial appearance of the polyhedra was variable among shrimp potentially corresponding with different levels of body lipid used for energy storage. We measured preinfection and postinfection TAG and protein levels during a series of experimental BP exposures using shrimp from a variety of sources with different inherent protein and TAG levels. Mysis-stage or early post-larval protein levels measured either preexposure or postexposure to BP showed no consistent relationship with BP infections. However, a viral prevalence of 86–100 % occurred at 72 h when initial TAG levels measured above 3.5 μg/mg in contrast to 35 % or less in shrimp when initial TAG levels <2.0 μg/mg. TAG levels in mysis stage-1 larva rose rapidly from 1 to about 11 μg/mg in 2 days for both infected and negative control populations, then dipped to about 2 μg/mg at day 9, and subsequently again rose in both populations but was about 3 μg/mg higher in the infected shrimp at day 15. Most shrimp died before the end of a week. Two other experiments starting with 9-day postlarvae (PL9s) were conducted when initial TAG levels were either about 1 or 7 μg/mg, and substantial mortalities did not occur in either. All individuals in the infected groups became infected, at about day 10 in the initial low TAG group and at day 3 in the initial high group. In the initial low TAG group, the TAG level of the infected group stayed about the same at day 20, but that of the control groups increased to about 4 μg/mg. In contrast, in the initial high TAG experiment in which the TAG level for both infected and control groups dipped from 7 to about 1.5 μg/mg at day 3, the level increased to about 10 μg/mg at day 20 and about 15 μg/mg at day 25. Stress caused by starving postlarvae (PL18s) reduced TAG but not protein reserves. Detection of the BP infection was delayed for 30 h when the group to be infected was starved for 48 h compared with continuously fed controls, but the prevalence of infection in both groups increased rapidly to similar high prevalence values above 80 % between 72 and 192 h PI.

Genetics Offspring of male and female crosses of high-growth and low-growth "families" of a well-defined population at GCRL Consortium at OI, Hawaii, were fed BP as 15-day-old postlarvae (PL15s) (Alcivar-Warren et al. 1997). The high × high-growth and low × high-growth (female × male) postlarvae, respectively, had a 77 and 85 % survival rate at 18 days postexposure compared with 19 and 24 % survival for the low × low-growth and high × low-growth offspring. All but the high × low-growth offspring, with a 68 % prevalence of infection, had an 88–100 % prevalence at day 4. The low × low-growth cross of 3-month-old shrimp fed the virus IHHNV (infectious hypodermal and hematopoietic necrosis virus, which has been classified as *Penaeus stylirostris* densovirus) exhibited the highest prevalence (48 %) at 30 days compared with the lowest (6 %) in the high × low-growth cross. Random amplified polymorphic DNA polymorphisms for the four crosses showed no clear relationship between the prevalence values of IHHNV and BP. Even with similar mtDNA haplotypes included in the initial crosses, the offspring of those crosses exhibited major differences in both steady-state levels and patterns of expression of mitochondrial 12s rRNA at various early developmental stages of the resulting offspring of the different crosses. Even though specific genetic markers and differences could not be associated with specific differences in susceptibility to different infections, shrimp growth, or regulation of gene expression, the genetics of the shrimp play an important role in triggering infections and mortality from infections.

Concurrent agents influencing infections Reo-like viruses in crustaceans often occur concurrently with other disease-causing viruses producing a synergistic effect. We (Krol et al. 1990) found such a virus infecting the anterior midgut epithelium and the R- and F-cells from the HP in larval specimens of *Litopenaeus vannamei*. The occurrence of the reo-like virus was observed only in BP-infected shrimp; however, shrimp with BP seldom had the reo-like virus. Other agents often associated with stress also occur in shrimp (Overstreet 1994).

Susceptibility to viral infections appears to be enhanced by crowding of host individuals. We determined that the prevalence and intensity of infections in shrimp crowded in commercial bait tanks were higher than in those collected from the wild shrimp population that was used to stock the bait tanks.

In summary for the endemic virus BP, there are a variety of susceptibility factors dealing with the environment, virus, and host, including the nutritional and molting state of the host, which can trigger a relatively harmless infection to develop into a severely pathogenic condition. In nature, this condition is seldom recognized because only microscopic-sized larvae and early postlarvae usually die from the infection, dead animals cannot be seen because they are small or readily eaten, and the opportunity for the agent and the larvae to come into close proximity is relatively rare. In aquaculture, the virus and larvae can and do come into contact. Hatcheries in the multibillion-dollar shrimp industry have prompted research to solve the previous major threat that collapsed hatcheries in the USA and elsewhere in the Western Hemisphere during the early days of the industry when brood stock was obtained from the wild. The related MBV in the Eastern Hemisphere also caused a similar major problem. Now that the industry can rear most penaeid species; detect BP and other infectious agents with PCR, gene probes, and other methods in routine monitoring; produce its own brood stock relative to disease resistance and growth potential; and accommodate ambient temperatures, infections of BP in aquaculture are rare. If they do occur, an entire spawn can be destroyed and disposed and the system decontaminated, and there would be a time loss of only a few days before a new spawn could be produced.

2.3.5.3 Exotic, or Introduced, Viruses: The Emergence of Viral Pathogens of Shrimp Aquaculture

With the exception of baculovirus species, most or all the pathogenic penaeid viruses are not native to the shrimp species being cultured or perhaps to any commercial shrimp. Consequently, the cultured species may be highly susceptible to and die from one or more of the viral agents. Most of the presently known 25 or so shrimp viruses cause mortality in subadult animals that have been reared for a few months. The high cost of feed and labor plus the relatively long animal growing period results in a high economic cost. Therefore, it is necessary to incorporate the costs of an inability to quickly replace the entire infected pond-reared adult or subadult stock as can be done for larvae in hatcheries infected with BP infections. Production losses associated with the exotic viruses white spot syndrome virus (WSSV) and Taura syndrome virus (TSV) provide excellent examples for examining the emergence of pathogenicity in aquaculture on a large scale.

The history of the shrimp aquaculture industry is a case study of the emergence of severe pathogens in hosts where none was known previously. For about two decades after the industry began in the 1970s, few viral pathogens were known. Beginning in the early 1990s, several severely pathogenic viral diseases emerged. We will consider two of the pathogens in some detail as they inform our understanding of the appearance of highly pathogenic symbionts through host switching.

White spot syndrome virus (WSSV) was first reported in 1992 from Taiwan. The virus belongs to the order Baculovirales. Although once considered a member of the Nudiviridae or Baculoviridae, WSSV is more closely related to the recently erected family Hytrosaviridae (Wu et al. 2009; Wang and Jehle 2009). WSSV is now considered a member of the genus *Whispovirus* in the family Nimaviridae. The virus is a rod-shaped, enveloped virion and among the largest viruses known at 120–150-nm wide by 270–290-nm long. The genome is double-stranded DNA of 295 kbp. The host range is among the widest of crustacean viruses and infects all decapod crustaceans that have been tested.

Taura syndrome virus (TSV) was first observed from Ecuador also in 1992. The virus belongs to the order Picornavirales and is a member of the family Dicistroviridae and the genus *Aparavirus*. The virion is icosahedral, nonenveloped, and 32 nm in diameter. The genome contains positive sense, single-stranded RNA of 10 kb. The host range is restricted mostly to members of the genus *Penaeus* sensu lato, although recently Overstreet et al. (2009) reported experimental infections with replication in at least some symbiotic barnacles. Pathogenicity varies with host species being most virulent in *Litopenaeus vannamei* and less so in other species of penaeid shrimp (Overstreet et al. 1997).

Although it is possible that the viruses were symbionts of wild shrimp prior to being detected in shrimp aquaculture, we wish to explore the hypothesis that they emerged by host transfer from insects or other terrestrial arthropods during the development and expansion of shrimp aquaculture. Roekring et al. (2002) perhaps first suggested this tantalizing hypothesis in their study of three shrimp parvoviruses whose closest known relatives are parasites of insects. Host switching is a common mechanism for the emergence of pathogenic pathogens. For example, in humans, 75 % of today's emerging human viral diseases are zoonotic, e.g., HIV-1 and HIV-2, influenza virus, Ebola virus, hantaviruses, Nipah virus, Zika virus, and the SARS coronavirus that causes severe acute respiratory syndrome, and many of the rest of human infectious diseases have their closest relatives residing in nonhuman animal hosts (Parrish et al. 2008; Pearce-Duvet 2006).

Most theories of the emergence of new diseases through host switching recognize several steps: cross-species transmission, establishment, and spread (Parrish et al. 2008; Domingo 2010; Antia et al. 2003). The first step in the process is an increase in the number of contact events between the recipient host and the donor host. The increase in contacts is usually due to an ecological change or disturbance. As the number of contacts between the donor and recipient hosts increases, there may be a concomitant increase in successful transfer of the pathogen. In the case of shrimp viruses, we suggest below that there was an increase in contact between penaeid shrimp and other arthropods during the early 1990s, and that facilitated cross-species transmission and the emergence of pathogens in penaeids.

Although cross-species transmission may occur, permanent establishment within the new host requires enough within-species transmission to maintain the pathogen in the new host. It is unknown how many times the ancestor of these two symbionts was transmitted to penaeid shrimp only to be eliminated because shrimp-to-shrimp transmission was not sufficient. However, at some point, either prior to cross-species transmission or after initial colonization, a mutation of the ancestor resulted in penaeid shrimp being highly susceptible to them, resulting in the maintenance of the infection locally in shrimp aquaculture.

A third set of factors allows for the pandemic spread. Sharp and Hahn (2010) review the successful establishments of HIV from primates to humans. Although there may have been as many as seven successful transmissions, most are restricted to Africa, and only one of the seven establishments of HIV (HIV-1-M) is responsible for the pandemic of HIV in the 1980s (Sharp and Hahn 2010; Pepin 2011). One or more factors other than trans-species transmission allowed for the amplification and spread after a transmission. In the case of HIV, it is thought that the increased use of vaccination in colonial Africa, chance, and increased world travel may have been the contributing factors (Pepin 2011). In the case of shrimp viruses, it may be that the greater number and higher shrimp densities in aquaculture settings coupled with the long-distance transport of live and fresh frozen shrimp has allowed worldwide viral spread after cross-species transmission.

The rapid expansion of the shrimp aquaculture industry in the 1980s and 1990s brought with it conditions conducive to increasing the rate of contact between shrimp and terrestrial arthropods, the majority of which are insects. According to

FAO statistics (FAO 2009a, b) in the late 1970s, most of the penaeid shrimp were coming from capture fisheries (Fig. 2.11). However, between 1992 and 1993 production of penaeid shrimp from aquaculture surpassed 50 % of all penaeid shrimp production worldwide (Fig. 2.12). This local peak occurred near the time that the two viruses appeared in shrimp aquaculture. Subsequently, culture production declined as a percentage of total penaeid production and remained below 50 % throughout the 1990s. However, the first decade of the twenty-first century saw a substantial increase and by 2010 the contribution from aquaculture to world's commercial shrimp increased supply by nearly 75 %.

Clearly, in the early 1990s, when viral pathogens emerged, there were twice as many penaeid shrimp in the market place as there had been a decade before. Although this does not mean that the total number of shrimp in the world doubled (not all of the shrimp in the world were captured), it does suggest that aquaculture substantially increased the total number of penaeid shrimp in the world and continues to do so. It is also clear that the increased number of aquacultured shrimp were at higher densities than wild shrimp. This increased population size and density are conducive to the spread of infectious diseases.

Perhaps more important for the cross-species transmission of the viruses from insect to shrimp is the habitat of those new shrimp. For the most part, they were in earthen ponds in the coastal zone. However, many of the ponds were located a distance from the coastline, and seawater was pumped over a considerable distance to provide the necessary culture medium. The point is that there may be more shrimp that were likely to have contacted terrestrial insects and other terrestrial arthropods, after the expansion of shrimp aquaculture than before it.

The second point in building a case for the origin of the two shrimp viruses from insects or other terrestrial arthropods is that the hosts of the viruses most closely related to shrimp viruses are insect or terrestrial arthropod hosts. Roekring

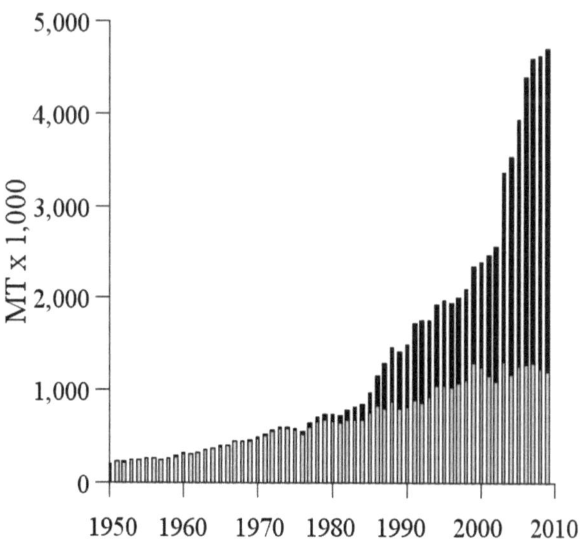

Fig. 2.11 Total world's supply of penaeid shrimp in metric tonnes. Production from the capture fishery is *light bars*. Production from aquaculture is in *dark bars*. Source FAO statistics (FAO 2009a, b)

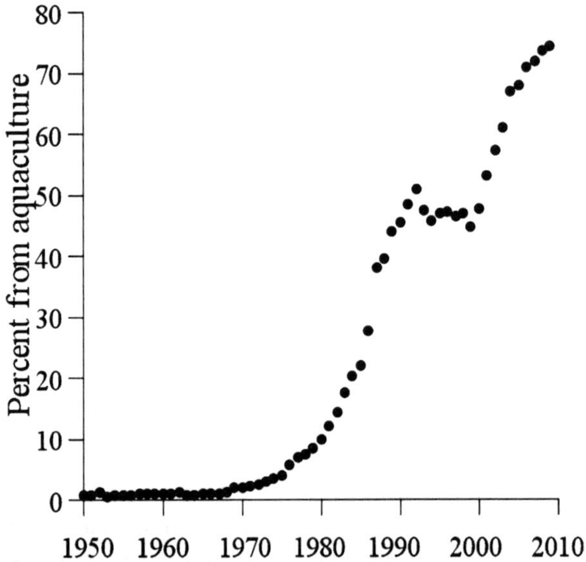

Fig. 2.12 Proportion of world's penaeid shrimp from aquaculture. Note that the proportion of penaeid shrimp production saw a local peak in the early 1990s above 0.5 and then declined during the emergence of Taura syndrome virus (TSV) and white spot syndrome virus (WSSV). Source FAO statistics (FAO 2009a, b)

et al. (2002) indicate that shrimp parvoviruses do not form a single clade but are distributed among clades that include insect parvoviruses. Roekring et al. (2002) suggested that viral transfers between crustaceans and insects occurred.

Most members of the Dicistroviridae are found in insects (Baker and Schroeder 2008). Guo et al. (2013) noted that the exceptions are TSV and the recently-described *mud crab dicistrovirus-1* (MCDV-1). MCDV-1 groups with TSV which could indicate that TSV in shrimp aquaculture arose from other crustaceans. It should be noted that the first reports of TSV were from the New World in the 1990s and that MCDV-1 was described from Asia long after TSV was transferred from the New World into Asia in 2000, suggesting that MCDV-1 arose after TSV was introduced into Asia.

Similarly, WSSV as a member of the order Baculovirales has as its closest relatives symbionts of insects. In particular, Jehle et al. (2013) group WSSV most closely with two insect viruses causing salivary gland hypertrophy, *Musca domestica salivary gland hypertrophy virus* (MdSGHV) and *Glossina pallidipes salivary gland hypertrophy virus* (GpSGHV).

We have provided a number of examples and triggers of the switch from one quality of a symbiotic relationship to another, specifically from commensal to pathogen. Examples of triggers include environmental ones like temperature, toxic chemicals (dose), chemotherapeutics, dietary changes, and geographic habits; internal ones include host physical site, host resistance or susceptibility, host modifications, and host switching; and combinations of these and other conditions.

2.3.6 Evolutionary Context for Changing Pathogenicity

The question arises, "Is becoming a pathogen an adaptation for a symbiont?" In an evolutionary context, which is where the concept of adaptation lies, why would a symbiont switch to being pathogenic, either at some stage in its life cycle or in some hosts at the same stage but not in others?

In some cases, symbionts are pathogenic as a clear adaptation, e.g., the *Levinseniella byrdi*, referenced above causes damage or even death of the intermediate host to complete its life cycle. In particular, *L. byrdi* causes a phenotypic change in the host that increases the host's chances of becoming a prey. In other cases, it is not so clear. Facultative symbionts, such as entomophagous nematodes, are both free living and symbiotic (Sudhaus 2008). The adaptation of being symbiotic is that it increases the habitat breadth of the symbiont, and inhabitation of a living organism removes the organism from competition with other organisms encountered living freely in the habitat. Thereby, the relationship allows the symbiont an escape from competition. So that may be the reason for a life history strategy in which both symbiotic and free-living phases exist. However, we are most interested in answering the question of why would a symbiont become pathogenic in some hosts and not in others from an evolutionary perspective.

There is a body of theory that addresses virulence evolution in symbiotic organisms. Traditionally, the virulence of a pathogen was thought to be a reflection of a recent host–symbiont association and that over time there would result a diminishment in virulence of the pathogen (May and Anderson 1983; Anderson and May 1992; Ewald 1994). It was argued that no symbiont would benefit from killing its host (destroying its habitat and thereby destroying itself). However, this thinking fails to recognize that natural selection maximizes reproductive success of the organism. Reproductive success is a composite of births (new infections) and survival (loss of infection) not just survival.

More recent theoretical studies suggest the evolution of an intermediate level of virulence. The new conclusion results from the recognition that there is often a trade-off between pathogen transmission (births) and virulence (pathogen survival) (Anderson and May 1982; May and Anderson 1983; Antia et al. 1994; Frank 1996; Koella and Restif 2001).

One indicator of reproductive success of a symbiont is the basic reproduction number (R_0), which is the mean number of new infections produced by a single infection (Anderson and May 1992; Diekmann et al. 1990; Lotz et al. 2003). It is calculated as the mean life span of an infection times the number of transmissions (equivalent to births) that would occur over that infectious period (Mollison 1995). It is very much the population growth rate of an infection and is analogous to the net reproductive rate (R_0) of free-living organisms. R_0 for both free-living and symbiotic organisms indicates population growth and therefore reproductive success (fitness). And as such, it should be maximized by natural selection.

For simplicity, we can represent transmission as β and life span as the reciprocal of the virulence α or pathogen-induced mortality. In this case, $R_0 = \frac{\beta}{\alpha}$ and R_0

increases as the infected hosts live longer (α decreases), and the infection is highly transmissible (β increases). However, there often occurs a trade-off between transmission and virulence. For example, in the case of TSV, the greater the load or intensity of the virus in a shrimp host, the greater the transmission (β increases). However, the greater the viral intensity, the greater the chance of mortality of the host (α increases) (Lotz 2010). Therefore, the trade-off is mediated by viral intensity. As a result, a balance exists between infectivity and virulence, and the maximum R_0 is obtained at intermediate levels of virulence (Fig. 2.13). This much of the theory assumes that the source of new infections is living infected hosts and that the death of the host ends infectivity (Ewald 1994; Hochberg 1998). What if transmission occurs not so much from living infected hosts but from hosts after they die? This is likely to be the case for several shrimp pathogens. Soto and Lotz (2001) and Lotz et al. (2003) demonstrated for WSSV and for TSV that transmission from dead infected shrimp is considerably greater than from living infected shrimp. This indicates that the transmissibility of a shrimp pathogen does not end with the life of the shrimp.

If transmission occurs from dead infected hosts, then R_0 is determined by the time to death of an infected host and the length of time that a dead infected host remains infectious. The infectious time of a dead shrimp depends on two factors, the ingestion of dead shrimp by other shrimp and the decay of the carcass (Soto and Lotz 2001; Lotz et al. 2003). The rate at which dead shrimp infectivity declines, whether by cannibalism or by carcass decay, is unrelated to the viral load; however, transmission rate (β) is. In this case, the load does not affect the time of infectivity of a dead infected shrimp, and the trade-off between load and time of infectivity disappears. In fact, increased virulence in live infected hosts causes a high pathogen load in dead animals and therefore increased infectivity of dead hosts without

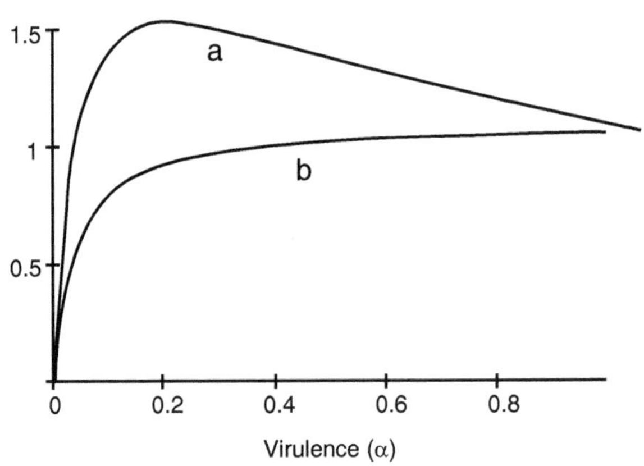

Fig. 2.13 Graph of relationship between two sources of infection (or two habitats) or two hosts assessed as pathogen-induced mortality. Curve *a* is the relationship between R_0 virulence when the source is a living host; *b* is the relationship between R_0 and virulence when the source is a dead host. Which curve obtained depends on the relative contribution of both living and dead hosts to the overall R_0

affecting the time of infectivity. So as virulence increases, R_0 also increases, and we expect virulence to increase over time if transmission is from dead shrimp (Fig. 2.13). Lotz (2010) did not consider the outcome if both living and dead hosts contribute; however, overall R_0 is $\frac{\beta_a}{\alpha} + \frac{\beta_d}{\delta}$, where β_a is the transmission rate from a living host, β_d is the transmission rate from dead hosts, α is the virulence of a pathogen to the living hosts, and δ is the infectivity decay of a dead infected host. The relative contributions of $\frac{\beta_a}{\alpha}$ and $\frac{\beta_a}{\delta}$ to R_0 are what will determine the final virulence and whether or not virulence will increase or decrease over evolutionary time. Although the above reasoning and Fig. 2.13 have been applied specifically to TSV and living and dead shrimp, they apply more generally to any two states that contribute to an overall R_0. The two states instead can be two habitats or two species of host. The state (habitat or species) that contributes the greatest to R_0 will dominate setting the optimal virulence. For species of symbiont that have both free-living and symbiotic relationships, the free-living habitat will contribute to the net reproductive rate (R_0), and the symbiotic habitat will contribute to the basic reproduction number (R_0). The two R_0s contribute to the overall growth (and thereby fitness) of the symbiont. If the two states represent different host species, then the host species that is responsible for the greatest contribution to the basic reproduction number will predominate. The conclusion is that what happens in the lesser contributing state will not be as important, and therefore very high virulence could be obtained in a host with little contribution to fitness of the symbiont. The observed virulence is a coincidental by-product of the adaptation to other habitats (Brown et al. 2012; Adiba et al. 2010). In particular, for some human bacterial pathogens, it has been postulated that bacterial pathogenicity evolved from antipredator selection by free-living bacteria and their predatory protists (Brüssow 2007; Adiba et al. 2010; Brown et al. 2012; Erken et al. 2013).

We have covered two evolutionary hypotheses to explain increased virulence of a symbiont: one contributes directly to the fitness of the symbiont, and the other is a coincidental outcome of selection for a trait important in another habitat.

Acknowledgments We thank Janet Wright, Jean Jovonovich, and Kim Overstreet for assistance. Some funding was provided by a grant from BP Exploration & Production, Inc.

References

Aanen DK, Eggleton P, Rouland-Lefèvre C, Guldberg-Frøslev T, Rosendahl S, Boomsma JJ (2002) The evolution of fungus-growing termites and their mutualistic fungal symbionts. Proc Natl Acad Sci USA 99(23):14887–14892

Adamo SA (2012) The strings of the puppet master: how parasites change host behaviour. In: Hughes D, Thomas F (eds) Parasitic manipulation. Oxford University Press, Oxford, UK, pp 36–51

Adamo SA, Webster JP (2013) Neural parasitology: how parasites manipulate host behaviour. J Exp Biol 216(1):1–2

Adiba S, Nizak C, van Baalen M, Denamur E, Depaulis F (2010) From grazing resistance to pathogenesis: the coincidental evolution of virulence factors. PLoS One 5(8):e11882
Alcivar-Warren A, Overstreet RM, Dhar AK, Astrofsky K, Carr WH, Sweeney J, Lotz JM (1997) Genetic susceptibility of cultured shrimp (*Penaeus vannamei*) to infectious hypodermal and hematopoietic necrosis virus and *Baculovirus penaei*: possible relationship with growth status and metabolic gene expression. J Invertebr Pathol 70(3):190–197
Anderson RM, May RM (1982) Coevolution of hosts and parasites. Parasitology 85(2):411–426
Anderson RM, May RM (1992) Infectious diseases of humans: dynamics and control. Oxford University Press, Oxford, UK
Antia R, Levin BR, May RM (1994) Within-host population dynamics and the evolution and maintenance of microparasite virulence. Am Nat 144(3):457–472
Antia R, Regoes RR, Koella JC, Bergstrom CT (2003) The role of evolution in the emergence of infectious diseases. Nature 426(6967):658–661
Baker AC, Schroeder DC (2008) The use of RNA-dependent RNA polymerase for the taxonomic assignment of Picorna-like viruses (Order Picornavirales) infecting *Apis mellifera* L. populations. Virol J 5:10. doi:10.1186/1743-422X-5-10
Barnard CJ, Behnke JM (eds) (1990) Parasitism and host behaviour. Taylor & Francis, London
Barrows EM (2011) Animal behavior desk reference, 3rd edn. CRC Press, Boca Raton, FL
Bates AE, Stickle WB, Harley CDG (2010) Impact of temperature on an emerging parasitic association between a sperm-feeding scuticociliate and Northeast Pacific sea stars. J Exper Mar Biol Ecol 384(1–2):44–50
Beaver PC, Jung RC, Cupp EW (1984) Clinical parasitology, 9th edn. Lea & Febiger, Philadelphia, PA
Bignell DE, Oskarsson H, Anderson JM (1980) Distribution and abundance of bacteria in the gut of a soil-feeding termite *Procubitermes aburiensis* (Termitidae, Termitinae). J Gen Microbiol 117(2):393–403
Bousfield EL, Heard RW (1986) Systematics, distribution ecology, and some host-parasite relationships of *Uhlorchestia uhleri* (Shoemaker) and *U. spartinophila*, new species (Crustacea: Amphipoda), endemic to salt marshes of the Atlantic coast of North America. J Crustacean Biol 6(2):264–274
Brown SP, Cornforth DM, Mideo N (2012) Evolution of virulence in opportunistic pathogens: generalism, plasticity, and control. Trends Microbiol 20(7):336–342
Brüssow H (2007) Bacteria between protists and phages: from antipredation strategies to the evolution of pathogenicity. Mol Microbiol 65(3):583–589
Bush AO, Heard RW Jr, Overstreet RM (1993) Intermediate hosts as source communities. Can J Zool 71(7):1358–1363
Bush AO, Lafferty KD, Lotz JM, Shostak AW (1997) Parasitology meets ecology on its own terms: Margolis et al. revisited. J Parasitol 83(4):575–583
Chouvenc T, Efstathion CA, Elliott ML, Su N-Y (2013) Extended disease resistance emerging from the faecal nest of a subterranean termite. Proc R Soc B 280(1770):20131885. doi:10.1098/rsbp.2013.1885
Cleveland LR (1926) Symbiosis among animals with special reference to termites and their intestinal flagellates. Q Rev Biol 1(1):51–60
Couch JA (1976) Attempts to increase *Baculovirus* prevalence in shrimp by chemical exposure. Prog Exp Tumor Res 20:304–314
de Bary HA (1879) Die Erscheinung der Symbiose. Karl J Tübner, Strassburg
Deardorff TL, Overstreet RM (1980a) Taxonomy and biology of species of *Goezia* (Nematoda: Anisakidae) from fishes, including three new species. Proc Helminthol Soc Wash 47(2): 192–217
Deardorff TL, Overstreet RM (1980b) *Contracaecum multipapillatum* (=*C. robustum*) from fishes and birds in the northern Gulf of Mexico. J Parasitol 66(5):853–856
Deardorff TL, Overstreet RM, Okihiro M, Tam R (1986) Piscine adult nematode invading an open lesion in a human hand. Am J Trop Med Hyg 35(4):827–830

Diekmann O, Heesterbeek JAP, Metz JAJ (1990) On the definition and the computation of the basic reproduction ratio R_0 in models for infectious diseases in heterogeneous populations. J Math Biol 28(4):365–382
Domingo E (2010) Mechanisms of viral emergence. Vet Res 41(6):38–52
Dyková I, Lom J (2007) Histopathology of protistan and myxozoan infections in fishes: an atlas. Academia, Praha
Erken M, Lutz C, McDougald D (2013) The rise of pathogens: predation as a factor driving the evolution of human pathogens in the environment. Microb Ecol 65(4):860–868
Ewald PW (1994) Evolution of infectious disease. Oxford University Press, Oxford, UK
FAO (2009a) Global aquaculture production (1950–2009). http://www.fao.org/fishery/statistics/global-aquaculture-production/query/en. Accessed 7 Aug 2014
FAO (2009b) Global capture production (1950–2009). http://www.fao.org/fishery/statistics/global-capture-production/query/en. Accessed 7 Aug 2014
Fournie JW, Overstreet RM (1993) Host specificity of *Calyptospora funduli* (Apicomplexa: Calyptosporidae) in atheriniform fishes. J Parasitol 79(5):720–727
Frank SA (1996) Models of parasite virulence. Q Rev Biol 71(1):37–78
Gaines JL Jr, Rogers WA (1972) Fish mortalities associated with *Goezia* sp. (Nematoda: Ascaridoidae) in central Florida. In: Proceedings of 25th annual conference of the Southeastern Association of Game Fish Commissioners, pp 496–497
Guo ZX, He JG, Xu HD, Weng SP (2013) Pathogenicity and complete genome sequence analysis of the mud crab dicistrovirus-1. Virus Res 171(1):8–14
Hammer HS, Stuck KC, Overstreet RM (1998) Infectivity and pathogenicity of *Baculovirus penaei* (BP) in cultured larval and postlarval Pacific white shrimp, *Penaeus vannamei*, related to the stage of viral development. J Invertebr Pathol 72(1):38–43
Hawkins WE, Solangi MA, Overstreet RM (1981) Ultrastructural effects of the coccidium, *Eimeria funduli* Duszynski, Solangi and Overstreet, 1979 on the liver of killifishes. J Fish Dis 4(4):281–295
Hochberg ME (1998) Establishing genetic correlations involving parasite virulence. Evolution 52(6):1865–1868
Hoffman GL, Landolt M, Camper JE, Coats DW, Stookey JL, Burek JD (1975) Disease of freshwater fishes caused by *Tetrahymena corlissi* Thompson, 1955, and a key for identification of holotrich ciliates of freshwater fishes. J Parasitol 61(2):217–223
Hose JE, Lightner DV, Redman RM, Danald DA (1984) Observations on the pathogenesis of the imperfect fungus, *Fusarium solani*, in the California brown shrimp, *Penaeus californiensis*. J Invertebr Pathol 44(3):292–303
Jehle JA, Abd-Alla AM, Wang Y (2013) Phylogeny and evolution of Hytrosaviridae. J Invertebr Pathol 112(Suppl 1):S62–S67
Johnson DS, Fleeger JW, Deegan LA (2009) Large-scale manipulations reveal that top-down and bottom-up controls interact to alter habitat utilization by saltmarsh fauna. Mar Ecol Prog Ser 377:33–41
Kelly DC (1982) *Baculovirus* replication. J Gen Virol 63(1):1–13
Ko RC (1976) Experimental infection of mammals with larval *Echinocephalus sinensis* (Nematoda: Gnathostomatidae) from oysters (*Crassostrea gigas*). Can J Zool 54(4):597–609
Ko RC (1977) Effects of temperature acclimation on infection of *Echinocephalus sinensis* (Nematoda: Gnathostomatidae) from oysters to kittens. Can J Zool 55(7):1129–1132
Koella JC, Restif O (2001) Coevolution of parasite virulence and host life history. Ecol Lett 4(3): 207–214
Krol RM, Hawkins WE, Overstreet RM (1990) Reo-like virus in white shrimp *Penaeus vannamei* (Crustacea: Decapoda): co-occurrence with *Baculovirus penaei* in experimental infections. Dis Aquat Organ 8(1):45–49
Labrie L, Komar C, Terhune J, Camus A, Wise D (2004) Effect of sublethal exposure to the trematode *Bolbophorus* spp. on the severity of enteric septicemia of catfish in channel catfish fingerlings. J Aquat Anim Health 14(4):231–237

LeBlanc BD, Overstreet RM (1990) Prevalence of *Baculovirus penaei* in experimentally infected white shrimp (*Penaeus vannamei*) relative to age. Aquaculture 87(3):237–242
LeBlanc BD, Overstreet RM (1991a) Effect of desiccation, pH, heat, and ultraviolet irradiation on viability of *Baculovirus penaei*. J Invertebr Pathol 57(2):277–286
LeBlanc BD, Overstreet RM (1991b) Efficacy of calcium hypochlorite as a disinfectant against the shrimp virus *Baculovirus penaei*. J Aquat Anim Health 3(2):141–145
Lefèvre T, Adamo SA, Biron DG, Missé D, Hughes D, Thomas F (2009) Invasion of the body snatchers: the diversity and evolution of manipulative strategies in host-parasite interactions. Adv Parasitol 68:45–83
Lewis EE, Campbell JF, Sukhdeo MVK (2002) The behavioural ecology of parasites. CABI Publishing, Wallingford, UK
Li L, Fröhlich J, Pfeiffer P, König H (2003) Termite gut symbiotic Archaezoa are becoming living metabolic fossils. Eukaryot Cell 2(5):1091–1098
Lightner DV (1996) A Handbook of shrimp pathology and diagnostic procedures for diseases of cultured penaeid shrimp. World Aquaculture Society, Baton Rouge, LA
Lightner DV, Redman RM (1998) Shrimp diseases and current diagnostic methods. Aquaculture 164(1):201–220
Lotz JM (2010) Evolutionary principles applied to disease control and health management in shrimp aquaculture. In: Alday-Sanz V (ed) The shrimp book. Nottingham University Press, Nottingham, UK, pp 679–694
Lotz JM, Flowers AM, Breland V (2003) A model of *Taura syndrome virus* (TSV) epidemics in *Litopenaeus vannamei*. J Invertebr Pathol 83(2):168–176
Maggenti MA, Maggenti AR, Gardner SL (2005) Online dictionary of invertebrate zoology: complete work, DigitalCommons @ University of Nebraska—Lincoln. http://digitalcommons.unl.edu/onlinedictinvertzoology/2. Accessed 7 Aug 2014
May RM, Anderson RM (1983) Epidemiology and genetics in the coevolution of parasites and hosts. Proc R Soc Lond Ser B 219(1216):281–313
McGaw IJ (2005) The decapod crustacean circulatory system: a case that is neither open nor closed. Microsc Microanal 11(1):18–36
Middlebrooks BL, Ellender RD, Lee YM (1989) Monthly assessment of lectin levels in *Penaeus aztecus* from the area of Biloxi Bay in Mississippi. J World Aquacult Soc 20(1):57A
Miller TL, Small HJ, Peemoeller B-J, Gibbs DA, Shields JD (2013) Experimental infections of *Orchitophyra stellarum* (Scuticociliata) in American blue crabs (*Callinectes sapidus*) and fiddler crabs (*Uca minax*). J Invertebr Path 114(3):346–355
Mollison D (1995) The structure of epidemic models. In: Mollison D (ed) Epidemic models: their structure and relation to data. Cambridge University Press, New York, pp 17–33
Moore J (2002) Parasites and the behavior of animals. Oxford series in ecology and evolution. Oxford University Press, New York
Noble ER, Noble GA, Schad GA, MacInnes AJ (1989) Parasitology: the biology of animal parasites, 6th edn. Lea & Febiger, Philadelphia, PA
Olson RE (1981) Effects of low temperature on the development of the microsporidan *Glugea stephani* in English sole (*Parophrys vetulus*). J Wildl Dis 17(4):559–562
Overstreet RM (1968) Parasites of the inshore lizardfish, *Synodus foetens*, from South Florida, including a description of a new genus of Cestoda. Bull Mar Sci 18(2):444–470
Overstreet RM (1982) Abiotic factors affecting marine parasitism. In: Mettrick DF, Desser SS (eds) Parasites, their world and ours: proceedings of the Fifth International Congress of Parasitology. Elsevier Biomedical Press, Amsterdam, pp 36–39
Overstreet RM (1983) Metazoan symbionts of crustaceans, Chapter 4. In: Provenzano AJ (ed) The biology of Crustacea: pathobiology, vol 6. Academic Press, New York, pp 155–250
Overstreet RM (1994) BP (*Baculovirus penaei*) in penaeid shrimps. USMSFP 10th Anniversary Review GCRL Special Publication 1:97–106
Overstreet RM (1997) Parasitological data as monitors of environmental health. Parassitologia 39(3):169–175

Overstreet RM (2013) Waterborne parasitic diseases in ocean. In: Kanki P, Grimes DJ (eds) Infectious diseases: selected entries from the Encyclopedia of sustainability science and technology. Springer Science + Business Media, New York, pp 431–496. doi:10.1007/978-1-4614-5719-0_15

Overstreet RM, Curran SS (2004) Defeating diplostomoid dangers in USA catfish aquaculture. Folia Parasitol 51(2/3):153–165

Overstreet RM, Safford S (1980) Diatoms in the gills of the commercial white shrimp. Gulf Res Rep 6(4):421–422

Overstreet RM, Stuck KC, Krol RA, Hawkins WE (1988) Experimental infections with *Baculovirus penaei* in the white shrimp, *Penaeus vannamei* (Crustacea: Decapoda), as a bioassay. J World Aquacult Soc 19(4):175–187

Overstreet RM, Hawkins WE, Deardorff TL (1995) The western mosquitofish as an environmental sentinel: parasites and histological lesions. In: Servos MR, Munkittrick KR, Carey JH, Van Der Kraak GJ (eds) Environmental fate and effects of pulp and paper mill effluents. St. Lucie Press, Delray Beach, FL, pp 495–509

Overstreet RM, Lightner DV, Hasson KW, McIlwain S, Lotz JM (1997) Susceptibility to Taura syndrome virus of some penaeid shrimp species native to the Gulf of Mexico and southeastern United States. J Invertebr Pathol 69(2):165–176

Overstreet RM, Curran SS, Pote LM, King DT, Blend CK, Grater WD (2002) *Bolbophorus damnificus* n. sp. (Digenea: Bolbophoridae) from the channel catfish *Ictalurus punctatus* and American white pelican *Pelecanus erythrorhynchos* in the USA based on life-cycle and molecular data. Syst Parasitol 52(1):81–96

Overstreet RM, Jovonovich J, Ma H (2009) Parasitic crustaceans as vectors of viruses, with an emphasis on three penaeid viruses. Integr Comp Biol 49(2):127–141

Parker JD, Spall RD, Warner MC (1971) Two new Myxosporida, *Henneguya gambusi* sp. n. and *Myxosoma pharyngeus* sp. n., in the mosquitofish, *Gambusia affinis* (Baird and Girard). J Parasitol 57(6):1297–1301

Parrish CR, Holmes EC, Moren DM, Park EC, Burke DS, Calisher CH, Laughlin CA, Saif LJ, Daszak P (2008) Cross-species viral transmission and the emergence of new epidemic diseases. Microbiol Mol Biol Rev 72(3):457–470

Pearce-Duvet JMC (2006) The origin of human pathogens: evaluating the role of agriculture and domestic animals in the evolution of human disease. Biol Rev 81(3):369–382

Pepin J (2011) The origin of AIDS. Cambridge University Press, Cambridge

Roberts LS, Janovy J Jr (2009) Gerald D Schmidt & Larry S Roberts' foundations of parasitology, 8th edn. McGraw-Hill, Boston, MA

Roekring S, Nielsen L, Owens L, Pattanakitsakul SN, Malasit P, Flegel TW (2002) Comparison of penaeid shrimp and insect parvoviruses suggests that viral transfers may occur between two distantly related arthropod groups. Virus Res 87(1):79–87

Sharp PM, Hahn BH (2010) The evolution of HIV-1 and the origin of AIDS. Philos Trans R Soc Lond B Biol Sci 365(1552):2487–2494

Shields JD, Overstreet RM (2007) Diseases, parasites, and other symbionts. Chapter 8. In: Kennedy VS, Cronin LE (eds) The blue crab, Callinectes sapidus. Maryland Sea Grant, College Park, MD, pp 299–417

Shoemaker CA, Martins ML, Xu D-H, Klesius PH (2012) Effect of *Ichthyophthirius multifiliis* parasitism on the survival, hematology and bacterial load in channel catfish previously exposed to *Edwardsiella ictaluri*. Parasitol Res 111(5):2223–2228

Small HJ, Miller TL, Coffey AH, Delaney KL, Schott E, Shields JD (2013) Discovery of an opportunistic starfish pathogen, *Ochitophrya stellarum*, in captive blue crabs, *Callinectes sapidus*. J Invertebr Pathol 114(2):178–185

Smith SJ (1981) The trematode fauna of a brackish coastal lagoon in Tasmania. PhD Dissertation, University of Tasmania, Hobart

Solangi MA, Lightner DV (1976) Cellular inflammatory response of *Penaeus aztecus* and *P. setiferus* to the pathogenic fungus, *Fusarium* sp., isolated from the California brown shrimp, *P. californiensis*. J Invertebr Pathol 27(1):77–86
Solangi MA, Overstreet RM (1980) Biology and pathogenesis of the coccidium *Eimeria funduli* infecting killifishes. J Parasitol 66(3):513–526
Solangi MA, Overstreet RM, Fournie JW (1982) Effect of low temperature on development of the coccidium *Eimeria funduli* in the Gulf killifish. Parasitology 84(1):31–39
Soto MA, Lotz JM (2001) Epidemiological parameters of *white spot syndrome virus* (WSSV) infections in *Litopenaeus vannamei* and *L. setiferus*. J Invertebr Pathol 78(1):9–15
Stickle WB, Kozloff EN, Story S (2007) Physiology of the ciliate *Orchitophrya stellarum* and its experimental infection of *Leptasterias* spp. Can J Zool 85(2):201–206
Stuck KC, Overstreet RM (1994) Effect of *Baculovirus penaei* on growth and survival of experimentally infected postlarvae of the Pacific white shrimp, *Penaeus vannamei*. J Invertebr Pathol 64(1):18–25
Stuck KC, Stuck LM, Overstreet RM, Wang SY (1996) Relationship between BP (*Baculovirus penaei*) and energy reserves in larval and postlarval Pacific white shrimp (*Penaeus vannamei*). Dis Aquat Organ 24(3):191–198
Sudhaus W (2008) Evolution of insect parasitism in rhabditid and diplogastrid nematodes. In: Makarov SE, Dimitrijević RN (eds) Advances in arachnology and developmental biology. Vienna-Belgrade-Sofia: SASA, Monographs 12, pp 143–161
Summers MD (1977) Characterization of shrimp baculovirus. EPA-600 3-77-130, Ecological research series. Environmental Research Laboratory, Gulf Breeze, FL
Thomas F, De Meeûs T, Renaud F (1996) Pattern of infection of *Gammarus aequicauda* (Amphipoda) with metacercariae of *Levinseniella tridigitata* (Trematoda: Microphallidae). J Helminthol Soc Wash 63(1):8–11
Trager W (1970) Symbiosis. Van Nostrand Reinhold Company, New York
U.S. Department of Health and Human Services, Centers for Disease Control and Prevention (2013) Intestinal parasite guidelines for domestic medical examination for newly arrived refugees. http://www.cdc.gov/immigrantrefugeehealth/guidelines/domestic/domestic-guidelines.html. Accessed 7 Aug 2014
Vidal-Martínez VM, Osorio-Sarabia D, Overstreet RM (1994) Experimental infection of *Contracaecum multipapillatum* (Nematoda: Anisakinae) from Mexico in the domestic cat. J Parasitol 80(4):576–579
Wang Y, Jehle JA (2009) Nudiviruses and other large, double-stranded circular DNA viruses of invertebrates: new insights on an old topic. J Invertebr Pathol 101(3):187–193
Wu GA, Juna SR, Sims GE, Kima SH (2009) Whole-proteome phylogeny of large dsDNA virus families by an alignment-free method. Proc Natl Acad Sci USA 106(31):12826–12831

Part II
Opportunistic Pathogenicity in the Aquatic World

Chapter 3
Opportunistic Infections in Elasmobranchs

Joanna Borucinska

Abstract This chapter describes factors potentially contributing to immunosuppression and thus opportunistic infections in captive and free-ranging cartilaginous fishes. The immune system in sharks is briefly discussed to show the similarities with higher vertebrates with the implications for comparable mechanisms governing immunosuppression. A brief description of stress as a factor inducing immunosuppression is provided. Potentially opportunistic infections with viruses, bacteria, algae, and parasites are discussed separately for elasmobranchs in captivity and in the wild, and their links with stress-inducing stimuli are suggested.

3.1 Introduction

Opportunistic infections have been defined as infections that are more frequent or more severe because of immunosuppression (www.cdc.gov). Although without doubt such infections were quietly affecting all life forms since time immemorial, the human immunodeficiency virus (HIV) pandemic erupting in the 1980s dramatically increased our awareness of their existence. Since then, our knowledge on the causes of diminished immunity has expanded beyond viral infections to encompass the indirect effects of various forms of stress and direct effects of environmental pollutants with immunosuppressive and immunomodulatory properties (Fries 1986). Some of the toxic effects of pollutants were heralded by Rachel Carson in her sentinel book *Silent Spring* (Carson 1962), but it took several decades to fully acknowledge the spectrum of their toxic effects. Despite the wealth of laboratory data, the mechanisms of their individual and cumulative toxicity are yet not fully understood. In addition, their causal link to the health of animals in the natural environment is still difficult to document (Corsolini et al. 1995). This is especially true to aquatic organisms that are chronically exposed to small albeit readily

J. Borucinska (✉)
Department of Biology, University of Hartford, West Hartford, CT, USA
e-mail: borucinsk@hartford.edu

© Springer International Publishing Switzerland 2016

C.J. Hurst (ed.), *The Rasputin Effect: When Commensals and Symbionts Become Parasitic*, Advances in Environmental Microbiology 3,
DOI 10.1007/978-3-319-28170-4_3

available and potentially biologically significant concentrations of various toxic compounds which can penetrate both the gill and mucocutaneous surfaces (Anderson 1996). Due to the insidious nature of such exposure, it is quite impossible for affected organisms to defend against the direct effects of pollution with immunosuppressant compounds, and error on the safe side seems to be the wiser choice in inferring their potentially deleterious effects (Fries 1986). Extrapolation of findings from experimental animals in which immunosuppression has been studied can be useful in this regard.

It is equally difficult to show the effects of stress, especially neurogenic stress, on the immune system of fish; luckily, the evolutionary conservation of the general adaptation syndrome (GAS) as defined by Hans Selye (1955) provided tools to identify stress in fishes (Harper and Wolf 2009) including sharks (Terrell 2004). This chapter will suggest the existence of opportunistic infections in wild and captive sharks, and it seems prudent to first briefly discuss the immune system and possible causes of immunosuppression in cartilaginous fishes.

3.2 The Immune System in Cartilaginous Fishes

3.2.1 Adaptive Immunity

An overview of the immune system of elasmobranchs has been provided by Luer et al. (2004), and since then, our knowledge on the immune mechanisms involved in innate and acquired immunity in sharks has expanded dramatically (Walsh et al. 2006a; Criscitiello 2014). Accordingly, elasmobranchs fill a unique immunological niche in that they are phylogenetically the oldest group of vertebrates to possess all of the components necessary for an adaptive immune system including immunoglobulins (Ig), T-cell receptors (TCR), the major histocompatibility complex (MHC), RAG-mediated rearrangement, somatic hypermutation, T and B lymphocytes, and the presence of primary and secondary lymphoid tissues (Marchalonis et al. 1998; Dooley and Flajnik 2006). Several classes of immunoglobulins including monomeric and pentameric IgM, IgW, and IgNAR have been described in elasmobranchs, each having unique features in their T-cell interactions and mechanisms of actions (Dooley and Flajnik 2006; Smith et al. 2012). Similarly to other vertebrates, elasmobranch fishes possess a thymus and spleen, but in the absence of bone marrow and lymph nodes, these fish have evolved unique lymphomyeloid tissues, namely, epigonal and Leydig organs, that are associated with the gonads and the esophagus, respectively. The epigonal and Leydig organs are thought to be vital in generating adaptive immunity in sharks. As in higher vertebrates, the regulation of elasmobranch immune responses likely involves intercellular signaling molecules such as cytokines. The cytokines identified so far in elasmobranches include IL-1b and IL-8, but this list is expanding and it points to high levels of evolutionary continuity of their form and function (Walsh et al. 2006a).

3.2.2 Innate Immunity

The innate immunity of sharks is very well developed, and again, it shares many components with other vertebrates including inflammation, phagocytosis, lysozyme, interferon, inducible NO-synthetase, nonspecific cytotoxicity, and the complement system (Walsh et al. 2006b). Sharks have a well-developed system of phagocytic, possible antigen-binding and processing cells termed the melanomacrophages that serve as an anatomical prelude to lymph nodes (Borucinska et al. 2009). The outstanding capacity of sharks to heal physical injury and combat infections includes their ability to mount a marked proliferation of immune cells and high phagocytic activity within their connective tissues and mesothelium to confine the spread of bacteria or delineate foreign objects (Bird 1978; Reif 1978; Borucinska et al. 2001, 2002). There is a claim that sharks can naturally harbor microorganisms within internal organs and tissues, which in other vertebrates would consistently be sterile, including blood (Terrell 2004; Mylniczenko et al. 2007). Some attribute this capability to physiological peculiarities in elasmobranchs including high levels of blood urea (Grimes 1990) and possibly some unique features of their lymphocytes including an isotype of light chainless antibodies (Criscitiello 2014). Despite the above claims, bacteria are actually the most common infectious diseases reported both from captive and wild-caught sharks (Camus et al. 2013; Garner 2013).

Some of the physiological peculiarities mentioned above have led to serious misconceptions regarding the resistance of sharks against both infections and cancer along with their perception of pain. This has contributed to at least two significant negative outcomes: a dramatic increase in shark-directed fisheries that led to worldwide decline in shark populations and the use of shark-derived pseudomedications that represent a diversion of patients from effective cancer treatments (Ostrander et al. 2004).

3.2.3 Stress and Immune Functions

It has been documented that in fish, similarly to other vertebrates, acute stress induces shifts in peripheral blood leukocyte distribution expressed as higher granulocyte to lymphocyte ratio (Harper and Wolf 2009; Van Rijn and Reina 2010). In prolonged stress, the brain–chromaffin and hypothalamus–pituitary–suprarenal organ (corpora axillaria in sharks) axis responds with high levels of ACTH and glucocorticoids. This leads to deleterious metabolic and physiologic effects, among which are immunosuppression via lymphoid cell apoptosis, lymphopenia, decreased phagocytosis, and disturbance in connective tissue metabolism and delayed wound healing (Fries 1986; Schreck 1996; Barton 2002; Walsh et al. 2002). Because it mostly is the T lymphocytes that are affected in the above listed processes, the most prominent clinical effect of chronic stress is increased susceptibility to intracellular pathogens and cancer (Van Rijn and Reina 2010).

From the above information, one can postulate, that similar to other vertebrates, sharks possess a highly functional immune system that can be challenged through toxins and stress-induced mechanisms conserved among all Chordata. Accordingly, opportunistic infections can be the result of stressful, immunosuppressive factors operating both in the marine and captive environments (Fries 1986).

3.3 Opportunistic Infections in Captive Elasmobranchs

3.3.1 Stress and Immunosuppression in Captivity

A true prevalence of infections in wild shark populations is impossible to establish for obvious technical reasons, and thus reports of infectious diseases in the wild elasmobranchs are uncommon. On the contrary, most of our knowledge on this subject comes from studying captive sharks exhibited in aquaria. Curiously, only a few of the reports of sharks dying in exhibits due to infections have addressed the causal relationships of the stress of captivity along with accompanying immunosuppression. Such stress begins with capture and handling and relocation to circumstances that are new to the animal and frequently inadequate with respect to environment and nutrition (Smith et al. 2004). In addition, captivity results in proximity to pathogens that sharks normally would not encounter in the wild but to which they are exposed due to the ecologically artificial mixing of diverse fish species and invertebrates which occurs in exhibits (Smith et al. 2004; Van Rijn and Reina 2010; Marancik et al. 2012; Stidworthy et al. 2014). An additional problem of captivity that potentially could exacerbate immune dysfunction is the very common condition of goiter, i.e., hypothyroidism described in sharks kept in exhibit aquaria (Blum and Luer 1985; Crow et al. 1998; Garner 2013). Experimentally induced hypothyroidism has resulted in significantly decreased white blood cell counts (WBC) in bony fishes, and similar mechanisms could be operating in sharks (Fries 1986). Accordingly, many if not most infections described in captive sharks have the potential of being opportunistic. What follows is a discussion of the infectious diseases that have been described in captive sharks.

3.3.2 Viral Infections in Captive Elasmobranchs

An enzootic viral infection of the smooth dogfish, *Mustelus canis* (Mitchill) (Leibovitz and Lebouitz 1985), has been described in wild-caught, laboratory-maintained, and aquaria-displayed populations of smooth dogfish that had been collected from the Woods Hole area of Massachusetts during the course of a

2.5-year period. The disease usually followed a stressful event and was characterized by the progressive development of gross and microscopic herpes-like skin lesions (Leibovitz and Lebouitz 1985). Virus particles (presumably either herpesvirus or adenovirus) have been reported in conjunction with ciliated protozoa and bacteria in cutaneous lesions of captive dusky smooth-hounds (Garner 2013), but the contribution of these lesions to morbidity was not discussed. The same author reported papillomatosis seen as multicentric dermal raised plaques in several shark species, and although no virus particles were isolated and the lesions regressed spontaneously, they may have been a "wart"-like condition associated with papillomavirus species typically infecting immunocompromised hosts.

3.3.3 Bacterial Infections in Captive Elasmobranchs

In a recent review of diseases reported from free-ranging and captive sharks, Garner (2013) listed that 33.5 % of all diseases diagnosed in his facility (receiving autopsy cases predominantly from aquaria) were infectious, and of these, bacterial infections were the most common. A bacterial infection that was most likely of opportunistic nature has been observed in a public aquarium-kept shark (Camus et al. 2013). It involved *Serratia marcescens* and was associated with cellulitis of the ampullary system and septicemia in a bonnethead shark, *Sphyrna tiburo* (L.). *Serratia marcescens* is a moisture-preferring, biofilm-forming environmental saprophyte that has been associated with several cases of opportunistic nosocomial infections in humans and does as well cause opportunistic infections both in mammals and in fish taken from waters polluted by human sewage (Camus et al. 2013).

Other examples of possibly opportunistic bacterial isolates from sharks housed in aquaria are mycobacteria. In one case, multiple granulomas from an epaulette shark (*Hemiscyllium ocellatum*) yielded *Mycobacterium avium* (Janse and Kik 2012); in the second case, splenic granulomas in an Atlantic guitarfish, *Rhinobatos lentiginosus*, were associated with *Mycobacterium chelonae* (Anderson et al. 2012). The source of these mycobacterial infections could include other exhibited animals as well as their human caretakers or visitors.

Bacterial infections in captive sharks can produce septicemias including those which have been attributed to *Aeromonas salmonicida* (Briones et al. 1998) and various *Vibrio* spp. (Grimes et al. 1984, 1985, 1989). Also, meningitis due to *Vibrio carchariae* (Stoskopf 1993) has been reported, and *Dermophthirius* sp. has been implicated as a vector of *Vibrio carchariae* transmitting the disease from shark to shark in aquaria (Grimes et al. 1984). Because *Vibrio* spp. are frequently isolated from healthy sharks in the wild (Knight et al. 1987) and in captivity (Bertone et al. 1996), one may speculate that in some cases captivity leads to an opportunity for the physiological colonization to progress to infection and disease. Direct evidence of stress-induced vibriosis in sharks has been provided by the reporting of a variety of *Vibrio* spp. having been isolated from Port Jackson (*Heterodontus portusjacksoni*) and epaulette (*Hemiscyllium ocellatum*) sharks and southern fiddler rays (*Trygonorrhina fasciata*) that died following a change in aquatic salinity (Callinan 1988).

3.3.4 Infections with Fungi and Algae Described from Captive Sharks

An example of a most likely opportunistic fungal infection in captive sharks is a condition known as "bonnethead shark disease." The etiological agent, *Fusarium solani*, was originally isolated from two newborn bonnethead sharks *Sphyrna tiburo* (Muhvich et al. 1989) that died in captivity; since then, the infection has been also described in captive scalloped hammerhead sharks *Sphyrna lewini*. According to Terrell (2004), the disease typically follows some environmental change and has been seen when shallow, warm-water sharks have been kept in deeper, cold water aquarium systems. The disease is progressive and refractory to treatment, with affected sharks dying with evidence of deep invasion of the fungus into skin, muscle, cartilage, and, occasionally, internal organs (Terrell 2004). More recently, lethal, systemic infections with two fungal organisms were described in a captive great hammerhead *Sphyrna mokarran* and a juvenile zebra shark *Stegostoma fasciatum*; the fungal organisms included *Exophiala pisciphila* and *Mucor circinelloides*, both of which are considered uncommon pathogens in human and veterinary medicine (Marancik et al. 2011).

Lesions in captive sharks elicited by histozoic algae were described in sevengill sharks *Notorynchus maculatus* (Blasiola and Turnier 1979). This case involved seven sevengill sharks and no other shark species were affected.

3.3.5 Parasitic Infections in Captive Elasmobranchs

Reports of protozoal diseases in elasmobranchs appear to be relatively uncommon with Cheung (1993) providing a review of reported protozoa including Microsporidia, amoeba, and ciliates. Most of reports found in the literature refer to infections in wild-caught individuals and rarely indicated more than a light infection (Clewley et al. 2002). In captive elasmobranchs, parasitic protozoa are usually associated with teleost fishes from the same exhibit and can become an overwhelming challenge. Such was the case of Atlantic stingrays *Dasyatis sabina* succumbing to a massive infection of the dinoflagellate *Amyloodinium ocellatum*, after being exposed to teleosts carrying the parasite (Lawler 1980 cited in Goerz 2004). Similarly, systemic scuticocilliatosis manifesting as a combination of necrotizing bronchitis, hepatitis, and meningoencephalitis due to infection with *Philasterides dicentrarchi* has been reported to cause acute death in several species of wild-caught juvenile and adult sharks of different species exhibited in display aquaria (Stidworthy et al. 2014). Both of the outbreaks studied by Stidworthy et al. occurred in mixed teleost-elasmobranch exhibits and presented as peracute to acute lethal infections affecting only sharks. Interestingly, one of the sharks had marked lymphoid depletion in the spleen. It also is noteworthy that invasive ciliates of the subclass Scuticociliatida have been reported in several autopsy reports from cases of captive sharks that died of either necrotizing bronchitis, hepatitis, or encephalitis (Garner 2013). Although they have been experimentally induced in sharks, ciliate infections in wild elasmobranchs are practically unknown (Goerz 2004), such that the above instances of death seem to represent classical examples of circumstances in which a combination of stress and proximity of bony fish with their microbiome results in opportunistic infections in sharks.

Lastly, infestation with the marine leech *Branchellion torpedinis* causes disease and mortality in captive elasmobranchs. It has been suggested that there may be immunomodulatory substances contained in the saliva of the leech which contribute both to severity of the associated diseases and difficulties in controlling it (Marancik et al. 2012).

3.4 Opportunistic Infections in Free-Ranging Elasmobranchs

3.4.1 Possible Causes of Immunosuppression in Marine Elasmobranchs

In wild sharks, habitat destruction and pollution, food source depletion, and disturbance by fishing activities including catch-and-release can all lead to elevated stress responses (Fries 1986; Anderson 1996). In addition, environmental toxins with immunosuppressive properties such as PCBs and heavy metals could increase the susceptibility to infections (Fries 1986). Most sharks being long-lived top predators will bioaccumulate and biomagnify the levels of environmental toxins in their tissues (Lyle 1986; Hornung et al. 1993; Corsolini et al. 1995; Nam et al. 2011a). It is anticipated that global pollution of oceans will become with time only a bigger problem as exemplified by the sequel of global climate changes, eutrophication and its accompanying increase in toxic algal blooms, extreme weather events, and spillage of oil and its concomitant usage of chemical dispersants and by the ocean plumes of microplastics that are becoming well incorporated into the marine food web. These effectors reach even to the highest trophic levels, including top predators like sharks (Law and Thompson 2014). A direct effect of the above list of environmental changes upon stress levels and immunosuppression in fish would be very difficult to document and has to be inferred from careful field observations.

3.4.2 Stranding Events Involving Sharks

It may well be that regularly observed stranding events, although much less frequent in sharks than bony fish, are a phenomenon whose occurrence is compounded by many of the above environmental factors. Published reports have linked stranding events occurring regularly along the Californian coast and involving juvenile salmon sharks *Lamna ditropis* to bacterial meningoencephalitis (Schaffer et al. 2013), but the immune functions in these sharks were not examined. Additional stranding in sharks has been linked to meningoencephalitis of unknown etiology (Dagleish et al. 2010), cold kills, fishing discards, and harmful algal blooms (Flewelling et al. 2010) and, lastly, to retained fishing gear (Adams et al. 2015). Marine unicellular algae can produce several toxic substances including brevetoxins, which are persistent, bioaccumulative, lipophilic polyether neurotoxins synthesized by the harmful algal bloom (HAB) dinoflagellate *Karenia brevis*. Brevetoxin exposure in wild sharks has been shown to induce marked neurochemical alterations that could induce not only abnormal behavior leading to stranding but also produce other biological effects including acute gill hemorrhages and embryo mortality (Flewelling et al. 2010; Nam et al. 2011b).

3.4.3 Retained Fishing Gear and Infections in Free-Ranging Sharks

Retained fishing gear exemplifies a wider problem of prolonged stress in sharks captured and then released due to their low commercial value or to comply with fisheries regulations. Both regular "J" and "circle" hooks can be retained anywhere between the esophagus and stomach of the shark or lodged in the body wall (Borucinska et al. 2001, 2002; Adams et al. 2015). Many of these hooks penetrate into the pericardial or abdominal cavities and cause chronic lesions due to bacterial, algal, or foreign body-induced inflammation that in some cases results in neoplastic processes (Borucinska et al. 2003). There is no question that many of the microorganisms introduced with the hooks could be considered opportunists that use the immunosuppression due to the stress of capture, release, and trauma, to become pathogenic. The paucity of intralesional microorganisms in some of these sharks would be consistent with suggestions that sharks are capable of clearing or retaining bacteria by immune phenomena and fibroproliferative tissue reactions (Bird 1978; Reif 1978; Stoskopf 1993; Heupel and Benett 1997; Heupel et al. 1998; Van Rijn and Reina 2010). In one of the reports of lesions associated with retained fishing hooks, *Corynebacterium* sp. and *Pseudomonas putrefaciens* were cultured from the affected shark, and histological evidence of intralesional bacteria (Fig. 3.1) and algal organisms (Fig. 3.2) was present (Borucinska et al. 2001).

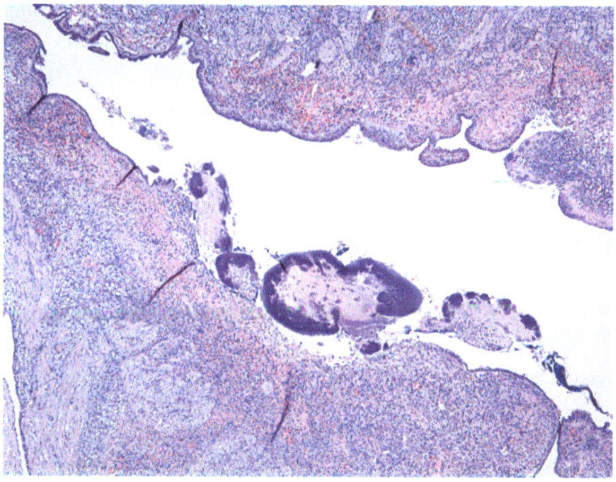

Fig. 3.1 Blue shark with gastritis, peritonitis, and pericarditis associated with retained fishing hook. Bacterial colonies within a fistula surrounding fishing hook embedded within the gastric wall. Tissue gram stain

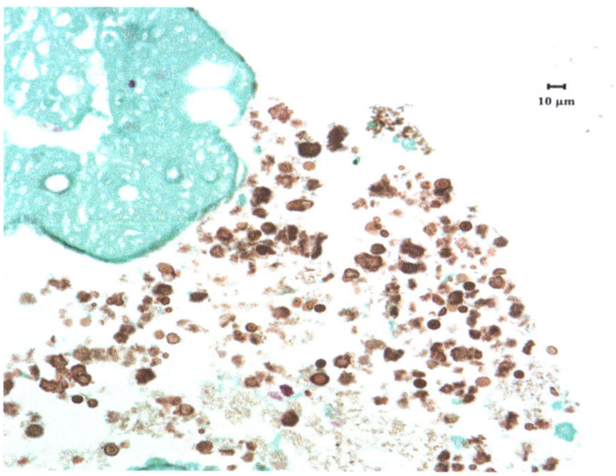

Fig. 3.2 Blue shark with gastritis, peritonitis, and pericarditis associated with retained fishing hook. Algal organisms (visible as black round structures) on the pericardial surface of the atrium with severe proliferative epicarditis. Grocott's silver stain

3.4.4 Metazoan Parasites in Free-Ranging Sharks

The last but perhaps not the least group of infectious agents that should be studied in relation to immunosuppression of elasmobranchs is the metazoan parasites. Although reports of parasitic metazoan infections in both wild and captive sharks are numerous (Cheung 1993; Stoskopf 1993), the elicited pathology is usually not addressed. Our observations indicate that the degree in pathological reactions to the presence of parasites can quite vary in their severity. For example, microfilarial-like nematodes in sharks can be confined in granulomas within multiple organs with no apparent harm (Borucinska and Heger 1999; Credille et al. 1993; Borucinska and Frasca 2002), and in other cases, nematodes and their larvae can cause severe granulomatous meningitis and encephalitis (Credille et al. 1993) or metritis with multisystemic granulomatous lesions (Borucinska and Adams 2013). Future studies of tissue responses to such parasites including morphological assessment of the lymphoid organs in affected sharks could help us to understand the many differences in host response and possibly link those differences to the immune status of the host.

References

Adams DH, Borucinska JD, Maillett K, Whitburn K, Sander TE (2015) Mortality due to a retained circle hook in a longfin mako shark *Isurus paucus* (Guitart-Manday). J Fish Dis 38:621–628

Anderson DP (1996) Environmental factors in fish health: immunological aspects. In: Iwama G, Nakanishi T (eds) The fish immune system. Academic, London, pp 289–305

Anderson ET, Frasca S Jr, Asakawa M, Fatzinger MH, Johnson J, Marchettere K, Goodale L, Risatti GR, Harms CA (2012) Splenic mycobacteriosis in an Atlantic guitarfish, *Rhinobatos lentiginosus* (Garman). J Fish Dis 35:541–544

Barton BA (2002) Stress in fishes: a diversity of responses with particular reference to changes in circulating corticosteroids. Integr Comp Biol 42:517–525

Bertone S, Gili C, Moizo A, Calegari L (1996) *Vibrio carchariae* associated with a chronic skin ulcer on a shark, *Carcharhinus plumbeus* (Nardo). J Fish Dis 19:429–434

Bird PM (1978) Tissue regeneration in three carcharhinid sharks encircled by embedded straps. Copeia 2:345–349

Blasiola GC Jr, Turnier JC (1979) Algal infection of the sevengill shark *Notorynchus maculatus*. J Fish Dis 2:161–163

Blum P, Luer CA (1985) Observations on the occurrence of thyroid hyperplasia in captive sharks. Fla Sci 48:32

Borucinska JD, Adams DH (2013) Parasitic metritis and oophoritis associated with nematode larvae in an Atlantic sharpnose shark (*Rhizoprionodon terraenovae*) collected in the western Atlantic off Florida. J Fish Dis 36:811–817

Borucinska JD, Frasca S Jr (2002) Naturally occurring lesions and microorganisms in two species of free-living sharks: the spiny (*Squalus acanthias*) and the less spotted (*Mustelus canis*) dogfish from the northwestern Atlantic. J Fish Dis 25:1–11

Borucinska JD, Heger K (1999) Disseminated granulomas in a shortfin mako shark associated with nematode larvae. J Wildl Dis 35:98–100

Borucinska JD, Martin J, Skomal G (2001) Peritonitis and pericarditis associated with gastric perforation by a retained fishing hook in a blue shark (*Prionace glauca*). J Aquat Anim Health 13:347–354

Borucinska JD, Kohler N, Natanson L, Skomal G (2002) Pathology associated with retained fishing hooks in blue sharks (*Prionace glauca*) (Linnaeus, 1758) with implications for their conservation. J Fish Dis 25:515–521

Borucinska JD, Harshbarger JC, Bogicevic T (2003) Hepatic cholangiocarcinoma and testicular mesothelioma in a wild-caught blue shark (*Prionace glauca*) (Linnaeus, 1758). J Fish Dis 26:43–49

Borucinska JD, Kotran K, Shackett M, Barker T (2009) Melanomacrophage cells in three species of free-ranging sharks from the northwestern Atlantic, the blue shark *Prionacae glauca*, the shortfin mako *Isurus oxyrhinchus*, and the thresher *Alopias vulpinus*. J Fish Dis 32:883–891

Briones V, Fernandez A, Blanco M, Ramiro F, Garcia J, Mendez JL, Goyache J, De Vincente ML (1998) Hemorrhagic septicemia by *Aeromonas salmonicida subsp. salmonicida* in a black-tip reef shark (*Carcharhinus melanopterus*). Zentralblatt der Veterinärmedizin 45:443–445

Callinan RB (1988) Diseases of Australian native fishes. In: Fish diseases: refresher course for veterinarians, Proceeding 106, pp 459–472. Post Graduate Committee of Veterinary Science, University of Sydney, Australia

Camus A, Berliner A, Clauss T, Hatcher N, Marancik D (2013) *Serratia marcescens* associated ampullary system infection and septicaemia in a bonnethead shark, *Sphyrna tiburo* (L.). J Wildl Dis 36:891–895
Carson RL (1962) Silent spring. Riverside, Cambridge
Cheung P (1993) Parasitic diseases of elasmobranchs. In: Stoskopf MK (ed) Fish medicine. W.B. Saunders, Philadelphia, PA, pp 782–808
Clewley A, Kocan RM, Kocan AA (2002) An intraerythrocytic parasite from the spiny dogfish, *Squalus acanthias* L., from the Pacific Northwest. J Wildl Dis 25:693–696
Corsolini S, Focardi S, Kannan K, Tanabe S, Borrell A, Tatsukawa R (1995) Congener profile and toxicity assessment of polychlorinated biphenyls in dolphins, sharks and tuna collected from Italian coastal waters. Mar Environ Res 40:33–53
Credille KM, Johnson L, Reimschuessel R (1993) Parasitic meningoencephalitis in nurse sharks (*Ginglymostoma cirratum*). J Wildl Dis 29:502–506
Criscitiello MF (2014) What the shark immune system can and cannot provide for the expanding landscape of immune therapy. Expert Opin Drug Discov 9:725–739
Crow GL, Atkinson MJ, Ron B, Atkinson S, Skillman AD, Wong GTF (1998) Relationship of water chemistry to serum thyroid hormones in captive sharks with goiters. Aquat Geochem 4:469–480
Dagleish MP, Baily JL, Foster G, Reid LJ, Barley J (2010) The first report of disease in a basking shark (*Cetorhinus maximus*). J Comp Pathol 143:284–288
Dooley H, Flajnik MF (2006) Antibody repertoire development in cartilaginous fish. Dev Comp Immunol 30:43–56
Flewelling LJ, Adams DH, Naar JP, Atwood KE, Granholm AA, O'Dea SN, Landsberg JH (2010) Brevetoxins in sharks and rays (Chondrichthyes, Elasmobranchii) from Florida coastal waters. Mar Biol 157:1937–1953
Fries CR (1986) Effects of environmental stressors and immunosuppressants on immunity in *Fundulus heteroclitus*. Am Zool 26:271–282
Garner MM (2013) A retrospective study of disease in elasmobranchs. Vet Pathol 50:377–389
Goerz CEC (2004) Protozoal diseases of elasmobranchs. In: Smith M, Warmolts D, Thoney D, Hueter R (eds) The elasmobranch husbandry manual: captive care of sharks, rays and their relatives. Ohio Biological Survey, Columbus, OH, pp 417–426
Grimes DJ (1990) Review of human pathogenic bacteria in marine animals with emphasis on sharks. In: Elasmobranchs as living resources: advances in the biology, ecology, systematics, and the status of fisheries. NOAA Technical Report NMFS 90
Grimes DJ, Stemmler J, Hada H, Maneval D, Hetrick FM, May EB, Jones RT, Stoskopf MK, Colwell RR (1984) Vibrio species associated with mortality of sharks held in captivity. Microb Ecol 10:271–282
Grimes DJ, Gruber SH, May EB (1985) Experimental infection of lemon sharks, *Negaprion brevirostris* (Poey), with Vibrio species. J Fish Dis 8:173–180
Grimes DJ, Burgess J, Crunkleton JA Jr, Brayton PR, Colwell RR (1989) Potential invasive factors associated with *Vibrio carchariae*, an opportunistic pathogen for sharks. J Fish Dis 12:69–72
Harper C, Wolf JC (2009) Morphological effects of the stress response in fish. ILAR J 50:387–396
Heupel MR, Benett MB (1997) Histology of dart tag insertion sites in the epaulette shark. J Fish Biol 50:1034–1041
Heupel MR, Simpfendorfer CA, Bennett MB (1998) Analysis of tissue responses to fin tagging in Australian carcharhinids. J Fish Biol 52:610–620
Hornung H, Krom MD, Cohen Y, Bernhard M (1993) Trace metal content in deep-water sharks from the eastern Mediterranean Sea. Mar Biol 115:331–338
Janse M, Kik MJL (2012) *Mycobacterium avium* granulomas in a captive epaulette shark (*Hemiscyllium ocellatum*). J Fish Dis 35:935–940
Knight II, Grimes DJ, Colwell RR (1987) Bacterial hydrolysis of urea in the tissues of carcharhinid sharks. Can J Fish Aquat Sci 45:357–360
Law LK, Thompson RC (2014) Microplastics in the seas. Science 345:144–145

Leibovitz L, Lebouitz SS (1985) A viral dermatitis of the smooth dogfish, *Mustelus canis* (Mitchill). J Fish Dis 8:273–279
Luer CA, Walsh CJ, Bodine AB (2004) The immune system of sharks, skates and rays. In: Carrier JC, Musick JA, Heithaus MR (eds) Biology of sharks and their relatives. CRC, Washington, DC, pp 369–389
Lyle JM (1986) Mercury and selenium concentrations in sharks from northern Australian waters. Aust J Mar Freshwat Res 37:309
Marancik DP, Berliner AL, Cavin JM, Clauss TM, Dove AD, Sutton DA, Wickes BL, Camus AC (2011) Disseminated fungal infection in two species of captive sharks. J Zoo Wildl Med 42:686–693
Marancik DP, Leary JH, Fast MM, Flajnik MF, Camus AC (2012) Humoral response of captive zebra sharks *Stegostoma fasciatum* to salivary gland proteins of the leech *Branchellion torpedinis*. Fish Shellfish Immunol 33:1000–1007
Marchalonis J, Schluter S, Bernstein RM, Hohman VS (1998) Antibodies of sharks: revolution and evolution. Immunol Rev 166:103–122
Muhvich AG, Reimschuessel R, Lipsky MM, Bennett RO (1989) *Fusarium solani* isolated from newborn bonnethead sharks, *Sphyrna tiburo* (L.). J Fish Dis 12:57–62
Mylniczenko ND, Harris B, Wiborn RE, Young FA (2007) Blood culture results from healthy captive and free-ranging elasmobranchs. J Aquat Anim Health 19:159–167
Nam D, Adams DH, Reyier EA, Basu N (2011a) Mercury and selenium levels in lemon sharks (*Negaprion brevirostris*) in relation to a harmful red tide event. Environ Monit Assess 176:549–559
Nam D, Adams DH, Flewelling LJ, Basu N (2011b) Neurochemical alterations in lemon shark (*Negaprion brevirostris*) brains associated with brevetoxin exposure. Aquat Toxicol 99:351–359
Ostrander GK, Cheng KC, Wolf JC, Wolfe MJ (2004) Shark cartilage, cancer and the growing threat of pseudoscience. Cancer Res 64:8485–8491
Reif WE (1978) Wound healing in sharks. Zoomorphologie 90:101–111
Schaffer PA, Lifland B, Van Sommeran S, Casper DR, Davis CR (2013) Meningoencephalitis associated with *Carnobacterium maltaromaticum*-like bacteria in stranded juvenile salmon sharks (*Lamna ditropis*). Vet Pathol 50:412–417
Schreck CB (1996) Immunomodulation: endogenous factors. In: Iwama G, Nakanishi T (eds) The fish immune system. Academic, London, pp 311–327
Selye H (1955) Stress and disease. Science 122:625–631
Smith M, Warmolts D, Thoney D, Heuter R (2004) The elasmobranch husbandry manual: captive care of sharks, rays, and their relatives. Ohio Biological Survey, Columbus, OH, pp 427–431
Smith LE, Crouch K, Cao W, Müller MR, Wu L, Steven J, Lee M, Liang M, Flajnik MF, Shih HH, Barelle CJ, Paulsen J, Gill DS, Dooley H (2012) Characterization of the antibody repertoire of the spiny dogfish (*Squalus acanthias*). Dev Comp Immunol 36:665–679
Stidworthy MF, Garner MM, Bradway DS, Westfall BD, Joseph B, Repetto S, Guglielmi E, Schmidt-Posthaus H, Thornton SM (2014) Systemic scuticocilliatosis (*Philasterides dicentrarchi*) in sharks. Vet Pathol 51:628–632
Stoskopf MK (1993) Fish medicine. W. B. Saunders, Philadelphia, PA
Terrell SP (2004) An introduction to viral, bacterial, and fungal diseases of elasmobranchs. In: Smith M, Warmolts D, Thoney D, Hueter R (eds) The elasmobranch husbandry manual: captive care of sharks, rays, and their relatives. Ohio Biological Survey, Columbus, OH, pp 427–431
Van Rijn JA, Reina R (2010) Distribution of leukocytes as indicators of stress in the Australian swellshark, *Cephaloscyllium laticeps*. Fish Shellfish Immunol 29:534–538
Walsh CJ, Wyffels JT, Bodine AB, Luer CA (2002) Dexamethasone-induced apoptosis in immune cells from peripheral circulation and lymphomyeloid tissues of juvenile clearnose skates, *Raja eglanteria*. Dev Comp Immunol 26:623–633

Walsh CJ, Luer CA, Bodine AB, Smith CA, Cox HL, Noyes DR, Gasparetto M (2006a) Elasmobranch immune cells as a source of novel tumor cell inhibitors: implications for public health. Integr Comp Biol 46:1072–1081

Walsh CJ, Toranto JD, Gilliland CT, Noyes DR, Bodine AB, Luer CA (2006b) Nitric oxide production in nurse sharks (*Ginglymostoma cirratum*) and clearnose skate (*Raja eglanteria*) peripheral blood leucocytes. Fish Shellfish Immunol 20:40–46

Chapter 4
Bacterial Opportunistic Pathogens of Fish

Nicolas Derome, Jeff Gauthier, Sébastien Boutin, and Martin Llewellyn

Abstract Bacterial opportunistic pathogens are defined as microorganisms causing disease in hosts experiencing atypical environmental stressors or having impaired immune function. In intensive aquacultural rearing, stress factors (such as hypoxia, abnormal pH, and high population density) generate an optimal setting for such pathogens to thrive. The status of these organisms—either as natural components of a healthy microbiome, or a latent step in disease establishment, or both—is still not entirely clear. In this chapter, we outline the current understanding (i.e., taxonomy, biology, disease impact, and current treatment options) of major opportunist bacterial genera of special interest in aquaculture: *Aeromonas*, *Flavobacterium*, and *Vibrio*. On a broader scale, we consider the importance of host/microbiota/environment interactions in opportunistic infections of teleost fish. Not only does this cross talk play a crucial role in defining disease, but their importance also reveals novel strategies to prevent and cure opportunistic diseases. As such, preventive measures to reduce host stress, along with active interventions

N. Derome (✉) • J. Gauthier
Institut de Biologie Intégrative et des Systèmes (IBIS), Université Laval, 1030, av. de la Médecine, Québec, QC, Canada, G1V0A6

Département de Biologie, Université Laval, Québec, QC, Canada
e-mail: nicolas.derome@bio.ulaval.ca

S. Boutin
Institut de Biologie Intégrative et des Systèmes (IBIS), Université Laval, 1030, av. de la Médecine, Québec, QC, Canada, G1V0A6

Département de Biologie, Université Laval, Québec, QC, Canada

Department of Infectious Disease, Medical Microbiology and Hygiene, Universität Heidelberg, Im Neuenheimer Feld 324, 69120 Heidelberg, Germany

M. Llewellyn
Institut de Biologie Intégrative et des Systèmes (IBIS), Université Laval, 1030, av. de la Médecine, Québec, QC, Canada, G1V0A6

Département de Biologie, Université Laval, Québec, QC, Canada

Molecular Ecology and Fisheries Genetics Laboratory, School of Biological Sciences, Bangor University, Gwynedd LL57 2UW, UK

© Springer International Publishing Switzerland 2016

C.J. Hurst (ed.), *The Rasputin Effect: When Commensals and Symbionts Become Parasitic*, Advances in Environmental Microbiology 3,
DOI 10.1007/978-3-319-28170-4_4

to enhance (or restore) the protective effect of the microbiome (i.e., prebiotics, probiotics, synbiotics), can mitigate bacterial opportunistic diseases.

4.1 Introduction

In recent decades, we have developed a better understanding of the role of microorganisms in beneficial long-term interactions with their host. Fish mucosae (skin, gills, and gut) and associated microbiota are an important primary defense barrier against pathogens (Trivedi 2012). Endogenous fish bacteria contribute to host defense in several ways:

1. Colonization resistance (CR) which prevents pathogen growth with competitive use of resources (Dillon and Charnley 2002)
2. Friction-preventing polymers
3. Stimulation of the innate immune response
4. Production of inhibitory compounds (Austin 2006; Ramsey and Whiteley 2009)

Disturbance of the relative taxonomic abundance of commensal microbiota (i.e., dysbiosis, often representing broken integrity of the microbiome (Dillon and Charnley 2002, see Sect. 4.2.3) is linked to specific human diseases (Frank et al. 2011). The decrease of commensal nonpathogenic bacteria in fish mucosae is correlated to increased abundance of potential pathogens (Boutin et al. 2013b). In many cases, pathogens may be isolated from healthy fish (Cahill 1990; Austin 2006). It is not clear whether asymptomatic carriers are merely a latent step of a disease cycle. Nonetheless, known pathogenic organisms present as a component of healthy fish microbiota are termed "opportunistic pathogens" (Austin and Austin 2007). Here, we review bacterial opportunistic pathogens involved in fish disease.

4.2 Overview

Opportunistic pathogens are defined as microorganisms associated with disease only in host individuals experiencing atypical environmental stressors or having impaired immune function (Boutin et al. 2013b). Most pathogens that impact aquaculture are deemed opportunistic. Important bacterial pathogens are represented by multiple species and several genera (Austin 2006). In this chapter, we focus on three genera with special reference to aquaculture: *Aeromonas*, *Flavobacterium*, and *Vibrio*.

4.2.1 The Genus Aeromonas

Aeromonas is a genus belonging to the Gammaproteobacteria that is ubiquitous in freshwater (Cahill 1990) and commonly found in the commensal microbiota of aquatic or terrestrial animals, plants, and natural soils (Abbott et al. 1992). Species among this genus are mainly pathogenic to fish and, to some extent, mammals and reptiles (Janda and Abbott 2010). Aeromonads are separated into two major groups, according to their morphological traits and optimum growth conditions (Janda and Abbott 2010): motile aeromonads, which tend to be mesophilic, and nonmotile aeromonads, which develop in psychrophilic conditions (Janda and Duffey 1988). The main diseases associated with motile aeromonads vary from opportunistic infections of fish to bacterial gastroenteritis in humans leading to bacteremia in immunocompromised patients (Austin and Austin 2012a, b). Due to their low salinity and cold temperature requirements, nonmotile aeromonads are almost exclusively pathogenic to freshwater fish. Especially in industrial aquaculture, their contribution to economic losses is worldwide and substantial (Nielsen et al. 2001). Solely in the province of Quebec, Canada, aeromonads are responsible for 30–60 % of all infection diagnoses in farmed salmonids each year (Morin 2010). The conditions to which farmed fish are exposed have been shown to disturb their natural microbiota (Boutin et al. 2013b), leaving them more vulnerable to opportunistic pathogens (Stecher et al. 2013). Aeromonads apparently exploit dysbiosis in aquaculture to emerge as virulent and damaging.

4.2.1.1 *Aeromonas hydrophila*

This bacterium is a ubiquitous and potent opportunistic pathogen of fish and humans. Zoonosis is unlikely in healthy individuals, but may occur via contact with mucus or tissue from a carrier fish (Lowry and Smith 2007). In farmed freshwater fish, *A. hydrophila* causes motile aeromonas septicemia (MAS), found in two main forms (Cipriano et al. 1984): (1) The acute form induces internal hemorrhages and generalized bacteremia, without apparent external symptoms, except for darkening of the skin and erratic behavior. (2) The chronic form is characterized by skin ulcers and underlying necrosis of the musculature. Inflammatory cells are often absent in necrotized tissue, but abundant in the adjacent epidermis. *Aeromonas hydrophila* is naturally present in the microflora of healthy fish (Trust et al. 1979). However, it takes advantage of environmental stresses to which fish are exposed. High-nutrient diets, combined with the lack of oxygen-producing plankton, lead to faster oxygen depletion, which facilitates *A. hydrophila* outbreaks in farmed channel catfish (Plumb et al. 1976).

Aeromonas hydrophila consists of heterotrophic facultative–anaerobic Gram-negative bacteria. They are rod shaped and motile, and their size is from 0.3 to 1.0 μm wide by 1.0–3.5 μm long (Horneman et al. 2007). Layered cell surface proteins (S-layers) facilitate osmotolerance, resistance to bactericidal compounds, and increased adherence (Sara and Sleytr 2000). Other virulence factors include fimbriae for increased adherence to the host tissues, cytotoxic enterotoxin, and hemolysin. The AH-3 strain also expresses effector proteins secreted by a molecular nanosyringe complex known as the type III secretion system (T3SS). The expression of this complex is activated by low extracellular calcium concentrations and allows specific injection of effector proteins in eukaryotic cells of the host

(Vilches et al. 2009). As for many Gammaproteobacteria, the T3SS is a key determinant in virulence (Coburn et al. 2007). Secreted effectors cause direct toxicity to infected cells by disrupting actin polymerization and disruption of intracellular signaling.

4.2.1.2 *Aeromonas salmonicida*

Aeromonas salmonicida subsp. *salmonicida* is the etiological agent of furunculosis, a ravaging opportunistic disease affecting salmonids in industrial aquaculture. Its role as the causative agent of salmonid furunculosis was first established near the end of the nineteenth century (Emmerich and Weibel 1890). In the province of Quebec, Canada, where the brook trout accounts for 95 % of farmed salmonids, furunculosis represents 30–60 % of all diagnoses of fish infections each year (Morin 2010). This disease is highly contagious and can decimate an experimental sea trout (*Salmo trutta* morpha *trutta*) and induce 75 % mortality in a group of brown trout (*Salmo trutta* morpha *fario*) in less than 7 days after challenge (Scott 1968). Zoonoses are not a concern, since the growth temperature range of this bacterium is not compatible with the internal temperature of humans. The concern is, however, not only on the high pathogenic potential of *A. salmonicida* between fish but on how its proliferation is facilitated in industrial fish culture systems. Elevated temperature makes optimum growth of *A. salmonicida* easier to achieve, and the high density of fish stocks allows rapid and proximal transmission of the bacteria between fish. Even outside aquaculture basins, high temperature (more than 18 °C) is correlated to increased abundance of *A. salmonicida* in the wild. Climate change has been shown to accelerate the prevalence of furunculosis in wild fish of the James Bay, Quebec, Canada (Tam et al. 2011).

Salmonid furunculosis manifests itself in three major forms (Cipriano and Bullock 2001): (1) The peracute form of furunculosis mostly affects fingerlings. Clinical indications are not obvious, since the only evident marks are darkening of the skin, slight exophthalmia, and premature death. (2) Acute infections affect mostly juvenile and adult salmonids. This form is notorious for the variety of physiological and behavioral damages induced in the infected fish. First signs include color darkening and hemorrhage at the base of fins and the oral cavity. This condition worsens with hemorrhages in internal and reproductive organs, necroses of the kidney and spleen, and also gross intestine congestion (Scott 1968). *Aeromonas salmonicida* then gains access to the bloodstream via the multiple internal lesions, which leads to generalized septicemia and, ultimately, death of the infected fish. (3) The chronic form of furunculosis primarily affects older fish or species which have greater innate resistance to infection by *A. salmonicida*, such as the rainbow trout (*Oncorhynchus mykiss*) (Scott 1968). It is this form of furunculosis which is characterized by boil-like lesions in varying numbers on the skin of infected fish. The underlying ulcers may extend deep into the musculature.

Aeromonas salmonicida are Gram-negative rod-shaped bacteria of about 1×1.5 μm in size, with cells arranged in bunches (Loch and Faisal 2010). They are facultative anaerobes whose optimum growth temperature ranges between

22 and 25 °C. A vast range of virulence factors allow rapid and efficient colonization of eukaryotic cells. These include an extracellular matrix of lipopolysaccharide (LPS) and a surface protein array (A-layer) which, respectively, enhance autoaggregation (Johnson et al. 1985) and adherence to eukaryotic cells (Garduño and Kay 1992). An arsenal of effector products are also secreted specifically to eukaryotic hosts by a molecular nanosyringe complex known as the type III secretion system (T3SS). The T3SS is the main virulence factor of *A. salmonicida*, and loss-of-function (LOF) mutations can drastically reduce virulence. Compared to the fully virulent wild-type (wt) strain JF5054, 2×10^5 times more CFU per fish of T3SS-negative mutants must be administered in order to achieve similar fish mortality (Vanden Bergh and Frey 2013). Genome sequencing of the reference strain A449 (Reith et al. 2008) has revealed major insights about other secretion routes such as the T2SS or T6SS. However, the T3SS genes are located on a large conjugative plasmid (pAsa5), while the others are on chromosomal loci. Previous studies have revealed that loss of virulence in *A. salmonicida* is due to thermolability of this plasmid (Stuber et al. 2003). High temperature (more than 25 °C) and stressful growth conditions lead to rearrangement of the T3SS-bearing plasmid by DNA recombination events between insertion sequences (IS) (Tanaka et al. 2012).

4.2.2 *The Genus* Flavobacterium

The genus *Flavobacterium* occurs in most aquatic ecosystems. Some species belonging to this genus are known as pathogenic agents of flavobacteriosis in fish. The four most important etiological agents of flavobacteriosis are *F. psychrophilum*, *F. columnare*, *F. branchiophilum*, and *F. johnsoniae* (Bernardet and Bowman 2006). Flavobacteriosis is a major problem in Scandinavian aquaculture, but it also affects fisheries worldwide (Decostere et al. 1998; Durborow et al. 1998; Madetoja et al. 2002; Bernardet and Bowman 2006; Mohamed and Ahmed Refat 2011). *Flavobacterium* sp. are naturally present in the fish microbiota. When the immune defenses of their host weaken, they undergo a positive shift in their virulence (Cahill 1990; Nematollahi et al. 2003; Austin 2006; Bernardet and Bowman 2006; Boutin et al. 2013b). Common symptoms of Flavobacteriosis include the erosion of external tissues of the fish (e.g., gills, fin, jaws, and tails). Evidence shows that *Flavobacterium* sp. is transmitted vertically as well as horizontally through water and physical contact (Bernardet and Bowman 2006).

4.2.2.1 *Flavobacterium psychrophilum*

Flavobacterium psychrophilum is the etiological agent of the (1) peduncle disease, the (2) cold-water disease, and the (3) rainbow trout fry syndrome. Those diseases occur mostly in salmonids, but its prevalence as a pathogen has also been demonstrated in other fish families (Amita et al. 2000; Liu et al. 2001). The symptoms of those diseases are characteristic lesions and necrosis occurring on, or near, the peduncle (Davis 1946); skin lesions located near the dorsal fin are associated in severe cases with a systemic infection causing anorexia, distended abdomen, and darkened pigmentation in the region of the caudal peduncle. In those severe cases, the bacterium may be isolated from internal organs such as the spleen and kidney. In rainbow trout fry syndrome, the several symptoms include lethargy and abnormal spiral swimming behavior. Subsequently, darkening of the gills, swelling of the abdomen, and reddening of the vent may be observed prior to death.

Flavobacterium psychrophilum are strictly aerobic, Gram-negative, slender, flexible rods. Their typical growth temperature ranges from 4 to 23 °C and is supported in water of low salinity but not high salinity. Strains of *F. psychrophilum* showed genetic heterogeneity with four serotypes identified in Japan, the United States, and Europe and the presence of diverse plasmids (Chakroun et al. 1998) (Wakabayashi et al. 1994; Lorenzen and Olesen 1997; Chakroun et al. 1998; Izumi et al. 2003). As the bacteria can be detected in sexual products from both female and male fish, parental transmission could likewise constitute a route of colonization (Holt 1993; Ekman et al. 1999; Amita et al. 2000; Vatsos et al. 2001, 2006; Cipriano 2005). This transmission route may lead to the development of disease in fry (Rangdale et al. 1996; Brown et al. 1997; Rangdale et al. 1997a, b; Kumagai et al. 2000; Cipriano 2005). Furthermore, *F. psychrophilum* is part of the natural microbiota, which makes fish a main reservoir for horizontal transmission (Bullock and Snieszko 1981; Holt 1993; Lorenzen 1994; Madetoja et al. 2002). Transmission via water streams is also demonstrated as a possible cross-contamination route—a claim that is supported by the ability of this bacterium to survive in streaming water and sediments for several months (Rangdale 1995; Brown et al. 1997; Madsen and Dalsgaard 1999; Vatsos et al. 2001; Madetoja et al. 2002; Sundell and Wiklund 2011).

However, even if *F. psychrophilum* is widely transmissible, the disease can occur only in immune-suppressed or stressed fish. Some studies proved that contact of the bacterium even at high concentrations is not sufficient to trigger an outbreak in healthy fish (Iida and Mizokami 1996; Ostland et al. 1997; Decostere et al. 2000). When the fish are stressed, the bacteria can outcompete the immune system and the natural microflora to invade the organisms via the gills, skin injuries, or gut (Lorenzen 1994; Liu et al. 2001).

4.2.2.2 *Flavobacterium columnare*

Flavobacterium columnare is an aerobic Gram-negative rod responsible for the columnaris disease. Phylogenetic analysis of the 16S rRNA gene has revealed strong heterogeneity among *F. columnare* strains. However, the high percentage of DNA–DNA hybridization further leads to a consideration of all those strains as being one species.

The virulence of this bacterium is strain dependent and is effective against a large range of host families, such as the Salmonidae, Cyprinidae, and Anguillidae. As is the case for many pathogens, a phylogenetic relationship between strain genotype and host species range is observed (Darwish and Ismaiel 2005). In fingerlings, death occurs early and no external lesions are observed. In the Salmonidae, specific lesions located near the dorsal fin have led to the naming of this condition as "saddleback disease." *Flavobacterium columnare* is also an etiological agent of gill disease, along with *F. branchiophilum*. Symptoms of this disease include (1) white spots on the side of the head, the gills, and fins, (2) erosion of the mouth and fins, (3) necrosis, and (4) hemorrhagic lesions. In severe cases, the disease becomes systemic, and the bacteria colonize internal organs (Durborow et al. 1998). Stressful conditions have been shown to increase the probability of infections. Once fish are infected, they become a reservoir for the pathogen, which facilitates epizootic transmission of the disease. Presence of *F. columnare* favors colonization by other secondary pathogens like *Saprolegnia*. The disease is mostly transmitted via the surrounding waters and via moribund fish. Survival of *F. columnare* in water was estimated to be (1) 77 h in freshwater at 20 °C and (2) up to 16 days in alkaline water at 25 °C (Fijan 1967). The bacterial agent is able to colonize and infect fish without interindividual contact via the water column (Welker et al. 2005).

4.2.2.3 *Flavobacterium branchiophilum*

This bacterial species was first isolated from gills of salmonids in Japan, Ontario, and Oregon, as well as rainbow trout, sheatfish, and silver carp in Hungary. The etiological agent of gill disease is a Gram-negative rod which grows at temperatures ranging from 5 to 30 °C in low salinity (0–0.1 %).

Bacterial gill disease is one of the most important conditions affecting the salmonid aquaculture industry in Ontario (Ferguson et al. 1991; Turnbull 1993; Ostland et al. 1994). This disease is characterized by colonization of the gills at the lamellar surface by *F. branchiophilum*, which induces lamellar epithelial necrosis (Speare and Ferguson 1989; Speare et al. 1991). All the isolated strains so far are able to attach to the gills; however, only the most virulent ones colonize the gills substantially (Turnbull 1993; Ostland et al. 1995). Virulence of *F. branchiophilum* is not as pronounced as *F. columnare* or *F. psychrophilum*. Even in severe cases, the disease does not become systemic and usually does not invade the inner organs. In fact, the disease causes mortality only in juveniles (Kimura et al. 1978; Wakabayashi et al. 1980). The virulence of *F. branchiophilum* is attributed to various enzymatic and hemagglutinating activities detected in the extracellular products. Touchon et al. showed that the genome of *F. branchiophilum* presents the first observation of a cholera-like toxin gene acquired by a non-*Proteobacteria* (Touchon et al., 2011). Several adhesin-encoding genes are present, which may be linked to the virulence of the pathogen (Touchon et al. 2011). Few studies have focused on the transmission route of *F. branchiophilum* in aquaculture systems.

Evidence suggests, however, that it probably occurs horizontally from the water supply. Indeed, bath exposure experiments reveal the presence of the pathogen in the water during outbreaks (Heo et al. 1990; Ostland et al. 1995).

4.2.3 The Genus Vibrio

The link between the genus *Vibrio* and pathogenicity dates from the beginning of modern microbiology. Several marine *Vibrio*, such as *V. anguillarum*, *V. harveyi*, and *V. vulnificus*, have proven to be potent pathogens of marine fish, migrating freshwater fish, bivalves, and crustaceans (Austin 2010). In an aquacultural context, high stock density facilitates epizootic transmission and augments the incidence of zoonoses affecting human consumers (Ghittino et al. 2003).

4.2.3.1 *Vibrio anguillarum*

The 1718's outbreak of *pestis rubra anguillarum* (Lat. "red pest of eels") in Italy was the first reference to a bacterial fish disease in the scientific literature (Bonaveri 1761 (quoted by Drouin de Bouville 1908)). However, it was not until the end of the nineteenth century that the causal agent of this disease was determined. Following an epizootic episode dating back to 1817 in migrating eels, Canestrini isolated the causal agent, which he named *Bacterium anguillarum* (1893). Sixteen years later, the etiological agent of the "red pest" in the Baltic Sea was described by Bergman (1909), who proposed the name *Vibrio anguillarum* from his findings. Subsequently, it was established that both researchers had described the same etiological agent, whose latter name has been retained. The disease, now generally referred to as "vibriosis," causes fin and mouth rot, hemorrhages, generalized septicemia, and superficial skin lesions (Egidius 1987). Currently, it is widely known that vibriosis affects not only eels but also benthopelagic or benthic fish such as cod (Bagge and Bagge 1956), halibut (Hoare et al. 2002), and turbot (Grisez et al. 1996). Migrating freshwater fish, like Pacific and Atlantic salmon, also are susceptible targets of vibriosis (Evelyn 1971; Arkoosh et al. 1998; Frans et al. 2011). To a lesser extent, bivalves and crustaceans are occasionally infected (Paillard et al. 2004; Aguirre-Guzmán et al. 2004).

Vibrio anguillarum are Gram-negative, curved-rod-shaped bacteria. Their motility is guided by chemotaxis and mediated by a monotrichous polar flagellum (Larsen and Boesen 2001). Colonies growing on 2 % NaCl blood agar at 22 °C typically are yellowish, low, convex, and shiny and have unguent consistency and are sized from 3 to 5 mm after 48 h (Myhr et al. 1991).

The first pillar of bacterial virulence is proper recognition of the host. *Vibrio anguillarum* is not an exception and exhibits motility driven by chemotactic recognition of skin or gut mucus components (O'Toole et al. 1996, 1999). Active motility is required in the early stages of infection. A decrease in virulence and persistence in systemic infections is observed when the flagellin genes *flaA*, *flaD*, and *flaE* undergo deleterious mutations (Milton et al. 1996; McGee et al. 1996). Following movement toward the host, virulent bacteria "grip" to the host epithelial cells. In *V. anguillarum*, bacterial adherence is facilitated by a myriad of adherence factors (i.e., adhesins) such as fimbriae, outer membrane proteins (OMP), and extracellular polysaccharides (Pizarro-Cerdá and Cossart 2006). Invasion of host tissue is followed by degradation of epithelial mucus due to extracellular metalloprotease EmpA (Han et al. 2011). Once *V. anguillarum* has reached the blood vessels, the activity of hemolysins causes erythrocyte lysis and release of intracellular heme. Chelated iron ions are then captured by a siderophore-dependent iron acquisition system, which is encoded on pJM1, a ~65 kb plasmid (Stork et al. 2002).

4.3 Factors Triggering Opportunistic Infections

Examples of pathogenic outbreaks in aquaculture present some common trends:

- Opportunistic pathogens seem to be present at low prevalence in healthy fish, indicating that under normal conditions, the immune system can maintain these organisms at low abundance.
- Practices from aquaculture modify the environmental conditions that greatly impact the virulence or diversification of the pathogens (virus or bacteria).

In the following sections, we will explore the three main factors triggering opportunistic infections: (1) environmental factors, (2) relationships between pathogens and commensals, and (3) host–microbe interactions.

4.3.1 Environmental Factors

Fish welfare is known to be influenced by environmental parameters, as for all animals. There are now strong indications that circadian and seasonal variation of temperature and light can affect health (Zapata et al. 1992; Nelson 2004; Bowden et al. 2007). Such factors can affect fish susceptibility to infectious diseases due to increased prevalence of the pathogen or due to increased susceptibility in the host (Revie et al. 2002; Lillehaug et al. 2003; Hjelm et al. 2004; Ondračková et al. 2004). Temperature is one of the most influential environmental factors as fish are ectotherms and their inability to regulate their internal temperature influences their immune response (Baras 1995). For example, the columnaris disease occurs when temperature rises higher than 14 °C and most infections are observed between 20 and 30 °C (Durborow et al. 1998). Temperature may therefore influence both innate and adaptive immune responses in many different fish species (Bly and Clem 1992; Ellis 2001). Usually, higher temperature enhances the immune response in fish whereas lower temperature exerts an adverse effect (Bly and Clem 1992). Opportunistic diseases that appear at higher temperature are thus either due to (1) autoimmune damage triggering the infections or to (2) a direct effect on pathogen growth.

A second influential environmental parameter triggering opportunistic infections in fish is stress. The growing demand for fish is answered by the industry with extensive production, which often negatively affects the welfare of farmed fish (Ashley 2007). Based on health, physiological, and behavioral indicators, animal stress is a widely used indicator of welfare. Stress is a response reaction to a stimulus that can alter the homeostatic state of an individual (Barton and Iwama 1991). In aquaculture, stress results from handling, sorting, grading, transporting, and stocking. When these stress factors overwhelm the adaptive capacity of the fish, the stress response becomes detrimental to the fish's health. Adams considered that stress can be divided into two different effects (Adams 1990):

- Direct effects influence the organisms by altering their physiological, hormonal, or cellular functions.
- Indirect effects operate at a higher level integrating the population or community effect by affecting the energy resources available for fish.

At the interface between direct and indirect effects, we can now append the effect of stress on the associated microbiome of fishes. Infectious diseases are known to be triggered by stress (Snieszko 1974; Wakabayashi 1991; Freestone et al. 2008; Littman et al. 2010; Boutin et al. 2012; Moloney et al. 2013). It is widely accepted that opportunistic pathogens, among which some are present in healthy fish microbiota, become infectious when hosts are stressed (Durborow et al. 1998; Le Moullac et al. 1998; Starliper 2011). However, it is nearly impossible to draw a global framework of how and when opportunistic diseases occur following stressful conditions. Fortunately, the number of studies on infectious outbreaks in aquaculture is growing. Recent studies highlight some patterns which increase the infectivity and spread of an infection, such as the number of host species cultured together and the intensity of culture.

Population density is the other factor that clearly influences the spread of the infection. In aquaculture settings, density is often a thousand times higher than in nature (Pulkkinen et al. 2010). This increases the probability of horizontal pathogen transmission through fish-to-fish contact. Indeed, host limitation (the frequency of host encounter) is one of the major limiting factors in pathogen population dynamics. The probability of transmission increases greatly when hosts are numerous and homogeneously distributed, especially if hosts are stressed by crowding or other factors (Snieszko 1974; Lipsitch et al. 1995; Ebert 1998; Pulkkinen et al. 2010). This disappearance of host range limitations also changes how the dynamics between hosts and pathogens evolve. The classical trade-off in this dynamics is that pathogens will reduce their virulence in order to avoid depletion of available hosts (Anderson and May 1982; Ebert and Mangin 1997). However, in aquaculture settings, frequency of host contact no longer is a limiting resource. Consequently, the removal of host encounter as a selection pressure may lead to increased virulence and transmission, resulting in increased pathogen fitness (Day 2002). A notable example illustrating this assumption is *F. columnare*. Virulence assays showed that *F. columnare* is not only able to survive in water and in carcasses but is also able to increase its dispersal rate. This results in more efficient transmission from dead to living fish, rather than between two living fish (Pulkkinen et al. 2010).

4.3.2 Microbe–Microbe Interactions

Interactions between microbial organisms including opportunistic pathogens in the environment as part of the host's commensal microbiome can also influence disease outcome.

4.3.2.1 Horizontal Gene Transfer

High microbial density correlates to the occurrence of horizontal gene transfer (HGT), especially of resistance and virulence genes. HGT can occur by three different ways:

1. Transformation, which is the acquisition of free DNA from the environment
2. Transduction, involving transfer of genetic material via phage infection
3. Conjugation, which is gene transfer through a type IV pilus connecting two cells (Sorensen et al. 2005)

Acquisition of antimicrobial resistance genes is mostly due to mutations, but the spreading of resistance is linked to HGT. The spread of virulence factors via HGT may potentially explain the emergence of some opportunistic pathogens.

4.3.2.2 Agonism and Antagonism

Different types of interactions occur between bacterial species. One of them, mutualism, consists of a beneficial relationship between one or more different individuals or species. In bacteria, mutualism can induce a specialization of certain taxa which efficiently produce select gene products consumed by the microbial community. Such mutualism in return favors the loss of genes by inefficient producers. Gene loss, leading to interdependency between taxa, is tentatively explained by the Black Queen Hypothesis (Morris et al. 2012). On the other hand, bacterial species also compete against each other. Indeed, competition is one of the most abundant interactions between not only bacteria but also most of living organisms (Begon et al. 1990). Patterns of agonism and antagonism explain most of the network of interactions that exist in the bacterial meta-community.

Changes among these interactions that result in dysbiosis are also likely to trigger an opportunistic pathogen emergence.

4.3.2.3 Quorum Sensing

Microbes have developed a process to communicate between cells. This process, named quorum sensing (QS), changed the vision that bacteria regulate their biological processes independently. This system was first discovered in *Aliivibrio fischeri*, which expresses its bioluminescence in mid-exponential growth phase (Nealson et al. 1970). The signal molecule was found to be acyl-homoserine lactone (AHL). However, several other types of cell–cell communication signals have also been characterized in recent years: furanosyl borate (AI-2), cyclic thiolactone (AIP), hydroxyl palmitic acid methyl ester (PAME), methyl dodecanoic acid which acts as a diffusible signal factor (DSF), and farnesoic acid (FA) (Miller and Bassler 2001; Whitehead et al. 2001; Fuqua and Greenberg 2002). Quorum sensing is involved in the regulation of various biological processes, such as bioluminescence, antimicrobial compound production, plasmid conjugation, motility, virulence, and biofilm formation (Whitehead et al. 2001; Federle and Bassler 2003). It may also influence the emergence of opportunistic pathogens.

4.3.2.4 Quorum Quenching

As a response to this interspecies communication, some bacteria had developed a way to interfere to outcompete their rivals. This process, quorum quenching, consists of inhibiting the QS pathways by two different ways: (1) binding of non-signal molecules to the receptor (noncognate AHLs, intermediates of the AHL biosynthetic pathway, dicyclic peptides) or (2) degrading the QS signal with different enzymes (AHL-lactonases, decarboxylases, AHL-acylase, and deaminase) (Kalia 2013). Many opportunistic pathogens of fish have based their capacities to respond to environmental factors on QS. Indeed, *Aeromonas hydrophila*, *A. salmonicida*, *Vibrio anguillarum*, *V. harveyi*, and *Yersinia ruckeri* express their pathogenicity through QS systems (Bruhn et al. 2005).

4.3.3 Host–Microbe Interactions

Fish are in permanent interaction with microbes. This leads them to develop efficient processes to recognize potential partners. The most evolved system in that field is the immune system, based on molecular recognition of microbes. Recognition induces a signal which activates the effective immune functions. Two components of the immune system have emerged:

1. The innate immune system, which deals with the recognition of conserved patterns, accessible and shared by many pathogens
2. The adaptive immune response, based on receptors targeting specific structures of a given pathogen

4.3.3.1 Innate Immunity

The innate receptors are present on the periphery of all cells of the immune system (e.g., macrophages, dendritic cells). This process is important for the maintaining of homeostasis (i.e., stability in the structure of the commensal microbiome). To maintain homeostasis, receptors and signals of innate immunity are constitutively expressed by the host (Dixon et al. 2004). Innate immunity receptors are named pattern recognition receptors (PRR) (Akira et al. 2006). There are different types of those receptors: toll-like receptors (TLRs) and their coreceptor CD14, scavenger receptors, mannose receptors, integrins CD11b-c/CD18, and complement receptors CR1,2,3. Those cell surface receptors bind conserved structures of microbes named pathogen-associated molecular patterns (PAMPs) (Medzhitov and Janeway 1999). These surface molecules are specifically produced by bacteria as peptidoglycans (PGN), lipopolysaccharides (LPS), or lipoteichoic acid (LTA) and are not specific to pathogenic bacteria (Delneste et al. 2007).

4.3.3.2 Adaptive Immunity

The adaptive response will target specific pathogens via the activation of B and T cells. This response is characterized by immunological memory, which allows quicker recognition of the pathogen after a first infection. The B cells can recognize antigens in their native form. However, T cells recognize only those peptides which have been presented by the major histocompatibility complex (MHC), located at the surface of antigen-presenting cells.

Innate and adaptive immunity were considered as separate for a long time. However, in recent decades, studies have shown that both systems are interconnected, with innate immunity shaping the adaptive response (Medzhitov and Janeway 1999). Furthermore, the role of commensal bacteria was unknown. It is now acknowledged that commensal bacteria are involved in the maturation of both immune responses. Such bacteria even interfere with colonization of the host by pathogenic agents (Rakoff-Nahoum et al. 2004; Kelly et al. 2005; Mazmanian and Kasper 2006; O'Mahony et al. 2008). Protection against pathogens is then influenced by both host-mediated immunity and by the microbiome, but also by the dialog between those two components.

When the integrity of the microbiome is broken, this state is called dysbiosis. Stress on the host, changes in the environmental factors, or antibiotic use can disturb the natural microflora of fish (i.e., dysbiosis). This results in a weakening of the primary defense barrier of the host (Boutin et al. 2013b; Stecher et al. 2013).

As we previously discussed, endogenous fish bacteria contribute to the immune function by such factors as colonization resistance (CR), immune response stimulation, and production of microbially inhibitory compounds (Dillon and Charnley 2002; Austin 2006; Ramsey and Whiteley 2009; Maslanik et al. 2012; Naik et al. 2012). The importance of commensal bacteria to fish health is starting to be understood. Furthermore, their role in disease prevention explains why probiotics are successful in aquaculture against many of the previously discussed fish pathogens (Moriarty 1998; Nikoskelainen et al. 2003; Farzanfar 2006; Nakayama et al. 2009; Nayak 2010; Burbank et al. 2011; Boutin et al. 2012, 2013a).

4.4 Prevention and Treatment of Opportunistic Infections

4.4.1 Antimicrobial Compounds

Aquaculture is a growing industry that encounters many issues with disease. The way to prevent and control those outbreaks was once exclusively the use of antimicrobial agents (Armstrong et al. 2005; Cole et al. 2009; Burridge et al. 2010). These chemicals were not only used to treat diseases but also as prophylaxis against bacterial infections. Beyond disease control, their side effects are exploited by the industry. For example, in the United States, 70 % of the annual production of antibiotics is used to increase the growth of the livestock (SCAN and the European Commission Health and Consumer Protection Directorate-General 2003; Balcazar et al. 2006). The extensive use of those chemicals in aquaculture leads to certain drawbacks for human and animal health (Witte 1998; Phillips et al. 2004; Marshall and Levy 2011). One of these is the persistence of antimicrobials in the environment, which induces a selective pressure on resistant bacteria. The results likely will include an increase in multiresistant microbial strains which could possibly be pathogenic for humans or livestock (Gold and Moellering 1996; Rangdale et al. 1997b; Levy 1998; Waldvogel 1999; Lindsay and Holden 2006; Arias and Murray 2009). In this context, the new methods being considered to prevent bacterial diseases are (1) the research of new antimicrobials acting upon secreted products of bacteria, instead of the cellular machinery itself (Chapra et al. 1997; Stephens and Shapiro 1997), and (2) alternative approaches such as vaccines, phages, probiotics, or genetic selection.

4.4.2 Vaccines

Vaccines exploit the adaptative immune response of fish (Foott and Hedrick 1987; Kent et al. 1999). Vaccines mimic a first infection by introducing antigens from the pathogens into the organism. Three types of inoculation are available in aquaculture: oral vaccination with antigen included in the food, immersion in diluted solution of the antigens, or intraperitoneal injection in the body cavity. Oral vaccination is seen as the easiest way to vaccinate fish because it does not require many efforts to administer. However, the decrease in fish mortality resulting from exposure to pathogens by this route is also significantly lower than what is traditionally associated with other methods, perhaps due to antigen destruction in the gut (Hart et al. 1988; Nakanishi and Ototake 1997; Sommerset et al. 2005). Vaccinations by immersion and by injection are the two main ways used so far in aquaculture. Vaccines for most viral and bacterial pathogens are available for salmonids and other fish species of significant economical value (Biering et al. 2004; Håstein et al. 2004). Vaccines have proven efficient in preventing many major bacterial diseases. However, for some pathogens, a single exposure to the vaccine is not enough to induce a long-term protection, and a second vaccination is needed during the production period. Most available virus vaccines, based on either inactivated virus or recombinant subunit proteins, are generally not efficient unless delivered by injection. Live viral vaccines have shown interesting results (Benmansour and De Kinkelin 1996) but are not yet considered safe for widespread use (Sommerset et al. 2005). Furthermore, the cost of developing new vaccines for virus and parasites is an important drawback to their commercial usage.

4.4.3 Phage Therapy

The use of bacteriophage as a therapy was developed as an alternative to antibiotics. As phage particles are very specific to their bacterial hosts, they do not target both pathogens and the normal flora, and thus, their use may minimize the chance of secondary infections following antibiotic-induced dysbiosis. Furthermore, phage particles replicate at the site of infection; thus, curative doses can be fairly small. Moreover, although bacteria can become resistant to phage, these viral organisms can mutate and therefore evolve to counter phage-resistant bacteria (Matsuzaki et al. 2005). In aquaculture, phage therapy is a new field, but the growing number of phage types isolated in the last decades is promising (Matsuzaki et al. 2005; Vinod et al. 2006; Shivu et al. 2007; Stenholm et al. 2008; Crothers-Stomps et al. 2010; Defoirdt et al. 2011). The most important advantage of phage is that they might kill planktonic pathogens living in the surrounding water in addition to pathogens proliferating in carrier fish. Possible drawbacks of phage therapy include the possible transduction of virulence factors between bacteria, as well as the fact that the vertebrate host may mount an immune response against the phage itself.

4.4.4 Probiotics

The use of probiotics is probably the most widespread alternative to antibiotics so far. Probiotics are defined as a "live microbial culture added to feed or environment (water) to increase viability (survival) of the host" (Gram and Ringø 2005). This definition was recently modified by Merrifield et al. (2010) who defined a probiotic in aquaculture as:

> A live, dead or component of a microbial cell that, when administered via the feed or to the rearing water, benefits the host by improving either disease resistance, health status, growth performance, feed utilisation, stress response, which is achieved at least in part via improving the hosts or the environmental microbial balance.

Although the mechanisms by which probiotics exert their beneficial effects require further investigation, probiotic administration showed promising results on growth performance and general health of teleost fish (Gatesoupe 2010). Despite the aforementioned advantages of probiotics, the viability of live bacteria during large-scale production of food (i.e., commercial diets) and during transition through the gastrointestinal tract is not always reliable (Ringø et al. 2014).

4.4.5 Prebiotics

To resolve the problem of viability, the prebiotic concept has been suggested and developed (Mahious and Ollevier 2005). A prebiotic is defined by Roberfroid (2007) as:

> A non digestible food ingredient that beneficially affects the host by selectively stimulating the growth and/or activity of one or a limited number of bacteria in the colon, that can improve the host health.

To be classified as a prebiotic, a dietary ingredient needs to follow one of several different criteria: (1) resist gastric acidity, hydrolysis by digestive enzymes, and gastrointestinal absorption; (2) be fermented by the intestinal microbiota; and (3) be able to selectively stimulate the growth and activity of beneficial bacteria (Gibson 2004). First used in 1995 by Hanley, the number of prebiotics is increasing and includes many compounds, such as inulin and oligosaccharide compounds (Hanley et al. 1995; Ringø et al. 2014). Prebiotics are used as nutrients by probiotic bacteria (Geraylou et al. 2012). Many prebiotics are fermented by the gut microbiota into short-chain fatty acids (SCFA), which are the main energy source for colonic epithelial cells. Prebiotics are thus associated with maintenance of the epithelium (Cummings and Macfarlane 2002; Maslowski and Mackay 2010). The SCFA also modulate lipid synthesis (Marcil et al. 2002) as well as stimulate the immune system and aid with host resistance against pathogens (Maslowski and Mackay 2010).

4.4.6 Synbiotics

Synbiotics refer to nutritional complements combining probiotics and prebiotics (Cerezuela et al. 2011). Synbiotics aim to simultaneously seed and maintain probiotic strains as the dominant species in the gut after treatment cessation (Rurangwa et al. 2009). Despite recent progress in the field of synbiotics in aquaculture, there is limited information available on different aspects of synbiotic effects on fish (Cerezuela et al. 2011).

4.4.7 Selection Genetics

Selection of livestock to improve the growth and resistance is an old method used for centuries. However, selection programs in aquaculture represent a relatively recent field. Historically, the focus has been on fish growth rate. However, with the growing threat of infectious disease, pathogen resistance became a major phenotype for breeding programs. With the substantial advances in genetics, breeding programs are now based on genetic loci, which are quantitatively linked to the resistance (Baerwald et al. 2011; Langevin et al. 2012; Massault et al. 2011; Ozaki et al. 2001; Rodríguez-Ramilo et al. 2011; Verrier et al. 2013). These regions, named quantitative trait loci (QTL), are becoming interesting markers for the breeding program.

A new way assess the genetic basis of resistance is to observe QTL associated with the relationship between host and microbiome. There is evidence that QTL correlate with an influence of host genetic variation on fecal microbiome composition in mice (Benson et al. 2010; McKnite et al. 2012). Those taxa under host genetic control correspond with species and genera thought to interact with host immunity (Benson et al. 2010; McKnite et al. 2012). The QTL analysis of skin microbiome composition has recently been undertaken in the salmonid *Salvelinus fontinalis* (Boutin et al. 2014). Standing genetic variation among components of the teleost adaptive immune system is increasingly well characterized (Dionne et al. 2009; Pavey et al. 2013). While toll-like receptors (TLRs) are present in multiple teleost species (Palti 2011), there has been no work to date to correlate genetic diversity at these innate immune loci (inter- or intraspecies) with commensal microbiome diversity. Experiments in zebra fish highlight the role that TLRs play in modulating the intestinal microbiota, whereby alkaline phosphatase is produced via a TLR-4-myD88-controlled pathway to inhibit an inflammatory responses to gut microbiota (Bates et al. 2007). Given that desirable microbiome characteristics may exist from an aquaculture perspective (e.g., disease resistance, nutrient absorption, stress resilience), it is encouraging that a host genetic basis may exist for selection of such traits.

4.5 Conclusion

In this chapter, we have defined opportunistic teleost pathogens as pathogen species whose presence can be detected from apparently healthy teleost hosts. The status of these organisms—either as a natural component of a healthy commensal microbiome, or a latent step in disease establishment, or both—is still not entirely clear. In our discussion, we have limited ourselves to bacterial pathogens. We excluded viruses, which make up a significant proportion of known teleost pathogens, because these cannot safely be classed as opportunists owing to their obligate parasitic lifestyle. We outlined the major opportunist bacterial genera with special emphasis on disease in aquaculture. In doing so, we discussed the current understanding of bacterial pathogen taxonomy, biology, disease impact, and treatment options.

Unlike directly transmitted pathogens, understanding disease evolution and transmission caused by opportunistic pathogens necessitates a holistic view. Thus, in the second section of this chapter, we consider the importance of environment–host, pathogen–microbiome, and host–pathogen interactions in the context of opportunistic pathogens in teleosts. We demonstrate that not only do these factors play a crucial role in defining disease, but their importance opens up exciting new avenues to treat disease. As such, preventive measures to reduce stress, along with active interventions to enhance the protective effect of the microbiome (prebiotic, probiotics), can all mitigate the impact and prevalence of infectious disease.

The global demand for fish protein grows daily. Meanwhile, wild stocks are dwindling. Furthermore, anthropogenic activities have resulted in the epizootic dispersal of disease agents between farmed and wild individuals, as well as between wild populations via the introduction of invasive host species. A clear understanding of the drivers of disease caused by opportunistic pathogens is thus critical, not only to guarantee safe and sustainable aquaculture but also to protect existing wild fish species.

References

Abbott SL, Cheung WK, Kroske-Bystrom S et al (1992) Identification of Aeromonas strains to the genospecies level in the clinical laboratory. J Clin Microbiol 30:1262–1266

Adams SM (1990) Status and use of biological indicators for evaluating the effects of stress on fish. Am Fish Soc Symp 8:1–8

Aguirre-Guzmán G, Ruíz HM, Ascencio F (2004) A review of extracellular virulence product of Vibrio species important in diseases of cultivated shrimp. Aquacult Res 35:1395–1404. doi:10.1111/j.1365-2109.2004.01165.x

Akira S, Uematsu S, Takeuchi O (2006) Pathogen recognition and innate immunity. Cell 124:783–801. doi:10.1016/j.cell.2006.02.015

Amita K, Hoshino M, Honma T, Wakabayashi H (2000) An investigation on the distribution of Flavobacterium psychrophilum in the Umikawa River. Fish Pathol 35:193–197

Anderson RM, May RM (1982) Coevolution of hosts and parasites. Parasitology 85:411–426

Arias CA, Murray BE (2009) Antibiotic-resistant bugs in the 21st century—a clinical super-challenge. N Engl J Med 360:439–443. doi:10.1056/NEJMp0804651

Arkoosh MR, Casillas E, Huffman P et al (1998) Increased susceptibility of juvenile Chinook salmon from a contaminated estuary to Vibrio anguillarum. Trans Am Fish Soc 127:360–374. doi:10.1577/1548-8659(1998)127<0360:ISOJCS>2.0.CO;2

Armstrong SM, Hargrave BT, Haya K (2005) Antibiotic use in finfish aquaculture: modes of action, environmental fate, and microbial resistance. In: Hargrave BT (ed) Environmental effects of marine finfish aquaculture. Springer, Berlin, pp 341–357

Ashley PJ (2007) Fish welfare: current issues in aquaculture. Appl Anim Behav Sci 104:199–235

Austin B (2006) The bacterial microflora of fish, revised. ScientificWorldJournal 6:931–945. doi:10.1100/tsw.2006.181

Austin B (2010) Vibrios as causal agents of zoonoses. Vet Microbiol 140:310–317. doi:10.1016/j.vetmic.2009.03.015

Austin B, Austin DA (2007) Bacterial fish pathogens: diseases of farmed and wild fish. Praxis, Chichester

Austin B, Austin DA (2012a) Aeromonadaceae Representative (Aeromonas salmonicida). In: Austin B, Austin DA (eds) Bacterial fish pathogens. Springer, Dordrecht, pp 147–228

Austin B, Austin DA (2012b) Aeromonadaceae representatives (motile aeromonads). In: Austin B, Austin DA (eds) Bacterial fish pathogens. Springer, Dordrecht, pp 147–228

Baerwald MR, Petersen JL, Hedrick RP, Schisler GJ, May B (2011) A major effect quantitative trait locus for whirling disease resistance identified in rainbow trout (*Oncorhynchus mykiss*). Heredity 106:920–926

Bagge J, Bagge O (1956) Vibrio anguillarum som årsak til ulcus sykdom hos torsk (Gadus callarias Linné). Nord Vet Med 8:481–492

Balcazar J, Blas I, Ruizzarzuela I et al (2006) The role of probiotics in aquaculture. Vet Microbiol 114:173–186. doi:10.1016/j.vetmic.2006.01.009

Baras E (1995) Seasonal activities of Barbus barbus: effect of temperature on time-budgeting. J Fish Biol 46:806–818

Barton BA, Iwama GK (1991) Physiological changes in fish from stress in aquaculture with emphasis on the response and effects of corticosteroids. Annu Rev Fish Dis 1:3–26. doi:10.1016/0959-8030(91)90019-g

Bates JM, Akerlund J, Mittge E, Guillemin K (2007) Intestinal alkaline phosphatase detoxifies lipopolysaccharide and prevents inflammation in zebrafish in response to the gut microbiota. Cell Host Microbe 2:371–382. doi:10.1016/j.chom.2007.10.010

Begon M, Harper JL, Townsend CR (1990) Ecology: individuals populations and communities, 2nd edn. Blackwell, Oxford

Benmansour A, De Kinkelin P (1996) Live fish vaccines: history and perspectives. Dev Biol Stand 90:279–289

Benson AK, Kelly SA, Legge R et al (2010) Individuality in gut microbiota composition is a complex polygenic trait shaped by multiple environmental and host genetic factors. Proc Natl Acad Sci U S A 107:18933–18938. doi:10.1073/pnas.1007028107

Bergman AM (1909) Die rote Beulenkrankheit des Aals. Ber Aus K Bayer Vers 2:10–54

Bernardet J-F, Bowman J (2006) The genus *Flavobacterium*. In: Dworkin M, Falkow S, Rosenberg E et al (eds) The prokaryotes. Springer, New York, pp 481–531

Biering E, Villoing S, Sommerset I, Christie KE (2004) Update on viral vaccines for fish. Dev Biol 121:97–113

Bly JE, Clem LW (1992) Temperature and teleost immune functions. Fish Shellfish Immunol 2:159–171

Boutin S, Bernatchez L, Audet C, Derôme N (2012) Antagonistic effect of indigenous skin bacteria of brook charr (*Salvelinus fontinalis*) against *Flavobacterium columnare* and *F. psychrophilum*. Vet Microbiol 155:355–361. doi:10.1016/j.vetmic.2011.09.002

Boutin S, Audet C, Derôme N (2013a) Probiotic treatment by indigenous bacteria decreases mortality without disturbing the natural microbiota of Salvelinus fontinalis. Can J Microbiol 59:662–670
Boutin S, Bernatchez L, Audet C, Derôme N (2013b) Network analysis highlights complex interactions between pathogen, host and commensal microbiota. PLoS One 8:e84772. doi:10.1371/journal.pone.0084772
Boutin S, Sauvage C, Bernatchez L et al (2014) Inter individual variations of the fish skin microbiota: host genetics basis of mutualism? PLoS One 9:e102649. doi:10.1371/journal.pone.0102649
Bowden TJ, Thompson KD, Morgan AL et al (2007) Seasonal variation and the immune response: a fish perspective. Fish Shellfish Immunol 22:695–706
Brown LL, Cox WT, Levine RP (1997) Evidence that the causal agent of bacterial cold-water disease Flavobacterium psychrophilum is transmitted within salmonid eggs. Dis Aquat Organ 29:213–218
Bruhn JB, Dalsgaard I, Nielsen KF et al (2005) Quorum sensing signal molecules (acylated homoserine lactones) in gram-negative fish pathogenic bacteria. Dis Aquat Organ 65:43–52
Bullock GL, Snieszko SF (1981) Fin rot, coldwater disease, and peduncle disease of salmonid fishes. U.S. Fish and Wildlife Service, Fish Disease Leaflet No. 25
Burbank DR, Shah DH, LaPatra SE et al (2011) Enhanced resistance to coldwater disease following feeding of probiotic bacterial strains to rainbow trout (Oncorhynchus mykiss). Aquaculture 321:185–190. doi:10.1016/j.aquaculture.2011.09.004
Burridge L, Weis JS, Cabello F et al (2010) Chemical use in salmon aquaculture: a review of current practices and possible environmental effects. Aquaculture 306:7–23
Cahill MM (1990) Bacterial flora of fishes: a review. Microb Ecol 19:21–41. doi:10.1007/bf02015051
Canestrini G (1893) La malattia dominante delle anguille: richerche batteriologiche. Atti R Ist Veneto Sci Lett Ed Arti 7:809–814
Cerezuela R, Meseguer J, Esteban M (2011) Current knowledge in synbiotic use for fish aquaculture: a review. J Aquacult Res Dev 1:1–7
Chakroun C, Grimont F, Urdaci MC, Bernardet J-F (1998) Fingerprinting of Flavobacterium psychrophilum isolates by ribotyping and plasmid profiling. Dis Aquat Organ 33:167–177
Chapra I, Hodgson J, Metcalf B, Poste G (1997) The search for antimicrobial agents effective against bacteria resistant to multiple antibiotics. Antimicrob Agents Chemother 41:497–503
Cipriano RC (2005) Intraovum infection caused by Flavobacterium psychrophilum among eggs from captive Atlantic salmon broodfish. J Aquat Anim Health 17:275–283
Cipriano RC, Bullock GL (2001) Furunculosis and other diseases caused by Aeromonas salmonicida. United States Geological Survey
Cipriano R, Bullock G, Pyle S (1984) Aeromonas hydrophila and other septicemias of fish. U.S. Fish and Wildlife Service Publications
Coburn B, Sekirov I, Finlay BB (2007) Type III secretion systems and disease. Clin Microbiol Rev 20:535–549. doi:10.1128/CMR.00013-07
Cole DW, Cole R, Gaydos SJ et al (2009) Aquaculture: environmental, toxicological, and health issues. Int J Hyg Environ Health 212:369–377
Crothers-Stomps C, Høj L, Bourne DG et al (2010) Isolation of lytic bacteriophage against Vibrio harveyi. J Appl Microbiol 108:1744–1750
Cummings JH, Macfarlane GT (2002) Gastrointestinal effects of prebiotics. Br J Nutr 87:S145–S151
Darwish AM, Ismaiel AA (2005) Genetic diversity of *Flavobacterium columnare* examined by restriction fragment length polymorphism and sequencing of the 16S ribosomal RNA gene and the 16S–23S rDNA spacer. Mol Cell Probes 19:267–274
Davis HS (1946) Care and diseases of trout. US Department of Interior Research Report No. 12
Day T (2002) Virulence evolution via host exploitation and toxin production in spore-producing pathogens. Ecol Lett 5:471–476

Decostere A, Haesebrouck F, Devriese LA (1998) Characterization of four Flavobacterium columnare (Flexibacter columnaris) strains isolated from tropical fish. Vet Microbiol 62:35–45. doi:10.1016/s0378-1135(98)00196-5

Decostere A, Lammens M, Haesebrouck F (2000) Difficulties in experimental infection studies with *Flavobacterium psychrophilum* in rainbow trout (*Oncorhynchus mykiss*) using immersion, oral and anal challenges. Res Vet Sci 69:165–169

Defoirdt T, Sorgeloos P, Bossier P (2011) Alternatives to antibiotics for the control of bacterial disease in aquaculture. Curr Opin Microbiol 14:251–258, http://dx.doi.org/10.1016/j.mib.2011.03.004

Delneste Y, Beauvillain C, Jeannin P (2007) Immunité naturelle: Structure et fonction des toll-like receptors. Med Paris 23:67–74

Dillon R, Charnley K (2002) Mutualism between the desert locust Schistocerca gregaria and its gut microbiota. Res Microbiol 153:503–509, http://dx.doi.org/10.1016/S0923-2508(02)01361-X

Dionne M, Miller KM, Dodson JJ, Bernatchez L (2009) MHC standing genetic variation and pathogen resistance in wild Atlantic salmon. Philos Trans R Soc Lond B Biol Sci 364:1555–1565. doi:10.1098/rstb.2009.0011

Dixon DR, Bainbridge BW, Darveau RP (2004) Modulation of the innate immune response within the periodontium. Periodontol 2000 35:53–74. doi:10.1111/j.0906-6713.2004.003556.x

Drouin de Bouville R de (1908) Les maladies des poissons d'eau douce d'Europe: d'après les travaux des divers icthyo-pathologistes et le traité du professeur Hofer (Deuxième édition, revue et augmentée)/par R. de Drouin de Bouville,... Berger-Levrault (Paris)

Durborow RM, Thune RL, Hawke JP, Camus AC (1998) Columnaris disease: a bacterial infection caused by *Flavobacterium columnare*. Southern Regional Aquaculture Center (SRAC) Publication No. 479

Ebert D (1998) Experimental evolution of parasites. Science 282:1432–1436

Ebert D, Mangin KL (1997) The influence of host demography on the evolution of virulence of a microsporidian gut parasite. Evolution 51:1828–1837

Egidius E (1987) Vibriosis: pathogenicity and pathology. A review. Aquaculture 67:15–28. doi:10.1016/0044-8486(87)90004-4

Ekman E, Börjeson H, Johansson N (1999) Flavobacterium psychrophilum in Baltic salmon Salmo salar brood fish and their offspring. Dis Aquat Organ 37:159–163

Ellis AE (2001) Innate host defense mechanisms of fish against viruses and bacteria. Dev Comp Immunol 25:827–839. doi:10.1016/s0145-305x(01)00038-6

Emmerich R, Weibel F (1890) Über eine durch Bakterien verursachte Infektionskrankheit der Forellen. Allg Fisch-Ztg 15:73–77, 85–92

Evelyn TPT (1971) First records of vibriosis in pacific salmon cultured in Canada, and taxonomic status of the responsible bacterium, Vibrio anguillarum. J Fish Res Board Can 28:517–525. doi:10.1139/f71-073

Farzanfar A (2006) The use of probiotics in shrimp aquaculture. FEMS Immunol Med Microbiol 48:149–158. doi:10.1111/j.1574-695X.2006.00116.x

Federle MJ, Bassler BL (2003) Interspecies communication in bacteria. J Clin Invest 112:1291–1299

Ferguson HW, Ostland VE, Byrne P, Lumsdsen JS (1991) Experimental production of bacterial gill disease in trout by horizontal transmission and by bath challenge. J Aquat Anim Health 3:118–123

Fijan N (1967) The survival of Chondrococcus columnaris in waters of different quality. Bull Off Int Epizoot 69:1159–1166

Foott JS, Hedrick RP (1987) Seasonal occurrence of the infectious stage of proliferative kidney disease (PKD) and resistance of rainbow trout, Salmo gairdneri Richardson, to reinfection. J Fish Biol 30:477–483

Frank DN, Zhu W, Sartor RB, Li E (2011) Investigating the biological and clinical significance of human dysbioses. Trends Microbiol 19:427–434. doi:10.1016/j.tim.2011.06.005

Frans I, Michiels CW, Bossier P et al (2011) Vibrio anguillarum as a fish pathogen: virulence factors, diagnosis and prevention. J Fish Dis 34:643–661. doi:10.1111/j.1365-2761.2011.01279.x

Freestone PP, Sandrini SM, Haigh RD, Lyte M (2008) Microbial endocrinology: how stress influences susceptibility to infection. Trends Microbiol 16:55–64

Fuqua C, Greenberg EP (2002) Listening in on bacteria: acyl-homoserine lactone signalling. Nat Rev Mol Cell Biol 3:685–695

Garduño RA, Kay WW (1992) Interaction of the fish pathogen Aeromonas salmonicida with rainbow trout macrophages. Infect Immun 60:4612–4620

Gatesoupe FJ (2010) Probiotics and other microbial manipulations in fish feeds: prospective health benefits. In: Watson RR, Preedy VR (eds) Bioactive foods in promoting health. Probiotics and prebiotics. Academic, San Diego, CA, pp 541–552

Geraylou Z, Souffreau C, Rurangwa E et al (2012) Effects of arabinoxylan-oligosaccharides (AXOS) on juvenile Siberian sturgeon (Acipenser baerii) performance, immune responses and gastrointestinal microbial community. Fish Shellfish Immunol 33:718–724. doi:10.1016/j.fsi.2012.06.010

Ghittino C, Latini M, Agnetti F et al (2003) Emerging pathologies in aquaculture: effects on production and food safety. Vet Res Commun 27:471–479. doi:10.1023/B:VERC.0000014204.37722.b6

Gibson GR (2004) Fibre and effects on probiotics (the prebiotic concept). Clin Nutr Suppl 1:25–31

Gold HS, Moellering RC (1996) Antimicrobial-drug resistance. N Engl J Med 335:144–1453

Gram L, Ringø E (2005) Prospects of fish probiotics. In: Holzapfel WH, Naughton PJ (eds) Microbial ecology of the growing animal. Elsevier, Amsterdam, pp 379–417

Grisez L, Chair M, Sorgeloos P, Ollevier F (1996) Mode of infection and spread of Vibrio anguillarum in turbot Scophthalmus maximus larvae after oral challenge through live feed. Dis Aquat Organ 26:181–187

Han Y, Mo Z, Xiao P et al (2011) Characterization of EmpA protease in Vibrio anguillarum M3. J Ocean Univ Chin 10:379–384. doi:10.1007/s11802-011-1781-x

Hanley F, Brown H, Carberry J (1995) First observations on the effects of mannan oligosaccharide added to the hatchery diets for warmwater Hybrid Red Tilapia. In: Nutritional biotechnology in the feed and food industries. Proceedings of Alltech's 11th annual symposium, Lexington, KY

Hart S, Wrathmell AB, Harris JE, Grayson TH (1988) Gut immunology in fish: a review. Dev Comp Immunol 12:453–480

Håstein T, Gudding R, Evensen O (2004) Bacterial vaccines for fish—an update of the current situation worldwide. Dev Biol 121:55–74

Heo G-J, Kasai K, Wakabayashi H (1990) Occurrence of Flavobacterium branchiophila associated with bacterial gill disease at a trout hatchery. Fish Pathol 25:99–105

Hjelm M, Riaza A, Formoso F et al (2004) Seasonal incidence of autochthonous antagonistic Roseobacter spp. and Vibrionaceae strains in a turbot larva (Scophthalmus maximus) rearing system. Appl Environ Microbiol 70:7288–7294. doi:10.1128/aem.70.12.7288-7294.2004

Hoare R, Hovland H, Langston AL et al (2002) Susceptibility of three different strains of juvenile Atlantic halibut (Hippoglossus hippoglossus L.) cultured at two different temperatures to Vibrio anguillarum and temperature effect on antibody response. Fish Shellfish Immunol 13:111–123

Holt RA, Rohovec JS, Fryer JL (1993) Bacterial coldwater disease. In: Inglis V, Roberts RJ, Bromage NR (eds) Bacterial disease of fish. Blackwell, Oxford, pp 3–22

Horneman AJ, Ali A, Abbott SL (2007) Aeromonas. In: Murray PR, Baron EJ, Landry ML, Jorgensen JH, Pfaller MA (eds) Manual of clinical microbiology, 9th edn. ASM, Washington, DC, pp 715–722

Iida Y, Mizokami A (1996) Outbreaks of coldwater disease in wild Ayu and Pale Chub. Fish Pathol 31:157–164. doi:10.3147/jsfp.31.157

Izumi S, Liu H, Aranishi F, Wakabayashi H (2003) A novel serotype of Flavobacterium psychrophilum detected using antiserum against an isolate from amago, Oncorhynchus

masou rhodurus Jordan & Gilbert, in Japan. J Fish Dis 26:677–680. doi:10.1046/j.1365-2761.2003.00502.x
Janda JM, Abbott SL (2010) The genus Aeromonas: taxonomy, pathogenicity, and infection. Clin Microbiol Rev 23:35–73. doi:10.1128/CMR.00039-09
Janda JM, Duffey PS (1988) Mesophilic aeromonads in human disease: current taxonomy, laboratory identification, and infectious disease spectrum. Rev Infect Dis 10:980–997
Johnson CM, Tatner MF, Horne MT (1985) Autoaggregation and extracellular A-layer protein in Aeromonas salmonicida. Aquaculture 46:163–166. doi:10.1016/0044-8486(85)90200-5
Kalia VC (2013) Quorum sensing inhibitors: an overview. Biotechnol Adv 31:224–245
Kelly D, Conway S, Aminov R (2005) Commensal gut bacteria: mechanisms of immune modulation. Trends Immunol 26:326–333. doi:10.1016/j.it.2005.04.008
Kent ML, Dawe SC, Speare DJ (1999) Resistance to reinfection in chinook salmon Oncorhynchus tshawytscha to Loma salmonae (Microsporidia). Dis Aquat Organ 37:205
Kimura N, Wakabayashi H, Kudo S (1978) Studies on bacterial gill disease in salmonids, 1: Selection of bacterium transmitting gill disease. Fish Pathol 12:233–242
Kumagai A, Yamaoka S, Takahashi K et al (2000) Waterborne transmission of Flavobacterium psychrophilum in Coho salmon eggs. Gyobyo Kenkyu Fish Pathol 35:25–28
Langevin C, Blanco M, Martin SAM, Jouneau L, Bernardet J-F, Houel A, Lunazzi A, Duchaud E, Michel C, Quillet E, Boudinot P (2012) Transcriptional responses of resistant and susceptible fish clones to the bacterial pathogen *Flavobacterium psychrophilum*. PLoS One 7, e39126. doi:10.1371/journal.pone.0039126
Larsen MH, Boesen HT (2001) Role of flagellum and chemotactic motility of Vibrio anguillarum for phagocytosis by and intracellular survival in fish macrophages. FEMS Microbiol Lett 203:149–152. doi:10.1111/j.1574-6968.2001.tb10833.x
Le Moullac G, Soyez C, Saulnier D, Ansquer D, Avarre JC, Levy P (1998) Effect of hypoxic stress on the immune response and the resistance to vibriosis of the shrimp *Penaeus stylirostris*. Fish Shellfish Immunol 8:621–629. doi:10.1006/fsim.1998.0166
Levy SB (1998) Multidrug resistance—a sign of the times. N Engl J Med 338:1376–1378
Lillehaug A, Lunestad BT, Grave K (2003) Epidemiology of bacterial diseases in Norwegian aquaculture—a description based on antibiotic prescription data for the ten-year period 1991 to 2000. Dis Aquat Organ 53:115–125
Lindsay J, Holden M (2006) Understanding the rise of the superbug: investigation of the evolution and genomic variation of *Staphylococcus aureus*. Funct Integr Genomics 6:186–201. doi:10.1007/s10142-005-0019-7
Lipsitch M, Nowak MA, Ebert D, May RM (1995) The population dynamics of vertically and horizontally-transmitted parasites. Proc Biol Sci 260:321–327. doi:10.1098/rspb.1995.0099
Littman RA, Bourne DG, Willis BL (2010) Responses of coral-associated bacterial communities to heat stress differ with Symbiodinium type on the same coral host. Mol Ecol 19:1978–1990. doi:10.1111/j.1365-294X.2010.04620.x
Liu H, Izumi S, Wakabayashi H (2001) Detection of Flavobacterium psychrophilum in various organs of ayu Plecoglossus altivelis by in situ hybridization. Fish Pathol 36:7–11
Loch TP, Faisal M (2010) Isolation of Aeromonas salmonicida subspecies salmonicida from Lake Whitefish (Coregonus clupeaformis) inhabiting Lakes Michigan and Huron. J Great Lakes Res 36(Suppl 1):13–17. doi:10.1016/j.jglr.2009.07.002
Lorenzen E (1994) Studies on Flexibacter psychrophilus in relation to rainbow trout fry syndrome (RTFS). Royal Veterinary and Agricultural University, Copenhagen
Lorenzen E, Olesen NJ (1997) Characterization of isolates of Flavobacterium psychrophilum associated with coldwater disease or rainbow trout fry syndrome II: serological studies. Dis Aquat Organ 31:209–220. doi:10.3354/dao031209
Lowry T, Smith SA (2007) Aquatic zoonoses associated with food, bait, ornamental, and tropical fish. J Am Vet Med Assoc 231:876–880. doi:10.2460/javma.231.6.876
Madetoja J, Dalsgaard I, Wiklund T (2002) Occurrence of Flavobacterium psychrophilum in fish-farming environments. Dis Aquat Organ 52:109–118

Madsen L, Dalsgaard I (1999) Reproducible methods for experimental infection with Flavobacterium psychrophilum in rainbow trout Oncorhynchus mykiss. Dis Aquat Organ 36:169

Mahious A, Ollevier F (2005) Probiotics and prebiotics in aquaculture. In: 1st Regional workshop on techniques for enrichment for use in larviculture. Urmia, Iran

Marcil V, Delvin E, Seidman E et al (2002) Modulation of lipid synthesis, apolipoprotein biogenesis, and lipoprotein assembly by butyrate. Am J Physiol Gastrointest Liver Physiol 283:G340–G346

Marshall BM, Levy SB (2011) Food animals and antimicrobials: impacts on human health. Clin Microbiol Rev 24:718–733

Maslanik T, Tannura K, Mahaffey L et al (2012) Commensal bacteria and MAMPs are necessary for stress-induced increases in IL-1β and IL-18 but not IL-6, IL-10 or MCP-1. PLoS One 7: e50636

Maslowski KM, Mackay CR (2010) Diet, gut microbiota and immune responses. Nat Immunol 12:5–9

Massault C, Franch R, Haley C, De Koning DJ, Bovenhuis H, Pellizzari C, Patarnello T, Bargelloni L (2011) Quantitative trait loci for resistance to fish pasteurellosis in gilthead sea bream (*Sparus aurata*). Anim Genet 42:191–203

Matsuzaki S, Rashel M, Uchiyama J et al (2005) Bacteriophage therapy: a revitalized therapy against bacterial infectious diseases. J Infect Chemother 11:211–219

Mazmanian SK, Kasper DL (2006) The love-hate relationship between bacterial polysaccharides and the host immune system. Nat Rev Immunol 6:849–858

McGee K, Horstedt P, Milton DL (1996) Identification and characterization of additional flagellin genes from Vibrio anguillarum. J Bacteriol 178:5188–5198

McKnite AM, Perez-Munoz ME, Lu L et al (2012) Murine gut microbiota is defined by host genetics and modulates variation of metabolic traits. PLoS One 7:e39191. doi:10.1371/journal.pone.0039191

Medzhitov R, Janeway CA (1999) Innate immune induction of the adaptive immune response. Cold Spring Harb Symp Quant Biol 64:429–436. doi:10.1101/sqb.1999.64.429

Merrifield DL, Dimitroglou A, Foey A et al (2010) The current status and future focus of probiotic and prebiotic applications for salmonids. Aquaculture 302:1–18

Miller MB, Bassler BL (2001) Quorum sensing in bacteria. Annu Rev Microbiol 55:165–199

Milton DL, O'Toole R, Horstedt P, Wolf-Watz H (1996) Flagellin A is essential for the virulence of Vibrio anguillarum. J Bacteriol 178:1310–1319

Mohamed MH, Ahmed Refat NAG (2011) Pathological evaluation of probiotic, *Bacillus subtilis*, against *Flavobacterium columnare* in tilapia nilotica (*Oreochromis niloticus*) fish in Sharkia governorate, Egypt. J Am Sci 7:244–256

Moloney R, Desbonnet L, Clarke G et al (2013) The microbiome: stress, health and disease. Mamm Genome 25:49–74. doi:10.1007/s00335-013-9488-5

Moriarty DJW (1998) Control of luminous *Vibrio* species in penaeid aquaculture ponds. Aquaculture 164:351–358. doi:10.1016/s0044-8486(98)00199-9

Morin R (2010) L'utilisation des antibiotiques pour combattre la furonculose chez l'omble de fontaine génère de l'antibiorésistance chez Aeromonas salmonicida. Bull L'Association Aquaculteurs Qué 15:3–6

Morris JJ, Lenski RE, Zinser ER (2012) The black queen hypothesis: evolution of dependencies through adaptive gene loss. mBio. doi:10.1128/mBio.00036-12

Myhr E, Larsen JL, Lillehaug A et al (1991) Characterization of Vibrio anguillarum and closely related species isolated from farmed fish in Norway. Appl Environ Microbiol 57:2750–2757

Naik S, Bouladoux N, Wilhelm C et al (2012) Compartmentalized control of skin immunity by resident commensals. Science 337:1115–1119. doi:10.1126/science.1225152

Nakanishi T, Ototake M (1997) Antigen uptake and immune responses after immersion vaccination. Dev Biol Stand 90:59

Nakayama T, Lu H, Nomura N (2009) Inhibitory effects of Bacillus probionts on growth and toxin production of Vibrio harveyi pathogens of shrimp. Lett Appl Microbiol 49:679–684. doi:10.1111/j.1472-765X.2009.02725.x

Nayak SK (2010) Probiotics and immunity: a fish perspective. Fish Shellfish Immunol 29:2–14. doi:10.1016/j.fsi.2010.02.017

Nealson KH, Platt T, Hastings JW (1970) Cellular control of the synthesis and activity of the bacterial luminescent system. J Bacteriol 104:313–322

Nelson RJ (2004) Seasonal immune function and sickness responses. Trends Immunol 25:187–192

Nematollahi A, Decostere A, Pasmans F, Haesebrouck F (2003) *Flavobacterium psychrophilum* infections in salmonid fish. J Fish Dis 26:563–574. doi:10.1046/j.1365-2761.2003.00488.x

Nielsen ME, Høi L, Schmidt AS et al (2001) Is Aeromonas hydrophila the dominant motile Aeromonas species that causes disease outbreaks in aquaculture production in the Zhejiang Province of China? Dis Aquat Organ 46:23–29. doi:10.3354/dao046023

Nikoskelainen S, Ouwehand AC, Bylund G et al (2003) Immune enhancement in rainbow trout (*Oncorhynchus mykiss*) by potential probiotic bacteria (*Lactobacillus rhamnosus*). Fish Shellfish Immunol 15:443–452. doi:10.1016/s1050-4648(03)00023-8

O'Mahony C, Scully P, O'Mahony D et al (2008) Commensal-induced regulatory T cells mediate protection against pathogen-stimulated NF-κB activation. PLoS Pathog 4:e1000112. doi:10.1371/journal.ppat.1000112

O'Toole R, Milton DL, Wolf-Watz H (1996) Chemotactic motility is required for invasion of the host by the fish pathogen Vibrio anguillarum. Mol Microbiol 19:625–637

O'Toole R, Lundberg S, Fredriksson SA et al (1999) The chemotactic response of Vibrio anguillarum to fish intestinal mucus is mediated by a combination of multiple mucus components. J Bacteriol 181:4308–4317

Ondračková M, Reichard M, Jurajda P, Gelnar M (2004) Seasonal dynamics of Posthodiplostomum cuticola (Digenea, Diplostomatidae) metacercariae and parasite-enhanced growth of juvenile host fish. Parasitol Res 93:131–136

Ostland VE, Lumsden JS, MacPhee DD, Ferguson HW (1994) Characteristics of Flavobacterium branchiophilum, the cause of salmonid bacterial gill disease in Ontario. J Aquat Anim Health 6:13–26

Ostland VE, MacPhee DD, Lumsden JS, Ferguson HW (1995) Virulence of Flavobacterium branchiophilum in experimentally infected salmonids. J Fish Dis 18:249–262. doi:10.1111/j.1365-2761.1995.tb00300.x

Ostland VE, McGrogan DG, Ferguson HW (1997) Cephalic osteochondritis and necrotic scleritis in intensively reared salmonids associated with Flexibacter psychrophilus. J Fish Dis 20:443–451. doi:10.1046/j.1365-2761.1997.00323.x

Ozaki A, Sakamoto T, Khoo S, Nakamura K, Coimbra MR, Akutsu T, Okamoto N (2001) Quantitative trait loci (QTLs) associated with resistance/susceptibility to infectious pancreatic necrosis virus (IPNV) in rainbow trout (*Oncorhynchus mykiss*). Mol Genet Genomics 265:23–31

Paillard C, Le Roux F, Borrego JJ (2004) Bacterial disease in marine bivalves, a review of recent studies: trends and evolution. Aquat Living Resour 17:477–498. doi:10.1051/alr:2004054

Palti Y (2011) Toll-like receptors in bony fish: from genomics to function. Dev Comp Immunol 35:1263–1272. doi:10.1016/j.dci.2011.03.006

Pavey SA, Sevellec M, Adam W et al (2013) Nonparallelism in MHCIIbeta diversity accompanies nonparallelism in pathogen infection of lake whitefish (Coregonus clupeaformis) species pairs as revealed by next-generation sequencing. Mol Ecol 22:3833–3849. doi:10.1111/mec.12358

Phillips I, Casewell M, Cox T et al (2004) Does the use of antibiotics in food animals pose a risk to human health? A critical review of published data. J Antimicrob Chemother 53:28–52. doi:10.1093/jac/dkg483

Pizarro-Cerdá J, Cossart P (2006) Bacterial adhesion and entry into host cells. Cell 124:715–727. doi:10.1016/j.cell.2006.02.012

Plumb JA, Grizzle JM, Defigueiredo J (1976) Necrosis and bacterial infection in channel catfish (Ictalurus punctatus) following hypoxia. J Wildl Dis 12:247–253

Pulkkinen K, Suomalainen L-R, Read AF et al (2010) Intensive fish farming and the evolution of pathogen virulence: the case of columnaris disease in Finland. Proc Biol Sci 277:593–600. doi:10.1098/rspb.2009.1659

Rakoff-Nahoum S, Paglino J, Eslami-Varzaneh F et al (2004) Recognition of commensal microflora by toll-like receptors is required for intestinal homeostasis. Cell 118:229–241. doi:10.1016/j.cell.2004.07.002

Ramsey MM, Whiteley M (2009) Polymicrobial interactions stimulate resistance to host innate immunity through metabolite perception. Proc Natl Acad Sci USA 106:1578–1583. doi:10.1073/pnas.0809533106

Rangdale RE (1995) Studies on rainbow trout fry syndrome (RTFS). PhD thesis, University of Stirling, Stirling

Rangdale RE, Richards RE, Alderman DJ (1996) Isolation of Cytophaga psychrophila, causal agent of rainbow trout fry syndrome (RTFS) from reproductive fluids and egg surfaces of rainbow trout (Oncorhynchus mykiss). Bull Eur Assoc Fish Pathol 16:63–67

Rangdale RE, Richards RH, Alderman DJ (1997a) Colonisation of eyed rainbow trout ova with Flavobacterium psychrophilum leads to rainbow trout fry syndrome in fry. Bull Eur Assoc Fish Pathol 17:108–111

Rangdale RE, Richards RH, Alderman DJ (1997b) Minimum inhibitory concentrations of selected antimicrobial compounds against *Flavobacterium psychrophilum* the causal agent of rainbow trout fry syndrome (RTFS). Aquaculture 158:193–201. doi:10.1016/s0044-8486(97)00202-0

Reith ME, Singh RK, Curtis B et al (2008) The genome of Aeromonas salmonicida subsp. salmonicida A449: insights into the evolution of a fish pathogen. BMC Genomics 9:427. doi:10.1186/1471-2164-9-427

Revie CW, Gettinby G, Treasurer JW et al (2002) Temporal, environmental and management factors influencing the epidemiological patterns of sea lice (Lepeophtheirus salmonis) infestations on farmed Atlantic salmon (Salmo salar) in Scotland. Pest Manag Sci 58:576–584

Ringø E, Dimitroglou A, Hoseinifar SH, Davies SJ (2014) Prebiotics in finfish: an update. In: Merrifield DL, Ringø E (eds) Aquaculture nutrition: gut health, probiotics and prebiotics. Wiley-Blackwell, Oxford, pp 410–418

Roberfroid M (2007) Prebiotics: the concept revisited. J Nutr 137:830S–837S

Rodríguez-Ramilo ST, Toro MA, Bouza C, Hermida M, Pardo BG, Cabaleiro S, Martínez P, Fernández J (2011) QTL detection for Aeromonas salmonicida resistance related traits in turbot (*Scophthalmus maximus*). BMC Genomics 12:541. doi:10.1186/1471-2164-12-541

Rurangwa E, Laranja JL, Van Houdt R et al (2009) Selected nondigestible carbohydrates and prebiotics support the growth of probiotic fish bacteria mono cultures in vitro. J Appl Microbiol 106:932–940

Sara M, Sleytr UB (2000) S-layer proteins. J Bacteriol 182:859–868

SCAN, European Commission Health and Consumer Protection Directorate-General (2003) Opinion of the scientific committee on Animal Nutrition on the criteria for assessing the safety of microorganisms resistant to antibiotics of human clinical and veterinary importance

Scott M (1968) The pathogenicity of Aeromonas salmonicida (Griffin) in sea and brackish waters. J Gen Microbiol 50:321–327. doi:10.1099/00221287-50-2-321

Shivu MM, Rajeeva BC, Girisha SK et al (2007) Molecular characterization of Vibrio harveyi bacteriophages isolated from aquaculture environments along the coast of India. Environ Microbiol 9:322–331

Snieszko SF (1974) The effects of environmental stress on outbreaks of infectious diseases of fishes. J Fish Biol 6:197–208. doi:10.1111/j.1095-8649.1974.tb04537.x

Sommerset I, Krossøy B, Biering E, Frost P (2005) Vaccines for fish in aquaculture. Expert Rev Vaccines 4:89–101. doi:10.1586/14760584.4.1.89

Sorensen SJ, Bailey M, Hansen LH et al (2005) Studying plasmid horizontal transfer in situ: a critical review. Nat Rev Microbiol 3:700–710

Speare DJ, Ferguson HW (1989) Clinical and pathological features of common gill diseases of cultured salmonids in Ontario. Can Vet J 30:882
Speare DJ, Ferguson HW, Beamish FWM et al (1991) Pathology of bacterial gill disease: ultrastructure of branchial lesions. J Fish Dis 14:1–20
Starliper CE (2011) Bacterial coldwater disease of fishes caused by *Flavobacterium psychrophilum*. J Adv Res 2:97–108. doi:10.1016/j.jare.2010.04.001
Stecher B, Maier L, Hardt W-D (2013) "Blooming" in the gut: how dysbiosis might contribute to pathogen evolution. Nat Rev Microbiol 11:277–284. doi:10.1038/nrmicro2989
Stenholm AR, Dalsgaard I, Middelboe M (2008) Isolation and characterization of bacteriophages infecting the fish pathogen Flavobacterium psychrophilum. Appl Environ Microbiol 74:4070–4078
Stephens C, Shapiro L (1997) Bacterial protein secretion—a target for new antibiotics? Chem Biol 4:637–641
Stork M, Di Lorenzo M, Welch TJ et al (2002) Plasmid-mediated iron uptake and virulence in Vibrio anguillarum. Plasmid 48:222–228
Stuber K, Burr SE, Braun M et al (2003) Type III secretion genes in Aeromonas salmonicida subsp. salmonicida are located on a large thermolabile virulence plasmid. J Clin Microbiol 41:3854–3856. doi:10.1128/JCM.41.8.3854-3856.2003
Sundell K, Wiklund T (2011) Effect of biofilm formation on antimicrobial tolerance of *Flavobacterium psychrophilum*. J Fish Dis 34:373–383. doi:10.1111/j.1365-2761.2011.01250.x
Tam B, Gough WA, Tsuji L (2011) The impact of warming on the appearance of furunculosis in fish of the James Bay region, Quebec, Canada. Reg Environ Change 11:123–132. doi:10.1007/s10113-010-0122-8
Tanaka KH, Dallaire-Dufresne S, Daher RK et al (2012) An insertion sequence-dependent plasmid rearrangement in Aeromonas salmonicida causes the loss of the type three secretion system. PLoS One 7:e33725. doi:10.1371/journal.pone.0033725
Touchon M, Barbier P, Bernardet J-F et al (2011) Complete genome sequence of the fish pathogen Flavobacterium branchiophilum. Appl Environ Microbiol 77:7656–7662
Trivedi B (2012) Microbiome: the surface brigade. Nature 492:S60–S61
Trust TJ, Bull LM, Currie BR, Buckley JT (1979) Obligate anaerobic bacteria in the gastrointestinal microflora of the grass carp (Ctenopharyngodon idella), goldfish (Carassius auratus), and rainbow trout (Salmo gairdneri). J Fish Res Board Can 36:1174–1179. doi:10.1139/f79-169
Turnbull JF (1993) Bacterial gill disease and fin rot. In: Inglis V, Roberts RJ, Bromage NR (eds) Bacterial disease fish. Halsted, New York, pp 40–58
Vanden Bergh P, Frey J (2013) *Aeromonas salmonicida* subsp. *salmonicida* in the light of its type-three secretion system. Microb Biotechnol. doi:10.1111/1751-7915.12091
Verrier ER, Dorson M, Mauger S, Torhy C, Ciobotaru C, Hervet C, Dechamp N, Genet C, Boudinot P, Quillet E (2013) Resistance to a rhabdovirus (VHSV) in rainbow trout: identification of a major QTL related to innate mechanisms. PLoS One 8, e55302
Vatsos IN, Thompson KD, Adams A (2001) Adhesion of the fish pathogen Flavobacterium psychrophilum to unfertilized eggs of rainbow trout (Oncorhynchus mykiss) and n-hexadecane. Lett Appl Microbiol 33:178–182
Vatsos IN, Thompson KD, Adams A (2006) Colonization of rainbow trout, Oncorhynchus mykiss (Walbaum), eggs by Flavobacterium psychrophilum, the causative agent of rainbow trout fry syndrome. J Fish Dis 29:441–444
Vilches S, Jimenez N, Tomas JM, Merino S (2009) Aeromonas hydrophila AH-3 Type III secretion system expression and regulatory network. Appl Environ Microbiol 75:6382–6392. doi:10.1128/AEM.00222-09
Vinod MG, Shivu MM, Umesha KR et al (2006) Isolation of *Vibrio harveyi* bacteriophage with a potential for biocontrol of luminous vibriosis in hatchery environments. Aquaculture 255:117–124

Wakabayashi H (1991) Effect of environmental conditions on the infectivity of *Flexibacter columnaris* to fish. J Fish Dis 14:279–290. doi:10.1111/j.1365-2761.1991.tb00825.x
Wakabayashi H, Egusa S, Fryer JL (1980) Characteristics of filamentous bacteria isolated from a gill disease of salmonids. Can J Fish Aquat Sci 37:1499–1504
Wakabayashi H, Toyama T, Iida T (1994) A study on serotyping of Cytophaga-psychrophila isolated from fishes in Japan. Fish Pathol 29:101–104
Waldvogel FA (1999) New resistance in Staphylococcus aureus. N Engl J Med 340:556–557
Welker TL, Shoemaker CA, Arias CR, Klesius PH (2005) Transmission and detection of Flavobacterium columnare in channel catfish Ictalurus punctatus. Dis Aquat Organ 63:129–138
Whitehead NA, Barnard AM, Slater H et al (2001) Quorum-sensing in Gram-negative bacteria. FEMS Microbiol Rev 25:365–404
Witte W (1998) Medical consequences of antibiotic use in agriculture. Science 279:996–997
Zapata AG, Varas A, Torroba M (1992) Seasonal variations in the immune system of lower vertebrates. Immunol Today 13:142–147

Chapter 5
Fungal Secondary Invaders of Fish

Nicolas Derome, Jeff Gauthier, Sébastien Boutin, and Martin Llewellyn

Abstract Fungal and fungal-like opportunistic pathogens are of primary concern in aquaculture, because they affect a wide range of hosts (both vertebrate and invertebrate) and are very difficult to diagnose, control, or treat. One of the reasons for this difficulty is that some diseases caused by fungal and fungal-like pathogens have no clear external symptoms. Furthermore, the increasing emergence of fungal and fungal-like opportunistic pathogens is correlated with modern production cycles, relying on intensive techniques that often lead to poor water quality and high population density. This combination of stress factors is known as a catalyst for opportunistic infections in aquaculture due to the direct and indirect effects on the host immune response. Therefore, those suboptimal rearing conditions may lead to homeostatic imbalance in favor of secondary invaders. We have therefore covered fungal and fungal-like fish pathogens with significant impact in worldwide aquaculture (black yeasts, *Oomycetes*, and Microsporidia), as well as control and prevention strategies for pathogens belonging to these groups.

N. Derome (✉) • J. Gauthier
Institut de Biologie Intégrative et des Systèmes (IBIS), Université Laval, 1030, av. de la Médecine, Québec, QC, Canada, G1V0A6

Département de Biologie, Université Laval, Québec, QC, Canada
e-mail: nicolas.derome@bio.ulaval.ca

S. Boutin
Institut de Biologie Intégrative et des Systèmes (IBIS), Université Laval, 1030, av. de la Médecine, Québec, QC, Canada, G1V0A6

Département de Biologie, Université Laval, Québec, QC, Canada

Department of Infectious Disease, Medical Microbiology and Hygiene, Universität Heidelberg, Im Neuenheimer Feld 324, 69120 Heidelberg, Germany

M. Llewellyn
Institut de Biologie Intégrative et des Systèmes (IBIS), Université Laval, 1030, av. de la Médecine, Québec, QC, Canada, G1V0A6

Département de Biologie, Université Laval, Québec, QC, Canada

Molecular Ecology and Fisheries Genetics Laboratory, School of Biological Sciences, Bangor University, Gwynedd LL57 2UW, UK

© Springer International Publishing Switzerland 2016

C.J. Hurst (ed.), *The Rasputin Effect: When Commensals and Symbionts Become Parasitic*, Advances in Environmental Microbiology 3,
DOI 10.1007/978-3-319-28170-4_5

5.1 Introduction

The traditional "Fungi" kingdom encompasses an enormous diversity of species in terms of morphology, physiology, sexual states, ecology, and life cycle strategies, ranging from unicellular aquatic molds to large mushrooms, all of those heterotrophic organisms possessing a chitinous cell wall. Such a huge biological diversity led classical taxonomists to group similar-looking species, thus building highly paraphyletic or even polygenic groups species that were later identified, upon molecular characters, as distantly related species (Hibbett et al. 2007; Wang et al. 2009). Thus, recent molecular phylogenetic studies have helped to reshape thoroughly the classification of the kingdom *Fungi* to get a more comprehensive interpretation of evolutionary relationship among fungal lineages. The currently accepted classification by Hibbett et al. (2007) recognizes one kingdom (*Fungi*) and one subkingdom (*Dikarya*), which includes two phyla (*Basidiomycota* and *Ascomycota*); however, relationships with and among the four other phyla are not yet resolved (*Chytridiomycota*, *Neocallimastigomycota*, *Blastocladiomycota*, *Glomeromycota*, and *Microsporidia*).

This classification is restricted to organisms that belong in the monophyletic kingdom *Fungi*. It does not consider other organisms formerly included in the kingdom but which are now termed as fungal-like eukaryotic microorganisms (such as the *Oomycota*, *Mesomycetozoea*, and *Gymnomycota* (slime molds)), even if those are still studied by mycologists. The following are three examples of the most dramatic shifts in the classification relative to previous works, which concern taxonomic groups that have traditionally been included in the *Fungi* kingdom. First, one of the most famous "fungal" pathogens, *Saprolegnia parasitica*, belongs to the *Oomycota* class, which now forms a distinct phylogenetic lineage. Then, the *Microsporidia*, which are mostly obligate intracellular parasites, were included in the *Fungi* kingdom until recently (Keeling et al. 2000; Gill and Fast 2006; Liu et al. 2006; James et al. 2006). Then, Voigt et al. (2013), based on recent molecular phylogenomic studies, concluded that the *Microsporidia* were the sister group of *Fungi*. Finally, previously classified into the paraphyletic *Zygomycota* (*Fungi*), members of the *Ichthyophonus* genus, which are mostly parasites of fish, are now considered as fungal-like organisms and thus ranked as a monophyletic group into the *Mesomycetozoea*, also known as the DRIP clade (*Dermocystidium*, "rosette agent," *Ichthyophonus*, *and Psorospermium*) (Paps et al. 2013). According to Baldauf (2003), higher phylogenetic relationships among true fungi and fungi-like organisms are as follows: the true *Fungi*, *Microsporidia*, and *Mesomycetozoea* are placed along with the animals in the *Opisthokonta* supergroup (which includes animals), whereas all of the *Oomycota* are placed into the *Heterokonta* supergroup (Baldauf 2003). Then, the *Opisthokonta* supergroup divides into two groups: Holozoa (which includes animals) and *Holomycota* which includes, among others, true *Fungi* and the *Microsporidia* (Torruella et al. 2012).

Therefore, the main focus of this chapter is to give an overview on commonly reported "fungal" disease caused by both true fungal and fungal-like pathogens and the currently available methods for prevention and cure in an aquacultural context. An increasing number of other environmental fungi or amoeboid fungal-like pathogens are being diagnosed as the causative agent of fish death, further assessing their opportunistic nature.

Indeed, fungal diseases are of primary concern in aquaculture because they affect a wide range of hosts, both vertebrate and invertebrate, and are very difficult to control or treat. One of the reasons for this difficulty, especially for freshwater systems, is that some diseases caused by fungal and fungal-like pathogens have no clear external symptoms (Gozlan et al. 2005; Kocan and Hershberger 2006; Andreou et al. 2011, 2012). Although both fisheries and aquaculture represent a key protein supply for an increasing human population, our appreciation of the disease risk associated with fungal pathogen emergence remains incomplete (Gozlan et al. 2006).

The general cause of the observed worldwide spread of opportunistic pathogens is unclear, but they are thought to benefit from recent intensification of global trade (Grünwald et al. 2012). Thus, increased volume of fish translocations for farming and recreational fishing, escapes of domestic strains (i.e., farmed salmons), and invasive healthy fish carriers, such as *Pseudorasbora parva* originating from China—which recently allowed the rosette agent *Sphaerothecum destruens* to threaten European fish diversity and wild stocks of Chinook and Atlantic salmon in the USA (Andreou et al. 2012)—are all potentially to blame.

We have therefore covered some of the most significant fungal and fungal-like pathogens with significant impact in worldwide aquaculture (black yeasts, *Oomycetes*, and *Microsporidia*), as well as control and prevention strategies for pathogens belonging to these groups.

5.2 Overview

5.2.1 The Ascomycota *Phylum*

Phaeohyphomycosis is an umbrella term covering subcutaneous and systemic infections caused by pigmented fungi, where the agents are present in host tissue with brownish to olivaceous hyphal elements (McGinnis 1983; Ajello 1986). The dark coloration of these fungi is due to the presence of melanin in the cell wall. The so-called black yeasts (i.e., *Phaeohyphomycetes*) and their relatives are often identified as the causative agents of disease in cold-blooded vertebrates including fish (de Hoog et al. 2011). For instance, species belonging to the *Exophiala* genus are opportunistic pathogens that may infect a broad range of warm- and cold-blooded animals, including salmonids and the Atlantic cod (Gjessing et al. 2011). The term "black yeasts" indicates those melanized fungi that are able to reproduce in culture by unilateral budding (de Hoog 1993), although not all members of the *Phaeohyphomycota* group causing infection have this ability. *Phaeohyphomycetes* belong to a limited number but phylogenetically dispersed orders of *Ascomycota* (de Hoog and McGinnis 1987) and show diverse adaptations to extreme conditions (Ruibal et al. 2009), including the bodies of animal hosts. Systemic black yeast infections in fish have been described on many occasions (Yanong 2003). The infections are generally considered opportunistic as they generally spread following stress of captivity (poor water quality, handling, or aggression) (Otis et al. 1985; Silphaduang et al. 2000), which in turn may contribute to reduced immune function and, thus, trigger bacterial or parasitic infections. Of special interest are members of the order *Chaetothyriales*, including the *Exophiala* genus.

5.2.1.1 The Genus *Exophiala*

Localized and systemic *Exophiala* infections have been reported in several freshwater and marine teleosts (those ray-finned fish with a mobile jaw—e.g., Perciformes, Salmonidae, etc.), resulting in a variety of targeted tissues and inflammatory responses. Single tissue infections by *Exophiala* were observed in whitings (*Sillaginodes punctata*), in which lesions were confined to the skin (Reuter et al. 2003). By contrast, infections caused by *E. aquamarina*, *E. angulospora*, and other *Exophiala* in both leafy sea dragons (*Phycodurus eques*) and weedy sea dragons (*Phyllopteryx taeniolatus*) targeted the blood vessels, thus allowing systemic necrotizing lesions, and those have been observed mostly in the skeletal muscle, skin, kidney, gills, swim bladder, extradural sinus, and spinal cord and more sporadically in the heart, liver, spleen, muscle coats and serosa of the intestine, and mesentery (Nyaoke et al. 2009).

In salmonids, *Exophiala* target cranial tissues, both in cutthroat trout (*Oncorhynchus clarkii*) and Atlantic salmon (*Salmo salar*) (Otis et al. 1985). In cutthroat trout, lesions are characterized by granuloma formation with numerous giant cells in the brain and cranial tissues. The infection extends peripherally to include surrounding cranial structures, such as eyes and gills (Otis et al. 1985). In Atlantic salmon, lesions are systemic and involve the brain and kidneys. The host's inflammatory response varies from granulomatous to granulocytic, with the formation of microabscesses. *Exophiala salmonis* has been identified as a cause of up to 40 % mortality in Atlantic salmon hatcheries (Otis et al. 1985). Similarly, *Exophiala pisciphila* infection is associated with high mortality in Atlantic salmon. Hyphae invade cranial structures, including semicircular canals, and the lateral body line, accompanied by a granulomatous inflammatory reaction (Langdon and McDonald 1987).

Interestingly, *E. pisciphila* has been observed to also target cranial tissues (and the skin) in a non-salmonid fish, the smooth dogfish (*Mustelus canis*) (McGinnis and Ajello 1974). *Exophiala pisciphila* has been identified in cutaneous and visceral lesions in channel catfish (*Ictalurus punctatus*) (Gaskins and Cheung 1986). *Exophiala* strains have been observed to threaten Pleuronectidae; *E. pisciphila* was identified as the causative agent of muscle lesions in captive American sole (*Hippoglossoides platessoides*) (Strongman et al. 1997); another species of

Exophiala caused ulcerative lesions in Japanese flounder (*Paralichthys olivaceus*) (Kurata et al. 2008).

5.2.1.2 Other *Ascomycetes*

Systemic phaeohyphomycosis has also been associated with *Phoma herbarum* from a variety of fish species, including hatchery-reared Chinook salmon juveniles (*Oncorhynchus tshawytscha*) (Faisal et al. 2007); those exhibited abnormal swimming behavior, exophthalmia, multiple rounded areas of muscle softening, protruded hemorrhagic vents, and abdominal swelling. The targeted tissues were swim bladders, kidneys, and musculature. Finally, sporadic phaeohyphomycosis, caused by two species of *Ochroconis* genus: *O. humicola* and *O. tshawytscha*, was also reported in fish. *Ochroconis humicola* is a rare pathogen of cold-blooded vertebrates, particularly fish such as coho salmon (*Oncorhynchus kisutch*) (Ross and Yasutake 1973), Atlantic salmon (*Salmo salar*) (Schaumann and Priebe 1994), rainbow trout (*Oncorhynchus mykiss*) (Ross and Yasutake 1973), scorpion fish (*Inimicus japonica*) (Ajello et al. 1977), and walking catfish (*Clarias batrachus*) (Bhattacharya 1988). A case of infection in Chinook salmon due to *Ochroconis tshawytscha* was also reported (Doty and Slater 1946).

Another fungal disease, called aspergillomycosis, was observed in the African fish tilapia (*Oreochromis* sp.), and the causative agents of this disease are species such as *Aspergillus flavus*, *Aspergillus terreus*, and *Aspergillus japonicus*; also, *Aspergillus niger* was observed to cause internal and external infections in common carp (Willoughby 1994) and other fish species (Firouzbakhsh et al. 2005). *Aspergillus* species are observed to be part of the mycoflora of healthy fish such as *Oreochromis* sp., *Clarias gariepinas* (Mohamed and Refai 2010), and *Channa punctata* (Malathi and Rajendran 2012). Such observations confirm further the opportunistic nature of many fungal pathogens.

5.2.2 *The* Oomycota *Class*

The class of *Oomycota* includes the *Saprolegniacae* family (i.e., *Saprolegnia* sp.) and other typical water molds that are the “classic” secondary invaders, infecting mostly superficial areas of the body and requiring compromise of the exterior of the fish, poor water quality, or general immunosuppression.

5.2.2.1 The Genus *Saprolegnia*

Saprolegnia sp. can have multiple hosts and are thus capable of infecting different aquatic organisms (Sarowar et al. 2013). Thus, Saprolegniasis is a major disease problem in many aquatic organisms, including different wild and farmed fish species, causing huge losses in salmonid species such as Atlantic salmon, rainbow and brown trout, and non-salmonid species including eels, catfish (*Clarias gariepinas*), perch, and tilapia (Mohamed and Refai 2010; Bruno et al. 2011). Symptoms are characterized by white and gray patches of mycelial growth on the skin and fins of adult fish and cotton-like filamentous mycelium on eggs. In fish, death often occurs from disruption of the osmotic balance (hemodilution) following destruction of large areas of the epidermis by massive, invasive hyphal growth (Bruno et al. 2011). Fish eggs, on the other hand, are thought to be killed by hyphal breaching of the chorionic membrane regulating the osmosis of the embryo. *Saprolegnia* species can produce flagellated zoospores to disperse in the aquatic environment. *Saprolegnia* also forms secondary zoospores and secondary cysts that, for some species, contain boathooks that are presumed to aid in attachment to the fish skin or promote floating (buoyancy) in water (van den Berg et al. 2013). Losses resulting from saprolegniasis average 10 % in eggs and young fish, but losses of up to 50 % have been reported (Hatai and Hoshiai 1992; Bruno et al. 2011; van den Berg et al. 2013).

5.2.2.2 *Aphanomyces invadans*

Aphanomyces invadans is the etiological agent of epizootic ulcerative syndrome (EUS) (Egusa and Masuda 1971), which is a destructive disease affecting over 100 species of wild and cultured freshwater and estuarine finfish across the Asia-Pacific region. Unlike other Oomycoses, EUS symptoms include the presence of distinctive mycotic granulomas in internal tissues (Egusa and Masuda 1971; McKenzie and Hall 1976; Callinan et al. 1989; Roberts 1994; Vishwanath 1997). Isolates of *Aphanomyces* have been recovered from affected fish in Japan (Hatai et al. 1977), Australia (Fraser et al. 1992), the Philippines (Callinan et al. 1995), Thailand, Bangladesh (Willoughby et al. 1995), and Indonesia, all of them being recognized as a single species, *A. invadans* (David and Kirk 1997).

The infective life stage of *A. invadans* is the free-swimming zoospore that attaches to a fish host, encysts, and germinates to develop vegetative aseptate hyphae invading and ramifying through host tissues: skin, muscular tissue, and internal organs (Lilley et al. 1998; Kiryu et al. 2003). The formation of reniform and biflagellated zoospores, which can actively swim in water until they find a new fish host, follows and thus close the life cycle. Alternatively, the zoospore may encyst. This is a typical life cycle described in numerous *Oomycota*.

A very limited number of studies have used bath challenge to expose fish to *A. invadans* spores. Kiryu et al. (2003) challenged by aqueous exposure the Atlantic menhaden (*Brevoortia tyrannus*), a species highly susceptible to EUS, and confirmed that *A. invadans* was pathogenic by this route. Salinity of the rearing water for the fish had been decreased from 2 to 0.1 % over a 30-day period prior to challenge. The authors challenged Atlantic menhaden with *A. invadans* spores at 100 spores ml^{-1} for 5.5 h. Over the 27-day observation period, 9 of 14 net-stressed fish (64 %) developed focal ulcers, whereas among the fish presumed free of mechanical skin damage at the beginning of the challenge, only 5 of 36 fish (14 %) developed ulcerous lesions. Mortality was 64 % in the net-stressed group and 11 % in the group without skin damage. The drop of salinity levels prior to challenge is likely to have introduced an additional stressor that may have affected fish susceptibility.

5.2.2.3 The Genus *Ichthyophonus*

Ichthyophonus is an obligate parasite with a wide host spectrum, including some freshwater or anadromous fish (those which migrate upstream to spawn) and numerous marine fish (McVicar 1999). *Ichthyophonus* and other related microbes constitute a phylogenetic group at the boundaries of the animal–fungal divergence, which has been referred to as the DRIP clade, which represents an acronym of its original members *Dermocystidium*, rosette agent, *Ichthyophonus* and *Psorospermium* (Spanggaard et al. 1996; Ragan et al. 1996). *Ichthyophonus* was later ascribed to the class *Mesomycetozoea* (Mendoza et al. 2002), and this class was included within the *Opisthokonta* and not in *Fungi* in the new classification of Adl et al. (2005). Overall, a paucity of empirical investigations has resulted in a lack of understanding regarding *Opisthokonta*'s basic biological and epidemiological characteristics.

5.2.3 *The* Microsporidia *Phylum*

5.2.3.1 *Microsporidia*

Microsporidia are generally obligate intracellular parasites with 58 genera known to infect aquatic organisms: 35 of these genera are aquatic arthropod pathogens, 17 infect fish, and the remaining 6 are associated with various non-arthropod invertebrates, hyperparasites, and protists. Such a wide pathogenic spectrum certainly results from an extreme morphological plasticity and an ability to infect all known organ and tissue systems. Thus, despite only 58 genera being known to infect aquatic organisms, it is highly likely that many thousands of pathogenic microsporidian taxa remain undescribed in aquatic hosts.

Microsporidioses have thus been included in the emerging and opportunistic disease category (Didier 2005). Disease emergence is argued to be particularly likely in freshwater habitats as a result of their links with terrestrial and marine habitats and as a result of anthropogenic pressures (Okamura and Feist 2011). Taken together, balance shifts among host, pathogen, and environmental factors (e.g., due to pollution or climate change) are likely to be critical in progressing latent infection to patent microsporidiosis. In this way, the prevalence and severity of microsporidiosis in hosts residing in aquatic habitats may indirectly reflect immunocompetence, encompassing the host immune system, endogenous microbiota, and other biotic factors such as environmental microbial communities.

Microsporidians associated with bony fish occur in hosts inhabiting freshwater (lakes, rivers), brackish (estuaries, mangroves), and marine environments (rocky shorelines, open ocean, deep floor). Among the diseases attributed to *Microsporidia*, the microsporidial gill disease of salmonids (MGDS) is among the most significant infectious diseases affecting aquaculture-raised salmonids, including rainbow trout (*Oncorhynchus mykiss*) and Chinook salmon (*O. tshawytscha*). In Canada, this disease most often affects salmon in their second summer of marine cultivation when they are nearing market weight, and outbreaks can lead to a cumulative mortality rate exceeding 30 % (Kent and Speare 2005). In MGDS, sporogony occurs within the pillar and endothelial cells of the gill. These cells undergo hypertrophy, forming a xenoma composed primarily of parasite spores. The rupture of the xenoma releases spores and causes a severe, persistent gill inflammation; affected fish display signs of anoxia (Kent and Speare 2005). Xenomas are caused by various microsporidian genera including *Alloglugea*, *Amazonspora*, *Glugea*, *Ichthyosporidium*, *Loma*, *Microfilum*, *Microgemma*, *Neonosemoides*, *Pseudoloma*, *Spraguea*, and *Tetramicra* (Lom and Dyková 2005).

5.3 Control and Prevention

5.3.1 Phaeohyphomycosis

Diseases caused by black fungi are difficult to treat for two main reasons: first, diagnosis still requires direct microscopy, culture, histopathology, and mostly also molecular analysis for reliable identification down to the species level (Ajello 1986; Balajee et al. 2007; Ferrer et al. 2001; Zeng et al. 2007). *Fungi* with similar morphology may be phylogenetically remote from each other, which has a large impact on evaluation of the clinical course, appropriate therapy, and prophylaxis. Second, their recalcitrant nature is possibly enhanced by the presence of melanin in the cell wall (Jacobson 2000). Therefore, culture and susceptibility testing may provide useful information for selecting appropriate treatment protocols. Black fungi belonging to genera such as *Exophiala*, *Alternaria*, *Cladophialophora*, *Phialophora*, and *Cyphellophora* are susceptible to most currently used antifungals (Badali et al. 2009, 2011; Vitale et al. 2009). However, to our knowledge, no infection was reported in fish with these species, except for *Exophiala*.

To conclude, black fungi are not yet considered as clinically important zoonotic fungi. In fish, the main epidemiological situation is pseudoepidemics, i.e., opportunistic infection of a large host population due to a common source of stress. In intensive production context, management practices such as handling, crowding, transporting, fluctuating temperatures, poor and unstable water quality, and other husbandry deficiencies generate stress, which predisposes for overall microbial dysbiosis, which in turn creates biotic conditions favoring opportunistic diseases

(Boutin et al. 2012, 2013a, b). Therefore, it appears clearly enough that prophylaxis in fish should further target prevention of stress.

5.3.2 *Saprolegniasis*

Despite the common occurrence of epizootic ulcerative syndrome (EUS) in aquaculture, there are very few studies that have investigated its control. For example, lime or hydrated lime and salt were tested to minimize fish losses in infected fishponds (Lilley et al. 1998). Also, several plant-derived products such as ash, turmeric, neem seeds, and dried banana leaves were tested for prophylactic and therapeutic treatments, but have shown variable results (Campbell et al. 2001). Vaccination against the primary pathogen *A. invadans* is one of the most acceptable options to counteract the disease. Different antigenic preparations from *A. invadans* were evaluated as vaccine candidates in Indian carp (*Catla catla*), using respectively a fungal extract with and without adjuvant, and extracellular products (ECP) as antigens (Saikia and Kamilya 2012). Although in vivo tests of the three vaccine preparations in pathogen challenges showed a reduction of mortality in all the vaccinated groups, this reduction was not statistically significant. For saprolegniasis, no effective immunization is currently available, and nonspecific chemical agents such as malachite green and formalin have been, or will be, banned for use in aquaculture or wild settings, leading to dramatic reemergences of saprolegniasis. There is an urgent need to develop alternative preventive and curative strategies. To date, there are numerous general or more specific antifungal agents that were tested against *Saprolegnia*, such as chemicals, antibiotics, vaccines, humic substances, plant essential oils, and probiotics (Ali et al. 2014).

Exposure of chum salmon eggs to 1,3;1,6-β-D-glucans with a molecular mass of more than 2 kDa increased the survival of embryos and juveniles and their resistance to *Saprolegnia* infection by up to 2.5-fold when compared with the control chum salmons (Kiseleva et al. 2014). Three hours exposure of β-glucan suspension (10 and 15 mg l^{-1}) was observed to reduce the mortality of *Anabas testudineus* (climbing perch) spawns challenged with 3×10^5 CFU ml^{-1} of *Saprolegnia parasitica* (Das et al. 2013). Potassium permanganate ($KMnO_4$) and Freund's complete adjuvant (FCA) have been tested with tilapia (*O. niloticus*) challenged with *Saprolegnia ferax*. Treatment with $KMnO_4$ exhibited a protective role against oxidative stress response; furthermore FCA was observed to modulate the oxidative stress response and enhance fish immune response (Zahran and Risha 2013).

Antifungal activities of bronopol and 2-methyl-4-isothiazolin-3-one (MT) against members of the genus *Saprolegnia* were tested in vitro (Oono and Hatai 2007) and in vivo on rainbow trout brood stock (Branson 2002). Treatment with both molecules was effective in vitro to kill hyphae, vegetative cells and zoospores, at concentrations ranging from 20 to 200 mg l^{-1} for 30–60 min (Branson 2002; Oono and Hatai 2007). Overall, bronopol appears to qualify as a safe and effective replacement for malachite green and formalin in the prevention of fungal infections in the aquaculture environment.

Skin mucus is known to contain a number of nonspecific antimicrobial factors such as lysozymes, transferrin, and C-reactive protein, which are critically important in fish innate immunity (Van Muiswinkel and Nakao 2014). The majority of these factors are low molecular weight peptides or proteins, such as defensins, and are often active against a wide range of pathogenic microorganisms. Histone 2B-like protein isolated from channel catfish (*Ictalurus punctatus*) skin was inhibitory to *Saprolegnia* spp. These findings suggest that histones may be important defensive molecules in fish (Robinette et al. 1998).

As well as chemotherapeutics, several vaccine candidates have been proposed for *Saprolegnia* spp. However, given the low prevalence of antibodies that has been observed in *Saprolegnia*-infected trout (18.0 %) when compared to healthy relatives, it suggests possible immune suppression and the lack of an effective specific immune response in fish with saprolegniasis (Fregeneda-Grandes et al. 2009).

More recently, tests with endogenous probiotics were conducted in vivo in rainbow trout. Isolates identified as *Aeromonas sobria*, *Pantoea agglomerans*, *Pseudomonas fluorescens*, *Serratia fonticola*, *Xanthomonas retroflexus*, and *Yersinia kristensenii* exhibited significant antagonism against *Saprolegnia* in salmon eggs (Carbajal-González et al. 2011; Valenzuela et al. 2012; Carbajal-González et al. 2013). Some of these isolates are promising probiotic candidates for biological control of saprolegniasis. In their study, Liu and colleagues observed that lower incidence of saprolegniasis was strongly correlated with a high richness and abundance of specific commensal *Actinobacteria*, featuring genus *Frondihabitans* (*Microbacteriaceae*), which effectively inhibits attachment of *Saprolegnia* to salmon eggs (Liu et al. 2014). These results highlight that fundamental insights into factors impairing the microbial homeostasis of fish eggs may provide a new sustainable approach to mitigate emerging diseases.

5.3.3 Ichthyophonosis

Information on treatment and control methods for ichthyophonosis is very scarce. Some treatments, such as fenoxetol and paraclorofenoxetol or antibiotics, suggested for the first infection steps proved ineffective (Van Duijn 1956). Therefore, prophylactic measures, such as food pasteurization (McVicar 1982) or disinfection (Hershberger et al. 2008), were suggested. For example, chlorine and iodine solutions were successfully tested in vitro as spore-inactivating agents (Hershberger et al. 2008). Inactivation in seawater increased directly with halide concentration and exposure duration (1.5–13.3 ppm for 1–60 min) and at most iodine concentrations and exposure durations tested (5.9–10.7 ppm for 1–60 min). Among the available broad-spectrum antifungals against both superficial and systemic mycosis, the ketoconazole (KZ), an azole-derivative antifungal agent, demonstrated a significant inhibitory activity on *Ichthyophonus* growth in vitro (Chauvier and

Mortier-Gabet 1982; Franco-Sierra 1994; Hontoria et al. 2009). Azole antifungal action is based on the inhibition of the cytochrome P-450-dependent demethylation of lanosterol in fungi (Vanden Bossche et al. 1988). However, azoles also affect a number of mammalian cytochromes and thus can cause liver damage through mitochondrial activity inhibition (Rodriguez and Acosta 1996).

5.3.4 *Microsporidiosis*

Therapeutic drugs, such as antiprotozoals and antibiotics, are rarely successful against *Microsporidia*, and none are licensed for use in aquaculture. However, fumagillin was reported to be effective for controlling various microsporean infections in fish (Higgins et al. 1998) with most of the testing having been against *Loma salmonae* and *Nucleospora salmonis* in salmonids (Chinook salmon and rainbow trout). The compounds tested against *Loma salmonae* in juvenile rainbow trout were antibiotics (fumagillin) and antiprotozoals (quinine hydrochloride, sulfaquinoxaline, pyrimethamine, albendazole, amprolium, metronidazole), although of these only fumagillin (high or low dose) and albendazole have been shown to reduce the number of xenomas present 10 weeks after infection ($p < 0.01$) (Speare et al. 1999). Quinine treatment was effective at reducing xenomas when compared to control ones (57 % vs. 100 %) after 8 weeks of exposure to the pathogen ($p < 0.0001$), and branchial lesions did not appear in treated individuals until week 14 (Speare et al. 1998).

Recent findings have demonstrated that a protective cell-mediated immune response can be evoked against microsporidian parasites, opening the door to possible vaccine development (Rodriguez-Tovar et al. 2006). An intraperitoneal vaccine, consisting of a preparation of a low-virulence strain of *Loma salmonae*, was tested on *O. mykiss* (Speare et al. 2007). Fish receiving 10^3–10^5 killed spores had the best protection against experimental infection, with 85 % fewer xenomas in their gills than in the controls. Another intraperitoneal vaccine preparation with whole viable spores of the microsporidians *Glugea anomala* or *Glugea hertwigi* reduced the numbers of branchial xenomas by 80 % and 91 %, respectively, after a standard experimental infection of juvenile rainbow trout with the microsporidian *Loma salmonae*. Similar significant results were obtained when killed-spore preparations were used (Harkness et al. 2013). Interestingly, for *Loma salmonae*, a microsporidian gill parasite observed in rainbow trout, vaccinated trout were completely resistant to a *L. salmonae* challenge 6 weeks after vaccination (Rodríguez-Tovar et al. 2006).

5.4 Conclusion

"Clearly" enough, fish need to be reared under pristine conditions to avoid opportunistic diseases, including those caused by *Fungi*. Unfortunately, modern production cycles, relying on intensive techniques, often lead to poor water quality and high stocking densities. This combination is known as a catalyst for opportunistic infections in aquaculture, as suboptimal rearing conditions may lead to homeostatic imbalance (Barton and Iwama 1991) in favor of pathogenic agents (Boutin et al. 2013b; Stecher et al. 2013). Consequently, management practices have to consider the optimum conditions for parameters such as feeding rate, dissolved oxygen concentration, population density, and even temperature.

Several conservation and disease mitigation strategies have been proposed to control and prevent the spread of emerging fungal and fungal-like pathogens. In aquaculture, immunization and chemical control are among the preferred approaches to mitigate diseases (Bruno et al. 2011; van den Berg et al. 2013). In the last two decades, significant progress in the development of antifungal agents has been made, and the number of available antifungal drugs has increased by 30 % since 2000 (Foy and Trepanier 2010). Those that are available are controlled by legislation, but they are also used outside therapeutic boundaries, e.g., for prophylactic use or growth promotion.

However, prevention is the best medicine, and increased knowledge of basic biology and evolutionary relationships among such a huge diversity of fungal and fungal-like parasites will guide treatment and control methods. Additional control measures such as good husbandry, adequate feed composition, movement restrictions, immunostimulants, and biological control could contribute to reduced antimicrobial usage throughout the aquaculture industry.

To our best knowledge, no probiotic approach has been tested against fungal and fungal-like secondary invaders of fish. However, since opportunistic pathogens and parasites undergo a shift in virulence in response to host stress, the probiotic approach is worth testing, at least to prevent dysbiosis, and potentially for antagonism against such pathogens.

References

Adl SM, Simpson AGB, Farmer MA et al (2005) The new higher level classification of eukaryotes with emphasis on the taxonomy of protists. J Eukaryot Microbiol 52:399–451. doi:10.1111/j.1550-7408.2005.00053.x

Ajello L (1986) Hyalohyphomycosis and phaeohyphomycosis: two global disease entities of public health importance. Eur J Epidemiol 2:243–251

Ajello L, McGinnis MR, Camper J (1977) An outbreak of phaeohyphomycosis in rainbow trout caused by Scolecobasidium humicola. Mycopathologia 62:15–22

Ali SE, Thoen E, Evensen Ø, Skaar I (2014) Boric acid inhibits germination and colonization of Saprolegnia spores in vitro and in vivo. PLoS One 9:e91878. doi:10.1371/journal.pone.0091878

Andreou D, Gozlan RE, Stone D et al (2011) Sphaerothecum destruens pathology in cyprinids. Dis Aquat Organ 95:145–151. doi:10.3354/dao02356

Andreou D, Arkush KD, Guégan J-F, Gozlan RE (2012) Introduced pathogens and native freshwater biodiversity: a case study of Sphaerothecum destruens. PLoS One 7:e36998. doi:10.1371/journal.pone.0036998

Badali H, De Hoog GS, Curfs-Breuker I et al (2009) In vitro activities of eight antifungal drugs against 70 clinical and environmental isolates of Alternaria species. J Antimicrob Chemother 63:1295–1297. doi:10.1093/jac/dkp109

Badali H, de Hoog GS, Sudhadham M, Meis JF (2011) Microdilution in vitro antifungal susceptibility of Exophiala dermatitidis, a systemic opportunist. Med Mycol 49:819–824. doi:10.3109/13693786.2011.583285

Balajee SA, Sigler L, Brandt ME (2007) DNA and the classical way: identification of medically important molds in the 21st century. Med Mycol 45:475–490. doi:10.1080/13693780701449425

Baldauf SL (2003) The deep roots of eukaryotes. Science 300:1703–1706. doi:10.1126/science.1085544

Barton BA, Iwama GK (1991) Physiological changes in fish from stress in aquaculture with emphasis on the response and effects of corticosteroids. Annu Rev Fish Dis 1:3–26. doi:10.1016/0959-8030(91)90019-G

Bhattacharya U (1988) Scolebasidium humicola, a new fungal fish infection from India. Environ Ecol 6:532–3

Boutin S, Bernatchez L, Audet C, Derôme N (2012) Antagonistic effect of indigenous skin bacteria of brook charr (*Salvelinus fontinalis*) against *Flavobacterium columnare* and *F. psychrophilum*. Vet Microbiol 155:355–361. doi:10.1016/j.vetmic.2011.09.002

Boutin S, Audet C, Derome N (2013a) Probiotic treatment by indigenous bacteria decreases mortality without disturbing the natural microbiota of *Salvelinus fontinalis*. Can J Microbiol 59:662–670. doi:10.1139/cjm-2013-0443

Boutin S, Bernatchez L, Audet C, Derôme N (2013b) Network analysis highlights complex interactions between pathogen, host and commensal microbiota. PLoS One 8, e84772. doi:10.1371/journal.pone.0084772

Branson E (2002) Efficacy of bronopol against infection of rainbow trout (Oncorhynchus mykiss) with the fungus Saprolegnia species. Vet Rec 151:539–541

Bruno D, West PV, Beakes G (2011) Saprolegnia and other oomycetes. In: Woo P, Bruno D (eds) Fish diseases and disorders: viral, bacterial and fungal infections, 2nd edn. CABI, Wallingford, pp 669–720

Callinan RB, Fraser GC, Virgona JL (1989) Pathology of red spot disease in sea mullet, Mugil cephalus L., from eastern Australia. J Fish Dis 12:467–479. doi:10.1111/j.1365-2761.1989.tb00558.x

Callinan R, Paclibare J, Bondad-Reantaso M et al (1995) Aphanomyces species associated with epizootic ulcerative syndrome (EUS) in the Philippines and red spot disease (RSD) in Australia: preliminary comparative studies. Dis Aquat Organ 21:233–238. doi:10.3354/dao021233

Campbell RE, Lilley JH, Taukhid V et al (2001) In vitro screening of novel treatments for Aphanomyces invadans. Aquacult Res 32:223–233. doi:10.1046/j.1365-2109.2001.00551.x

Carbajal-González MT, Fregeneda-Grandes JM, Suárez-Ramos S et al (2011) Bacterial skin flora variation and in vitro inhibitory activity against Saprolegnia parasitica in brown and rainbow trout. Dis Aquat Organ 96:125–135. doi:10.3354/dao02391

Carbajal-González MT, Fregeneda-Grandes JM, González-Palacios C, Aller-Gancedo JM (2013) Adhesion to brown trout skin mucus, antagonism against cyst adhesion and pathogenicity to rainbow trout of some inhibitory bacteria against Saprolegnia parasitica. Dis Aquat Organ 104:35–44. doi:10.3354/dao02582

Chauvier G, Mortier-Gabet J (1982) Recherches sur les possibilités de traitement chimique de l'ichthyophonose. Rev Fr AquarolNancy 9:90

Das BK, Pattnaik P, Debnath C et al (2013) Effect of β-glucan on the immune response of early stage of Anabas testudineus (Bloch) challenged with fungus Saprolegnia parasitica. SpringerPlus 2:197. doi:10.1186/2193-1801-2-197
David JC, Kirk P (1997) Index of fungi. In: Ainsworth and Bisby's dictionary of the Fungi, 8th edn. CAB International, Wallingford, UK, p 706
De Hoog GS (1993) Evolution of black yeasts: possible adaptation to the human host. Antonie Van Leeuwenhoek 63:105–109
De Hoog GS, McGinnis MR (1987) Ascomycetous black yeasts. Stud Mycol 30:187–199
De Hoog GS, Vicente VA, Najafzadeh MJ et al (2011) Waterborne Exophiala species causing disease in cold-blooded animals. Persoonia 27:46–72. doi:10.3767/003158511X614258
Didier ES (2005) Microsporidiosis: an emerging and opportunistic infection in humans and animals. Acta Trop 94:61–76. doi:10.1016/j.actatropica.2005.01.010
Doty MS, Slater DW (1946) A new species of heterosporium pathogenic on young chinook salmon. Am Midl Nat 36:663. doi:10.2307/2421459
Egusa S, Masuda N (1971) A new fungal of Plecoglossus altivelis. Fish Pathol 6:41–46. doi:10.3147/jsfp.6.41
Faisal M, Elsayed E, Fitzgerald SD et al (2007) Outbreaks of phaeohyphomycosis in the chinook salmon (Oncorhynchus tshawytscha) caused by Phoma herbarum. Mycopathologia 163:41–48. doi:10.1007/s11046-006-0084-z
Ferrer C, Colom F, Frasés S, Mulet E, Abad JL, Alió JL (2001) Detection and identification of fungal pathogens by PCR and by ITS2 and 5.8S ribosomal DNA typing in ocular infections. J Clin Microbiol 39:2873–2879. doi:10.1128/JCM.39.8.2873-2879.2001
Firouzbakhsh F, Ebrahimzadeh Mousavi H, Khosravi A (2005) Isolation and identification of pathogenic and saprophytic fungi from gill lesion in cultivated cyprinids (common carp, silver carp and grass carp). J Vet Res 60:15–9
Foy DS, Trepanier LA (2010) Antifungal treatment of small animal veterinary patients. Vet Clin North Am Small Anim Pract 40:1171–1188. doi:10.1016/j.cvsm.2010.07.006
Franco-Sierra A (1994) Estudio de las infecciones por el hongo Ichthyophonus sp. en peces de interés comercial. Universitat Autónoma de Barcelona
Fraser GC, Callinan RB, Calder LM (1992) Aphanomyces species associated with red spot disease: an ulcerative disease of estuarine fish from eastern Australia. J Fish Dis 15:173–181. doi:10.1111/j.1365-2761.1992.tb00651.x
Fregeneda-Grandes JM, Carbajal-González MT, Aller-Gancedo JM (2009) Prevalence of serum antibodies against Saprolegnia parasitica in wild and farmed brown trout Salmo trutta. Dis Aquat Organ 83:17–22. doi:10.3354/dao01999
Gaskins JE, Cheung PJ (1986) Exophiala pisciphila. A study of its development. Mycopathologia 93:173–184
Gill EE, Fast NM (2006) Assessing the microsporidia-fungi relationship: combined phylogenetic analysis of eight genes. Gene 375:103–109. doi:10.1016/j.gene.2006.02.023
Gjessing MC, Davey M, Kvellestad A, Vrålstad T (2011) Exophiala angulospora causes systemic inflammation in atlantic cod Gadus morhua. Dis Aquat Organ 96:209–219. doi:10.3354/dao02381
Gozlan RE, St-Hilaire S, Feist SW et al (2005) Biodiversity: disease threat to European fish. Nature 435:1046–1046. doi:10.1038/4351046a
Gozlan RE, Peeler EJ, Longshaw M et al (2006) Effect of microbial pathogens on the diversity of aquatic populations, notably in Europe. Microbes Infect 8:1358–1364. doi:10.1016/j.micinf.2005.12.010
Grünwald NJ, Werres S, Goss EM et al (2012) Phytophthora obscura sp. nov., a new species of the novel Phytophthora subclade 8d. Plant Pathol 61:610–622. doi:10.1111/j.1365-3059.2011.02538.x
Harkness JE, Guselle NJ, Speare DJ (2013) Demonstrated efficacy of a pilot heterologous whole-spore vaccine against Microsporidial gill disease in rainbow trout. Clin Vaccine Immunol 20:1483–1484. doi:10.1128/CVI.00340-13

Hatai K, Hoshiai G (1992) Mass mortality in cultured coho salmon (Oncorhynchus kisutch) due to Saprolegnia parasitica coker. J Wildl Dis 28:532–536. doi:10.7589/0090-3558-28.4.532
Hatai K, Egusa S, Takahashi S, Ooe K (1977) Study on the pathogenic fungus of mycotic granulomatosis—I. Fish Pathol 12:129–133. doi:10.3147/jsfp.12.129
Hershberger PK, Pacheco CA, Gregg JL (2008) Inactivation of *Ichthyophonus* spores using sodium hypochlorite and polyvinyl pyrrolidone iodine. J Fish Dis 31:853–858. doi:10.1111/j.1365-2761.2008.00959.x
Hibbett DS, Binder M, Bischoff JF et al (2007) A higher-level phylogenetic classification of the Fungi. Mycol Res 111:509–547. doi:10.1016/j.mycres.2007.03.004
Higgins MJ, Kent ML, Moran JD et al (1998) Efficacy of the fumagillin analog TNP-470 for Nucleospora salmonis and Loma salmonae infections in chinook salmon Oncorhynchus tshawytscha. Dis Aquat Organ 34:45–49. doi:10.3354/dao034045
Hontoria F, González MA, Sitjà-Bobadilla A et al (2009) Ketoconazole inhibits the growth and development of Ichthyophonus sp. (Mesomycetozoa) in vitro. J Eukaryot Microbiol 56:484–491. doi:10.1111/j.1550-7408.2009.00427.x
Jacobson ES (2000) Pathogenic roles for fungal melanins. Clin Microbiol Rev 13:708–717
James TY, Kauff F, Schoch CL et al (2006) Reconstructing the early evolution of Fungi using a six-gene phylogeny. Nature 443:818–822. doi:10.1038/nature05110
Keeling PJ, Luker MA, Palmer JD (2000) Evidence from beta-tubulin phylogeny that microsporidia evolved from within the fungi. Mol Biol Evol 17:23–31
Kent ML, Speare DJ (2005) Review of the sequential development of Loma salmonae (Microsporidia) based on experimental infections of rainbow trout (Oncorhynchus mykiss) and Chinook salmon (O. tshawytscha). Folia Parasitol (Praha) 52:63–68
Kiryu Y, Shields JD, Vogelbein WK et al (2003) Infectivity and pathogenicity of the oomycete Aphanomyces invadans in Atlantic menhaden Brevoortia tyrannus. Dis Aquat Organ 54:135–146. doi:10.3354/dao054135
Kiseleva M, Balabanova L, Elyakova L et al (2014) Effect of treatment of chum salmon Oncorhynchus keta (Walbaum) eggs with 1,3;1,6-β-D-glucans on their development and susceptibility to Saprolegnia infection. J Fish Dis 37:3–10. doi:10.1111/jfd.12043
Kocan R, Hershberger P (2006) Differences in Ichthyophonus prevalence and infection severity between upper Yukon River and Tanana River chinook salmon, Oncorhynchus tshawytscha (Walbaum), stocks. J Fish Dis 29:497–503. doi:10.1111/j.1365-2761.2006.00743.x
Kurata O, Munchan C, Wada S et al (2008) Novel *Exophiala* infection involving ulcerative skin lesions in Japanese flounder *Paralichthys olivaceus*. Fish Pathol 43:35–44. doi:10.3147/jsfp.43.35
Langdon J, McDonald W (1987) Cranial Exophiala pisciphila infection in Salmo salar in Australia. Bull Eur Assoc Fish Pathol 7:35–7
Lilley JH, Callinan R, Chinabut S et al (1998) Epizootic ulcerative syndrome (EUS) technical handbook. Aquatic Animal Health Research Institute, Bangkok
Liu YJ, Hodson MC, Hall BD (2006) Loss of the flagellum happened only once in the fungal lineage: phylogenetic structure of Kingdom Fungi inferred from RNA polymerase II subunit genes. BMC Evol Biol 6:74. doi:10.1186/1471-2148-6-74
Liu Y, de Bruijn I, Jack AL et al (2014) Deciphering microbial landscapes of fish eggs to mitigate emerging diseases. ISME J. doi:10.1038/ismej.2014.44
Lom J, Dyková I (2005) Microsporidian xenomas in fish seen in wider perspective. Folia Parasitol (Praha) 52:69–81
Malathi K, Rajendran K (2012) Isolation of fungi and bacteria from various tissues of freshwater fish *Channa Punctatus* (BLOCH). Asian J Microbiol Biotechnol Environ Sci 14:553–556
McGinnis MR (1983) Chromoblastomycosis and phaeohyphomycosis: new concepts, diagnosis, and mycology. J Am Acad Dermatol 8:1–16
McGinnis MR, Ajello L (1974) A new species of Exophiala isolated from channel catfish. Mycologia 66:518–520

McKenzie RA, Hall WTK (1976) Dermal ulceration of mullet (mugil Cephalus). Aust Vet J 52:230–231. doi:10.1111/j.1751-0813.1976.tb00076.x

McVicar AH (1982) *Ichthyophonus* infections in fish. Microbial Dis Fish 243–269

McVicar AH (1999) Ichthyophonus and related organisms. In: Woo PTK, Bruno DW (eds) Fish diseases and disorders: viral, bacterial and fungal infections. CAB, Wallingford

Mendoza L, Taylor JW, Ajello L (2002) The class mesomycetozoea: a heterogeneous group of microorganisms at the animal-fungal boundary. Annu Rev Microbiol 56:315–344. doi:10.1146/annurev.micro.56.012302.160950

Mohamed A, Refai MK (2010) The assessment of mycotic settlement of freshwater fishes in Egypt. J Am Sci 6:595–602

Nyaoke A, Weber ES, Innis C et al (2009) Disseminated phaeohyphomycosis in weedy seadragons (Phyllopteryx taeniolatus) and leafy seadragons (Phycodurus eques) caused by species of Exophiala, including a novel species. J Vet Diagn Invest 21:69–79

Okamura B, Feist SW (2011) Emerging diseases in freshwater systems. Freshw Biol 56:627–637. doi:10.1111/j.1365-2427.2011.02578.x

Oono H, Hatai K (2007) Antifungal activities of bronopol and 2-methyl-4-isothiazolin-3-one (MT) against saprolegnia. Biocontrol Sci 12:145–148

Otis EJ, Wolke RE, Blazer VS (1985) Infection of Exophiala salmonis in Atlantic salmon (Salmo salar L.). J Wildl Dis 21:61–64

Paps J, Medina-Chacón LA, Marshall W et al (2013) Molecular phylogeny of unikonts: new insights into the position of apusomonads and ancyromonads and the internal relationships of opisthokonts. Protist 164:2–12. doi:10.1016/j.protis.2012.09.002

Ragan MA, Goggin CL, Cawthorn RJ et al (1996) A novel clade of protistan parasites near the animal-fungal divergence. Proc Natl Acad Sci USA 93:11907–11912

Reuter R, Ham W, Davis S (2003) Exophiala sp. infection in captured King George whiting, Sillaginodes punctata. Bull Eur Assoc Fish Pathol 23:128–34

Roberts RJ, Frerichs GN, Tonguthai K, Chinabut S (1994) Epizootic ulcerative syndrome of farmed and wild fishes—a review. In: Roberts RJ, Muir JF (eds) Recent advances in aquaculture, vol 5. Blackwell, Oxford, pp 207–239

Robinette D, Wada S, Arroll T, Levy MG, Miller WL, Noga EJ (1998) Antimicrobial activity in the skin of the channel catfish *Ictalurus punctatus*: characterization of broad-spectrum histone-like antimicrobial proteins. Cell Mol Life Sci 54:467–475

Rodriguez RJ, Acosta D (1996) Inhibition of mitochondrial function in isolated rat liver mitochondria by azole antifungals. J Biochem Toxicol 11:127–131. doi:10.1002/(SICI)1522-7146(1996)11:3<127::AID-JBT4>3.0.CO;2-M

Rodríguez-Tovar LE, Becker JA, Markham RJF, Speare DJ (2006) Induction time for resistance to microsporidial gill disease caused by Loma salmonae following vaccination of rainbow trout (Oncorhynchus mykiss) with a spore-based vaccine. Fish Shellfish Immunol 21:170–175. doi:10.1016/j.fsi.2005.11.009

Rodriguez-Tovar LE, Markham RJF, Speare DJ, Sheppard J (2006) Cellular immunity in salmonids infected with the microsporidial parasite Loma salmonae or exposed to non-viable spores. Vet Immunol Immunopathol 114:72–83. doi:10.1016/j.vetimm.2006.07.006

Ross AJ, Yasutake WT (1973) Scolecobasidium humicola, a fungal pathogen of fish. J Fish Res Board Can 30:994–995. doi:10.1139/f73-161

Ruibal C, Gueidan C, Selbmann L et al (2009) Phylogeny of rock-inhabiting fungi related to Dothideomycetes. Stud Mycol 64:123–133S7. doi:10.3114/sim.2009.64.06

Saikia D, Kamilya D (2012) Immune responses and protection in catla (Catla catla) vaccinated against epizootic ulcerative syndrome. Fish Shellfish Immunol 32:353–359. doi:10.1016/j.fsi.2011.11.030

Sarowar MN, van den Berg AH, McLaggan D et al (2013) Saprolegnia strains isolated from river insects and amphipods are broad spectrum pathogens. Fungal Biol 117:752–763. doi:10.1016/j.funbio.2013.09.002

Schaumann K, Priebe K (1994) Ochroconis humicola causing muscular black spot disease of Atlantic salmon (Salmo salar). Can J Bot 72:1629–1634. doi:10.1139/b94-200
Silphaduang U, Hatai K, Wada S, Noga E (2000) Cladosporiosis in a tomato clownfish (Amphiprion frenatus). J Zoo Wildl Med 31:259–261
Spanggaard B, Skouboe P, Rossen L, Taylor JW (1996) Phylogenetic relationships of the intercellular fish pathogen Ichthyophonus hoferi and fungi, choanoflagellates and the rosette agent. Mar Biol 126:109–115. doi:10.1007/BF00571382
Speare DJ, Ritter G, Schmidt H (1998) Quinine hydrochloride treatment delays xenoma formation and dissolution in rainbow trout challenged with *Loma salmonae*. J Comp Pathol 119:459–465
Speare DJ, Athanassopoulou F, Daley J, Sanchez JG (1999) A preliminary investigation of alternatives to fumagillin for the treatment of Loma salmonae infection in rainbow trout. J Comp Pathol 121:241–248. doi:10.1053/jcpa.1999.0325
Speare DJ, Markham RJF, Guselle NJ (2007) Development of an effective whole-spore vaccine to protect against microsporidial gill disease in rainbow trout (Oncorhynchus mykiss) by using a low-virulence strain of Loma salmonae. Clin Vaccine Immunol 14:1652–1654. doi:10.1128/CVI.00365-07
Stecher B, Maier L, Hardt W-D (2013) "Blooming" in the gut: how dysbiosis might contribute to pathogen evolution. Nat Rev Microbiol 11:277–284. doi:10.1038/nrmicro2989
Strongman D, Morrison C, McClelland G (1997) Lesions in the musculature of captive American plaice Hippoglossoides platessoides caused by the fungus Hormoconis resinae (Deuteromycetes). Dis Aquat Organ 28:107–113. doi:10.3354/dao028107
Torruella G, Derelle R, Paps J et al (2012) Phylogenetic relationships within the Opisthokonta based on phylogenomic analyses of conserved single-copy protein domains. Mol Biol Evol 29:531–544. doi:10.1093/molbev/msr185
Valenzuela A, Campos V, Yañez F et al (2012) Application of artificial photoperiod in fish: a factor that increases susceptibility to infectious diseases? Fish Physiol Biochem 38:943–950. doi:10.1007/s10695-011-9580-2
Van den Berg AH, McLaggan D, Diéguez-Uribeondo J, van West P (2013) The impact of the water moulds Saprolegnia diclina and Saprolegnia parasitica on natural ecosystems and the aquaculture industry. Fungal Biol Rev 27:33–42. doi:10.1016/j.fbr.2013.05.001
Van Duijn C (1956) Diseases of fishes, 1st edn. Water Life, London, UK
Vanden Bossche H, Marichal P, Gorrens J et al (1988) Mode of action studies: basis for the search of new antifungal drugs. Ann N Y Acad Sci 544:191–207. doi:10.1111/j.1749-6632.1988.tb40404.x
Van Muiswinkel WB, Nakao M (2014) A short history of research on immunity to infectious diseases in fish. Dev Comp Immunol 43:130–150. doi:10.1016/j.dci.2013.08.016
Vishwanath TS (1997) Mycotic granulomatosis and seasonality are the consistent features of epizootic ulcerative syndrome of fresh and brackish water fishes of Karnataka, India. Asian Fish Sci 10:155–60
Vitale RG, Perez-Blanco M, De Hoog GS (2009) In vitro activity of antifungal drugs against Cladophialophora species associated with human chromoblastomycosis. Med Mycol 47:35–40. doi:10.1080/13693780802566333
Voigt K, Marano AV, Gleason F (2013) Ecological and economical importance of parasitic and zoosporic true Fungi. In: Kempken F (ed) The Mycota vol XI: agricultural applications, 2nd edn. Springer, Berlin
Wang H, Xu Z, Gao L, Hao B (2009) A fungal phylogeny based on 82 complete genomes using the composition vector method. BMC Evol Biol 9:195. doi:10.1186/1471-2148-9-195
Willoughby LG (1994) Fungi and fish diseases. Pisces, Stirling
Willoughby LG, Roberts RJ, Chinabut S (1995) Aphanomyces invaderis sp. nov., the fungal pathogen of freshwater tropical fish affected by epizootic ulcerative syndrome. J Fish Dis 18:273–276. doi:10.1111/j.1365-2761.1995.tb00302.x
Yanong RPE (2003) Fungal diseases of fish. Vet Clin North Am Exot Anim Pract 6:377–400

Zahran E, Risha E (2013) Protective role of adjuvant and potassium permanganate on oxidative stress response of Nile tilapia (Oreochromis niloticus) challenged with Saprolegnia ferax. SpringerPlus 2:94. doi:10.1186/2193-1801-2-94

Zeng JS, Sutton DA, Fothergill AW, Rinaldi MG, Harrak MJ, de Hoog GS (2007) Spectrum of clinically relevant *Exophiala* species in the United States. J Clin Microbiol 45:3713–3720. doi:10.1128/JCM.02012-06

Chapter 6
Opportunistic Pathogens of Marine Mammals

Stephanie Venn-Watson

Abstract Despite the ocean's vast extent, knowledge of opportunistic pathogens in the marine environment has been limited. Marine mammal health is likely tied to the ocean's health, and proposed causes of increased host susceptibilities to opportunistic infections include climate change, decreased prey availability, and contaminant exposures. Known or possible marine mammal opportunistic infections include bacteria (marine *Brucella* species, *Pseudomonas aeruginosa*, and *Staphylococcus aureus*), fungi (*Aspergillus* species and *Lacazia loboi*), parasites (*Cryptosporidium parvum*, *Giardia duodenalis*, *Toxoplasma gondii*), and viruses (herpesviruses). Understanding how commensals become pathogens in marine mammals may improve our recognition of changing ocean health and its impacts on mammalian hosts.

6.1 Introduction

Marine mammals, including whales, dolphins, sea lions, seals, manatees, sea otters, and polar bears, acquire bacterial, mycotic, parasitic, and viral infections (Dubey et al. 2003; Gaydos et al. 2008; Higgins 2000; Munn 2006; Raga et al. 1997; Sweeney and Ridgway 1975). Reports on marine mammal microbes vary from incidental isolation of a bacterium species to discoveries of novel viruses causing mass mortality events. There is increasing evidence that pathogens readily move between terrestrial and marine environments, and a myriad of pathogens once considered terrestrial are now routinely isolated from marine mammals (Higgins 2000; Munn 2006; Waltzek et al. 2012).

There appears to be an overall rise of diseases in the ocean attributed to heavier coastal biological and chemical contaminant loads, decreased habitats, climate

S. Venn-Watson (✉)
Translational Medicine and Research Program, National Marine Mammal Foundation, San Diego, CA, USA
e-mail: stephanie.venn-watson@nmmpfoundation.org

© Springer International Publishing Switzerland 2016
C.J. Hurst (ed.), *The Rasputin Effect: When Commensals and Symbionts Become Parasitic*, Advances in Environmental Microbiology 3,
DOI 10.1007/978-3-319-28170-4_6

change, and decreased prey (Bossart 2006; Burek et al. 2008; Gulland and Hall 2007; Harvell et al. 2004; Miller et al. 2002; Ross 2002). In the Arctic, for example, when sea ice decreases drastically, so does the haul out area for ice seals (Harwood 2001). This decrease in habitat increases the concentration and variety of marine mammals in one place, in turn increasing opportunities for microbial exposure, physiological stress, and opportunistic infections (Burek et al. 2008; Kovacs et al. 2011). Understanding more about the epidemiology and pathophysiology of marine mammal opportunistic infections can provide valuable comparative insight into how diseases emerge and move between terrestrial and marine environments, how commensals evolve in land and water, and what impacts environmental changes have on host susceptibility to infectious diseases.

6.2 Host Susceptibility

Recognition of opportunistic pathogens is dependent upon knowing which microbes are commensals, how and when a host is susceptible to infection, and if a host without this susceptibility would otherwise be spared of illness from a commensal microbe's exposure. While there is some knowledge of marine mammal immunity, we are still far from full understanding of their normal immune response, changing immunity with age and reproduction status, and complex interactions among host, pathogens, and the environment. There are a few conditions in marine mammals, however, that are known or are likely to increase the susceptibility of some marine mammals to opportunistic pathogens. These conditions include morbillivirus infection, chemical contaminant exposure, advanced age, and late-term pregnancy.

Morbilliviruses are highly contagious paramyxoviruses that infect marine mammals and cause large-scale die-offs, primarily in cetaceans (bottlenose dolphins, striped dolphins, and long-finned pilot whales) and phocids (Baikal seals, grey seals, and harbor seals) (Di Guardo et al. 2013; Domingo et al. 1992; Fernandez et al. 2008; Lipscomb et al. 1996; Osterhaus 1989). Morbilliviruses that infect marine mammals include canine distemper virus, phocine distemper virus, dolphin morbillivirus, and porpoise morbillivirus (Saliki et al. 2002). Cetaceans and phocids with morbillivirus infections often die either from acute viral pneumonia, viral encephalitis, or infections secondary to pan-lymphoid depletion (Kennedy 1998). Opportunistic pathogens reported in marine mammals with morbillivirus infections include *Aspergillus fumigatus*, herpesvirus, *Staphylococcus aureus*, and *Toxoplasma gondii* (Di Guardo et al. 2013; Domingo et al. 1992; Lipscomb et al. 1996; Osterhaus 1989). Large-scale marine mammal epizootics caused by morbillivirus, which can last months over large geographic areas, continue to occur; as recently as 2013–2014, an Atlantic bottlenose dolphin unusual mortality event was caused by dolphin morbillivirus and involved much of the US east coast (T. Rowles, personal communication). With each of these morbillivirus epizootics, concurrent resurgence of opportunistic infections is expected.

Exposures to chemical contaminants, including dichlorodiphenyltrichloroethane, organochlorine pesticides, polybrominated diphenyl ethers, and polychlorinated biphenyls, released into the ocean from the terrestrial environment, have been well documented in marine mammals globally (Aguilar et al. 2002; Rotander et al. 2012). Contaminant burdens, especially in cetaceans, can vary based upon geographic location, proximity to the coast and known high-exposure areas, age, and sex (Ross et al. 2000; Fair et al. 2007). Contaminant loads in marine mammals can increase with age due to bioaccumulation, especially in blubber, and offloading has been associated with lactation as contaminants are passed to the fetus (Norstrom and Muir 1994). As such, it is not unusual for older males to have the highest contaminant loads (Ross et al. 2000; Fair et al. 2007).

Associations between contaminants and impaired immunity in marine mammals have been reported in multiple studies. Harbor seals fed organochlorine-contaminated fish developed immunotoxic responses, including suppressed natural killer cell and T-cell activity (de Swart et al. 1996). In vitro experiments exposing peripheral blood mononuclear cells from marine mammals, including bottlenose dolphins, Dall's porpoises, and a California sea lion, to butyltin compounds led to suppressed immune cell proliferation (Nakata et al. 2002). Bottlenose dolphins living off the US Georgia coast with higher levels of polychlorinated biphenyls had decreased T-lymphocyte proliferation, and incidental exposures of bottlenose dolphins to high levels of polycyclic aromatic hydrocarbons from wildfire smoke demonstrated apparent changes in neutrophil availability (Schwacke et al. 2012; Venn-Watson et al. 2013).

Due to high levels of contaminants documented in the environment and marine mammals, paired with demonstrated immune changes associated with contaminant exposures, immunotoxic contaminants have been proposed as important underlying causes of emerging infectious diseases, including opportunistic infections, in marine mammals (Ross 2002). Examples may include the high prevalence of *Lacazia loboi* infections in Florida's Indian River Lagoon dolphins and increased bacterial infections in harbor porpoises (Beineke et al. 2005; Hammond et al. 2005; Lahvis et al. 1995; Murdoch et al. 2008).

Marine mammals in managed collections are increasingly living to advanced age. Geriatric marine mammal populations are improving our knowledge of age-associated susceptibilities to opportunistic infections. The mean age at death of a healthy, free-ranging bottlenose dolphin population is approximately 20 years old (Wells et al. 2013). Similar to older humans, geriatric dolphins aging from 30 to 50 years old can develop chronic hyperlipidemia and chronic inflammation (Venn-Watson et al. 2011). Additionally, older dolphins have a higher risk of developing pneumonia from opportunistic fungal and bacterial infections; increasing susceptibility to opportunistic pathogens with advanced age, especially pneumonia, also occurs in humans (Fein 1999; Venn-Watson et al. 2013). Pneumonia-associated pathogens in dolphins include *Staphylococcus aureus*, *Pseudomonas aeruginosa*, and *Cryptococcus neoformans* (Venn-Watson et al. 2012b). Further understanding of how advanced age increases marine mammal susceptibilities to opportunistic infections may help identify means to decrease morbidity within this group.

Similar to cattle and many other mammals, late-term pregnancy can increase small cetaceans' susceptibilities to *Brucella*-associated abortions (Carmichael and Kenney 1968; Garin-Bastuji et al. 1998; Samartino and Enright 1993; Miller et al. 1999). Increased erythritol levels during late pregnancy were initially believed to encourage *Brucella* growth, but that hypothesis was disproven, and the reasons for third-trimester brucellosis remain unknown (Samartino and Enright 1993; Smith et al. 1962). *Brucella* is a common microbe in marine mammals, and its role as a late-term pregnancy opportunist will be further discussed in Sect. 6.3.

As with terrestrial mammals, there are a myriad of factors that can negatively impact host immune status in marine mammals, leading to increased susceptibilities to opportunistic infections. Detailed summaries of known and suspected marine mammal opportunistic pathogens are provided in subsequent sections.

6.3 Bacterial Pathogens

Marine *Brucella* species Possibly the most widely published microbe in marine mammals is *Brucella*, an intracellular facultative bacterium. *Brucella* species, including *B. ceti* and *B. pinnipedalis*, are common microbes to which marine mammals have evidence of exposure globally (Hernandez-Mora et al. 2013; Foster et al. 2007; Jepson et al. 1997; Guzman-Verri et al. 2012). In small cetaceans, especially dolphins, brucellosis can cause sporadic, late-term fetal losses in dolphins (Foster et al. 2007; Guzman-Verri et al. 2012; Hernandez-Mora et al. 2013; Jepson et al. 1997; Meegan et al. 2012; Miller et al. 1999). This disease manifestation is similar to late-term abortions in multiple terrestrial species, including cattle, goats, sheep, bison, and dogs (Carmichael and Kenney 1968; Garin-Bastuji et al. 1998; Samartino and Enright 1993). Other clinical manifestations of brucellosis in cetaceans include meningoencephalitis, osteomyelitis, and lung lesions (Dagleish et al. 2006, 2008; Goertz et al. 2011; Gonzalez et al. 2002). Despite the high prevalence of exposure to *Brucella* in marine mammals, its clinical relevance outside of small cetaceans appears to be limited, and to date, novel introduction of *Brucella* or point-source population exposures have not been identified as causes of epizootics in marine mammals.

Brucellosis is not typically considered an opportunistic infection. It does, however, take advantage of the late-term pregnancy state in small cetaceans, and *Brucella*-associated fetal losses can be secondary to poor maternal health, such as bison with bronchointerstitial pneumonia (Ryan et al. 2001). Further, compromises in targeted aspects of the mammalian innate and adaptive immune system, including T cells and gamma interferon, may increase an animal's susceptibility to opportunistic *Brucella* infection, including reactivation of quiescent infections (He et al. 2001; Zhan and Cheers 1993). Thus, due to the high prevalence of

Brucella in the marine environment, changing host susceptibilities could be potential drivers for population-wide increases in brucellosis.

Pseudomonas aeruginosa. *Pseudomonas aeruginosa* is an opportunistic pathogen commonly found in terrestrial and marine environments, and the open ocean has been proposed as a reservoir (Khan et al. 2007; Lyczak et al. 2000). *Pseudomonas aeruginosa* has been isolated while screening free-ranging marine mammals, as well as those in rehabilitation centers (Buck et al. 2006; Khan et al. 2008; Morris et al. 2011; Thornton et al. 1998; Vedros et al. 1982). Marine mammal species from which *P. aeruginosa* have been identified include bottlenose dolphins, California sea lions (*Zalophus californianus*), and northern fur seals (*Callorhinus ursinus*). Similar to humans, however, clinical disease associated with *P. aeruginosa* infections has been reported, including pneumonia in dolphins (Diamond et al. 1979; Venn-Watson et al. 2008, 2012a, b).

Staphylococcus aureus. *Staphylococcus aureus* can cause disease in marine mammals, including nephritis, myocarditis, pneumonia, and septicemia (Ketterer 1974; Power and Murphy 2002; Siebert et al. 2002; Venn-Watson et al. 2008, 2012a, b). While *S. aureus* has also been identified during screening studies in free-ranging and rehabilitation marine mammals, its prevalence in several studies was low; as such, it is not yet clear whether *S. aureus* is a commensal or a primary pathogen in marine mammals (Buck et al. 2006; Johnson et al. 1998; Morris et al. 2011; Thornton et al. 1998; Vedros et al. 1982).

Of increasing interest in both terrestrial and marine environments is the emergence of community-acquired methicillin-resistant *Staphylococcus aureus* (MRSA). Perhaps not surprisingly, human epidemic clones of MRSA have been reported in marine mammals from managed collections that have routine contact with humans (Faires et al. 2009). Long-term, 15-month colonization by the Canadian epidemic MRSA strain (CMRSA)2 (USA100) was identified in at least one of these bottlenose dolphins. More surprising has been the isolation of MRSA from wild, free-ranging bottlenose dolphins, such as those living in the estuarine waters of Charleston, South Carolina (Schaefer et al. 2009). *Staphylococcus aureus*, including MRSA, may be highly prevalent in seawater (Goodwin et al. 2012; Soge et al. 2009). Understanding the epidemiology and molecular profiles of *S. aureus* and MRSA in wild marine mammals can help determine whether these bacteria are successfully moving from humans and terrestrial animals directly into the ocean, if they independently develop in the ocean, and if they are emerging as opportunistic or primary infections in marine mammals.

6.4 Fungal Pathogens

***Aspergillus* species** *Aspergillus fumigatus* is an opportunistic fungal infection in humans that has been increasing in prevalence and severity, presumably due to increasing numbers of susceptible populations (Pfaller and Diekema 2004). Gliotoxin, a mycotoxin from *A. fumigatus*, has immunosuppressive effects by increasing monocyte cell death (Stanzani et al. 2005). Thus, while *A. fumigatus* is an opportunistic pathogen, it may in turn increase an animal's susceptibility to more severe infections. Marine strains of *A. fumigatus* survive in the ocean, can be concentrated in shellfish, and may have increased toxicity due to aggravation of seawater and subsequent higher excretion of gliotoxin (Grovel et al. 2003; Kerazon et al. 2008).

Aspergillus fumigatus has been reported as an opportunistic infection in cetaceans with immunosuppressive morbillivirus infections (Domingo et al. 1992; Lipscomb et al. 1996). *Aspergillus fumigatus* can cause encephalitis, tracheitis, otitis, and pneumonia in cetaceans, including bottlenose dolphins, harbor porpoises (*Phocoena phocoena*), and northern bottlenose whales (*Hyperoodon ampullatus*) (Dagleish et al. 2006, 2008; Delaney et al. 2012; Joseph et al. 1986; Prahl et al. 2011; Reidarson et al. 1998; Seibel et al. 2010; Zagzebski et al. 2006). Given the presence of terrestrial and marine strains of *A. fumigatus*, understanding whether or not these strains are exclusive to each environment, sources of marine *A. fumigatus* and differences in pathogenicity are of interest for this opportunistic fungal pathogen.

Lacazia loboi. Lobomycosis is an endemic fungal infection among humans in South America and Central America. The prevalence of this disease can be as high as 8.5 % in some remote populations (Rodriquez-Toro 1993), but only one human case has been reported in the United States (Burns et al. 2000). Among all terrestrial and marine animals, lobomycosis is known to occur only in humans and bottlenose dolphins (Haubold et al. 2000). The marine environment has been established as a likely habitat for *L. loboi* and a reservoir for infection (Burmudez et al. 2009).While there has been at least one case of potential transmission of *L. loboi* from a dolphin to a human, comparative morphology of this organism between dolphins and humans indicates that they may not be identical (Symmers 1983; Haubold et al. 2000).

Lobomycosis is an emerging disease among bottlenose dolphins living in the Indian River Lagoon in Florida, United States (Durden et al. 2009). There is a high prevalence of lobomycosis among these animals (17–30 %), and it is endemic in the south compared to the north Indian River Lagoon (Murdoch et al. 2008; Reif et al. 2006). Immune function testing indicates that dolphins with lobomycosis have immune dysfunction compared to dolphins without lobomycosis (Reif et al. 2009). There is a need to understand why lobomycosis is emerging in dolphins living in the Southeast United States to explain host susceptibilities or pathogenic changes occurring in a potentially zoonotic disease along the US coastline.

In addition, there would be much to benefit from understanding why humans and bottlenose dolphins are the only known hosts for *L. loboi*, despite diverse differences in marine and terrestrial habitats. Identifying common host susceptibilities to an infectious pathogen may open doors to previously unidentified physiological characteristics, therapeutic approaches, and the comparative value to humans by researching diseases in bottlenose dolphins.

6.5 Parasitic Pathogens

Cryptosporidium parvum* and *Giardia duodenalis Within the past decade, numerous studies have demonstrated the presence of *Cryptosporidium parvum* and *Giardia* (especially *G. duodenalis*) in marine mammals, raising questions regarding marine mammals' potential role as vectors for these opportunistic pathogens (Appelbee et al. 2005; Bogomolni et al. 2008). *Cryptosporidium parvum* and *Giardia* have been reported in multiple species, including bowhead whales (*Balaena mysticetus*), bearded seals (*Erignathus barbatus*), California sea lions, harbor seals (*Phoca vitulina richardsi*), North Atlantic right whales (*Eubalaena glacialis*), and ringed seals (*Phoca hispida*) (Deng et al. 2000; Dixon et al. 2008; Gaydos et al. 2008; Hughes-Hanks et al. 2005). In part due to the discovery of *G. duodenalis* in many marine mammal species, it appears that there is a higher genetic diversity for this parasite than once thought (Lasek-Nesselquist et al. 2008, 2010). Interestingly, there are no clear reports of clinical illness in marine mammals due to either *C. parvum* or *G. duodenalis*. Thus, the significance of understanding these opportunistic pathogens in the marine environment may be more relevant to human health than marine mammal health.

Toxoplasma gondii. *Toxoplasma gondii* is an opportunistic protozoan parasite that infects many species (Hill et al. 2005). Its best known roles as an opportunist in humans are during (1) pregnancy-associated novel exposures leading to congenital toxoplasmosis and (2) reactivation of latent infections in people with immunocompromised conditions, causing severe outcomes, including disseminated toxoplasmosis (Arnold et al. 1997; Kravetz and Federman 2005). While felids are the definite hosts of *T. gondii*, infective oocysts can survive well in seawater and shellfish, enabling *T. gondii* to infect marine mammals (Lindsay et al. 2004, Lindsay and Dubey 2009; Fayer et al. 2004). *Toxoplasma gondii* appears to be common in marine environments, and, as such, manifestation of toxoplasmosis in marine mammals may be opportunistic (Jessup et al. 2004).

Toxoplasma gondii causes illness in pinnipeds and small cetaceans, and it has been associated with sea otter deaths (*Enhydra lutris*) (Dubey et al. 2003; Miller et al. 2004). In addition to sea otters, *T. gondii* infections or evidence of infection (i.e., antibodies) has been reported in a variety of marine mammals, including bottlenose dolphins, common dolphins (*Delphinus delphis*), grey seals (*Halichoerus grypus*), harbor porpoises (*Phocoena phocoena*), harbor seals (*Phoca vitulina vitulina*), Hawaiian monk seals (*Monachus schauinslandi*), polar bears, striped dolphins, and walrus (*Odobenus rosmarus*) (Cabezon et al. 2004, 2011; Dubey et al. 2007, 2008, 2009; Honnold et al. 2005; Jensen et al. 2010;

Lambourn et al. 2001). A novel Type X *T. gondii*, once considered a marine type, has been associated with mortalities in sea otters (Miller et al. 2008).

Type X has been identified in mussels, terrestrial carnivores, runoff, and sea otters in California, easily linking how this emerging pathogen may move between terrestrial and marine mammals (Miller et al. 2008). There is increasing evidence of movement of Type X to human populations. Approximately 15 % of *T. gondii* congenitally infected children born in North America during 1981–2009 were infected with Type X, greatly increasing the relevance of Type X to human health (MacLeod et al. 2012). As *T. gondii* increases in importance as a marine mammal opportunistic pathogen, understanding its movement between land and ocean, as well as increasing coastal-specific risk factors to host immunity, may help lead to mitigation strategies within identified ecological hot spots.

6.6 Viral Pathogens

Herpesviruses Herpesviruses (alphaherpesviruses and gammaherpesviruses) are frequently identified in marine mammals including beaked whales, belugas, bottlenose dolphins, California sea lions, dusky dolphins, harbor porpoises, harbor seals, northern elephant seals, Pacific white-sided dolphins, Stellar sea lions, and striped dolphins (Arbelo et al. 2010; Barr et al. 1989; Blanchard et al. 2001; Borst et al. 1986; Burek et al. 2005; Goldstein et al. 2006; Gulland et al. 1997; Kennedy et al. 1992; Maness et al. 2011; Noguchi et al. 2012; Van Bressem et al. 1994). Infections by herpesviruses have been associated with adrenal necrosis, cutaneous and mucosal lesions, encephalitis, lymphoid necrosis, interstitial nephritis, and ocular disease (Arbelo et al. 2010, 2012; Benson et al. 2006; Gulland et al. 1999; Kadoi et al. 1992; Kennedy et al. 1992; Manire et al. 2006). Herpesvirus infections appear to be opportunistic in marine mammals, including during morbillivirus infections in harbor seals and striped dolphins (Belliere et al. 2010; Osterhaus 1989; Soto et al. 2012).

In humans, herpesviruses are opportunistic pathogens, either from primary infection or reactivation of latent infections. Changes in T-cell subsets and interferon-α deficiencies may increase the risk of herpesvirus infections in organ transplant recipients and people with human immunodeficiency virus (HIV) infections (Lopez et al. 1983; Preiksaitis et al. 1983; Schooley et al. 1983). Kaposi's sarcoma, increased in people with HIV infection, is associated with human herpesvirus 8 infection (Cesarman and Knowles 1997). Lymphoma is another neoplastic condition associated with herpesvirus infections in humans, though its role remains undetermined (Hermouet et al. 2003; Klein 1975). Herpesvirus infections in marine mammals have also been associated with neoplasia in marine mammals, including urogenital carcinoma and B-cell lymphoblastic lymphoma in California sea lions (Buckles et al. 2006; King et al. 2002; Venn-Watson et al. 2012a, b). The prevalence of neoplasia is particularly high in California sea lions, present in approximately one in five dead stranded adults (Gulland et al. 1996). In addition to herpesviruses, sea lions with carcinoma have significantly higher levels of polychlorinated biphenyls compared to those without carcinoma, suggesting contaminant exposure as a risk factor (Ylitalo et al. 2005). An apparent immunogenic component, including major histocompatibility genes, has been associated with carcinoma in sea lions (Bowen et al. 2005). As such, there may be links among contaminant exposures, impaired immunity, and opportunistic herpesvirus infections as causal factors for common carcinoma in sea lions.

6.7 One Ocean, One Health

This chapter summarizes known and possible opportunistic infections in marine mammals, many of which may be increasing due to a combination of increased numbers of marine mammals harboring opportunistic pathogens and increased number of susceptible hosts. Chemical contaminants, climate change, habitat loss, crowding, and prey loss can increase the incidence of opportunistic infections in marine mammals. As these risk factors increase, so may the importance of opportunistic infections.

The ocean has an incredibly diverse environment of bacteria and viruses, including 0.5–1 million bacteria and 10 million viral particles per milliliter of surface sea water, which fosters immense opportunities for genetic exchange and emergence of infectious diseases (Suttle 2007). Further, epizootics in the marine environment can spread at extremely rapid rates; while the highest epidemic spread rate in terrestrial environments is 1000 km/year, morbillivirus epizootics among seals and dolphins have spread at rates greater than 3000 km/year, and herpesviruses have spread at 10,000 km/year among phocids (McCallum et al. 2003). As such, the relevance of marine mammal opportunistic infections, most of which are considered zoonotic infections that can affect humans and agricultural animals, may overlap with terrestrial species. The following prioritized list of needs, developed by Harvell et al. (2004), continues to be important to address unsolved problems related to increasing ocean diseases, including opportunistic pathogens, and their relevance to ocean and terrestrial animal health:

- Detect origins and reservoirs for marine diseases and trace flow of new pathogens from land to sea.
- Document longevity and host range of infectious diseases.
- Evaluate effect of greater taxonomic diversity of marine relative to terrestrial hosts and pathogens.
- Pinpoint facilitating role of anthropogenic agents as incubators and conveyors of marine pathogens (including immune suppressors).
- Adapt epidemiology models to analyze marine disease.

Future work in areas such as those listed above may further elucidate how changes in the ocean's health, as well as other factors, may turn marine mammal commensals into pathogens.

References

Aguilar A, Borrell A, Reijinders PJH (2002) Geographical and temporal variation in levels of organochlorine contaminants in marine mammals. Mar Environ Res 53:425–452

Appelbee AJ, Thompson RCA, Olson ME (2005) *Giardia* and *Cryptosporidium* in mammalian wildlife—current status and future needs. Trends Parasitol 21:370–376

Arbelo M, Sierra E, Esperon F, Watanabe TTN, Belliere EN, de los Monteros AE, Fernandez A (2010) Herpesvirus infection with severe lymphoid necrosis affecting a beaked whale stranded in the Canary Islands. Dis Aquat Organ 89:261–264

Arbelo M, Belliere EN, Sierra E, Sacchini S, Esperon F, Andrada M, Rivero M, Diaz-Delgado J, Fernandez A (2012) Herpes virus infection associated with interstitial nephritis in a beaked whale (*Mesoplodon densirostris*). BMC Vet Res 8:243

Arnold SJ, Kinney MC, McCormick MS, Dummer S, Scott MA (1997) Disseminated toxoplasmosis: unusual presentations in the immunocompromised host. Arch Pathol Lab Med 121:869–873

Barr B, Dunn JL, Daniel MD, Banford A (1989) Herpes-like viral dermatitis in a beluga whale (*Delphinapterus leucas*). J Wildl Dis 25:608–611

Beineke A, Siebert U, McLachlan M, Bruhn R, Thron K, Failing K, Muller G, Baumgartner W (2005) Investigations of the potential influence of environmental contaminants on the thymus and spleen of harbor porpoises (*Phocoena phocoena*). Environ Sci Technol 39:3933–3938

Belliere EN, Esperon F, Arbelo M, Munoz MJ, Fernandez A, Sanchez-Vizcaino JM (2010) Presence of herpesvirus in striped dolphins stranded during the cetacean morbillivirus epizootic along the Mediterranean Spanish coast in 2007. Arch Virol 155:1307–1311

Benson KAS, Manire CA, Ewing RY, Saliki JT, Townsend FI, Ehlers B, Romero CH (2006) Identification of novel alpha- and gammaherpesviruses from cutaneous and mucosal lesions of dolphins and whales. J Virol Methods 136:261–266

Blanchard TW, Santiago NT, Lipscomb TP, Garber RL, McFee WE, Knowles S (2001) Two novel alphaherpesviruses associated with fatal disseminated infections in Atlantic bottlenose dolphins. J Wildl Dis 37:297–305

Bogomolni AL, Gast RJ, Ellis JC, Dennet M, Pugliares KR, Lentell BJ, Moore MJ (2008) Victims or vectors: a survey of marine vertebrate zoonoses from coastal waters of the Northwest Atlantic. Dis Aquat Organ 81:13–38

Borst GH, Walvoort HC, Reijnders PJ, van der Kam JS, Osterhaus AD (1986) An outbreak of a herpesvirus infection in harbor seals (*Phoca vitulina*). J Wildl Dis 22:1–6

Bossart GD (2006) Marine mammals as sentinel species for oceans and human health. Oceanography 19:134–137

Bowen L, Aldridge BM, DeLong R, Melin S, Buckles EL, Gulland F, Lowenstine LJ, Stott JL, Johnson ML (2005) An immunogenetic basis for the high prevalence of urogenital cancer in a free-ranging population of California sea lions (*Zalophus californianus*). Immunogenetics 56:846–848

Buck JD, Wells RS, Rhinehart HL, Hansen LJ (2006) Aerobic microorganisms associated with free-ranging bottlenose dolphins in coastal Gulf of Mexico and Atlantic Ocean waters. J Wildl Dis 42:536–544

Buckles EL, Lowenstine LJ, Funke C, Vittore RK, Wong HN, St. Leger JA, Greig DJ, Duerr RS, Gulland FMD, Stott JL (2006) Otarine herpesvirus-1, not papillomavirus, is associated with endemic tumours in California sea lions (*Zalophus californianus*). J Comp Pathol 135:183–189

Burek KA, Gulland FM, Sheffield G, Beckmen KB, Keyes E, Spraker TR, Smith AW, Skilling DE, Evermann JF, Stott JL, Saliki JT, Trites AW (2005) Infectious disease and the decline of Steller sea lions (*Eumetopias jubatus*) in Alaska, USA: insights from serologic data. J Wildl Dis 41:512–524

Burek KA, Gulland FMD, O'Hara TM (2008) Effects of climate change on arctic marine mammal health. Ecol Appl 18:S126–S134

Burmudez L, Van Bressem MF, Reyes-Jaimes O, Sayegh AJ, Paniz-Mondolfi AE (2009) Lobomycosis in man and lobomycosis-like disease in bottlenose dolphin, Venezuela. Emerg Infect Dis 15:1301–1303

Burns RA, Roy JS, Woods C, Padhye AA, Warnock DW (2000) Report of the first human case of lobomycosis in the United States. J Clin Microbiol 38:1283–1285

Cabezon O, Resendes AR, Domingo M, Raga JA, Agusti C, Alegre F, Mons JL, Dubey JP, Almeria S (2004) Seroprevalence of *Toxoplasma gondii* antibodies in wild dolphins from the Spanish Mediterranean coast. J Parasitol 90:643–644

Cabezon O, Hall JA, Vincent C, Pabon M, Garcia-Bocanegra I, Dubey JP, Almeria S (2011) Seroprevalence of *Toxoplasma gondii* in North-eastern Atlantic harbor seal (*Phoca vitulina vitulina*) and grey seal (*Halichoerus grypus*). Vet Parasitol 179:253–256

Carmichael LE, Kenney RM (1968) Canine abortion caused by Brucella canis. J Am Vet Med Assoc 152:605–616

Cesarman E, Knowles DM (1997) Kaposi's sarcoma-associated herpesvirus: a lymphotropic human herpesvirus associated with Kaposi's sarcoma, primary effusion lymphoma, and multicentric Castleman's disease. Semin Diagn Pathol 14:54–66

Dagleish MP, Patterson IAP, Foster G, Reid RJ, Linton C, Buxton D (2006) Intracranial granuloma caused by asporogenic *Aspergillus fumigatus* in a harbour porpoise (*Phocoena phocoena*). Vet Rec 159:458–460

Dagleish MP, Foster G, Howie FE, Reid RJ, Barley J (2008) Fatal mycotic encephalitis caused by *Aspergillus fumigatus* in a northern bottlenose whale (*Hyperoodon ampullatus*). Vet Rec 163:602–604

de Swart RL, Ross PS, Vos JG, Osterhaus AD (1996) Impaired immunity in harbor seals (*Phoca vitulina*) exposed to bioaccumulated environmental contaminants: review of a long-term feeding study. Environ Health Perspect 104:823–828

Delaney MA, Terio KA, Colegrove KM, Briggs MB, Kinsel MJ (2012) Occlusive fungal tracheitis in 4 captive bottlenose dolphins (*Tursiops truncatus*). Vet Pathol 50:172–176

Deng MQ, Peterson RP, Cliver DO (2000) First findings of *Cryptosporidium* and *Giardia* in California sea lions. J Parasitol 86:490–494

Di Guardo G, Di Francesco CE, Eleni C, Cocumelli C, Scholl F, Cacalone C, Peletto S, Mignone W, Tittarelli C, Di Nocera F, Leonardi L, Fernandez A, Marcer F, Mazzaeriol S (2013) Morbillivirus infection in cetaceans stranded along the Italian coastline: pathological, immunohistochemical and biomolecular findings. Res Vet Sci 94:132–137

Diamond SS, Ewing DE, Cadwell GA (1979) Fatal bronchopneumonia and dermatitis caused by *Pseudomonas aeruginosa* in an Atlantic bottlenosed dolphin. J Am Vet Med Assoc 175:984–987

Dixon BR, Parrington LJ, Parenteau M, Leclair D, Santín M, Fayer R (2008) *Giardia duodenalis* and *Cryptosporidium* spp. in the intestinal contents of ringed seals (*Phoca hispida*) and bearded seals (*Erignathus barbatus*) in Nunavik, Quebec, Canada. J Parasitol 94:1161–1163

Domingo M, Visa J, Pumarola M, Marco AJ, Ferrer L, Rabanal R, Kennedy S (1992) Pathologic and immunocytochemical studies of morbillivirus infection in striped dolphins (*Stenella coeruleoalba*). Vet Pathol 29:1–10

Dubey JP, Zarnke R, Thomas NJ, Wong SK, Van Bonn W, Briggs M, Davis JW, Ewing R, Mense M, Kwok OC, Romand S, Thulliez P (2003) *Toxoplasma gondii*, *Neospora caninum*, *Sarcocystis neurona*, and *Sarcocystis canis*-like infections in marine mammals. Vet Parasitol 116(4):275–296

Dubey JP, Morales N, Sundar N, Velmurugan GV, Gonzalez-Barrientos CR, Hernandez-Mora G, Su C (2007) Isolation and genetic characterization of *Toxoplasma gondii* from striped dolphin (*Stenella coerfuleoalba*) from Cost Rica. J Parasitol 93:710–711

Dubey JP, Fair PA, Sundar N, Velmurugan G, Kwok OCH, McFee WE, Majumdar D, Su C (2008) Isolation of Toxoplasma gondii from bottlenose dolphins (*Tursiops truncatus*). J Parasitol 94:821–823

Dubey JP, Mergl J, Gehring E, Sundar N, Velmurugan GV, Kwok OCH, Grigg ME, Su C, Martineau D (2009) Toxoplasmosis in captive dolphins (*Tursiops truncatus*) and walrus (*Odobenus rosmarus*). J Parasitol 95:82–85

Durden WN, St. Leger J, Stolen M, Mazza T, Londono C (2009) Lacaziosis in bottlenose dolphins (*Tursiops truncatus*) in the Indian River Lagoon, Florida, USA. J Wildl Dis 45:849–856

Fair PA, Mitchum G, Hulsey TC, Adams J, Zolman E, McFee W, Wirth E, Bossart GD (2007) Polybrominated diphenyl ethers (PBDEs) in blubber of free-ranging bottlenose dolphins (*Tursiops truncatus*) from two southeast Atlantic estuarine areas. Arch Environ Contam Toxicol 53:483–494

Faires MC, Gehring E, Mergl J, Weese JS (2009) Methicillin-resistant *Staphylococcus aureus* in marine mammals. Emerg Infect Dis 15:2071–2072

Fayer R, Dubey JP, Lindsay DS (2004) Zoonotic protozoa: from land to sea. Trends Parasitol 20:531–536

Fein AM (1999) Pneumonia in the elderly: overview of diagnostic and therapeutic approaches. Clin Infect Dis 28:726–729

Fernandez A, Esperon F, Herraez P, de Los Monteros AE, Clavel C, Bernabe A, Sanchez-Vizcaino JM, Verborgh P, DeStephanis R, Toledano F, Bayon A (2008) Morbillivirus and pilot whale deaths, Mediterranean Sea. Emerg Infect Dis 14:792–794

Foster G, Osterman BS, Godfroid J, Jacques I, Cloeckaeart A (2007) *Brucella ceti* sp. nov. and *Brucella pinnipedialis* sp. nov. for Brucella strains with cetaceans and seals as their preferred hosts. Int J Syst Evol Microbiol 57:2688–2693

Garin-Bastuji B, Blasco JM, Grayon M, Verger JM (1998) Brucella melitensis infection in sheep: present and future. Vet Res 29:255–274

Gaydos JK, Miller WA, Johnson C, Zornetzer H, Melli A, Packham A, Jeffries SJ, Lance MM, Conrad PA (2008) Novel and canine genotypes of *Giardia duodenalis* in harbor seals (*Phoca vitulina richardsi*). J Parasitol 94:1264–1268

Goertz CEC, Frasca S Jr, Bochach GA, Cowan DF, Buck JD, French RA, De Guise S, Maratea J, Hinckley L, Ewalt D, Schlievert PM, Karst SM, Deobald CF, St. Aubin DJ, Dunn JL (2011) *Brucella* sp. vertebral osteomyelitis with intercurrent fatal *Staphylococcus aureus* toxigenic enteritis in a bottlenose dolphin (*Tursiops truncatus*). J Vet Diagn Invest 23:845–851

Goldstein T, Lowenstein LJ, Lipscomb TP, Mazet JA, Novak J, Stott JL, Gulland FMD (2006) Infection with a novel gammaherpesvirus in northern elephant seals (Mirounga angustirostris). J Wildl Dis 42:830–835

Gonzalez L, Patterson IA, Reid RJ, Foster G, Barberan M, Blasco JM, Kennedy S, Howie FE, Godfroid J, MacMillan AP, Schock A, Buxton D (2002) Chronic meningoencephalitis associated with Brucella infection in live-stranded striped dolphins (*Stenella coeruleoalba*). J Comp Pathol 126:147–152

Goodwin KD, McNay M, Cao Y, Ebentier D, Madison M, Griffith JF (2012) A multi-beach study of *Staphylococcus aureus*, MRSA, and enterococci in seawater and beach sand. Water Res 46:4195–4207

Grovel O, Pouchus YF, Verbist JF (2003) Accumulation of gliotoxin, a cytotoxic mycotoxin from *Aspergillus fumigatus*, in blue mussel (*Mytilus edulis*). Toxicon 42:297–300

Gulland FMD, Hall AJ (2007) Is marine mammal health deteriorating? Trends in the global reporting of marine mammal disease. Ecohealth 4:135–150

Gulland FMD, Trupkiewicz JG, Spraker TR, Lowenstine LJ (1996) Metastatic carcinoma of probable transitional cell origin in 66 free-living California sea lions (*Zalophus californianus*), 1979 to 1994. J Wildl Dis 32:250–258

Gulland FM, Lowenstein LJ, Lapointe JM, Spraker T, King DP (1997) Herpesvirus infection in stranded Pacific harbor seals of coastal California. J Wildl Dis 33:450–458

Gulland FMD, Haulena M, Lowenstine LJ, Munro C, Graham PA, Bauman J, Harvey J (1999) Adrenal function in wild and rehabilitated Pacific harbor seals (*Phoca vitulina richardii*) and in seals with phocine herpesvirus-associated adrenal necrosis. Mar Mamm Sci 15:810–827

Guzman-Verri C, Gonzalez-Barrientos R, Hernandez-Mora G, Morales JA, Baquero-Calvo E, Chaves-Olarte E, Moreno E (2012) *Brucella ceti* and brucellosis in cetaceans. Front Cell Infect Microbiol 2:3. doi:10.3389/fcimb.2012.00003

Hammond JA, Hall AJ, Dyrynda EA (2005) Comparison of polychlorinated biphenyl (PCB) induced effects on innate immune functions in harbor and grey seals. Aquat Toxicol 74:126–138

Harvell D, Aronson R, Baron N, Connell J, Dobson A, Ellner S, Gerber L, Kim K, Kuris A, McCallum H, Lafferty K, McKay B, Porter J, Pascual M, Smith G, Sutherland K, Ward J (2004) The rising tide of ocean diseases: unsolved problems and research priorities. Front Ecol Environ 2:375–382

Harwood J (2001) Marine mammals and their environment in the twenty-first century. J Mammal 82:630–640

Haubold EM, Cooper CR, Wen JW, McGinnis MR, Cowan DF (2000) Comparative morphology of *Lacazia loboi* (syn. *Loboa loboi*) in dolphins and humans. Med Mycol 38:9–14

He Y, Vemulapalli R, Zeytun A, Schurig GG (2001) Induction of specific cytotoxic lymphocytes in mice vaccinated with *Brucella abortus* RB51. Infect Immun 69:5502–5508

Hernandez-Mora G, Palacios-Alfaro JD, Gonzalez-Barrientos R (2013) Wildlife reservoirs of brucellosis: Brucella in aquatic environments. Rev Sci Tech 32:89–103

Hermouet S, Sutton CA, Rose TM, Greenblatt RJ, Corre I, Garand R, Neves AM, Bataille R, Casey JW (2003) Qualitative and quantitative analysis of human herpesviruses in chronic and acute B cell lymphocytic leukemia and in multiple myeloma. Leukemia 17:185–195

Higgins R (2000) Bacteria and fungi of marine mammals: a review. Can Vet J 41:105–116

Hill DE, Chirukandoth S, Dubey JP (2005) Biology and epidemiology of *Toxoplasma gondii* in man and animals. Anim Health Res Rev 6:41–61

Honnold SP, Braun R, Scott DP, Sreekumar C, Dubey JP (2005) Toxoplasmosis in Hawaiian monk seal (*Monachus schauinslandi*). J Parasitol 91:695–697

Hughes-Hanks JM, Rickard LG, Panuska C, Saucier JR, O'Hara TM, Dehn L, Rolland RM (2005) Prevalence of *Cryptosporidium* spp. and *Giardia* spp. in five marine mammal species. J Parasitol 91:1225–1228

Jensen SK, Aars J, Lydersen C, Kovacs KM, Asbakk K (2010) The prevalence of *Toxoplasma gondii* in polar bears and their marine mammal prey: evidence for a marine transmission pathway? Polar Biol 33:599–606

Jepson PD, Brew S, MacMillan AP, Baker JR, Barnett J, Kirkwood JK, Kuiken T, Robinson IR, Simpson VR (1997) Antibodies to Brucella in marine mammals around the coast of England and Wales. Vet Rec 141:513–515

Jessup DA, Miller M, Ames J, Harris M, Kreuder C, Conrad PA, Mazet JAK (2004) Southern sea otter as a sentinel of marine ecosystem health. Ecohealth 1:239–245

Johnson SP, Nolan S, Gullland FM (1998) Antimicrobial susceptibility of bacteria isolated from pinnipeds stranded in central and northern California. J Zoo Wildl Med 29:288–294

Joseph BE, Cornell LH, Simpson JG, Migaki G, Griner L (1986) Pulmonary aspergillosis in three species of dolphin. Zoo Biol 5:301–308

Kadoi K, Mochizuki A, Ikeda T, Kamata H, Yukawa M, Inoue Y (1992) Susceptibility of a line of dolphin kidney cell culture to several herpesviruses. J Basic Microbiol 32:227–232

Kennedy S (1998) Morbillivirus infections in aquatic mammals. J Comp Pathol 119:201–225

Kennedy S, Lindstedt IJ, McAliskey MM, McConnell SA, McCullough SJ (1992) Herpesviral encephalitis in a harbor porpoise (*Phocoena phocoena*). J Zoo Wildl Med 23:374–379

Kerazon I, Grovel O, Du Pont TR, Le Pape P, Pouchus YF (2008) Effects of seawater on growth and gliotoxin excretion of marine strains of Aspergillus fumigatus Fres. Toxicon 51:398–405

Khan NH, Ishii Y, Kimata-Kino N, Esaki H, Nishino T, Nishimura M, Kogure K (2007) Isolation of *Pseudomonas aeruginosa* from open ocean and comparison with freshwater, clinical, and animal isolates. Microb Ecol 53:173–186

Ketterer PJ, Rosenfeld LE (1974) Septic embolic nephritis in a dolphin caused by *Staphylococcus aureus*. Aust Vet J 50:123

Khan NH, Ahsan M, Yoshizawa S, Hosoya S, Yokota A, Kogure K (2008) Multilocus sequence typing and phylogenetic analyses of *Pseudomonas aeruginosa* isolates from the ocean. Appl Environ Microbiol 74:6194–6205

King DP, Hure MC, Goldstein T, Aldridge BM, Gulland FMD, Saliki JT, Buckles EL, Lowenstein LJ, Stott JL (2002) Otarine herpesvirus-1: a novel gammaherpesvirus associated with urogenital carcinoma in California sea lions (*Zalophus californianus*). Vet Microbiol 86:131–137

Klein G (1975) The Epstein-Barr virus and neoplasia. N Engl J Med 293:1353–1357

Kovacs KM, Lydersen C, Overland JE, Moore SE (2011) Impacts of changing sea-ice conditions on Arctic marine mammals. Mar Biodivers 41:181–194

Kravetz JD, Federman DG (2005) Toxoplasmosis in pregnancy. Am J Med 118:212–216

Lahvis GP, Wells RS, Kuehl DW, Stewart JL, Rhinehart HL, Via CS (1995) Decreased lymphocyte responses in free-ranging bottlenose dolphins (*Tursiops truncatus*) are associated with increased concentrations of PCBs and DDT in peripheral blood. Environ Health Perspect 103:67–72

Lambourn DM, Jeffries SJ, Dubey JP (2001) Seroprevalence of *Toxoplasma gondii* in harbor seals (*Phoca vitulina*) in Southern Puget Sound, Washington. J Parasitol 87:1196–1197

Lasek-Nesselquist E, Bogomolni AL, Rebecca G, Welch M, Ellis JC, Mitchell S, Moore M (2008) Molecular characterization of *Giardia intestinalis* haplotypes in marine animals: variation and zoonotic potential. Dis Aquat Organ 81:39–51

Lasek-Nesselquist E, Welch DM, Sogin ML (2010) The identification of a new *Giardia duodenalis* assemblage in marine vertebrates and a preliminary analysis of *G. duodenalis* population biology in marine systems. Int J Parasitol 40:1063–1074

Lindsay DS, Dubey JP (2009) Long-term survival of *Toxoplasma gondii* sporulated oocysts in seawater. J Parasitol 95:1019–1020

Lindsay DS, Collins MV, Mitchell SM, Wetch CN, Rosypal AC, Flick GJ, Zajac AM, Alan Lindquist A, Dubey JP (2004) Survival of *Toxoplasma gondii* oocysts in Eastern oysters (*Crassostrea virginica*). J Parasitol 90:1054–1057

Lipscomb TP, Kennedy S, Moffett D, Krafft A, Klaunberg BA, Lichy JH, Regan GT, Worthy GAJ, Taubenberger JK (1996) Morbilliviral epizootic in bottlenose dolphins of the Gulf of Mexico. J Vet Diagn Invest 8:283–290

Lopez C, Fitzgerald PA, Siegal FP (1983) Severe acquired immune deficiency syndrome in male homosexuals: diminished capacity to make interferon-a in vitro associated with severe opportunistic infections. J Infect Dis 148:962–966

Lyczak JB, Cannon CL, Pier GB (2000) Establishment of *Pseudomonas aeruginosa* infection: lessons from a versatile opportunist. Microbes Infect 2:1051–1060

MacLeod R, Boyer KM, Lee D, Mui E, Wroblewski K, Karrison T, Noble AG, Withers S, Swisher CN, Heydemann PT, Sautter M, Babiarz J, Rabiah P, Meier P, Grigg ME, the Toxoplasmosis Study Group (2012) Prematurity and severity are associated with *Toxoplasma gondii* alleles (NCCCTS, 1981–2009). Clin Infect Dis 54:1595–1605

Maness HTD, Nollens HH, Jensen ED, Goldstein T, LaMere S, Childress A, Sykes J, St. Leger J, Lacave G, Latson FE, Wellehan JFX Jr (2011) Phylogenetic analysis of marine mammal herpesviruses. Vet Microbiol 149:23–29

Manire CA, Smolarek KA, Romero CH, Kinsel MJ, Clauss TM, Byrd L (2006) Proliferative dermatitis associated with a novel alphaherpesvirus in an Atlantic bottlenose dolphin (*Tursiops truncatus*). J Zoo Wildl Med 37:174–181

McCallum H, Harvell D, Dobson A (2003) Rates of spread of marine pathogens. Ecol Lett 6:1062–1067

Meegan J, Dunn JL, Venn-Watson SK, Smith CR, Sidor I, Jensen ED, Van Bonn WG, Pugh R, Ficht T, Adams LG, Nielsen K, Romano TA (2012) Serologic response in bottlenose dolphins Tursiops truncatus infected with Brucella sp. using a dolphin-specific indirect ELISA. Dis Aquat Org 102:73–85

Miller WG, Adams LG, Ficht TA, Cheville NF, Payeur JP, Harley DR, House C, Ridgway SH (1999) *Brucella*-induced abortions and infection in bottlenose dolphins (*Tursiops truncatus*). J Zoo Wildl Med 30:100–110

Miller MA, Gardner IA, Kreuder C, Paradies DM, Worcester KR, Jessup DA, Dodd E, Harris MD, Ames JA, Packham AE, Conrad PA (2002) Coastal freshwater runoff is a risk factor for *Toxoplasma gondii* infection of southern sea otters (*Enhydra lutris nereis*). Int J Parasitol 32:997–1006

Miller MA, Grigg ME, Kreuder C, James ER, Melli AC, Crosbie PR, Jessup DA, Boothroyd JC, Brownstein D, Conrad PA (2004) An unusual genotype of *Toxoplasma gondii* is common in California sea otters (Enhydra lutris nereis) and is a cause of mortality. Int J Parasitol 34:275–284

Miller MA, Miller WA, Conrad PA, James ER, Melli AC, Leutenegger CM, Dabritz HA, Packham AE, Paradies D, Harris M, Ames J, Jessup DA, Worcester K, Grigg ME (2008) Type X *Toxoplasma gondii* in a wild mussel and terrestrial carnivores from coastal California: new linkages between terrestrial mammals, runoff, and toxoplasmosis of sea otters. Int J Parasitol 35:1155–1168

Morris PJ, Johnson WR, Pisani J, Bossart GD, Adams J, Reif JS, Fair PA (2011) Isolation of culturable microorganisms from free-ranging bottlenose dolphins (*Tursiops truncatus*) from the southeastern United States. Vet Microbiol 148:440–447

Munn CB (2006) Viruses as pathogens of marine organisms—from bacteria to whales. J Mar Biol Assoc UK 86:453–467

Murdoch ME, Reif JS, Mazzoil M, McCulloch SD, Fair PA, Bossart G (2008) Lobomycosis in bottlenose dolphins (Tursiops truncatus) from the Indian River Lagoon, Florida: estimation of prevalence, temporal trends, and spatial distribution. Ecohealth 5:289–297

Nakata H, Sakakibara A, Kanoh M, Kudo S, Watanabe N, Nagai N, Miyazaki N, Asano Y, Tanabe S (2002) Evaluation of mitogen-induced responses to marine mammal and human lymphocytes by in-vitro exposure of butyltins and non-*ortho* coplanar PCBs. Environ Pollut 120:245–253

Noguchi K, Shimoda H, Tereda Y, Shimojima M, Kohyama K, Inoshima Y, Maeda K (2012) Isolation of a novel herpesvirus from a Pacific white-sided dolphin. Arch Virol 158:695–699

Norstrom RJ, Muir DCG (1994) Chlorinated hydrocarbon contaminants in arctic marine mammals. Sci Total Environ 154:107–128

Osterhaus A (1989) A morbillivirus causing mass mortality in seals. Vaccine 7:483

Pfaller MA, Diekema DJ (2004) Rare and emerging opportunistic fungal pathogens: concern for resistance beyond *Candida albicans* and *Aspergillus fumigatus*. J Clin Microbiol 42:4419–4431

Power E, Murphy S (2002) *Staphylococcus aureus* septicaemia in a killer whale. Vet Rec 150:819

Prahl S, Jepson PD, Sanchez-Hanke M, Deaville R, Siebert U (2011) Aspergillosis in the middle ear of a harbour porpoise (*Phocoena phocoena*): a case report. Mycoses 54:e260–e264

Preiksaitis JK, Rosno S, Grumet C, Merigan TC (1983) Infections due to herpesviruses in cardiac transplant recipients: role of the donor heart and immunosuppressive therapy. J Infect Dis 147:974–981

Raga JA, Balbuena JA, Aznar J, Fernandez M (1997) The impact of parasites on marine mammals: a review. Parassitologia 39:293–296

Reidarson TH, Harrell JH, Rinaldi MG, McBain J (1998) Bronchoscopic and serologic diagnosis of Aspergillus fumigatus pulmonary infection in a bottlenose dolphin (*Tursiops truncatus*). J Zoo Wildl Med 29:451–455

Reif J, Murdoch E, Mazzoil M, Fair P, Bossart G (2006) Lobomycosis in bottlenose dolphins: spatial aggregation of an emerging infectious disease. Epidemiology 17:S434–S435

Reif JS, Peden-Adams MM, Romano TA, Rice CD, Fair PA, Bossart GD (2009) Immune dysfunction in Atlantic bottlenose dolphin (*Tursiops truncatus*) with lobomycosis. Med Mycol 47:125–135

Rodriquez-Toro G (1993) Lobomycosis. Int J Dermatol 32:324–332

Ross PS (2002) The role of immunotoxic environmental contaminants in facilitating the emergence of infectious diseases in marine mammals. Hum Ecol Risk Assess 8:277–292

Ross PS, Ellis GM, Ikonomou MG, Barrett-Lennard LG, Addison RF (2000) High PCB concentrations in free-ranging Pacific killer whales, *Orcinus orca*: effects of age, sex and dietary preference. Mar Pollut Bull 40:504–515

Rotander A, van Bavel B, Polder A, Riget F, Auounsson GA, Gabrielsen GW, Vikingsson G, Bloch D, Dam M (2012) Polybrominated diphenyl ethers (PBDEs) in marine mammals from Arctic and North Atlantic regions, 1986–2009. Environ Int 40:102–109

Ryan JC, Gidlewski T, Ewalt DR, Hennager SG, Lambourne DM, Olsen SC (2001) Seroconverstion and abortion in cattle experimentally infected with *Brucella* sp isolated from a Pacific harbor seal (*Phoca vitulina* richardsi). J Vet Diagn Invest 13:379–382

Saliki JT, Cooper EJ, Gustavson JP (2002) Emerging morbillivirus infections of marine mammals. Ann N Y Acad Sci 969:51–59

Samartino LE, Enright FM (1993) Pathogenesis of abortion of bovine brucellosis. Comp Immunol Microbiol Infect Dis 16:95–101

Schaefer AM, Goldstein JD, Reif JS, Fair PA, Bossart GD (2009) Antibiotic-resistant organisms cultured from Atlantic bottlenose dolphins (*Tursiops truncatus*) inhabiting estuarine waters of Charleston, SC and Indian River Lagoon, FL. Ecohealth 6:33–41

Schooley RT, Hirsch MS, Colvin RB, Cosimi B, Tolkoff-Rubin NE, McCluskey RT, Burton RC, Russell PS, Herrin JT, Delmonico FL, Giorgi JV, Henle W, Rubin RH (1983) Association of herpesvirus infections with T-lymphocyte-subset alterations, glomerulopathy, and opportunistic infections after renal transplantation. N Engl J Med 308:107–313

Schwacke LH, Zolman ES, Balmer BC, De Guise S, George RC, Hoguet J, Hohn AA, Kucklick JR, Lamb S, Levin M, Litz JA, McFee WE, Place NJ, Townsend FI, Wells RS, Rowles TK (2012) Anaemia, hypothyroidism and immune suppression associated with polychlorinated biphenyl exposure in bottlenose dolphins (*Tursiops truncatus*). Proc Biol Sci 279:48–57

Seibel H, Beineke A, Siebert U (2010) Mycotic otitis media in a harbour porpoise (*Phocoena phocoena*). J Comp Pathol 143:294–296

Siebert U, Muller G, Desportes G, Weiss R, Hansen K, Baumgartner W (2002) Pyogranulomatous myocarditis due to *Staphylococcus aureus* septicaemia in two harbor porpoises (*Phocoena phocoena*). Vet Rec 150:273–277

Smith H, Williams AE, Pearce JH, Keppie J, Harris-Smith PW, Fitz-George RB, Witt K (1962) Foetal erythritol: a cause of the localization of *Brucella abortus* in bovine contagious abortion. Nature 193:47–49

Soge OO, Meschke JS, No DB, Roberts MC (2009) Characterization of methicillin-resistant *Staphylococcus aureus* and methicillin-resistant coagulase-negative *Staphylococcus* spp. Isolated from US West Coast public marine beaches. J Antimicrob Chemother 64:1148–1155

Soto S, Gonzalez B, Willoughby K, Maley M, Olvera A, Kennedy S, Marco A, Domingo M (2012) Systemic herpesvirus and morbillivirus co-infection in a striped dolphin (*Stenella coeruleoalba*). J Comp Pathol 146:269–273

Stanzani M, Orciuolo E, Lewis R, Kontoyiannia P, Martins SLR, St. John LS, Komanduri KV (2005) Aspergillus fumigatus suppresses the human cellular immune response via gliotoxin-mediated apoptosis of monocytes. Blood 105:2258–2265

Suttle CA (2007) Marine viruses—major players in the global ecosystem. Nat Rev Microbiol 5:801–812

Sweeney JC, Ridgway SH (1975) Common diseases of small cetaceans. J Am Vet Med Assoc 167:533–540

Symmers WS (1983) A possible case of Lobo's disease acquired in Europe from a bottle-nosed dolphin (*Tursiops truncatus*). Bull Soc Pathol Exot Filiales 76:777–784

Thornton SM, Nolan S, Gulland FMD (1998) Bacterial isolates from California sea lions (*Zalophus californianus*), harbor seals (*Phoca vitulina*), and northern elephant seals (*Mirounga angustirostris*) admitted to a rehabilitation center along the central California coast, 1994–1995. J Zoo Wildl Med 29:171–176

Van Bressem MF, Van Waerebeek K, Garcia-Godos A, Dekegel D, Pastoret PP (1994) Herpes-like virus in dusky dolphins, Lagenorhynchus obscurus, from Coastal Peru. Mar Mamm Sci 10:354–359

Vedros NA, Quinlivan J, Cranford R (1982) Bacterial and fungal flora of wild northern fur seals (Callorhinus ursinus). J Wildl Dis 18:446–456

Venn-Watson S, Smith CR, Jensen ED (2008) Primary bacterial pathogens in bottlenose dolphins *Tursiops truncatus*: needles in haystacks of commensal and environmental microbes. Dis Aquat Org 79:87–93

Venn-Watson S, Smith CR, Gomez F, Jensen ED (2011) Physiology of aging among healthy, older bottlenose dolphins (*Tursiops truncatus*): comparisons with aging humans. J Comp Physiol B 181:667–680

Venn-Watson S, Benham C, Gulland FMD, Smith CR, St. Leger J, Yochem P, Nollens H, Saliki JT, Colegrove K, Wellehan JFX, Rivera R (2012a) Clinical relevance of novel Otarine herpesvirus-3 in California sea lions (*Zalophus californianus*): association with juvenile strandings and B cell lymphoblastic lymphoma. Vet Res 43:85

Venn-Watson S, Daniels R, Smith C (2012b) Thirty year retrospective evaluation of pneumonia in a bottlenose dolphin population. Dis Aquat Organ 99:237–242

Venn-Watson S, Smith CR, Jensen ED, Rowles T (2013) Assessing the potential health impacts of the 2003 and 2007 firestorms on bottlenose dolphins (*Tursiops truncatus*) in San Diego Bay. Inhal Toxicol 25:481–491

Waltzek TB, Cortex-Hinojosa G, Wellehan JFX Jr, Gray GC (2012) Marine mammal zoonoses: a review of disease manifestations. Zoonoses Public Health 59:521–535

Wells RS, McHugh KA, Douglas DC, Shippee S, McCabe EB, Barros NB, Phillips GT (2013) Evaluation of potential protective factors against metabolic syndrome in bottlenose dolphins: feeding and activity patterns of dolphins in Sarasota Bay, Florida. Front Endocrinol (Lausanne) 4:139

Ylitalo GM, Stein JE, Hom T, Johnson LL, Tilbury KL, Hall AJ, Rowles T, Greig D, Lowenstine LJ, Gulland FMD (2005) The role of organochlorines in cancer-associated mortality in California sea lions (*Zalophus californianus*). Mar Pollut Bull 50:30–39

Zagzebski KA, Gulland FMD, Haulena M, Lander ME, Greig DJ, Gage LJ, Hanson MB, Yochem PK, Stewart BS (2006) Twenty-five years of rehabilitation of odontocetes stranded in Central and Northern California, 1977–2002. Aquat Mamm 32:334–345

Zhan Y, Cheers C (1993) Endogenous gamma interferon mediates resistance to *Brucella abortus* infection. Infect Immun 61:4899–4901

Part III
Opportunistic Pathogenicity in the Terrestrial World

Chapter 7
Opportunistic Pathogens of Terrestrial Plants

Amy O. Charkowski

Abstract Pathogens of terrestrial plants are categorized as biotrophic, hemi-biotrophic, or necrotrophic based on similarities in how they manipulate plant cells and based on when or if in the plant–microbe interaction plant cell death occurs. Biotrophic pathogens require live plant cells for infection, hemi-biotrophs require live plant cells to initiate colonization, but the plant cell eventually dies, and necrotrophs kill plant cells early in their interactions with their hosts. Pathogens in each of these categories that have arisen in multiple kingdoms and species within single genera may fall into multiple pathogen categories. The borders between these classifications are blurred and can be affected by environment and plant host species. Broad host range necrotrophic plant pathogens are sometimes referred to as opportunistic because some infect mainly damaged plants or only cause disease on harvested and stored fruits and vegetables. However, true opportunistic pathogens are rarely studied in plant pathology because they do not cause significant economic losses in agriculture. The broad host range necrotrophs may be the closest to true opportunistic pathogens. Recent mechanistic and genomic data shows that these pathogens are adept at manipulating plant defenses and subverting them to the benefit of their necrotrophic lifestyle. This subversion of conserved plant defenses likely contributes to the broad host range of these pathogens and the constant challenge of identifying resistant germplasm. Plants can also resist necrotrophic pathogens by producing antimicrobial compounds or plant cell wall modifications, but the genes encoding for these resistance mechanisms are little studied compared to canonical gene-for-gene resistance that act against biotrophic and hemi-biotrophic pathogens.

A.O. Charkowski (✉)
Department of Plant Pathology, University of Wisconsin at Madison, 1630 Linden Drive, Madison, WI 53706, USA
e-mail: acharkowski@wisc.edu

© Springer International Publishing Switzerland 2016

C.J. Hurst (ed.), *The Rasputin Effect: When Commensals and Symbionts Become Parasitic*, Advances in Environmental Microbiology 3,
DOI 10.1007/978-3-319-28170-4_7

7.1 Introduction

Terrestrial plants transform the energy of the sun into sugar by photosynthesis. Many species of plant-associated bacterial and eukaryotic microbes colonize plants and can harvest nutrients derived from photosynthesis or extracted from soil by the plant. These microbes can colonize plant surfaces; the interstitial spaces between plant root, leaf, flower, and stem cells; and inside the plant vascular system. The microbes, collectively denoted the phytobiome, may benefit through symbiotic, commensal, or pathogenic relationships with the plant or with other plant-colonizing microbes or invertebrates. Modern agriculture is essentially an exercise in reducing environmental diversity in an effort to have predictable and high yields from a few plant species bred to thrive in a disturbed environment.

Microbial relationships with plants can be described with a Venn diagram or disease triangle. Important interactions, such as disease or symbiotic nitrogen fixation, only occur when a microbe capable of colonizing a particular plant genotype is present in an appropriate environment (Fig. 7.1). The effect of environment varies in quality and importance for each type of interaction. For example, the fungal pathogen *Rhizoctonia solani* requires a cool damp environment to infect germinating seedlings. *Rhizoctonia* causes either a less severe or no disease when the soil is warm or when plants are mature. In contrast, as long as its insect vector transmits the bacterial pathogens *Xylella fastidiosa* into susceptible grape plants, the bacterium will likely cause disease. In this case, as long as the local environment supports the plant and the vector, the environment has relatively little effect on the pathogen. Understanding the components of these interaction diagrams helps us understand disease epidemiology and etiology and can lead to effective plant health management strategies.

7.2 What Is an Opportunistic Pathogen?

The definition of opportunistic pathogen is not agreed upon in pathogenesis literature. Most reviews on this topic consider only animal pathogens, completely ignoring all relevant discoveries in plant pathology [e.g., see Brown et al. (2012) and Casadevall and Pirofski (1999)]. And, as a result, the definitions provided in these reviews can be only poorly adapted to plant pathogens. The terms biotroph, hemi-biotroph, and necrotroph, each of which is described later in this chapter, are better suited to plant pathogenesis.

For example, opportunistic pathogens were recently defined as pathogens that are not specialists or not obligate pathogens (Brown et al. 2012). With this definition, Brown et al. divide opportunistic pathogens into three categories, including commensal, environmental, and parasitic opportunistic pathogens. Their goal was, in part, to show that classical virulence theories in evolution require that pathogens be both specialists and obligate pathogens, so this theory breaks down with most

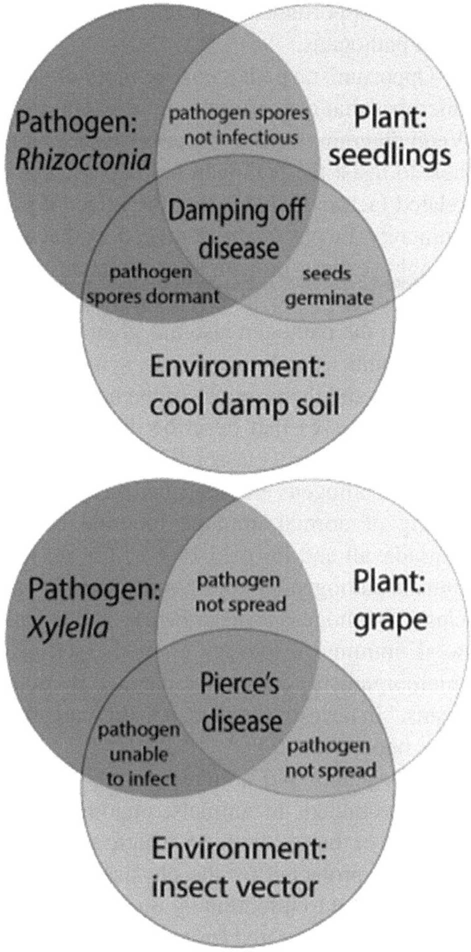

Fig. 7.1 Interaction diagrams for a necrotroph, *Rhizoctonia solani*, and a biotroph, *Xylella fastidiosa*

pathogens. Since this definition covers almost all plant pathogens, it is not particularly useful when discussing plant pathogens. Within their definition of opportunistic, commensal opportunists would include plant pathogens such as *Phytophthora infestans* and *Pseudomonas syringae*, environmental opportunists would include soft rot bacteria and decay fungi, and parasitic opportunists would include any obligate pathogen that could be considered a zoonotic infection, such as phytoplasma, *Liberibacter*, or *tomato spotted wilt virus*.

Brown et al. argue that virulence factors in opportunistic pathogens are "spandrels" or the secondary effect of factors that would provide a primary or direct fitness benefit in some other host or in another environment. This conjecture is not possible to defend within their definition of opportunistic pathogens in plants. There are numerous examples of virulence factors providing a clear fitness advantage for plant pathogens that fall into their definition of commensal, environmental, and parasitic opportunists. The remainder of their analysis is similarly poorly suited to plant pathogens.

Opportunistic pathogens are more often defined as commensal or environmental microbes that only cause disease in compromised hosts. Therefore, in terms of the Venn diagrams used to describe disease, it may be presumed that the ability of the host to resist the pathogen has become detrimentally affected. The list of possibly related factors includes that the affected host may lack key components of innate immunity, be physically damaged, or that unusual environmental conditions such as drought, excess moisture, or temperature extremes may have compromised the ability of the host to resist the pathogen. The subject of upregulated virulence genes in the pathogen also has great importance.

In plants as well as animals, symptoms occur both because of the action of the pathogen and the response of the host. Casadevall and Pirofski developed damage–response curves that describe disease progress in animal pathogens and divided pathogens into six classes (Casadevall and Pirofski 1999). In this scheme, opportunistic pathogens are described as microbes that "do not cause host damage in the setting of normal immune function because they have low intrinsic virulence (Casadevall and Pirofski 1999)." For reasons described below, this type of opportunistic pathogen remains almost entirely unexamined in terrestrial plants. The next class of pathogens is described as "pathogens that cause damage either in hosts with weak immune responses or in the setting of normal immune responses. These microorganisms cause host damage by both host- and pathogen-mediated mechanisms." In terrestrial plants, it is this category of pathogen that is most often referred to as opportunistic.

Both plants and animals may become infected with opportunistic pathogens when wounded. In animals, opportunistic pathogens occur after deep puncture wounds or burns, both of which allow microbes to access parts of the animal normally protected by the physical barrier of the skin. In plants, severe wounds also can lead to infection by opportunistic pathogens. For example, trees damaged by storms or harvested fruits, tubers, or roots that are bruised or punctured are likely to become diseased by pathogens not generally found on undamaged plants. Terrestrial plants constantly face smaller wounds caused by herbivorous birds, mammals, insects, and nematodes, and growing and otherwise healthy plants are generally able to effectively protect wounds from feeding damage, so feeding wounds rarely become the site of opportunistic infections.

Unlike animals, plants do not have circulating cells or adaptive immunity, but they do have innate immune systems capable of suppressing disease. With terrestrial plants, the role of innate immunity in suppression of opportunistic pathogens is evident in laboratory studies with plant mutants lacking key components of the immune response, and this will be described later in more detail. It is also apparent in plant breeding, where plant lines with compromised resistance to opportunistic pathogens are generally removed from breeding populations, often without intent. This occurs because when developing new plant varieties, the plant breeders must maintain seed, rhizomes, stem cuttings, or other propagative material from each line, and plant lines with compromised innate immunity will not produce healthy seed or other propagative material. Plants with mutations in genes required for innate immunity certainly occur in natural ecosystems as well, but based upon their relatively limited presence, it may be presumed likely that ecologically those plants do not compete effectively.

Opportunistic pathogens also cause disease in animals whose commensal microbial communities have been disrupted due to another disease or antibiotics. Commensal microbes can also protect plants from opportunistic pathogens, and this phenomenon is exploited by some commercially available biocontrol microbes widely used in agriculture. Commensal microbes and beneficial symbionts likely play an important role in protecting roots from opportunistic pathogens, but the complexities of soil make this difficult to study, so much remains to be learned about which microbes play important protective roles and the mechanisms that result in these beneficial interactions.

7.3 Opportunistic Is a Rarely Used Term in Plant Pathology

The term "opportunistic" is still widely used for animal pathogens, but it is now rarely used to describe plant pathogens. This difference in view of plant and animal pathogens has had little discussion in peer-reviewed literature. The rare use of the term opportunistic may represent important aspects of ethics and economics that lead to differences in how plant and animal pathogens are studied.

When studying plant disease, the individual plant is rarely important; rather, it is the population that matters. Farmers attempt to manage disease through manipulation of irrigation, soil health, crop rotations, plant variety choices, and many other means. They rarely worry about curing individual plants, although individual plants that remain healthy in the face of an epidemic may become important in plant breeding. Similarly, ecologists consider the effects of plant disease, such as chestnut blight, on populations and biological communities, but generally don't consider curing individuals. In addition, one of the general criteria for publishing on plant diseases is that the disease be economically significant. As a result, rare and usual opportunistic pathogens are not described or studied by plant pathologists. In contrast, we do care about and study individuals in human medicine, and the scientific literature is filled with reports of rare opportunistic infections of immune-compromised or injured patients.

Opportunistic infections in people are generally considered of limited communicability, meaning that they have a low probability of transmission. In human medicine, where individuals are important, the nature of a disease being described as having lessened transmissibility merits mention, but does not mean that the disease is not considered worthy of study. In plant disease, however, the ability of a disease to spread and cause economically important losses is required for the disease problem to be considered worth studying. Opportunistic infections in people may also be localized, causing significant damage only in a part of the person, such as a finger or foot. With animals, loss of a limb generally affects fitness more than loss of a leaf or branch affects plant health since plants can often replace the damaged part.

Conversely, in animal medicine, and particularly in human medicine, we are generally unable to confirm that microbes isolated from immune-compromised or injured patients are indeed pathogens. As a result, the literature is filled with examples of type species described as opportunistic pathogens that have never been reported again in any natural environment, leaving the ecological role of these species a mystery. With plants, completing Koch's postulates and establishing pathogenicity of suspected opportunistic pathogens are simple in comparison. As a result, we could test hundreds of microbes for the capacity to be an opportunistic pathogen on plants, if we wished, but this exercise has not been considered important among plant pathologists, particularly in light of the intense competition for agricultural and ecological research funding. These important ethical, economic, and dogmatic differences in the study of animal and plant diseases mean that we know little about opportunistic plant pathogens and especially little about those microbes that Casadevall and Pirofski classify as true opportunistic pathogens (Casadevall and Pirofski 1999).

7.4 Is Opportunistic Synonymous with Necrotrophic?

The most common general descriptors for plant pathogens are biotrophic, hemibiotrophic, and necrotrophic, and these are defined in the following paragraphs (Fig. 7.2). The delineation between categories are blurred, and how a pathogen is categorized may depend upon the host and environment. In the past, necrotrophs were often described as opportunistic or brute-force pathogens, but a more sophisticated understanding of how necrotrophic pathogens work shows that their virulence strategies are complex and not simply due to massive production of enzymes or toxins. Because of the difficulty in classifying microbes into such simple categories and because the boundaries of the categories are blurred, alternate classification schemes have been proposed. For example, Aguilar-Trigueros and colleagues proposed using a multidimensional trait-based approach to classifying root-associated fungi since it can be difficult to draw distinct lines between

Fig. 7.2 Diagrams of continuum of pathogen categories, from biotroph to saprotroph. Biotrophs can only grow on living hosts. Hemi-biotrophs obtain nutrients and grow on living hosts, but eventually kill the host cells during pathogenesis. Necrotrophs kill host cells during the early stages of disease and obtain nutrients primarily from dead host cells. Saprotrophs are not pathogens and obtain nutrients from dead cells. The borders between these categorizations are not fixed, and the placement of pathogenic species on this continuum can shift depending on host genotype, pathogen genotype, and environment

pathogens and saprotrophs (Aguilar-Trigueros et al. 2014). They identified several traits, such as production of enzymes involved with different types of plant colonization, among which are cell wall-degrading enzymes and invertases, and specific microbial structures, such as appressoria or resting structures, that might be important for characterizing fungi along what they describe as the "symbiotic and saprotrophic continuum." A similar approach could be used for all microbes in the phytobiome, but as yet, we do not know enough about the ecology and genomics of most microbes in the phytobiome to construct useful multidimensional classifications, so we still rely on simple one-dimensional categorizations.

7.4.1 Biotrophic Pathogens

Biotrophic pathogens colonize and obtain their nutrients only from living plant tissue. This class of pathogens has intimate interactions with their host plants and is generally located entirely within host cells, such as viruses or phloem-colonizing bacteria, or invades living host cells with specialized structures, such as haustoria produced by rust fungi or downy mildew fungi. Biotrophic pathogens can be beneficial. For example, a mild virus strain can protect a plant against more severe strains of the same virus, a phenomenon known as cross protection (Nishiguchi and Kobayashi 2011). Biotrophic pathogens are not generally considered opportunistic pathogens.

7.4.2 Hemi-biotrophic Pathogens

Hemi-biotrophic pathogens can multiply on living host cells for a portion of their life cycle, but may eventually kill the host cells. *Pseudomonas syringae*, a common colonist on leaf surfaces, is among the most widely studied and, at the genetic level, one of the best understood hemi-biotrophs (O'Brien et al. 2011). In laboratory experiments, it multiplies quickly after being inoculated into or on leaves. Necrotic lesions will form after a few days, at which point bacterial multiplication stops. Bacterial mutants unable to form lesions may not show fitness defects in laboratory settings, but the lesions appear to be important for fitness under field conditions (Hirano et al. 1997), perhaps because lesions provide a protected refuge for the pathogen cells.

The *Pseudomonas syringae* bacterium is easily isolated from angiosperm leaves in natural and agricultural systems (Morris et al. 2013). This bacterium does not usually cause disease when it is present on leaves and is not known to cause epidemics in natural systems. Rather, it is most problematic in agricultural systems where crop diversity within a field is low, such as in tomato production. And with some crops, even those with very low diversity, *P. syringae* is not an important pathogen. For example, in potato, *P. syringae* is easily isolated from leaves, but there are no reports of it causing disease. Since the most typical state of this pathogen seems to be as a colonist of healthy leaves, it suggests that *P. syringae* is often a commensal or, possibly even through competition with other microbes, serves as a beneficial bacterium on leaf surfaces.

Can we consider hemi-biotrophic pathogens that are common in the environment but that only cause disease under specific conditions, such as in a monocultured cropping system when rain is frequent, to be opportunistic pathogens? If so, then hemi-biotrophic pathogens such as *P. syringae* would sometimes fit this description. And yet, plant pathologists do not consider this pathogen to be opportunistic because of the highly specific way in which it can manipulate host plant cells. In order to grow in association with plant cells, *P. syringae* secretes toxins and effector proteins that have specific host cell targets (Lindeberg et al. 2012). These toxins and proteins disable plant defenses and promote bacterial growth adjacent to the cells being manipulated by the bacterial pathogen. This intimate highly regulated interaction is not what we would expect from an opportunistic pathogen.

7.4.3 Necrotrophic Pathogens

Necrotrophic pathogens kill plant cells early in their interaction with the plant and then digest the plant cell components. At a first glance, necrotrophs appear to be simply an impatient or aggressive saprotroph, and for this reason, they used to commonly be called brute-force or opportunistic pathogens. For example, necrotrophs and saprotrophs encode homologous plant cell wall-degrading enzymes, and some necrotrophs are capable of growing as saprotrophs, although in that role they are generally poor competitors. Necrotrophs are best known for causing significant losses in stored fruits or vegetables or in plants physically damaged by weather or farm equipment, which is another reason that this class of pathogens is referred to as opportunistic pathogens.

Over the past decade, as genome sequences became available for necrotrophic plant pathogens and we gained a more sophisticated understanding of pathogenesis and plant defenses against microbes, it became clear that at least some species of necrotrophs use intricate strategies to overcome plant defenses and are not simply brute-force pathogens [e.g., see Toth and Birch (2005)]. As a result, the term opportunistic has fallen mostly out of use when describing necrotrophic pathogens, and it fell out of use with little or no discussion of whether this term is appropriate for commonly studied plant pathogens. Therefore, the goal of this chapter will be to examine some of what is known about necrotrophic pathogens of terrestrial plants.

7.4.3.1 Broad Host Range Versus Narrow Host Range Necrotrophs

Necrotrophs can be divided into two categories based on their host ranges, and these categories generally reflect the pathogenicity strategy of the pathogen. Narrow host range necrotrophs tend to rely upon host-specific toxins that attack a target present only in some species or some subtypes of a particular species [e.g., see Lorang et al. (2012) and Condon et al. (2014)]. In contrast, broad host range pathogens encode toxins or enzymes that affect targets present in a wide range of or in all plant species. Broad host range pathogens have arisen within multiple kingdoms of biological families, and they use similar strategies to attack plants, including production of plant cell wall-degrading enzymes and toxins, manipulation of the local pH, and subversion of programmed plant cell death pathways. The border between hemi-biotroph and necrotroph can be blurry, and single genera, such as *Pseudomonas*, may include both hemi-biotrophic and necrotrophic pathogens. This chapter will focus on broad host range necrotrophs since opportunistic pathogens are most likely to be found within this category of plant pathogens.

7.4.3.2 How Many Times Has the Trait "Necrotroph" Arisen?

Many symbiotic categories have arisen numerous times in a myriad of environments. As a result, necrotrophic pathogens of terrestrial plants are found in multiple evolutionary branches in multiple phyla. It is not possible to predict from microbial phylogeny in which genera necrotrophs are likely to be found, and necrotrophs tend to have convergent traits, many of which will be described below. Similarly, plant resistance to necrotrophic pathogens cannot be predicted to the plant species or even genus level [e.g., see Chung et al. (2011)]. Therefore, individual plant genotypes must be assayed to identify plant lines resistant or tolerant to necrotrophs.

Necrotrophs tend to have similar life cycles, regardless of their phylogeny. Their life cycle is complex and involves multiple distinct stages. First, they must attach to and, if they are bacterial pathogens, invade their hosts through wounds or natural openings such as stomata or lenticels. In contrast, some necrotrophic fungi are able to directly penetrate the plant cuticle via specialized structures, such as appressoria. Once on or inside the plant, the necrotroph may enter a latent phase, with disease triggered by environmental stresses or host maturity. Lesion formation generally occurs concurrently with pathogen growth for both bacterial and fungal necrotrophs. This is followed by pathogen dispersal to new hosts via spores, agricultural equipment, or water. Some necrotrophs may have invertebrate vectors, such as fruit flies or snails, but this has been little explored. One clear example is the significant reduction of bacterial stalk rot of maize after the introduction of insect-resistant transgenic maize (Dalmacio et al. 2007).

7.4.3.3 How Do Broad Host Range Necrotrophs Cause Disease?

The broad host range necrotrophs are among the least commonly studied class of plant pathogens, perhaps because for many years, researchers thought their virulence strategy consisted of little more than massive production of plant cell wall-degrading enzymes or toxins. There is extensive literature on some of these pathogens, such as the widespread fungal pathogen *Sclerotinia sclerotiorum*, while others, such as pectolytic *Bacillus* and *Clostridium* species, remain almost entirely unexamined. The effort put into understanding the virulence mechanisms and epidemiology of necrotrophic plant pathogens, as with all other types of plant pathogens, generally has more to do with the ease of working with the pathogen and its hosts, and whether it is a recent invader into an agricultural system, than with the yearly losses or long-term potential for loss of the pathogen. For this chapter, three pathogens, which are common worldwide, will be used to illustrate broad host range necrotrophs as possible examples of opportunistic pathogens, including the oomycete pathogen *Pythium*, the fungal pathogen *Sclerotinia*, and the gram-negative bacterium *Pectobacterium*.

Broad host range necrotrophs tend to survive well outside their hosts. For example, *Pectobacterium* survives well in surface and ground water, and *Pythium* and *Sclerotinia* can both make durable survival structures that overwinter well in soil. They also tend to be common in the environment and easily isolated from soil, water, and apparently healthy plants. Because they can be easily isolated from healthy plants, the ecological role that these microbes play is unclear. Are they providing a benefit to the plant or are they waiting for the first sign of distress to overwhelm the plant defenses?

These necrotrophic pathogens tend to be capable of quiescent infections [for a recent review, see Prusky et al. (2013)]. Depending on the pathogen, they may enter a plant through natural openings, such as stomata or lenticels, through wounds, or through abscission scars on fruits. Some fungal necrotrophs produce structures known as appressoria that are capable of penetrating intact plant tissue. The pathogen may remain inside the plant for long periods of time in a quiescent state and not cause disease until weeks or months after infection. The signals and molecular mechanisms that trigger onset of disease are often not clear, but several mechanisms have been reported, including changes in plant cell wall structure and the reduction in defense compounds that occur as fruit ripening or changes in pH, temperature, or ethylene concentration. In the case of pathogens that attack stored fruit, we view these microbes as pathogens because we wish to use the fruit ourselves. They may, however, benefit the plant by releasing the plant seeds from the fruit and, through decay of the fruit, enriching the soil around the plant seed.

Pythium

The oomycete *Pythium* causes seedling and root rots in crops such as beans (Nzyngize et al. 2012) and can also rot fleshy vegetables, such as potato tubers (Fiers et al. 2012). Some species promote plant growth rather than cause disease (Benhamou 2012). Pathogenic *Pythium* species cause a watery decay of plant tissue. Young seedlings or plant cuttings will appear stunted, and if removed from soil or potting mix, root decay is usually obvious. In fleshy vegetables, *Pythium* kills the plant cells without causing extensive maceration, turning the vegetable into something resembling a wet sponge. This symptom gives the *Pythium* disease of potato its name, "leak," since the potato tubers remain intact, but when squeezed, the water in the tuber will leak out. Distinctive *Pythium* oospores or zoospores are usually present in the diseased tissue, making it possible to observe the presence of the oomycete. This pathogen is associated with seeds planted into wet soil and with stored vegetables. *Pythium* is common in the environment and is found in soil, water, and decaying plant material. There are numerous *Pythium* species, but only a few are well-characterized, with *Pythium ultimum* among those most clearly described.

Pythium encodes multiple classes of virulence genes including plant cell wall-degrading enzymes and secreted proteins, known as effectors, that likely disrupt plant defenses (Adhikari et al. 2013; Jiang and Tyler 2012). This genus lacks some virulence genes present in hemi-biotrophic oomycetes, such as the RXLR class of effector proteins named because they contain an arginine-X-leucine-arginine sequence at the amino terminus of the protein with X potentially being any amino acid, that are important for virulence of the hemi-biotrophic oomycete pathogen *Phytophthora*. *Pythium* may produce toxins and plant hormones as well (Ottmann et al. 2009; Rey 2001 #5594). Because the *Pythium* genome is large, complex, and diploid for most of its life cycle, and because the *Pythium* genome is relatively difficult to manipulate, much remains to be learned about how oomycete necrotrophs cause disease.

Sclerotinia

The fungal pathogen *Sclerotinia sclerotiorum* is a homothallic (self-fertile) fungus that infects over 400 dicot plant species and causes white mold disease. The fungus overwinters as sclerotia, which are melanized bundles of tightly packed differentiated hyphae (Chet et al. 1969). Sclerotia form apothecia that release ascospores capable of infecting plants, generally through flowers. Further spread during the growing season is caused by hyphae spread from plant to plant. Hyphae can also grow from sclerotia and infect host plants. The fungus causes necrosis on plant stems, killing plants and reducing overall productivity of the field. The hyphae are often visible on diseased or dead plants as white fuzz, and hard black sclerotia may be evident inside plant stems.

An important part of *S. sclerotiorum* pathogenicity is its ability to turn the plant defense response against the plant. It induces plant cells to initiate a programmed cell death and then metabolizes the dead plant cells. The fungus produces oxalic acid, which contributes to pathogenicity through multiple mechanisms, including enhancing plant cell wall degradation, de-regulating plant defenses, and inducing programmed plant cell death [Amselem et al. (2011) and references cited therein]. Monocots, which degrade oxalic acid, are resistant to this broad host range necrotroph. Transformation of dicot plants with the monocot (gene) confers resistance to white mold, demonstrating the key role that this single small molecule plays in *Sclerotinia sclerotiorum* pathogenicity. *Sclerotinia* encodes genes for production of other secondary metabolites and for plant cell wall-degrading enzymes that also likely contribute to virulence (Amselem et al. 2011).

Pectobacterium

The gram-negative bacterium *Pectobacterium* is a diverse genus that infects numerous angiosperm plants (Ma et al. 2007). Pectobacterium is commonly found in decaying potatoes along with pectolytic *Clostridium* and pectolytic *Bacillus*, both of which are generally referred to as secondary invaders. However, each of these three genera are capable of decaying plants on their own, so this combination of species might be more accurately referred to as a disease complex. However, to date, *Pectobacterium* is the only one of these three genera that has been extensively studied as a plant pathogen.

Pectobacterium relies upon a diverse array of plant cell wall-degrading enzymes (PCWDE), including pectinases, cellulases, xylanases, and proteases to macerate plant cell walls, which causes wilt and decay symptoms. As a genus, *Pectobacterium* is a broad host range necrotroph, but individual species sometimes appear to be limited to particular types of hosts. For example, *Pectobacterium atrosepticum* is rarely found on plants other than potato, and *Pectobacterium aroidearum* is mainly found on monocots, while other *Pectobacterium* species are generally found on dicots.

The genomic factors that affect *Pectobacterium* host range remain unknown, and it is not clear if narrow host range *Pectobacterium* lack key virulence genes to allow them to become broad host range or if these narrow host range species induce resistance on most host plants, thereby limiting their host range. This could occur if they encode genes that greatly increase their fitness on a few host species and if these same genes induce resistance on most other plant species, a phenomenon that is commonly seen in biotrophs and hemi-biotrophs.

Although virulence factors that expand or contract the host range of *Pectobacterium* are not yet described, the extensively studied type III secretion system (T3SS) may affect which tissues the pathogen can infect. This secretion system consists of a needlelike protein appendage used as a sensory probe. Secreted effector proteins translocated into host cells through the needle aid with infection of the eukaryotic target by disrupting host protein function or altering host gene expression. *Pectobacterium* strains encoding a T3SS and the single effector described in *Pectobacterium* to date can cause disease in plant leaves, while strains lacking the T3SS and effector, either naturally or through designed mutations, are impaired in causing disease in leaves.

7.4.3.4 Epidemiology and Management of Broad Host Range Necrotrophs

These pathogens are important in agricultural systems, but have been little studied in natural ecosystems. Unlike important invasive biotrophs or hemi-biotrophs, such as Dutch elm disease, this class of pathogens has not been described as making major changes in natural landscapes, although, as discussed later in this chapter, they are likely to influence the types of plants that grow in wetlands.

The broad host range necrotrophs tend to be common in soil and water and, like *Pectobacterium*, may also be associated with or spread by invertebrates. They tend to be able to infect plants at any point during growth or storage of either the plant or harvested fruit or vegetables, but they tend not to be able to affect seeds during storage. Farmers manage these diseases mainly through sanitation, crop rotation, exclusion of the pathogen through planting healthy seed, and careful management of field irrigation and fruit and vegetable storage conditions. Resistant plant varieties are rarely available to help manage these diseases, and, for the most part, how plants resist broad host range necrotrophs is poorly understood.

7.4.3.5 The Critical Role of Pectinases in Necrotrophic Pathogenesis

Plants cover most of the terrestrial surface and are abundant in some types of water bodies. Plant cell walls contain much carbon and other nutrients that can serve as a food source for bacterial growth, and numerous types of plant cell wall-degrading enzymes are encoded by microbes. Genes encoding these enzymes are exchanged among microbes, and the microbes encoding these enzymes are widespread, so at first glance, it is amazing that plants can survive this potential onslaught of microbes capable of digesting their cell walls. Clearly, there has been strong selection in plants for mechanisms to inhibit cell wall degradation.

Pectate, hemi-cellulose, and cellulose are the main polymers in the primary plant cell wall. These polymers are analogous to reinforced concrete used in buildings, with calcium-cross-linked pectate (polygalacturonate) and hemi-cellulose (polyose) polymers playing the role of concrete and cellulose (polyglucose) acting as steel bars to provide strength and rigidity. Pectate is commonly methylated to make pectin, and pectin is less susceptible to enzymatic degradation. The degree of pectin methylation correlates with plant resistance to broad host range necrotrophs [e.g., see Marty et al. (1997)].

Plant cells are under high turgor pressure, and the plant cell wall is important for maintaining cell integrity. If the cell wall becomes too damaged, the plant cell will burst, causing the plant to wilt or show decay symptoms. The pectin and hemi-cellulose in plant cell walls are good nutrient sources, and many necrotrophs and saprophytes can grow on pectin or hemi-cellulose as their sole carbon source.

Because the cellulose strands are protected by pectin, plant pathogens must first depolymerize pectin before being able to nutritionally access cellulose or hemi-cellulose; thus, pectinases are key in necrotrophic pathogenesis (Lionetti et al. 2007; Berlin et al. 2007). These pathogens often produce multiple types of pectinases, including pectin lyases and polygalacturonases. These enzymes can vary in cofactors, pH optima, and pI (isoelectric point), thereby altering their efficacy in different plant species (Roy et al. 1999; Nachin and Barras 2000; Prusky and Yakoby 2003). Plants also produce multiple types of plant cell wall modifying enzymes, and these enzymes are required for plant growth and maturation. Necrotrophic pathogens are able to modulate expression of these enzymes to affect efficacy of pathogen-produced pectinases. For example, both *Botrytis* and

Pectobacterium can induce an *Arabidopsis* enzyme that demethylates pectin, which presumably makes the pathogen-encoded pectinases more efficacious in pectate degradation (Raiola et al. 2011).

As mentioned above, methylation of pectate, to produce pectin, can protect the plant cell wall against some pectinases. Plants also produce molecules that inhibit pectinases (McMillan and Perombelon 1995; Jolie et al. 2010). The evolution of pectinases and their inhibitors likely fits the Red Queen hypothesis which proposes that the basic survival of a species requires it to constantly evolve, adapt, and proliferate against equally active opponents in order to simply hold its competitive ground, but this has been relatively little studied in plant pathogens. Overall, the study of evolution in plant pathogens has suffered because models tend to be built with a few lab strains of pathogens amenable to molecular biology and plants lines obtained from plant breeders rather than with randomly selected strains from either naturally occurring or agricultural field sites. When randomly selected environmental strains are examined (Kniskern et al. 2010), the findings tend not to agree with models derived from lab strains and plant-breeding lines [e.g., compare Kniskern et al. (2010) and Bent and Mackey (2007)].

7.4.3.6 Necrotrophs Can Turn Plant Defenses Against the Plant

A major difference across pathogenic lifestyles is which partner in the interaction controls when the host cell dies. Biotrophs require living cells, and so if the plant host can detect that it is being parasitized and kill those plant cells supporting the pathogen, then the plant can protect the rest of itself against disease. This response is generally called the hypersensitive response (HR), and it is considered a key part of the plant defense response. Hemi-biotrophs eventually kill plant cells, but premature plant cell death via the HR provides resistance against hemi-biotroph plant pathogens. In contrast, necrotrophs induce cell death and profit from this plant response [reviewed in Mengiste (2012)]. At least some necrotrophs, such as *Pectobacterium carotovorum*, cause no apparent suppression of plant defenses (Kim et al. 2011).

The HR and associated plant defenses, such as large reactive oxygen bursts and deposition of the polysaccharide callose, are triggered when a plant cell recognizes that it is being parasitized. Some biotrophic and hemi-biotrophic pathogens encode numerous proteins that are secreted into host cells to suppress these defensive responses. For example, the plant pathogen *P. syringae* translocates proteins into host cells that suppress callose formation (DebRoy et al. 2004). Plant recognition of either the pathogen proteins themselves or the effect of these pathogen proteins on host cells triggers the HR (Grant et al. 2006).

Since specific plant proteins often recognize specific pathogen proteins or their effects, this phenomenon is called the gene-for-gene interaction. In some cases, plant breeders have used these major dominant plant resistance genes to develop plant varieties that resist important pathogens. However, because pathogen effector repertoires tend to be diverse among strains within a pathogenic species, use of these types of resistance genes tends to result in a strong selection for pathogen strains that lack the effector protein targeted by the resistance gene, and this type of resistance tends not to be durable for long periods of time in agricultural settings.

Gene-for-gene responses are not commonly found for broad host range necrotrophic pathogens. This may be because these pathogens are using effector proteins or small molecules to target highly conserved resistance pathways to induce, rather than suppress, HR-induced plant cell death. For example, *Pectobacterium* encodes a single protein capable of eliciting plant defense, DspE, which is likely translocated directly into plant cells (Kim et al. 2011; Hogan et al. 2013). It has been suggested that DspE is among the most conserved family of plant defense eliciting proteins, which are known as effectors, found in bacterial pathogens. Unlike other DspE alleles, the *Pectobacterium* DspE is truncated, does not suppress plant defenses, and lacks all conserved cysteines, suggesting that it lacks the enzymatic activities often found in alleles from hemi-biotrophs. Its only function appears to be induction of plant cell death for the benefit of the *Pectobacterium* cells.

The plant cell death triggered by necrotrophic and biotrophic plant pathogens may not occur via the same pathway. The broad host range necrotroph *Sclerotinia sclerotiorum* triggers a little understood apoptotic-like plant cell death that is tied to fungal production of oxalic acid and that appears to benefit the fungus. Fungal mutants unable to produce oxalic acid are nonpathogenic, but they still induce plant cell death and apparently do so by eliciting the macromolecular degradation pathway termed autophagy (Kabbage et al. 2013). Conversely, the necrotroph *Botrytis cinerea* induces and benefits from autophagy (Lai et al. 2011). Therefore, it appears that pathogens have at least two pathways which they can use to elicit plant cell death and that these pathways and their associated phenomena may not be equally beneficial to necrotrophic pathogens.

Many additional phenomena occur along with plant cell death during plant resistance responses, and these other plant responses also affect biotrophs and necrotrophs differently. For example, when plants recognize an attacking pathogen, they release a burst of reactive oxygen species (ROS), such as hydrogen peroxide and oxygen radicals. This ROS burst inhibits growth of biotrophs and hemi-biotrophs, but may benefit necrotrophs, particularly if the ROS burst occurs later in disease development (Williams et al. 2011; Govrin and Levine 2000).

7.4.3.7 Plant Signaling Molecules and Hormones Affect Resistance to Broad Host Range Necrotrophs

Plants lack circulating immune cells and adaptive immunity, but they can still effectively protect themselves against disease. Plants resist microbial pathogens through preformed antimicrobial chemicals, and they also respond to pathogens by inducing defenses. Once a plant cell recognizes that it is being manipulated by a pathogen, multiple plant hormones signal plant disease defense pathways in plant cells. Some of these hormones are conserved across distantly related plant families, while others are found only in particular plant families.

Some of the most studied plant defense signal hormones include salicylic acid (SA), jasmonic acid (JA), abscisic acid (ABA), and ethylene (ET). In general, researchers describe SA-mediated defenses as protecting against biotrophs and the initial attack of hemi-biotrophs, while JA- and ET-mediated defenses protect against necrotrophs (Glazebrook 2005). However, there are numerous examples that do not fit this generalization, and it is clear that complex ties exist among hormone-mediated pathways that are probably affected differently depending upon the particular host–pathogen species that are interacting. As with much of biology, we try to describe these dynamic interactions with static two-dimensional models that poorly describe even simple interactions.

A common feature among necrotrophs is production of plant hormones. Both eukaryotic and prokaryotic necrotrophs produce plant hormones as well as molecules that interfere with hormone signaling. For example, necrotrophs may produce the defense hormones ethylene or abscisic acid or the ethylene inhibitors (Marquez-Villavincencio et al. 2011). These may also produce the plant growth hormones auxin or gibberellic acid, which likely affects the ability of the necrotroph to obtain nutrients from the plant host. Pathogens may also synthesize molecules such as auxin as host-manipulating hormones (Yang et al. 2007).

7.4.3.8 Limits to Our Understanding of Necrotroph Molecular Interactions with Plants

There are significant inconsistencies in data from experiments on necrotroph–plant interactions, and there are also large areas that remain relatively unexplored. To date, the majority of the work used to develop resistance models was completed in growth chambers, with a single plant species (*Arabidopsis*) grown in potting mix and with only a few strains of plant pathogens. Each of these important limitational aspects may affect the appropriate interpretation of data derived from this work. For example, interactions among signaling pathways are not necessarily conserved across plant families. Also, results from plants grown in growth chambers often do not reflect what occurs when plants are grown under field conditions. Finally, potting mix is both antimicrobial and relatively sterile compared to natural or agricultural soil, so that a large portion of the typical phytobiome is missing in growth chamber experiments and the phytobiome undoubtedly influences plant defenses. Therefore, what we think we know about these defense pathways must be considered with caution, and alternate explanations should be carefully considered.

Another significant consideration is the lack of available model systems for plant storage organs. Necrotrophic pathogens are a significant problem on storage organs of plants, such as tap roots, tubers, rhizomes, and fleshy fruits. Most of the common plant models, such as *Arabidopsis*, rice, and *Medicago*, lack these organs. Tomatoes, which are also used commonly as a plant model, produce fleshy fruit, but the majority of genetic and genomic work with tomato–necrotroph interactions is still completed with leaf, stem, or root pathogens.

7.4.3.9 Effects of Environment on Plant Resistance Responses

Disease caused by broad host range necrotrophs is often associated with extreme environmental conditions, particularly high levels of moisture, or with plant parts that have been removed from the plant, such as cut flowers or tomato fruits. Plant defense responses are dependent upon the environment, with multiple environmental factors affected plant defenses in complex ways [for a recent review, see Hua (2013)]. Most of this work has been done with biotrophic or hemi-biotrophic pathogens; the effect of environment on molecular defenses against broad host range necrotrophs remains almost entirely unexplored.

Where environment has been examined, there tend to be effects on both the plant and the pathogen. For example, UV light affects both defense responses in plants (Demkura and Ballaré 2012) and virulence gene expression in the necrotroph *Botrytis cinerea* (Canessa et al. 2013). In at least some cases, pathogen virulence genes are upregulated under conditions that suppress plant defenses. Soft rot bacterial pathogens upregulate pectate lyases under low-oxygen conditions (Babujee et al. 2012); plants are unable to heal wounds under these same conditions, making them susceptible to soft rot bacteria.

7.4.3.10 Latent Infections of Plants with Necrotrophs

Necrotrophs can often be isolated from healthy plants, suggesting that at least some necrotrophs also exist in a latent or quiescent phase in association with plants (Prusky et al. 2013). Both the environment and increasing maturity of the host can affect latent infections. Some environments inhibit plant defenses or metabolism, such as reduced oxygen levels or temperature and water stress, and these environments are considered conducive to disease. One important difference in pathogenicity in plant structures such as tubers or fruits compared to roots is the low-oxygen environment that often occurs inside the fruit or tuber during infection. Gene expression of necrotrophs under low-oxygen environments has only recently been examined (Babujee et al. 2012) and is still little understood.

Changes in plant maturity that result in changes in plant cell wall or cuticle structure, plant hormone homeostasis, or reductions in antimicrobial compounds also are correlated with the transition of necrotrophs from latent to active pathogens [reviewed for fungal necrotrophs in Prusky et al. (2013)]. Many of these changes are similar to what occur in other plant tissues. For example, the necrotroph *Colletotrichum* secretes ammonia which, in mature fruit, increases the local pH, induces activation of the salicylic acid (SA) production pathway, and eventually leads to plant cell leakage and death, which in turn favors necrotroph disease development (Alkan et al. 2012). Some bacterial pathogens also increase the local pH of plant tissues, but whether this affects plant gene expression remains unexplored (Marquez-Villavincencio et al. 2011). Other fungi acidify their environment in fruits, as occurs in leaves, by producing organic acids, with an example being the production of oxalic acid by *S. sclerotiorum* (Rollins and Dickman 2001).

7.4.3.11 Do Opportunistic Pathogens Control Ecosystems?

Plant pathogens are capable of reshaping entire landscapes. This effect is particularly evident when invasive pathogens kill trees across large regions and change the way our landscape both appears and functions. The disappearance of large chestnut trees and also of butternut and elm trees from most of North America in the twentieth century due to invasive fungal pathogens changed forest and urban ecosystems and destroyed economic opportunities. The bacterial pathogen *Liberibacter*, which causes citrus greening, now threatens to change our agroecosystem within the next few years by its destructive effects on citrus production.

Currently, much of the terrestrial landscape has been converted to modern agricultural production, with an estimated 40 % of all plant production going to support humans. Modern agriculture occurs on highly disturbed grounds, with crop plants bred to have many attributes of invasive species. For example, crop seeds germinate quickly when conditions are appropriate, often emerging from the soil in less than a week. Crop plants tend to be homogeneous within a field and only one or two crops covering large regions in some parts of the world.

Modern agriculture is essentially an exercise in reducing environmental diversity in an effort to have predictable and high yields from a few plant species bred to thrive in a disturbed environment. As a result, the disease and weed problems in modern agriculture are often predictable due to the homogeneous landscape we have created. Farmers can even use forecasting tools based on just a few environmental parameters, such as temperature and moisture level, to accurately predict when some diseases are likely to develop and will spray pesticides based on these forecasting systems (REF).

In agricultural systems, broad host range necrotrophic pathogens, such as *Pythium* and soft rot bacterial pathogens, tend to cause losses when there is excessive soil moisture. This suggests that *Pythium* and similar necrotrophic pathogens may affect the success of plant species in wetlands. Farmers take advantage of plant susceptibility to excess water and use water for weed control in some crops, such as cranberry, where fields are strategically flooded for a short time in the spring to manage weeds (Sandler and Mason 2010). Unfortunately, the role of broad host range necrotrophs in wetland ecosystems remains almost entirely unexplored. Examining this question would likely lead both to a better understanding of wetland ecosystems and to a better understanding of how plants resist broad host range necrotrophs.

7.4.3.12 Evolution of Necrotrophic Pathogens

Over the past 20 years, there has been a convergence in the study of the molecular mechanisms that contribute to plant and animal pathogenesis. Genetic and later genomic research showed that plant and animal pathogens use the same secretion systems and some of the same virulence strategies when manipulating host cells. In contrast, evolutionary theory describing plant and animal pathogenesis still seems to be developing separately. This latter point is evidenced in most work that involves classifying pathogen types, which generally considers only either plant or animal pathogens, and in theories that try to describe trade-offs in selective pressures on pathogens.

7.5 Conclusion

Opportunistic pathogenicity, as an ecological concept typically defined for animal pathogens, (Casadevall and Pirofski 1999), has not been examined for plant pathogens. A census of the proportion of soil-, water-, or plant-borne microbes capable of causing disease on a compromised plant would be of interest since it could highlight reservoirs of potential virulence genes for more aggressive plant pathogens. The class of microbes most closely associated with a true opportunistic pathogenicity in terrestrial plants are the broad host range necrotrophic pathogens. Broad host range necrotrophs are found in multiple kingdoms of life and in multiple distantly related families, demonstrating that this phenotype has arisen independently multiple times. The constraints and selection pressures acting upon evolution of broad host range necrotrophs are poorly understood. These pathogens kill host cells early in the disease process and obtain nutrients primarily from dead host cells. The causative pathogens do not rely upon a single toxin or enzyme for pathogenicity. Their ability to cause disease relies upon multiple plant cell wall-degrading enzymes, toxins and other secondary metabolites, as well as their capabilities to induce programmed plant cell death and to withstand the plant defense responses. In turn, plants resist broad host range pathogens with multiple approaches, none of which are thoroughly understood. Our understanding thus far is that antimicrobial peptides and metabolites, enzyme inhibitors, and plant cell wall modifications may all contribute toward resistance against broad host range necrotrophs.

References

Adhikari BN, Hamilton JP, Zerillo MM et al (2013) Comparative genomics reveals insight into virulence strategies of plant pathogenic oomycetes. PLoS One 8:e75072

Aguilar-Trigueros CA, Powell JR, Anderson IC et al (2014) Ecological understanding of root-infecting fungi using trait-based approaches. Trends Plant Sci 19:432–438

Alkan N, Fluhr R, Prusky D (2012) Ammonium secretion during *Colletotrichum coccodes* infection modulates salicylic and jasmonic acid pathways of ripe and unripe tomato fruit. Mol Plant Microbe Interact 25:85–96

Amselem J, Cuomo CA, van Kan JAL et al (2011) Genomic analysis of the necrotrophic fungal pathogens *Sclerotinia sclerotiorum* and *Botrytis cinerea*. PLoS Genet 7:e1002230

Babujee L, Apodaca J, Balakrishnan V (2012) Evolution of the metabolic and regulatory networks associated with oxygen availability in two phytopathogenic enterobacteria. BMC Genomics 13:110

Benhamou N, Le Floch G, Vallance J, Gerbore J, Grizard D, Rey P (2012) *Pythium oligandrum*: an example of opportunistic success. Microbiology 158:2679–2694. doi:10.1099/mic.0.061457-0

Bent AF, Mackey D (2007) Elicitors, effectors, and R genes: the new paradigm and a lifetime supply of questions. Annu Rev Phytopathol 45:399–436

Berlin A, Maximenko V, Gilkes N et al (2007) Optimization of enzyme complexes for lignocellulose hydrolysis. Biotechnol Bioeng 97:287–296

Brown SP, Cornforth DM, Mideo N (2012) Evolution of virulence in opportunistic pathogens: generalism, plasticity, and control. Trends Microbiol 20:336–342

Canessa P, Schumacher J, Hevia MA et al (2013) Assessing the effects of light on differentiation and virulence of the plant pathogen *Botrytis cinerea*: characterization of the white collar complex. PLoS One 8:e84223

Casadevall A, Pirofski LA (1999) Host-pathogen interactions: redefining the basic concepts of virulence and pathogenicity. Infect Immun 67:3703–3713

Chet I, Henis Y, Kislev N (1969) Ultrastructure of sclerotia and hyphae of *Sclerotium rolfsii* Sacc. J Gen Microbiol 57:143–147

Chung YS, Holmquist K, Spooner DM et al (2011) A test of taxonomic and biogeographic predictivity: resistance to soft rot in wild relatives of cultivated potato. Phytopathology 101:205–212

Condon BJ, Wu D, Kraševec N et al (2014) Comparative genomics of *Cochliobolus* phytopathogens. In: Dean R, Lichens-Park A, Chitteranjan K (eds) Genomics of plant-associated fungi: monocot pathogens. Springer, New York, pp 41–67

Dalmacio SC, Lugod TR, Serrano EM et al (2007) Reduced incidence of bacterial rot on transgenic insect-resistant maize in the Philippines. Plant Dis 91:346–351

DebRoy S, Thilmony R, Kwack YB et al (2004) A family of conserved bacterial effectors inhibits salicylic acid-mediated basal immunity and promotes disease necrosis in plants. Proc Natl Acad Sci U S A 101:9927–9932

Demkura PV, Ballaré CL (2012) UVR8 mediates UV-B-induced Arabidopsis defense responses against *Botrytis cinerea* by controlling sinapate accumulation. Mol Plant 5:642–652

Fiers M, Edel-Hermann V, Chatot C et al (2012) Potato soil-borne diseases. A review. Agron Sustain Dev 32:93–132

Govrin EM, Levine A (2000) The hypersensitive response facilitates plant infection by the necrotrophic pathogen *Botrytis cinerea*. Curr Biol 10:751–757

Glazebrook J (2005) Contrasting mechanisms of defense against biotrophic and necrotrophic pathogens. Annu Rev Phytopathol 43:203–227

Grant SR, FIsher EJ, Chang JH et al (2006) Subterfuge and manipulation: type III effector proteins of phytopathogenic bacteria. Annu Rev Microbiol 60:425–449

Hirano SS, Ostertag EM, Savage SA (1997) Contribution of the regulatory gene *lemA* to field fitness of *Pseudomonas syringae* pv. syringae. Appl Environ Microbiol 63:4304–4312

Hogan CS, Mole BM, Grant SR et al (2013) The type III secreted effector DspE Is required early in *Solanum tuberosum* leaf infection by *Pectobacterium carotovorum* to cause cell death, and requires Wx((3-6))D/E motifs. PLoS One 8:e65534

Hua J (2013) Modulation of plant immunity by light, circadian rhythm, and temperature. Curr Opin Plant Biol 16:406–413

Jiang RHY, Tyler BM (2012) Mechanisms and evolution of virulence in oomycetes. Annu Rev Phytopathol 50:295–318

Jolie RP, Duvetter T, Van Loey AM et al (2010) Pectin methylesterase and its proteinaceous inhibitor: a review. Carbohydr Res 345:2593–2595
Kabbage M, Williams B, Dickman MB (2013) Cell death control: the interplay of apoptosis and autophagy in the pathogenicity of *Sclerotinia sclerotiorum*. PLoS Pathog 9:e1003287
Kim HS, Thammarat P, Lommel SA et al (2011) *Pectobacterium carotovorum* elicits plant cell death with DspE/F, but does not suppress callose or induce expression of plant genes early in plant-microbe interactions. Mol Plant Microbe Interact 24:773–786
Kniskern JM, Barret LG, Bergelson J (2010) Maladaptation in wild populations of the generalist plant pathogen *Pseudomonas syringae*. Evolution 65:818–830
Lai Z, Wang F, Zheng Z et al (2011) A critical role of autophagy in plant resistance to necrotrophic fungal pathogens. Plant J 66:953–968
Lindeberg M, Cunnac S, Collmer A (2012) *Pseudomonas syringae* type III effector repertoires: last words in endless arguments. Trends Microbiol 20:199–208
Lionetti V, Raiola A, Camardella L et al (2007) Overexpression of pectin methylesterase inhibitors in Arabidopsis restricts fungal infection by *Botrytis cinerea*. Plant Physiol 143:1871–1880
Lorang J, Kidarsa T, Bradford CS et al (2012) Tricking the guard: exploiting plant defense for disease susceptibility. Science 338:659–662
Ma B, Hibbing ME, Kim HS et al (2007) The host range and molecular phylogenies of the soft rot enterobacterial genera *Pectobacterium* and *Dickeya*. Phytopathology 97:1150–1163
Marquez-Villavincencio M, Weber B, Witherell RA et al (2011) The 3-hydroxy-2-butanone pathway is required for *Pectobacterium carotovorum* pathogenesis. PLoS One 6:e22974
Marty P, Jouan B, Bertheau Y et al (1997) Charge density in stem cell walls of *Solanum tuberosum* genotypes and susceptibility to blackleg. Phytochemistry 44:1435–1441
McMillan GP, Perombelon MCM (1995) Purification and characterization of a high pI pectin methyl esterase isoenzyme and its inhibitor from tubers of *Solanum tuberosum* subsp. *tuberosum* cv. Katahdin. Physiol Mol Plant Pathol 46:413–427
Mengiste T (2012) Plant immunity to necrotrophs. Annu Rev Phytopathol 50:267–294
Morris CE, Monteil CL, Berge O (2013) The life history of *Pseudomonas syringae*: linking agriculture to earth system processes. Annu Rev Phytopathol 51:85–104
Nachin L, Barras F (2000) External pH: an environmental signal that helps to rationalize pel gene duplication in *Erwinia chrysanthemi*. Mol Plant Microbe Interact 13:882–886
Nishiguchi M, Kobayashi K (2011) Attenuated plant viruses: preventing virus diseases and understanding the molecular mechanism. J Gen Plant Pathol 77:221–229
Nzyngize JR, Lyumugabe F, Busogoro JP et al (2012) Pythium root rot of common bean: biology and control methods. A review. Biotechnol Agron Soc Environ 16:405–413
O'Brien HE, Thakur S, Guttman DS (2011) Evolution of plant pathogenesis in *Pseudomonas syringae*: a genomics perspective. Annu Rev Phytopathol 49:269–289
Ottmann C, Luberacki B, Küfner I et al (2009) A common toxin fold mediates microbial attack and plant defense. Proc Natl Acad Sci U S A 106:10359–10364
Prusky D, Yakoby N (2003) Pathogenic fungi: leading or led by ambient pH? Mol Plant Pathol 4:509–516
Prusky D, Alkan N, Mengiste T et al (2013) Quiescent and necrotrophic lifestyle choice during postharvest disease development. Annu Rev Phytopathol 51:155–176
Raiola A, Lionetti V, Elmaghraby I et al (2011) Pectin methylesterase is induced in Arabidopsis upon infection and is necessary for a successful colonization by necrotrophic pathogens. Mol Plant Microbe Interact 24:432–440
Rey P, Leucart S, Desilets H, Belanger RR, Larue JP, Tirilly Y (2001) Production of indole-3-acetic acid and tryptophol by *Pythium ultimum* and Pythium group F: possible role in pathogenesis. Eur J Plant Pathol 107:895–904
Rollins JA, Dickman MB (2001) pH signaling in *Sclerotinia sclerotiorum*: identification of a *pacC*/RIM1 homolog. Appl Environ Microbiol 67:75–81
Roy C, Kester H, Visser J et al (1999) Modes of action of five different endopectate lyases from *Erwinia chrysanthemi* 3937. J Bacteriol 181:3705–3709

Sandler HA, Mason J (2010) Flooding to manage dodder (*Cuscuta gronovii*) and broad-leaved weed species in cranberry: an innovative use of a traditional strategy. Renew Agric Food Syst 25:257–262

Toth IK, Birch PRJ (2005) Rotting softly and stealthily. Curr Opin Plant Biol 8:424–429

Williams B, Kabbage M, Kim HJ et al (2011) Tipping the balance: *Sclerotinia sclerotiorum* secreted oxalic acid suppresses host defenses by manipulating the host redox environment. PLoS Pathog 7:e1002107

Yang S, Zhang Q, Guo J et al (2007) Global effect of indole-3-acetic acid biosynthesis on multiple virulence factors of *Erwinia chrysanthemi* 3937. Appl Environ Microbiol 73:1079–1088

Chapter 8
Non-spore-Forming Bacterial Entomopathogens: Their Toxins, Hosts and the Environment: Why Be a Pathogen

Mark R.H. Hurst

Abstract This chapter discusses the wide range of variables that can affect the virulence of a bacterial pathogen, as well as its ability to survive. These factors are diverse and operate both externally and internally to the bacterial cell. Similarly, the host is influenced by an array of environmental parameters. These processes in turn affect the efficacy of the pathogen. Factors that impart stress on the host will increase its susceptibility to a pathogen. When not in the presence of a host, the pathogen needs to be able to maintain itself, allowing it to persist and infect subsequent hosts. Pathogens are varied in both their ability to cause disease and in their host specificity. These factors affect pathogen dispersal and transfer to another host. In some instances, spread of a pathogen may be enabled by pathogen-induced changes in host behaviour. At the subcellular level, a pathogen is subjected to natural rates of gene mutation and/or large genomic changes resulting from processes such as horizontal gene transfer, whereby new virulence determinants might be acquired. Whether inside or outside of the host, selective biotic and abiotic forces act on the pathogen and its non-pathogenic counterparts. Competition and the process of natural selection may lead to the emergence of either more benign or more virulent strains.

This chapter focuses on non-spore-forming bacterial insect pathogens (entomopathogens) which, due to the absence of a spore-like survival structure, are more likely to be influenced by environmental changes. Entomopathogens may therefore provide greater clues for defining the driving forces behind pathogen evolution. A broad overview of external and subcellular variables that may influence both the pathogen and its host is given. The mechanisms of pathogen entry to the host and bacterial factors allowing a pathogen to survive within the host are then outlined. A range of pathogens, including those that are free-living, animal-vectored or plant-associated, are documented. Shared components of some

---

M.R.H. Hurst (✉)
Farm Systems & Environment, AgResearch, Lincoln Research Centre, Christchurch, New Zealand

Bio-Protection Research Centre, Lincoln, Christchurch, New Zealand
e-mail: mark.hurst@agresearch.co.nz

© Springer International Publishing Switzerland 2016
C.J. Hurst (ed.), *The Rasputin Effect: When Commensals and Symbionts Become Parasitic*, Advances in Environmental Microbiology 3,
DOI 10.1007/978-3-319-28170-4_8

multicomponent toxin systems that reside in both pathogens and symbionts are discussed, allowing us to propose a possible origin of the toxin transport machinery. In the final sections of the chapter, two case studies are presented: (1) the species-specific and chronic disease in grass grubs (*Costelytra zealandica*) caused by the bacterium *Serratia entomophila* and (2) the rapidly killing broad host-range pathogen *Yersinia entomophaga*. In both these instances the respective strategies of the pathogen are discussed in relation to their environment, applying the knowledge from earlier in the chapter to help define why these two different types of pathogen reside in the same environment and ask "why be a pathogen?"

8.1 Introduction

There is a wide range of terrestrial invertebrate species, many of which interact with other species belonging to similar trophic levels, enabling parallels to be drawn between different systems. Given the vast knowledge in this area, this chapter focuses on the class Insecta and its pathogens. Insects evolved as a separate class over 250 million years ago and have since developed a wide range of complex relationships with microorganisms, ranging from commensalism to parasitism or pathogenesis. The life cycle, diet and environment of the host place selective pressures on the relationship between the pathogen and its host, resulting in an evolutionary arms race and the formation of highly specialised host–pathogen relationships.

Insect pathogens are varied and include many prokaryotes (bacteria), eukaryotes (protists, nematodes, fungi and microsporidia) and insect-specific viruses that can be virulent under the appropriate conditions. Some of these pathogens, such as microsporidia, viral species and distant relatives of symbionts, are difficult to isolate or culture, meaning that their true roles in virulence within natural ecosystems remain largely unknown.

Most pathogens of insects need to be ingested to have an effect. However, entomopathogenic fungi can penetrate the cuticle using a combination of cuticle-piercing proteins and metabolites (Pedrini et al. 2013) and formation of a specialised germ tube structure called the aspersorium, which physically enables the fungus to enter through the insect cuticle. Other pathogens may require an animal vector such as a nematode to allow them to enter their target organism. Fungi and bacteria are the most frequently isolated entomopathogens and are used in areas such as biological control of insect pests. However, the active growth phase of a fungus is susceptible to changes in environmental conditions, such as temperature and water availability, limiting their potential effectiveness under drier environmental conditions (Jaronski 2010).

In this chapter, we emphasise Gram-negative pathogens, which in some circumstances mirror mammalian pathogens in their ability to infect and overwhelm the insect host. Gram-negative bacteria differ from Gram-positive bacteria in their inability to retain the dye crystal violet, because of a much thinner layer of peptidoglycan in their cell walls, giving the cells a red appearance following gram staining. Many Gram-positive bacteria can form spore-like structures, which provide a resting stage that allows survival of the bacterium in the absence of a potential host. For Gram-negative bacteria, which cannot produce survival structures, occupation of a host or its cadaver may provide an effective survival strategy that allows multiplication of the pathogen and protection of the resultant population from environmental extremes. This may extend the survival of the resident population before it once again comes into contact with its preferred live host, from which infection and dissemination of the bacteria can ensue. The cadaver may also be encountered by other soil invertebrates or ingested by higher animals, such as birds, resulting in dispersal of the bacteria in a new geographic location. In other systems, the cadaver may indirectly benefit not only the pathogen but also a pathogen-associated organism such as a plant endophytic entomopathogen. The plant-associated entomopathogenic fungal strains of *Metarhizium* can transfer nitrogen back to the host plant from the decaying insect (Behie and Bidochka 2014).

Spore-forming entomopathogens such as *Bacillus thuringiensis* have been extensively reviewed (Jurat-Fuentes and Jackson 2012), and because of their ability to form a survival structure, they will not be covered in any great detail in this chapter. Recently, several distinct non-spore-forming insect pathogens from the genera *Pseudomonas*, *Serratia* and *Yersinia* have been identified. These pathogens are discussed extensively in subsequent sections of this chapter.

8.2 The Bacterial Pathogen

To provide context for later sections of the chapter, we need to first understand how bacterial cells work. Bacteria replicate by an asexual process, in which a mother cell divides into two clonal daughter cells. To maintain cellular integrity and handle diverse environments while maintaining the ability to export and transport substances such as proteolytic enzymes or toxins, bacteria are typically surrounded by a rigid cell wall. In some instances, the cell wall can bud off in microcapsules called outer membrane vesicles (Kulkarni and Jagannadham 2014). In bacteria such as *Xenorhabdus nematophila*, these outer membrane vesicles contain insect-active toxins, other degradative enzymes such as chitinases, and the chaperonin protein GroEL, which is also insect active (Khandelwal and Banerjee-Bhatnagar 2003; Joshi et al. 2008).

Some cell wall components, such as fimbriae or pili, form appendages that can aid in cell adherence to the host and enhance virulence. Cell surface components such as lipopolysaccharides aid adhesion and can help mask the bacteria from detection by other bacteria or the immune system of higher animals or deter grazing by protozoa (Molmeret et al. 2005). The ability of a microbe to move through the environment or within the host intestine can be mediated by large bacterial appendages called flagella.

In vitro bacterial growth can be categorised into three phases. The lag phase is where the vegetative cells adjust to new environmental conditions, such as those found when encountering a new host. In the exponential or log phase, bacteria rapidly multiply and produce larger cells, thus maximising their surface area, which enables rapid assimilation of available nutrients. When nutrients are depleted, bacteria can enter stationary phase, in which the cells become smaller and may be more resistant to changes in environmental conditions. At this stage, population density-dependent signalling molecules are produced in a process called quorum sensing, potentially changing the rate or behaviour of bacterial growth (motile to sessile or vice versa), and in the metabolic profile of the bacterial population (Solano et al. 2014; Kostakioti et al. 2013).

Bacteria may form several types of relationships with other living organisms. Relationships in which at least one species benefits are called symbiotic. These include *mutualism*, in which both organisms benefit; *commensalism*, in which one organism benefits without affecting the other; and *parasitism*, in which one organism benefits at the expense of the other. Some bacteria are detrimental to the host and are referred to as pathogenic. Pathogenicity is the ability of an organism to enter a host and cause disease. The degree of pathogenicity, or the comparative ability to cause disease, is called virulence. This is influenced by the ability of the target organism to defend itself using both physical barriers and its inherent immune system. Colonisation is the process where a pathogen has invaded and started growing within the host. Virulence reflects the microbe's capacity for infection, such as the ability to colonise and invade the host, and the severity of the disease that it produces. Interestingly, virulence can differ significantly between strains of the same pathogenic bacterial species.

The various attributes of virulence and the basic concepts of pathogenicity are summarised by Casadevall and Pirofski (1999, 2001). Further to this, the various degrees of microbe–host interaction, ranging from antagonistic to cooperative symbiotic relationships, have been described as a "symbiosis continuum" (Dimijian 2000).

There are several forms of bacterial pathogenicity. *Obligate* bacterial pathogens complete their life cycles within the insect host and cannot survive outside of the host; they must cause disease to be transmitted from one host to another. An example of this is the Gram-positive scarab beetle pathogen *Paenibacillus popilliae*. *Facultative* pathogenic bacteria are those that can grow independently in the environment as well as within a host. They include the scarab beetle pathogens *Serratia entomophila* and *Serratia proteamaculans*. *Opportunistic* pathogens exploit specific conditions and may only infect their host after it has been compromised, either by stress, such as malnourishment, or wounding, which enables the bacterium to enter the haemocoelic cavity.

Opportunism is dependent on the genetic make-up of the pathogen and is generally a matter of whether the pathogen can multiply within the insect haemolymph (the open circulatory system of arthropods). Opportunistic pathogens cannot cause disease in a healthy host and can be transmitted from one host to another without causing disease. Opportunistic pathogens may spend most of their life cycle in the environment without meeting a host, so disease outbreaks rarely, if ever, occur under normal conditions. Due to their natural association with the host, commensal or endemic host organisms are more likely to become opportunistic pathogens.

Depending on the virulence of the bacterial strain and the immune status of the host, entry of bacteria into an insect can result in three distinct outcomes. *Toxaemia* occurs when the bacteria remain confined to the host gut lumen but produce toxins that are disseminated into the haemolymph. *Bacteraemia* is where bacteria invade the haemolymph but do not produce toxins or other harmful factors, which may in part reflect a secondary symbiotic relationship. Finally, *septicaemia* occurs when the bacteria multiply in high numbers and produce toxins and other degradative compounds that lead to host death. In such an instance, non-pathogenic opportunistic bacteria may also become involved in the sepsis.

8.2.1 The Need for a Pathogen to Spread

To ensure the long-term survival of a pathogen, it needs to self-transmit effectively. This represents a facet which in itself depends on the lifestyle and role of the host species in its ecosystem. A pathogen that kills rapidly unless vectored by another organism will be limited in its ability to spread beyond the host. This is particularly true in situations where rapid incapacitation of the host will limit its movement and potential contact with other hosts. During this rapid time course, the insect may not have sufficient chance to mount a defence or change behaviour, limiting the opportunity for developing resistance to the pathogen. In contrast, a pathogen that causes a chronic infection will have an increased probability of host-to-host contact through the movement of the host. Chronic infection also provides an opportunity for the pathogen to maintain itself in an environment that is relatively devoid of competitors, within the insect host, where it is protected from significant external environmental pressures. Some pathogens, such as *Yersinia entomophaga*, are indiscriminate in their host range, killing a wide range of insect species, while other bacterial pathogens have a narrow or specific host range, such as *S. entomophila*, which can only infect the larvae of the scarab beetle *Costelytra zealandica*. Acquisition of natural point mutations, or new virulence determinants through horizontal gene transfer (HGT, described later), may change the host range of a pathogen or cause it to become weaker, more potent, or even benign. This, in an extreme instance of a chronic relationship, may lead to the development of a possible symbiotic relationship or the formation of a highly virulent pathogen.

These scenarios raise several important questions, including what the benefits of being a pathogen are, whether pathogenic traits are a reflection of the environment that the pathogen naturally resides in or whether these traits enable a pathogen to survive in an environment to which it is not naturally adapted. It is plausible that a pathogen can establish a niche within the host, where it can grow to numbers not possible in a nutrient-poor environment. In this instance, the interior of animals represents a rich source of nutrients for pathogens, including sugars, amino acids and sources of nitrogen, such as urea and ammonia, whose presence promotes replication (Rohmer et al. 2011). Under these circumstances, bacteria may reach a growth rate that could not be achieved outside the host. Within a larger pathogen population, natural genomic point mutations may occur. Under the right conditions, and if certain other bacteria are present within the host, bacteria can transfer certain traits to each other through HGT. This genetic transfer may cause the recipient organism to become more competitive, either at that point in time or after release from the cadaver.

8.2.2 External Influences on the Pathogen and Its Host

The ability of the opportunist to cause disease also depends on a variety of other factors that impart stress on the host, affecting its susceptibility in either a positive or negative way. These factors include the health and population density of the host, host diet and whether various abiotic factors, including environmental conditions such as temperature or humidity, are conducive to growth of the pathogen. To aid in the prediction of disease and define optimal conditions for pathogen proliferation, the disease triangle (host, pathogen and environment) was initially proposed by McNew in 1960 for defining conditions conducive to plant pathogens. This concept has been expanded beyond these three factors to encompass more abiotic and biotic factors, among which are the relationship between the environment, the pathogen, its host and time (summarised in Scholthof 2007). Defined variables can also be incorporated, including inherent susceptibility of the host, relative reproductive rate of both the pathogen and the host, minimal inoculum potential of the pathogen, duration of infection, prevalence of pathogen or host, the age of the host, the presence of other pathogens and other trophic level interactions. The combination of these variables will dictate the resultant field efficacy of a pathogen (Fig. 8.1).

In recent times, it has become increasingly apparent that land-use change, where the current ecosystem is dramatically shifted through clearing of the land or changing farm-based systems (including chemical use), can directly affect the dynamics of both the pathogen and the host. This may favour the host in some instances through the introduction of more palatable exotic pastures or favour the pathogen through practices such as irrigation, where the subsequent moist conditions are more optimal for certain pathogens.

8.2.3 Virulence Is Influenced by the State of the Host

Poor health of the host increases the potential for disease, and may lead to greater pathogen accumulation, which in turn provides greater opportunity for the pathogen to infect other individuals, and may also give rise to new variant strains. Reduced health of the host will also lead to reduced fecundity or ability to mate. A well-nourished healthy insect is more likely to successfully fend off a pathogen. Honey

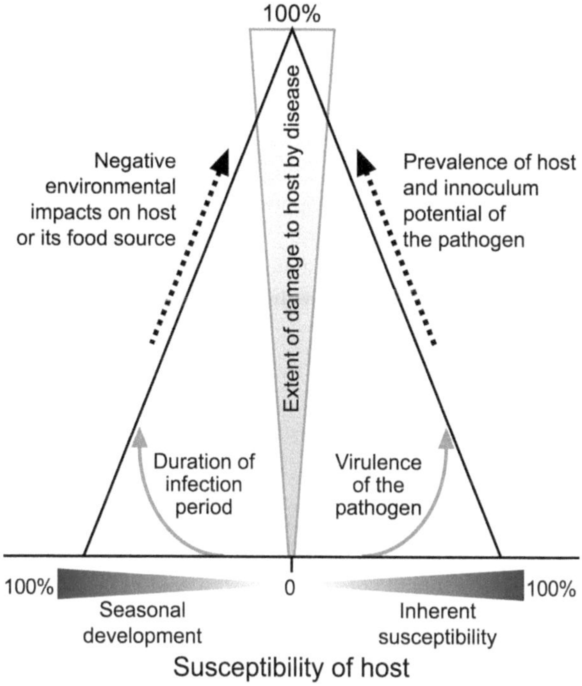

Fig. 8.1 Variation of the disease triangle: variables that can influence pathogen efficacy. Changes in host food source affect the health and well-being of the host, influencing growth and vigour. Negative effects on the host will increase host susceptibility to a pathogen. Population density of the host influences dispersal of the pathogen, where a greater resident host population will increase the opportunity for pathogen transfer. Certain life stages may render the host more vulnerable to infection (seasonal development), while a subset of the host population may be more genetically susceptible to a pathogen. In relation to the pathogen, a longer duration of infection will increase the opportunity to spread throughout the host population, while rapid onset of disease will reduce the opportunity for host resistance or recovery to occur (duration of infection period) and reduce the opportunity for the pathogen to spread

bees (*Apis mellifera*) that have a diverse diet, including pollen from different plants, have greater immune competence than those that feed on the same pollen type (Alaux et al. 2010). Larvae of the velvet bean caterpillar (*Anticarsia gemmatalis*) were more resilient to infection by the entomopathogenic fungus *Nomuraea rileyi*, when the larvae were reared on soybean leaves than on an artificial diet. The fungus was also found to be more efficacious against the younger larval stages (Boucias et al. 1984).

In agricultural systems, the presence of a monophytic or endophytic crop that is suboptimal for insect growth can lead to a greater population density of less fit insects that will be more likely to accumulate bacterial pathogens. Phytophagous insects have a significantly higher nitrogen and phosphorus content (~9 % N, ~0.5 % P) than their host plants (~1.5 % N ~0.05 % P) (Elser et al. 2000; Fagan et al. 2002). Using a model system composed of the plant hoppers, *Prokelisia dolus* and *Prokelisia marginata*, and their host plant, *Spartina* cordgrass, Huberty and Denno (2006) demonstrated that insects fed on plants high in nitrogen grew more rapidly and were healthier. Under nitrogen- and phosphorus-limiting conditions, *P. dolus* increased feeding activity, while *P. marginata* moved to higher-quality *Spartina*. Such dynamics in turn affect the efficacy of an insect pathogen either within or upon entering the combined ecological system. More frequent grazing and compromised health may increase the opportunity for *per os* entry of the pathogen into an insect host, which is more likely to cause disease in the stressed host. Alternatively, movement of a host in search of a more suitable food source may expose it to different or more virulent pathogens.

8.2.4 Bacteria Sense and Respond to Changes in the Environment

The wide range of biotic and abiotic factors that affect the host are exacerbated on a unicellular pathogen. The pathogen must rapidly respond in order to survive or compete with other neighbouring organisms. Bacteria are widely distributed in the environment, and many also coexist within, or pass benignly through, the intestinal tract of animals allowing their downstream release into the environment through processes such as defecation and decay. This potential movement of bacteria through different environments, even within the body of the host, means that they must adapt rapidly to survive and outcompete other bacteria in the different surroundings. To achieve this, many insect-associated bacterial species have developed complex regulatory networks, allowing the complete shutdown or initiation of metabolic processes which alleviates the metabolic burden on the bacterium. These metabolic strategies often reflect the requirement of microorganisms to survive in various challenging abiotic and biotic environments.

In addition to abiotic pressures placed on the host, the pathogen is also affected by abiotic factors, and variables such as temperature can influence the expression of many bacterial virulence determinants (Mekalanos 1992). Bacteria can sense and respond to these environmental changes using signal transduction pathways. This is typically mediated by cell wall-associated two-component regulators that can relay an external stimulus or signal internal to the cell, following which the bacterial cell can respond by differentially expressing target genes (Stock et al. 2000; Mascher et al. 2006). The PhoP-PhoQ two-component regulator of the entomopathogenic nematode-associated bacterium *Photorhabdus luminescens* senses and responds to the availability of magnesium. The *P. luminescens* PhoP-PhoQ mutant is avirulent when injected into Egyptian cotton leafworm *Spodoptera littoralis* larvae (Derzelle et al. 2004). Temperature can also trigger the expression of virulence determinants. Assessment of cell extracts derived from the bacterium *Yersinia enterocolitica* W22703 grown in vitro at either 10 or 30 °C showed that only those extracts derived from cultures grown at 10 °C were insect active following ingestion by the tobacco hornworm, *Manduca sexta* (Bresolin et al. 2006). In addition, under conditions of high humidity, fungal growth and efficacy are often increased (Jaronski 2010). For example, the mycelial development of *Beauveria bassiana* on an insect cadaver is optimal at 92 % humidity or higher (Ferron 1977).

8.3 Changes in Pathogen Genome Composition

In addition to these external factors, biotic processes are significantly influenced at the subcellular level by factors such as the natural mutation rate or through large-scale changes mediated by HGT. In the case of HGT, large regions of DNA can be transferred from one organism to another, influencing species diversity. In the environment, HGT occurs by a range of processes, including: *transformation*, which is the uptake of naked DNA from the environment (Mell and Redfield 2014); *transduction*, which is the incorporation and subsequent transfer of DNA by bacteriophage particles; and *conjugation*, which is the transfer of genetic material from one bacterium to another through a conjugative pilus.

The role of these mechanisms in gene transfer in the natural environment has been reviewed (Darmon and Leach 2014; Thomas and Nielsen 2005; Paz-Y-Mino and Espinosa 2010). When discussing HGT from an ecological perspective, it is important to understand the concepts of "sympatry" and "allopatry" that have been widely used by evolutionists to describe eukaryotic life forms (Mayr 1961). This terminology is now being used to describe the lifestyles of bacteria and viruses (Moliner et al. 2010; Diene et al. 2013). In this context, *sympatric* organisms are free-living generalist bacteria that coexist with other microbes, meaning that there is ample opportunity for HGT to occur (Georgiades 2012; Merhej et al. 2011). This may inadvertently lead to the production of large chimeric genomes with multiple ribosomal operons capable of translating DNA from a diverse range of species that differ in both DNA G+C composition and/or codon use (Georgiades and Raoult 2010; Raoult and Boyer 2010). In contrast, *allopatric* organisms, represented by obligate or intracellular bacteria, are specialists that are confined to a single environment, such as the host insect. Because of this mode of living, symbionts are more likely to be physically separated from other microbial communities. Limited contact in turn leads to a reduced frequency of genetic exchange and may result in a reduction in genome size (Georgiades 2012; Merhej et al. 2011; Moran 2002). This genetic isolation results in the generation of ribosomal operons that are less likely to translate DNA of an atypical (to that bacterium) codon composition.

8.3.1 Making Use of Acquired DNA

Diene et al. (2013) defined three essential steps in the formation of highly pathogenic bacteria, by their ability to utilise a new DNA resource. They coined the term "opportunity, power and usage" to describe a sympatric lifestyle, where *opportunity* describes the ability of a bacterium to interact with other bacteria and exchange foreign DNA (Georgiades and Raoult 2010); *power* is the ability of the bacterium to integrate foreign DNA sequences acquired by HGT (Bapteste et al. 2009); and *usage* signifies that the bacterium must also be able to maintain and use these sequences to produce biologically active transcripts, proteins or metabolites. Based on these criteria, it is apparent that having robust translational machinery is crucial for maintaining recently integrated sequences. DNA sequences with atypical or rare codon usage and/or G+C content are most likely to be lost as they are less likely to be effectively translated by the host genome (Medrano-Soto et al. 2004).

8.3.2 Large-Scale Horizontal Gene Transfer

Two of the most frequently transferred elements in HGT are plasmids and genomic islands, which, in the context of pathogens, result in "evolution in quantum leaps" (Groisman and Ochman 1996). The typical features of a genomic island are a DNA G+C content that differs from that of the core genome and the presence of DNA repeat elements such as a tRNA gene flanked by direct repeats. Active or remnant mobility and integration elements, such as insertion sequence elements, may also be present in genomic islands. Loss of function of these elements may denote an island that is in the final stages of integration into the host genome (Hacker and Carniel 2001). Genomic islands that encode virulence determinants are called pathogenicity islands (PAIs). The term PAI was first used by Hacker et al. (1990) to describe two large, unstable regions in the chromosome of uropathogenic *Escherichia coli* strain 536. Virulence determinants encoded by PAIs can include adhesins, iron uptake systems, secretion systems, invasins and toxins (Hacker and Kaper 2000). PAIs that encode entomocidal toxins have been reported in a diverse range of bacteria, including *P. luminescens* (Waterfield et al. 2002, 2004), *Xenorhabdus* (Sergeant et al. 2006), *Y. enterocolitica* W22703 (Bresolin et al. 2006) and *Y. entomophaga* (Hurst et al. 2011c). Plasmid-based PAIs have also been identified in *S. entomophila* and *S. proteamaculans* (Dodd et al. 2006; Hurst et al. 2011a).

Plasmids also play an important role in the transfer of virulence determinants. Plasmids are typically self-replicating circular DNA molecules that can be transferred between bacteria through a process termed conjugation. Due to evolutionary constraints and the metabolic cost of plasmid replication, plasmids typically encode genes that benefit the bacterial host in a particular circumstance. The key factor limiting the rate of conjugation is the ability of a donor cell to be in close proximity with a recipient cell. In nutrient-limiting conditions such as soil, where different microbial communities can be physically separated by particles within the soil, opportunities for conjugation may be limited. However, a nutrient-rich fluid environment, such as is found in the invertebrate gut, provides a habitat conducive to conjugation. In this latter instance, the diverse microflora of the host gut may further increase the opportunity for gene transfer. For example, rates of conjugation are twofold higher between *Salmonella enterica* Newport and *E. coli* in the gut of the lesser mealworm *Alphitobius diaperinus* than in vitro on filter paper (Poole and Crippen 2009). Watanabe and Sato (1998) reported the transfer of the plasmids pBPW1::Tn7 and RSF1010 between the insect-resident bacterium *Enterobacter cloacae* and the plant epiphytic bacterium *Erwinia herbicola* in the guts of *Bombyx mori* silkworm larvae. There are also several reports showing plasmid transfer between *Bacillus* species in lepidopteran and coleopteran larvae (Thomas et al. 2000, 2001; Jarrett and Stephenson 1990). Plasmid transfer was also noted between *E. coli* and *Yersinia pestis* within the flea, *Xenopsylla cheopis* (Hinnebusch et al. 2002).

Three other significant agents involved in gene transfer are transposons, insertion sequence elements and bacteriophages. Many transposons and insertion sequence elements can mobilise independent of the bacterial host recombination system, a factor which in combination with their high degree of nucleotide conservation increases the likelihood of DNA recombination, leading to gross genome DNA rearrangements such as deletions and inversions. In some instances, similar regions of DNA contained on two different plasmids can recombine to form a single plasmid termed a *cointegrate*, which may have a greater ability to replicate in other bacteria. Bacteriophages are the other significant agent in gene transfer. In some environments, numbers of bacteriophages reach an order of magnitude higher than bacteria (Weinbauer 2004. Wommack and Colwell 2000; McDaniel et al. 2010). Because of their broad host range and ability to transfer genetic information between bacterial populations, bacteriophages are often implicated in the fitness of bacterial genomes. Of note, many bacteriophages of animal pathogens transfer or encode virulence genes (reviewed by Brunder and Karch 2000).

8.3.3 Preserving Genetic Self-Identity

To help preserve genetic self-identity, many bacteria have developed mechanisms to protect against incoming foreign DNA. Protection is achieved through a variety of mechanisms, including the production of extracellular DNases that degrade the foreign DNA before it reaches the bacterial cell wall. However, if foreign DNA enters the cell, it is then faced with a subcellular immune response. Protective mechanisms include DNA restriction–modification systems composed of a DNA methyltransferase, which methylate the cell's own DNA, enabling self-recognition and protection of the host DNA, as well as a restriction endonuclease, which cleaves foreign unmodified (unmethylated) DNA (Pingoud et al. 2005). In certain circumstances, such as an aborted phage infection (Labrie et al. 2010), cells can undergo altruistic cell suicide to help prevent further infection of the bacterial population.

Some bacteria have a highly unique cellular memory system mediated by a specific set of elements, termed CRISPR-cas (clustered regulatory interspaced short palindromic repeat (CRISPR)-associated proteins) elements. These elements can recognise and capture specific sequences of incoming DNA, allowing their future recognition and destruction (Westra et al. 2012, 2014).

8.3.4 Defining Regions of Recently Acquired DNA

To help define HGT-acquired regions of DNA, terms such as a "core", "pan-genome" and "mobilome" have been defined. *Core-genome* genes are common to all strains within a species, typically comprising housekeeping genes that encode proteins essential for cell survival (Sarkar and Guttman 2004; Medini et al. 2005). These genes are most likely to be vertically inherited (Sharp and Li 1987). The *pan-genome* encompasses both a core genome and a dispensable genome, containing both genes that are present in the genomes of two or more strains, as well as genes unique to a single strain (Medini et al. 2005). The *mobilome* describes a mobile gene pool located on mobile genetic elements, such as plasmids and PAIs, which can be transferred by HGT between microorganisms (Frost et al. 2005; Siefert 2009; Medini et al. 2005). This transfer can be accomplished both vertically (mother to daughter) and horizontally (between non-relative strains).

Annotation of 20 pathogenic and commensal *E. coli* strains by Touchon et al. (2009) showed that the overall genomic organisation was highly conserved, with a core of ~2000 genes being similar between strains. The remaining genes were either shared by a subset of strains or were unique. It has also become evident that bacterial genomes are not discrete static entities but are instead amalgamations of multiple genes from different species. For example, assessment of 657 prokaryotic genomes for genes with aberrant DNA composition showed that approximately 21 % of genes in any genome will have been acquired from other sources (Popa et al. 2011).

8.3.5 Selective Forces on Acquired DNA

After a bacterial pathogen has acquired DNA and has gained the ability to colonise a new niche, the pre-existing core genome and the newly acquired virulence genes undergo further selection in a process called "pathoadaption". This may then mean that genes not essential for virulence ("anti-virulence genes") are either eliminated or their expression levels are reduced (Maurelli 2007). If a bacterium finds itself in a niche that is relatively nutrient rich, metabolically costly pathways and/or toxin production may be selected against and therefore lost over time. Reductions in gene load and metabolic output mean that the bacterium can multiply at a higher rate, enabling it to compete more efficiently with neighbouring bacteria. These competing bacteria may be non-pathogenic organisms of the same microbial species or strain (isogenic). Hence, maintaining virulence genes or other acquired genes must be of benefit to the host. In a further scenario, within the host, the pathogen may evolve to a point that limits life beyond the host, a process described as "short-sighted evolution" (Levin and Bull 1994).

8.3.6 *Subtle Genome Changes*

The external influences placed on unicellular organisms can place high selective pressures for survivors that are altered in other virulence-associated attributes that in some instances may enable the survival of the pathogen within a host. Subtle genome-based changes are likely to reflect the ability of *Pseudomonas aeruginosa* PAO1 bacteriophage-resistant variants to be more resistant to phagocytosis and produce greater amounts of virulence factors (Hosseinidoust et al. 2013). It is speculated (Levin and Bull 1994; Molmeret et al. 2005; Albert-Weissenberger et al. 2007) that the ability of a pathogen to survive in the insect haemocoel may have evolved from natural selection as a result of pressures occurring when bacteria are ingested by protists, which are frequent grazers of microorganisms. In this instance, these grazers may inadvertently select for bacterial survivors that either are subsequently resistant to lysis or more recalcitrant to ingestion. The survivors may then have a greater ability to tolerate complex systems such as the mammalian immune system (Molmeret et al. 2005). In support of this theory, *Pseudomonas* sp. CM10 strains that overproduce exopolysaccharides are selected for in the presence of grazing protists (Matz et al. 2002). It has been proposed that obligate pathogens such as *Rickettsia* species may have used amoeba or protozoa as an ancestral host (Ogata et al. 2006).

Alternatively, the pressure of a bacteriophage on the bacterial community may outweigh the need for a bacterium to be pathogenic, and accordingly the strain may lose virulence in favour of tolerating or gaining resistance to a phage. For example, in the presence of bacteriophages, the motility and pathogenicity of *Serratia marcescens* towards the wood tiger moth *Parasemia plantaginis* were reduced (Friman et al. 2011).

8.4 Pathogen–Host Interaction

8.4.1 *Passage of the Pathogen Through the Insect Gut*

If a pathogen survives the overarching biotic and abiotic variables that influence both its spread and vigour to come into contact with a suitable host, it must then be able to gain entry to the host to be able to cause an effect. There are several mechanisms by which pathogens can gain access to their host. The most effective and direct route is entry through the oral cavity by *per os* feeding. Upon entry, the ability of a pathogen to establish and maintain a niche within a host is an essential step in virulence. The insect alimentary tract can be crudely defined as a compartmentalised tube through which food material passes. There are, however, three defined regions, called the foregut, midgut and hindgut (Chapman 2012; Engel and Moran 2013; Fig. 8.2a).

Within the gut, physiochemical parameters such as dissolved oxygen content, pH, redox potential and ionic strength often differ between the various compartments of the gut, and there are also differences between insect species. The pH of the insect gut is often intrinsically related to the diet of the insect, where highly alkaline or acidic conditions may be required to facilitate the degradation of the more recalcitrant-ingested compounds. The pH through the gut of the rose chafer *Pachnoda ephippiata* varies between each compartment, with a pH of ~8 in the

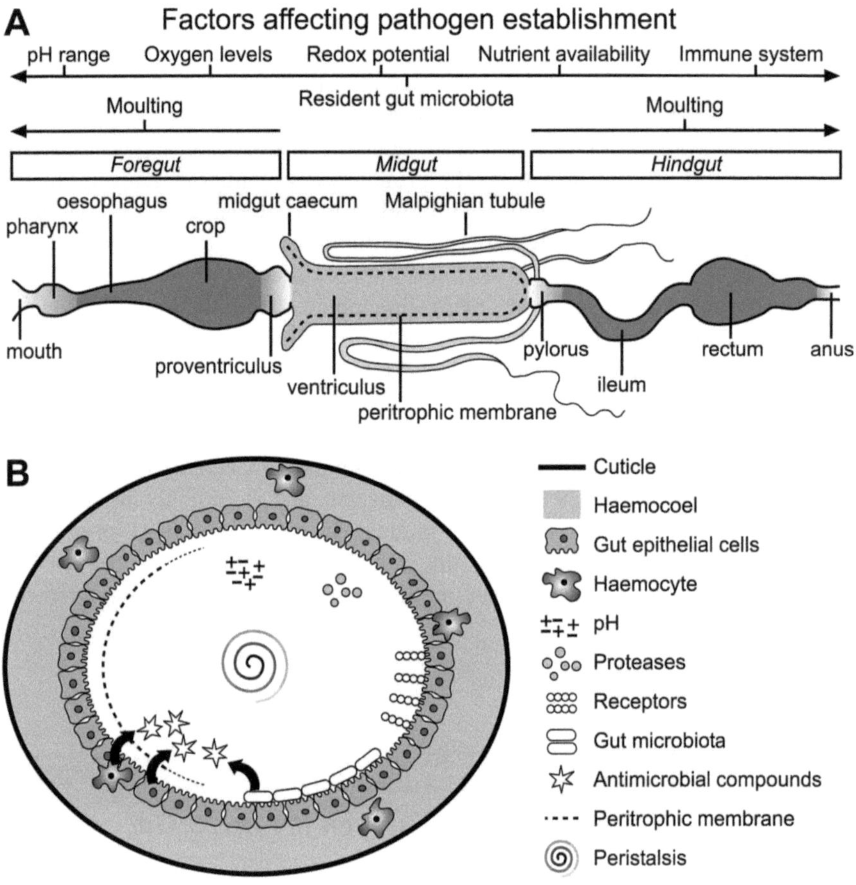

Fig. 8.2 Schematic of the insect intestinal tract. (**a**) Longitudinal section depicting the three main areas of the gut, the fore-, mid- and hindgut, as well as gut-based factors that influence pathogen establishment. (**b**) Cross section of insect intestinal tract depicting further variables that can influence the establishment of a pathogen, including the presence of resident microbiota, physical barriers such as the insect cuticle, the intestinal epithelial cell lining and the host immune system (haemocytes) located in the insect haemocoel

anterior midgut to >10 in the central midgut, while the hindgut is at the neutral pH of 7 (Lemke et al. 2003). These physiochemical factors, along with biotic factors such as the host immune system, limit the growth of the endemic saprophytic microorganisms within the insect.

For *per os* challenge to be effective, a sufficient number of pathogenic bacteria must be ingested so that a percentage of the bacteria can survive the initial insect defence systems. Subsequently, the bacteria may establish a site of colonisation, allowing persistence or a base from which toxins can be produced to incapacitate the host. However, it is more common that the bacteria move through the gut with the ingested food, which is compacted as a bolus that moves through the intestine via a retrograde muscle-driven process termed peristalsis.

Upon ingestion, the antimicrobial compounds in an insect's saliva may kill many of the ingested microbes. For example, the saliva of the cotton bollworm *Helicoverpa zea* reduces the infectivity and subsequent colonisation of *S. marcescens*, *P. aeruginosa* and *B. thuringiensis* (Musser et al. 2005). Once ingested, a pathogen must compete against, and preferably outnumber, the resident microbial population, before maintaining itself within the host. An inverse relationship was identified between gut microbial diversity and the ability of *S. marcescens* to colonise the intestine of the desert locust *Schistocerca gregaria* (Dillon et al. 2005). In an attempt to sequester the pathogen population, the host produces antimicrobial agents on recognition of a pathogen. Of note, antimicrobial genes in the larvae of the common fruit fly *Drosophila melanogaster* are up-regulated following ingestion of the bacterium *Pseudomonas entomophila* (Vodovar et al. 2005).

Similar to the external environment, within the insect the pathogen is exposed to a range of abiotic and biotic factors. When these changes occur, the pathogen is signalled by its two-component regulators to alter its gene expression, enabling it to effectively compete with its neighbours and/or host for resources. In some environments, elements such as iron may be significantly limited, which may necessitate that the bacterium activates complex machinery, such as siderophores, to enable the targeted capture of iron. Increased siderophore production may be particularly important when faced with a competing bacterium (Loper and Buyer 1991). While in the intestine, a pathogen may produce antimicrobials that are either limiting (bacteriostatic) or lethal (bactericidal) to other microbial cells. The bacterium itself may be influenced by factors such as temperature and population density. Throughout these processes, the physiochemical changes encountered by the bacterium or the host may provide the necessary cues to trigger the production of microbial virulence factors that are required to cause disease. Sometimes the metabolic work of a pathogen in producing toxins can be exploited by neighbouring non-toxin-producing cells, including isogenic strains, known as "cheats", which can replicatively outcompete the pathogenic strain because of their reduced metabolic load (Raymond et al. 2010; Rankin et al. 2011).

Another barrier to pathogenic bacteria establishing themselves in the insect digestive tract is a netlike structure called the peritrophic membrane (Fig. 8.2a). This matrix, which forms around the food bolus, is composed of chitin and glycoproteins. It serves as a mechanical barrier to ingested abrasive debris and is augmented by an array of antimicrobial compounds that reduce infection by microorganisms. The equivalent pore size of the net that forms the peritrophic membrane can vary from 7 to 36 nm, depending on the species (Lehane 1997; Terra 2001; Hegedus et al. 2009), rendering passage almost impossible to either a bacterial or a fungal pathogen. The peritrophic membrane divides the midgut into the ecto- (within the gut lumen) and endo-peritrophic spaces between the peritrophic membrane and the midgut epithelial lining. A *D. melanogaster dcy* mutant produced a peritrophic membrane with a reduced matrix width, making the net pore size slightly larger. The *dcy* mutant flies were more susceptible to infection by *P. entomophila* and its associated monolysin pore-forming toxin. In the latter instance, it was proposed that the peritrophic membrane has a role in limiting diffusion of the monolysin toxin (Kuraishi et al. 2011). Together, these findings provide evidence that the peritrophic membrane plays a role in defence against microorganisms.

Bacterial pathogens have developed several mechanisms to degrade the peritrophic matrix, including the production of chitinases (Wiwat et al. 2000; Thamthiankul et al. 2001), to gain entry through the peritrophic membrane to the "ectoperitrophic" space. Histological observations of *C. zealandica* larvae following *per os* challenge with the *Y. entomophaga* insect-active toxin complex (outlined later), termed the Yen-TC, which has associated chitinase activity, revealed a generalised dissolution and loss of the peritrophic membrane. TC-associated chitinases have been found in other bacteria, including *X. nematophila* and *P. luminescens* (Busby et al. 2013). It has also been shown that the combination of chitinases with gut-active toxins often synergistically enhances efficacy. Transmission electron microscopy showed that low levels of *S. marcescens* endochitinase ChiAII can perforate the peritrophic membrane. Combining low concentrations of ChiAII with the *B. thuringiensis* CryIC toxin increased bacterial *per os* activity against *S. littoralis* larvae (Regev et al. 1996). Fedhila et al. (2010) fed larva of the greater wax moth, *Galleria mellonella*, toxin-deficient strains of *B. thuringiensis* (407Cry^{-}) and the *cry*-deficient *Bacillus cereus* strain D23, both independently and in combination with Cry1C Bt toxin. The authors observed significantly greater mortality when both the bacterium and the toxin were administered in combination. It is therefore plausible that other bacterial enzymes such as chitinase may also enhance the bacterial activity of *B. thuringiensis* and *B. cereus*. Once in the ectoperitrophic space, bacterial toxins and degradative enzymes, such as chitinases, lecithinase and phospholipase C, are in closer proximity to the intestinal epithelial cell layer. This leads to a more rapid dissolution of the intestinal cells, providing possible entry points for the pathogen into the insect's haemocoelic cavity.

Many mammalian pathogens encode a complex type III secretion system that enables the bacteria to dock directly with the intestinal cells and actively transport toxin molecules into the target cell, resulting in their death (Tseng et al. 2009). Intestinally active type III secretion systems have yet to be identified in entomopathogenic bacteria. This absence with insect systems may be a result of the peritrophic membrane, which prevents the bacteria from making direct contact with the insect intestine (Vallet-Gely et al. 2008). In instances where type III systems have been identified in entomopathogenic bacteria, such as *P. luminescens* or *P. aeruginosa*, they are deployed against phagocytic cells (components of the later-described humeral response) located in the haemocoel, called haemocytes (Fauvarque et al. 2002).

Some bacteria have developed ways to deceive the insect cells, allowing their direct transport into the haemocoelic cavity. An example of this occurs in the Gram-positive bacteria *Paenibacillus lentimorbus* and *P. popilliae* which are obligate pathogens that infect a range of scarab beetles and require a host to replicate and produce spores. Once in the larval midgut, *P. popilliae* cells germinate, releasing vegetative cells that are absorbed into the midgut cells by a process called phagocytosis (Splittstoesser et al. 1973). From the midgut cells, the vegetative cells enter the haemocoel and multiply to high densities without generating a cellular defence response (Kawanishi et al. 1978). When this occurs, host-based defence responses such as melanisation are apparently not activated, and the larvae take on a milky or white appearance, leading to the term "milky disease". Subsequent death of the insect is attributed to depletion of nutrient-bearing fat reserves and weakening of the insect (Sharpe and Detroy 1979).

8.4.2 Non-intestinal Routes of Pathogen Entry

To circumvent the host defences, some pathogens have coevolved with their own host vector to allow active transmission directly into the insect haemocoelic cavity. These bacteria are typically not highly active by *per os* challenge and therefore often depend on their vector to initiate the disease process. Through this association with a vector, the genome of the pathogen may become more streamlined to contain only those genes that aid a symbiotic lifestyle and enable survival in the insect haemocoel. Nematode symbionts belonging to the genera *Photorhabdus* or *Xenorhabdus* typically need to be vectored by their associated nematode into the insect haemocoel to have an effect (ffrench-Constant et al. 2007; Owuama 2001). Once contact with an insect has been made, the nematode physically penetrates the insect cuticle, where it regurgitates the bacterial symbiont directly into the insect haemocoel. The rapid change in environment during the transition from nematode to insects triggers the large-scale release of bacterial metabolites and toxins. These molecules are required to enable the bacterium to survive the insect's immune humoral response.

8.4.3 Overcoming Host Haemocoelic Defences

The open circulatory system of the insect, termed the haemocoel, is a fluid-filled cavity with a neutral pH (Chapman 2012; Fig. 8.2b). The haemocoelic fluid enables the transfer of nutrients within the insect. It also harbours the insect immune system, the main components of which are phagocytic haemocytes. Upon entry of a foreign body into the haemocoelic cavity, the haemocytes respond by attempting to sequester the invader away from the rest of the haemocoelic cavity through encapsulation and compartmentalisation. In the resulting areas of haemocyte coagulation, high concentrations of antimicrobial peptides, cytotoxic molecules and phenoloxidases are produced by the host haemocytes (Bidla et al. 2005; Haine et al. 2007). The phenoloxidase cascade triggers the production of the black pigment, melanin (González-Santoyo and Córdoba-Aguilar 2012; Cerenius and Söderhäll 2004). Many pathogens that are active in the haemocoel encode a diverse range of toxins, enabling the rapid incapacitation of the host immune system and survival of the pathogen. This often necessitates a coordinated regulatory response upon the pathogen entering the haemocoel.

In *Xenorhabdus nematophila*, full insect-based virulence is controlled by the FlhDC master regulator, which in turn is involved in controlling the expression of genes required for motility (flagella synthesis) and for lipase, protease and haemolysin production (Givaudan and Lanois 2000; Park and Forst 2006). The *X. nematophila* FlhDC-dependent gene *xhbF*, which is involved in siderophore production, is up-regulated in the early stages of infection of Egyptian cotton leafworm *S. littoralis* larvae, at a time when the supply of iron in the insect haemocoelic cavity is limited (Jubelin et al. 2011). Interestingly, the iron-rich sites within the haemocoel are the first areas colonised by *X. nematophila* (Nichol et al. 2002). As to whether the invading bacteria actively require iron at this stage or are sequestering the available iron away from the host has yet to be ascertained. Limiting the iron supply stresses the insect, potentially lowering its immunity and favouring the pathogen. In contrast, *P. luminescens* uses a different strategy, called "phase variation", which results in gross physiological changes that significantly alter the behaviour of the bacterium. Phase variation is achieved through mutation or small-scale genome rearrangements, such as a DNA promoter inversion, allowing gene expression to be turned on or off. In response to sudden changes in environmental parameters, *P. luminescens* can switch between a pathogenic form (P-form) and a mutualistic form (M-form). The P-form produces an array of secondary metabolites, such as antibiotics and inhibitors of phenoloxidase, leading to growth of both the bacterium and its nematode in the insect. In contrast, the slower-growing M-forms are less motile and are more tolerant of antibiotics. This is also the form required for colonisation of the nematode through the production of bacterial Mad fimbriae (Somvanshi et al. 2010). The switch between these two forms has been linked to the inversion of a small region of DNA in the *Mad* fimbrial locus, mediated by an invertase. This switch in turn results in the production of a putative transcriptional activator, MadJ, which is thought to regulate the expression of genes involved in M-form generation (Somvanshi et al. 2012). It has been calculated that bacteria with 20 phase–variable loci can potentially exist in greater than a million different states, which increases the overall ability of the population to survive under atypical environmental conditions (Pallen and Wren 2007).

Several metabolites and toxins active in the haemocoel have been identified in both *Photorhabdus* and *Xenorhabdus* species. The *P. luminescens* genome contains several later-described anti-feeding prophage (Afp) variants, called *Photorhabdus* virulence cassettes (PVCs), which are composed of 16 conserved genes and two other toxin genes, some of which encode toxins similar to Mcf (makes caterpillars floppy) and LopT (Hurst et al. 2004; Yang et al. 2006). A PVC from the related species *Photorhabdus asymbiotica*, called PaPVCpnf, is lethal to *G. mellonella* larvae postinjection (Yang et al. 2006). In addition, there are several later-described insect-active TC clusters that were first identified in *P. luminescens* (Bowen et al. 1998).

Several other *Photorhabdus* and *Xenorhabdus* haemocoelically active toxins have been characterised. These include the *Photorhabdus* insect-related binary toxins (PirAB), which are active when injected into the haemocoelic cavity of *G. mellonella* (Waterfield et al. 2005b), or following *per os* challenge in the diamondback moth *Plutella xylostella* (Waterfield et al. 2005b; Blackburn et al. 2006). Another group of binary toxins, named XaxA and XaxB, from *X. nematophila*, exhibit cytolytic, haemolytic and proapoptotic effects in mammalian and insect cell lines (Vigneux et al. 2007). In addition, the *P. luminescens*-derived Mcf toxin enables *E. coli* cells to survive within the haemocoel of *M. sexta* larvae. The Mcf1 protein causes programmed cell death of haemocytes and midgut epithelial cells (Daborn et al. 2002) by apoptosis (Dowling et al. 2004). A second Mcf1-like protein, termed Mcf2, is located next to a type I exporter in a gene arrangement that is similar to that found in *Pseudomonas protegens* CHA0 (formally *P. fluorescens* CHA0) (Rodou et al. 2010). Members of the metalloprotease family of enzymes also exhibit insect-active properties, an example of which is the *P. luminescens* metalloprotease PrtS, which, when injected, causes melanisation of *G. mellonella* and *D. melanogaster* larvae (Held et al. 2007). In addition, Cabral et al. (2004) found that the metalloproteases PrtA and PrtS inhibit insect antibacterial factors. The bacterium *Xenorhabdus bovienii* produces a lecithinase that can incapacitate the *G. mellonella* larval host and fend off other microbial competitors (Pinyon et al. 1996). The *X. nematophila* proteins XciA (xenocin) and XimB are a cytotoxic RNase and an immunity protein, respectively. Under iron-limiting conditions, xenocin inhibited the growth of several bacterial species, including two other *Xenorhabdus* isolates. The production of antibiotics such as xenocin is thought to prevent putrification of the cadaver and provide greater resources for the pathogen through elimination of other bacteria (Singh and Banerjee 2008; Singh et al. 2013). Additional haemocoelically active toxins from *Photorhabdus, Xenorhabdus* and *B. thuringiensis*, such as haemolysins, cytolysins and proteases, are outlined in a review by Nielsen-LeRoux et al. (2012). Some of the bacterial metabolites in the haemocoel have a duel function. For example, the secreted product of *Photorhabdus*, hydroxystilbene, limits the growth of competing microorganisms in the dying insect and is also an inhibitor of the phenol oxidase produced by the insect's phagocytic haemocytes (Joyce et al. 2008; Waterfield et al. 2009). In a parallel system in *X. nematophila*, the bacterium produces the compound benzylideneacetone which, similarly to hydroxystilbene, has a dual role both as an antimicrobial compound and as an inhibitor of phenoloxidase activity (Song et al. 2011).

In the case of *P. luminescens*, the net result of this dramatic shift in bacterial behaviour during the transition from the nematode to the haemocoelic cavity means that as few as five *P. luminescens* bacterial cells are required to cause death of a *G. mellonella* larva (Clarke and Dowds 1991). The wide variety of toxins produced by both *P. luminescens* and its *Xenorhabdus* counterpart is likely to reflect the broad and non-discriminant nature of their associated nematode host, which is able to enter a wide range of different insect species. Once the nematode has gained entry into the insect host, the bacteria are delivered directly into the haemocoelic cavity, where they are immediately confronted by the insect's immune defence system. To enable survival, the pathogen needs to deploy a diverse range of haemocoelically active toxins, a subset of which may only be active against certain insect species.

8.4.4 Atypical Means of Host Entry

8.4.4.1 Haemocoelic Entry by Injection

Several non-*per os*-active bacteria are also lethal when injected directly into the haemocoelic cavity of *G. mellonella* larva. These bacteria include strains of *B. cereus* (Fedhila et al. 2006), *Campylobacter jejuni* (Senior et al. 2011), *Francisella tularensis* (Aperis et al. 2007), *Burkholderia mallei* (Schell et al. 2008) and *Yersinia* species (Fuchs et al. 2008). However, in these systems at least several hundred cells must be injected to cause an effect. Fuchs et al. (2008) identified that injection of at least 5×10^4 cells of *Y. enterocolitica*, *Yersinia mollaretii* or *Yersinia bercovieri* were required to cause >80 % mortality within 5 days of injection in *G. mellonella* larva. The ability of several mammalian pathogens, such as *C. jejuni*, to kill through haemocoelic injection is thought to relate to similarities between the immune systems of insects and mammals. The *G. mellonella* haemocytes are very similar to the macrophages of the mammalian immune system (Bergin et al. 2005; Mukherjee et al. 2011; Mylonakis et al. 2007; Lavine and Strand 2002). The high number of cells required to cause lethality relative to *P. luminescens* may either reflect subtle changes in toxin and receptor structures or that only a limited number of the injected bacteria-derived toxins cause an effect.

8.4.4.2 Opportunistic Routes of Pathogen Entry

Other routes that a pathogen may use to gain direct access to an insect's haemocoelic cavity are via the piercing mouthparts of arthropods, which serve as a vectoring mechanism for microorganisms. The digestive crop of the ant lion *Myrmeleon bore* contains *B. cereus*, *Morganella morganii* and *S. marcescens*, which are themselves able to cause significant mortality when injected into the haemolymph of the common cutworm, *Spodoptera litura*. Therefore, these bacteria, via their entry through the piercing mouth parts of the ant lion, are implicated in the death of the target insect (Nishiwaki et al. 2007). Under conditions of high host population, density insect combat can occur, and, although less targeted, the direct cuticular wounding by larval mouthparts enables a point of entry for an opportunistic pathogen. An alternate route is vectoring through the ovipositor of a parasitic wasp. Premature mortality of a recipient insect could be caused by the entry of pathogens through the puncture wounds made during parasitoid wasp *Microctonus hyperodae* oviposition in the host, the Argentine stem weevil *Listronotus bonariensis* (Jackson and McNeill 1998). In other instances, the act of reproduction can transmit disease. Surface inoculation of the male common fruit fly *D. melanogaster* genitalia with high numbers of *S. marcescens* infected the females, resulting in systemic infection and insect death (Miest and Bloch-Qazi 2008).

8.4.5 Growth of the Pathogen Within the Host

Throughout the disease process, a pathogen itself also undergoes age-related processes. Once inside the insect, the optimal (logarithmic) growth of the pathogen can either occur in the insect intestinal tract, such as with the grass grub pathogen *S. entomophila*, or in the haemocoelic cavity, such as with either *Paenibacillus* or the nematode-associated bacterium *P. luminescens*. However, it is unknown if all of the bacteria that have reached a peak in population growth are all metabolically active (i.e., secreting both toxins and degradative enzymes) or if instead only a subset of the population are in a metabolically active state and others are more passive in their communal role. It is also unknown how long the subpopulation of these cells may survive in vivo relative to the duration of the disease versus their longevity as measured in vitro.

In some instances, the bacteria may transition from their fluid-based motile phase to a more sedentary sessile phase through the formation of a complex structure called a biofilm that may facilitate their long-term survival. Microbial biofilms are a complex aggregate of bacteria embedded in an extracellular matrix that facilitates the self-attachment and binding of bacteria to biological or non-biological surfaces (Hall-Stoodley et al. 2004). Microbial biofilms not only afford the microbes with protection from extreme environmental conditions but also represent the formation of a population base that protects against competitors. Biofilms are also areas that are conducive to HGT process such as conjugation (Aminov 2011), because of the associated closer proximity and more sessile nature of the contained microbes. The bacterium *P. aeruginosa* can establish a biofilm in the crop of *D. melanogaster*, and yet a *P. aeruginosa* mutant defective in biofilm formation has been found to traverse the intestinal wall and was more virulent to the insect (Mulcahy et al. 2011). Further, a *P. aeruginosa* mutant that exhibited a greater ability to form a biofilm was less virulent towards *D. melanogaster* flies (de Bentzmann et al. 2012). Biofilms have been implicated in the population density-dependant process of quorum sensing in the alkaline insect gut. The high cell density in the microenvironment of a biofilm means that the pH at the immediate cell surface may differ from that of the greater surroundings, allowing quorum sensing to occur (Borlee et al. 2008). A *P. aeruginosa* PAO1 quorum-sensing mutant exhibited reduced *per os* virulence towards the white cabbage butterfly *Pieris rapae* than its wild-type counterpart (Borlee et al. 2008). *P. luminescens* also forms a biofilm, enabling it to adhere to the nematode crop via the Mad fimbriae (Somvanshi et al. 2010). Formation of a biofilm by *Y. pestis* allows the bacterium to be maintained in the crop of its vector, the flea *Xenopsylla cheopis*. In some of these *Y. pestis* strains, the TC-encoding loci have mutated, rendering them inactive, and thus it has been proposed that in these instances, the pathogen and vector are on the path to a commensal relationship (Chouikha and Hinnebusch 2012; Martínez 2013).

8.4.6 Pathogen Influences on Host Behaviour

Pathogen-derived metabolites, including toxins, can have a direct effect on host physiology and behaviour and can assist in maintenance of the pathogen within the host. In cases where the microbe triggers the cessation of feeding, it prevents the ingestion of food and therefore the movement of material through the gut. This may in turn prolong the initial presence of the ingested pathogen within the insect gut, allowing it to become fully established. Clearing of the intestinal tract, through vomiting or the increased production of frass, may remove the bulk of the resident bacterial population providing a greater opportunity for the pathogen to establish. It may also alter the physiochemical properties of the gut, such as pH. In both of these scenarios, depriving the host of a nutrient source will place it under stress, thus further reducing its ability to fend off the pathogen. The insect may also become paralysed; undergo changes in its appearance, such as colouration or shape; or become flaccid. The rate and timing of these pathogen-associated effects may depend on the stage of insect growth or on physical processes such as moulting. In addition to the exoskeleton, the exoskeletal lining of the insect fore- and hindgut areas is also moulted, effectively cleansing these regions of their associated microflora (Fig. 8.2b).

Host response processes such as vomiting or clearance of the gut may also inadvertently aid invasion by the pathogen. Voiding of the intestinal fluids may remove the endogenous microflora; physically disturb the intestinal cell layer, which provides points of entry for the pathogen; and/or reduce the fluid volume, thereby increasing both the speed at which virulence develops and the frequency with which the bacteria and/or its metabolites make contact with a target site.

Changes in the colouration of an insect, where the cadaver takes on a brown to black appearance, can reflect the production of melanin. In the instance of

Rickettsiella infection, melanisation is often localised, leading to a punctated or speckled appearance (Jurat-Fuentes and Jackson 2012).

In other cases the pathogen may cause the insect to undergo behavioural changes that facilitate the spread of the pathogen. For example, tropical carpenter ants, *Camponotus leonardi*, infected with the parasitic fungus *Ophiocordyceps unilateralis* undergo fungal-induced behavioural changes resulting in "zombie ants". These ants travel from the forest canopy down to the understory to die on sapling leaves, where environmental conditions are optimal for growth and subsequent dispersal of the fungal pathogen (Hughes et al. 2011). Locust species are known to thermoregulate to overcome fungal infections (Gardner and Thomas 2002). A phenomenon called behavioural fever can occur in response to a pathogen (Louis et al. 1986), during which an insect may seek warmer temperatures in an attempt to limit the growth of the pathogen while optimising its own metabolic processes. The heightened metabolism rate boosts the insect's immune responses. This behaviour is also dependent on the type of pathogen. In the case of the Moroccan locust, *Dociostaurus maroccanus*, fever was observed when the insect was infected with the fungus *Metarhizium anisopliae* but not when infected with *Beauveria bassiana* (Blanford and Thomas 1999).

Though not pathogens, symbionts can also influence the behaviour of the host. Parasitised pea aphids (*Acyrthosiphon pisum*) bearing the secondary symbiont *Hamiltonella defensa* produce significantly more offspring than unparasitised aphids, meaning that the symbiont may increase its own success through vertical transmission (Oliver et al. 2006). The symbiont can also protect the host through the production of toxins that prevent infection by the parasitoid wasp *Aphidius ervi*. In this instance, *H. defensa* strains harbour the bacteriophage APSE-3, which encodes a putative YD repeat toxin and causes greater than 85 % mortality in infecting parasitoid wasps (Oliver et al. 2005).

8.5 Possible Origins of Multicomponent Bacteria-Derived Toxins

Pathogens can encode a wide range of toxins that can also differ in their complexity. Bacterial toxins are typically categorised by their site of action in the host. Exotoxins and types III–VII secretion system cytotoxins modulate intracellular targets, while endotoxins, membrane-damaging toxins and superantigens act on the cell surface, and exoenzymes modulate targets in the extracellular matrix (Engleberg et al. 2012). Toxins that are internalised by the cell typically act to modify the subcellular skeleton, which is composed of a protein called actin. The physical alteration of the internal subcellular actin scaffold interferes with the ability of the cell to carry out subcellular functions, leading to death of the cell (Schiavo and van der Goot 2001; Aktories et al. 2012). Over time, many toxins have acquired different targeting or accessory molecules, leading to potential changes in both efficacy and/or host range. Some systems, such as the type III, type IV and type VI secretion machinery, are highly complex, comprising several proteins that form a functional complex (Tseng et al. 2009).

It has been proposed that the bacterial type III secretion machinery and the bacterial flagellum may share a common ancestor and have subsequently evolved independently of each other (Gophna et al. 2003). Type IV systems typically contain cell wall membrane-spanning regions and are either involved in processes such as DNA transfer via a pilus or in the transfer of effector (toxin) molecules (Alvarez-Martinez and Christie 2009). It is thought that some of these systems have been co-opted or evolved from functions that were in place before the origin of multicellular eukaryotic hosts (Frank et al. 2005; McCann and Guttman 2008; Leiman et al. 2009; Maezawa et al. 2006). Evidence supporting this theory is the recent finding that type VI secretion systems play a role in the "competitive duelling" of bacteria through a tit-for-tat exchange of toxin effector molecules (Russell et al. 2014). In these instances, effector molecules such as DNA- and cell wall-degrading enzymes target a recipient bacterium, where in the absence of an immunity protein will lead to cell death (Russell et al. 2014). A component protein of the type VI machinery contains a VgrG domain (valine–glycine repeat protein G) that forms a protein cylinder, called a beta-barrel. Such beta-barrel structures typify proteins involved in transport or pore formation (Galdiero et al. 2007; Wimley 2003). These VgrG domain proteins often encode a C-terminal effector, or, alternately, can bind to an independent effector protein (Shneider et al. 2013).

Similar to type VI machinery, some *Rhs* (recombination hot spot) elements encode toxic moieties that are actively delivered to a target cell to kill the recipient bacterium (Koskiniemi et al. 2013; Koskiniemi et al. 2014). *Rhs* proteins comprise an N-terminal region, a conserved "*Rhs* core" region and a C-terminal "tip" region of 100–200 amino acids that is highly divergent between different *rhs* elements. The *Rhs* core region contains multiple copies of the characteristic 15-amino acid sequence known as the YD repeat. Interestingly, the DNA that encodes the N-terminus has a high G+C content, while the C-terminus region is atypically AT-rich (Hill et al. 1994).

Similar to type VI machinery, *Rhs* elements may have initially been used to transfer toxin proteins between bacteria, facilitating clonal selection by outcompeting non-resistant variants. Over time, these systems have been co-opted by the bacteria, where the acquisition of alternate effector proteins leads to pathogenesis (Koskiniemi et al. 2014). Similar to VgrG, the structure TC-C(*Rhs*), which is part of the multicomponent insect-active TC, has also been shown to form a beta-barrel (Busby et al. 2013).

Of the Gram-negative entomopathogens, toxin complexes (TCs) and another group of multicomponent toxin the phage-like Afp are of relevance to latter parts of this chapter, which will briefly be covered in the following section.

8.5.1 Insect-Active Toxin Complexes

TCs were first identified in the bacterium *Photorhabdus luminescens* (Bowen et al. 1998). Aside from those in *Photorhabdus,* TCs have also been identified in *X. nematophilus* (Morgan et al. 2001), *Pseudomonas syringae* pv. tomato DC3000 (Buell et al. 2003), *Paenibacillus nematophila* (ffrench-Constant and Waterfield 2005), *B. thuringiensis* (Blackburn et al. 2011) and members of the *Yersinia* genus (Fuchs et al. 2008; Hurst et al. 2011c), among others. Typically, TCs are composed of three proteins designated as TC-A, TC-B and TC-C, which combine to form a complex with insecticidal activity (ffrench-Constant and Waterfield 2005; Fig. 8.3c). If the TC-BC subcomplex components are co-expressed in the same cell, they can combine with TC-A components from other species, or other TC clusters, to form a TC with altered host specificity. These combinations suggest that the predicted carriage molecule component, TC-A, varies relative to host range (Sergeant et al. 2003; Waterfield et al. 2005a, b). Further evidence for this was provided by Lee et al. (2007), who determined that the *X. nematophila* XptA1 TC-A component can bind to brush-border membrane vesicles (reside on the epithelial cells, Fig. 8.2b) of the white cabbage butterfly *Pieris brassicae*. Recent structural analysis has revealed that the *P. luminescens* TC-A molecule (TcdA1) (Gatsogiannis et al. 2013) and Yen-TC form a pentameric cage (Landsberg et al. 2011; Meusch et al. 2014), into which the TC-B and TC-C components dock.

The TC-C components are composed of two protein domains: the *Rhs* amino-terminal domain and a distinct carboxyl domain that encodes the main toxin/effector (Hurst et al. 2000). This effector molecule is enveloped by the TC-B component together with the amino-terminal region of the TC-C (*Rhs*), which forms the beta-barrel (Busby et al. 2013; Meusch et al. 2014). Orthologues of these carboxyl terminal-located effectors have been identified in other documented toxin systems and encode alternate transporting and targeting domains. This likely reflects a universally conserved role for the effectors and suggests that the limiting factor is the ability of the effector to gain entry into the cell. Depending on the host and/or mode of entry, such as *per os* application or direct entry via the circulatory system, different carrier vehicles are required to enable the toxin to withstand the different physiochemical parameters and to dock with the different receptor types. In the case of *Y. entomophaga*, the carboxyl terminus of TC-C YenC1 shares high amino acid similarity with the carboxyl terminus of the *E. coli*-derived cytotoxic necrotising factor 1 (Cnf1). The *E. coli* Cnf toxin is a single polypeptide that encodes its own amino-terminal receptor-targeting domain (Lemichez et al. 1997), which, in *Y. entomophaga*, is replaced at a functional level by TC-A (YenA1) and TC-B (YenB). These two components serve to protect and deliver the YenC1 effector to the target site.

Histologically, ingestion of the *P. luminescens* TC toxin, Tca, by larvae of the tobacco hornworm, *M. sexta*, results in apical swelling and blebbing (sloughing off) of large cytoplasmic vesicles by the intestinal columnar cells, leading to the eventual dissolution of the intestinal tract (Bowen et al. 1998; Bowen and Ensign

1998). The Tca toxin is active against *M. sexta* larvae by both *per os* challenge and intra-haemocoelic injection (Blackburn et al. 1998). In contrast, the American cockroach was killed only if the toxin was injected into the haemolymph, with no observed *per os* activity (Bowen and Ensign 1998). Based on this information, it would seem that TC toxins may interact with either a universal receptor or multiple different receptors present on both the apical and basal surfaces of the insect midgut cells (Blackburn et al. 1998). It is unknown why TC toxins are haemocoelically and *per os* active; however, this commonality may reflect the nature of their delivery. It is likely that the TC-A molecules are required to protect the TC-BC complex from the harsh environment of the insect gut, while the TC-BC subcomponents can act independently in places such as the haemocoel. In agreement with this, TC-B- and TC-C-related subcomplexes often reside as independent entities in the genomes of other microorganisms (ffrench-Constant and Waterfield 2005). In some instances, they are fused to form a single open reading frame (ORF) such as in case of *Burkholderia pseudomallei* (Busby et al. 2013).

In the context of entomopathogens, differences in host range may have been achieved through the acquisition of TC-C insect-active toxins and their associated proteins (TC-A and TC-B). Alternatively, the independent acquisition of the TC-B and TC-C (effector) components, which can complement a pre-existing TC-A (host–target specificity) component, can increase the diversification of a pathogen's own insect-active genes. Further to this, the *Rhs* amino-terminus provides the capacity ("power") to encapsulate recently acquired proteins that may have otherwise been detrimental to the bacterium. The DNA encoding the *Rhs*, or possibly VgrG, proteins may ensure their maintenance in the genome as a result of their ability to package potentially toxic effector molecules that confer a benefit to the host. In the instance of the *Rhs* AT-rich DNA that encodes the TC-C carboxyl domain, the less stringent binding of DNA may give greater flexibility in the accruement of natural mutations or in homologous recombination with less similar regions encoded by other *Rhs* elements. These gene acquisitions could be termed "modular" (or mosaic) evolution (Bornberg-Bauer and Albà 2013; Hachani et al. 2014).

In some instances, disparate systems that have toxic functions are used to fulfil a similar role in the transfer of an effector molecule or other component outside the cell. These processes require the production of morphologically similar transfer mechanisms. Hence, the assembled protein components are likely to share similar mechanical properties, even when they are derived from evolutionarily distinct organisms, in a process known as *convergent evolution*. Convergently evolved systems may share little, if any, DNA similarity, a phenomenon that is well documented in bacteriophage systems (Pedulla et al. 2003).

8.5.2 The Afp: A Mobile-Type VI Tailocin

An example of convergent evolution is evident following bioinformatic comparison of the Afp with its PVC orthologues. The Afp, is encoded by a prophage gene cluster of 18 ORFs. The translated products of 16 of these ORFs show similarity to the previously mentioned PVCs, while ORFs 17 and 18 of these systems have been found to encode different effector molecules (Hurst et al. 2004). Orthologues of the Afp have also been identified in the genomes of *Yersinia ruckeri* ATCC 29473 (Hurst et al. 2011a) and in marine bacteria such as *Vibrio campbellii* AND4.11 (Persson et al. 2009) and *Pseudoalteromonas luteoviolacea* (Shikuma et al. 2014). Electron microscopy of the Afp and the PVC orthologue PaPVC-Pnf showed that they are similar in morphology to R-type pyocins, which are bacteriophage tail-like particles originally identified in *P. aeruginosa* (Hurst et al. 2007a; Yang et al. 2006). The finding that expression of ORFs 1–16 of the Afp results in the formation of a bacteriophage tail-like structure (Rybakova et al. 2013) validated their role as a carrier vehicle for the toxin or effector molecules. Further, similarly to R-type pyocins, the Afp exhibits two morphological forms: a noncontracted or relaxed form where the outer sheath resembles a bullet-like structure and an extended or contracted form where the inner core of the noncontracted form protrudes. The R-type pyocins adsorb to a lipopolysaccharide on a target bacterial cell and cause lethality through the rapid contraction of the sheath component and subsequent penetration of the core through the outer bacterial membrane (Michel-Briand and Baysse 2002). Based on this, it is thought that the Afp and its orthologues undergo a similar contraction, enabling them to deliver the toxin to the target cell. Given that the puncturing of the eukaryotic target cell wall may in itself be lethal to the cell, it is tempting to speculate that the function of the Afp may be to induce cell wall-mediated endocytosis, whereby the target cell encapsulates and internalises the incoming toxin (Fig. 8.3f). Based on bioinformatics and protein structural data, along with consideration of their predicted function, it is likely that PVCs and the Afp are variants of the type VI secretion system (Bönemann et al. 2010; Zhang et al. 2012). Recently, the term tailocins, defined as defective phages that lack heads and contain no DNA, has been used to describe pyocins, PVC and Afp (Gill and Young 2011). Of further interest, distant orthologues of the Afp have been found in some bacterial symbionts (Penz et al. 2010, 2012). In the instance of the free-living marine bacterium *P. luteoviolacea,* the Afp orthologue induces metamorphosis of the marine tube worm *Hydroides elegans* through a yet to be defined mechanism (Shikuma et al. 2014). The ability of members of *P. luteoviolacea* to promote establishment and metamorphosis of various invertebrate and algal species (Bowman 2007) suggests that this bacterium is in some way associated with higher organisms.

8.5.3 Symbionts as Progenitors of Virulence Determinants

The transition between a symbiotic and a pathogenic lifestyle is of particular interest, as the previously mentioned symbionts can both manipulate host behaviour and protect the host from parasitism through the production of toxins. Many symbionts harbour known variants of key virulence determinants, such as types III, IV and VI secretion systems, including the Afp. Hence, symbionts represent an intermediate state between commensal and pathogenic microorganisms. Further to the definition above, symbionts can be categorised as primary *obligate symbionts* or secondary *facultative symbionts*. Obligate symbionts are often restricted to a specialised compartment within the insect, called the bacteriocyte. They are dependent on their host for certain metabolic factors because they have lost some metabolic functions, leading to a minimal genome (Bordenstein and Reznikoff 2005; Moran 2002). Because of their allopatric lifestyle, the opportunity for primary symbionts to make contact with other bacteria is limited. Facultative symbionts are more recently evolved and are not restricted to a specific region of the insect. They have a beneficial but not essential role in survival of their host insect, with some showing evidence of HGT (Oliver et al. 2010), and are more sympatric than allopatric.

Genomic analysis of several symbiont species has found that obligate symbionts of insects often lack dedicated protein secretion systems, while many facultative symbionts encode protein secretion systems, orthologues of which reside in known pathogens (Toft and Andersson 2010; Preston 2007; Coombes 2009; Dale and Moran 2006). It is thought that these systems may have developed to enable the establishment of a symbiont population, and possibly also the subsequent transfer of factors such as metabolites or nutrients, to and from the host for mutual benefit. It is speculated that the type III secretion systems of insect endosymbiotic bacteria may have undergone gene loss or gene divergence (Maezawa et al. 2006; Toft and Fares 2008) to mediate the invasion of host cells and thereby ensure transmission of bacteria to host offspring (Moya et al. 2008). In an alternate scenario, pathogens may have acquired virulence determinants such as type III systems from endosymbionts. Organisms that have acquired these new genes may then undergo events such as point mutations or HGT, from which selective pressure arises, leading to their further refinement to become either more or less virulent. These changes may be either to the protein itself or to its regulation. Ultimately, this process is influenced by the surrounding environment.

The nonmotile *Buchnera aphidicola* sp. strain APS, a symbiont of the pea aphid *A. pisum*, encodes a subset of flagella genes that form the flagellar hook–basal body that cover the bacterial cell surface (Maezawa et al. 2006). The authors suggest that the hook–basal body has been co-opted from the flagella system and is likely to be involved in the transport of yet to be defined molecules. In relation to the Afp, bioinformatic analysis has revealed variants of the Afp in the amoebal symbiont *Candidatus Amoebophilus asiaticus* (Penz et al. 2010) and the bacterium *Cardinium hertigii* cEper1, a symbiont of the parasitic wasp *Encarsia pergandiella* (Penz et al. 2012). Genome analysis has revealed that *C. hertigii* lacks all known protein secretion systems but does encode an Afp orthologue that shares DNA identity with the corresponding regions of the *A. asiaticus* genome. The authors suggest that at one time *C. hertigii* may have been a symbiont of amoebae or other protists, with a proposal that during these symbiotic associations, the type VI Afp-like system transports host-manipulating compounds directly into the symbiont host (Penz et al. 2012). Hence, the presence of Afp orthologues in these symbionts may in part define the possible origins of the orthologues.

In relation to the symbiotic relationship, it remains unknown as to which member evolved first, the pathogen or the symbiont. Recent phylogenetic analysis by Sarris et al. (2014) of Afp-like variants has identified relatives in a diverse range of bacterial species including members of the *Archaea*. The authors proposed that they are of an ancient origin and likely to have spread through HGT. This combined with the recent finding that bacteria partake in active warfare using various mechanical devices such *Rhs* and type VI systems, combined with the documented commonalities between type III and type IV secretions systems and bacterial appendages such as the conjugative pilus, provides strong support to the idea of a free-living sympatric origin.

8.5.4 Plant-Associated Entomopathogens

Further to insect symbionts, another likely reservoir of virulence determinants is plant-associated microorganisms. Plants evolved some 400 million years ago, much earlier than insects. Therefore, on the evolutionary timescale, plants and microorganisms have had a greater chance to develop the necessary machinery to allow the mutualistic sharing of nutrients and to evolve mechanisms for the protection of the plant from abiotic and biotic factors. There are many examples of complex endophytic partnerships of root-associated microbes, such as the insect-active root-associated fungal endophytes (Ball et al. 2006).

It is also known that several bacterial pathogens of humans and animals, such as *P. aeruginosa, Burkholderia* and *Salmonella enterica*, can also infect plants (Schikora et al. 2012). The opportunistic human pathogen *P. aeruginosa* has conserved virulence factors that are active towards diverse hosts, such as the roundworm *Caenorhabditis elegans,* the lepidopteran *G. mellonella* and the plant *Arabidopsis thaliana*. This host diversity suggests a possible evolutionary link to the emergence of mammalian pathogenicity (Mahajan-Miklos et al. 2000; Rahme et al. 1995, 2000). A similar scenario has also been suggested wherein invertebrate pathogens are the source of the emergence of mammalian pathogens (Waterfield et al. 2004). Genome-based similarities have also been found between the insect pathogen *P. luminescens* and the human pathogen *Y. enterocolitica* (Heermann and Fuchs 2008). Further to this, plant bacterial pathogen-associated type III secretion systems have been well documented (Büttner and He 2009).

In relation to plant-associated bacteria, root-associated *Pseudomonas* species can form a variety of mutualistic or commensal relationships. The close association of microorganisms with the plant, especially those in the root zone, can have beneficial effects on improving plant growth through the direct production of plant growth-promoting factors such as auxin. In addition, plant-associated bacteria can suppress some plant pathogens through the release of anti-protozoan nematode-active metabolites and, in some instances, insect-active compounds. Several root-associated insect-active pseudomonads have been identified, including *P. protegens, P. chlororaphis* and *P. entomophila* (Kupferschmied et al. 2013). The bacteria *P. protegens* Pf-5 (formally *P. fluorescens* Pf-5) and *P. protegens* CHA0 were isolated from so-called suppressive soils; soils in which disease does not occur under conditions optimal for plant pathogen development (Kinkel et al. 2011). The strain *P. protegens* CHA0 is *per os* active in a number of lepidopteran species, such as *S. littoralis,* the tobacco budworm *Heliothis virescens* and *P. xylostella* (Ruffner et al. 2013). Orthologues of the *Photorhabdus* insect-active toxin are present within the genomes of *P. protegens* Pf-5 and *P. protegens* CHA0, including the previously discussed *P. luminescens* Mcf toxin. In *P. protegens* CHA0, a Mcf orthologue encoded by the *fitD* gene is part of a larger gene cluster (*fitA-E*), components of which are proposed to mediate the transport of FitD (Mcf) out of the bacterial cell (Péchy-Tarr et al. 2013). Further *fit* orthologues have been identified in the genomes of *P. protegens* Pf-5 and *P. chlororaphis* (Loper et al. 2012; Ruffner et al. 2013; Shen et al. 2013). Both *P. protegens* and *P. chlororaphis* have *per os* activity against *S. littoralis, H. virescens* and *P. xylostella,* while *fit*-deficient strains of *P. protegens* CHA0 exhibited no *per os* activity (Ruffner et al. 2013).

The bacterium *P. protegens* CHA0 appears to be translocated directly into the haemocoel of the white cabbage butterfly *P. brassica* within 24 h of *per os* challenge by a yet to be defined mechanism (Kupferschmied et al. 2013). The bacterium is highly lethal even in low doses when injected into the haemocoel of *M. sexta* and *G. mellonella* larvae (Péchy-Tarr et al. 2008). The ability of the bacterium to survive within the haemocoel suggests that, similar to the nematode-associated bacteria *Photorhabdus* and *Xenorhabdus*, *P. protegens* CHA0 has the capacity to suppress, evade or overwhelm the insect immune system.

Another member of the *Pseudomonas* species, *P. entomophila*, is pathogenic at high doses in both larvae and adults of *D. melanogaster* (Vodovar et al. 2005). This bacterium contains a number of entomo-active virulence determinants (Vodovar et al. 2006), with the pore-forming protein toxin monalysin causing particularly significant damage to *D. melanogaster* gut cells (Opota et al. 2011). In addition to monalysin, *P. entomophila* also harbours TC-related toxins, including haemolysins and lipopeptides (Vodovar et al. 2006).

8.5.5 Free-Living Pathogens

So far, we have discussed pathogens that form an association with either an animal vector or a plant. In addition to this, there are a wide range of free-living pathogens, many of which have non-pathogenic counterparts. Certain members of the genus *Serratia* can infect a wide range of hosts, including plants, vertebrates and invertebrates (Hejazi and Falkiner 1997; Grimont and Grimont 1978; Inglis and Lawrence 2001). The earliest record of *Serratia* as an insect pathogen was from the reddened cadavers of dead silkworms (Steinhaus 1941). The causative bacterium, *S. marcescens*, produces large quantities of degradative enzymes such as lecithinase (Steinhaus 1959), chitinase (Lysenko 1976) and protease (Kaska 1976), which are likely to assist in invading the haemocoelic cavity, from where the bacteria rapidly multiply to kill the insect by septicaemia (Podgwaite and Cosenza 1976). Many strains of *S. marcescens* are *per os* active against a number of insect species, including tsetse flies *Glossina* spp. (Poinar et al. 1979), the corn earworm *Helicoverpa zea* (Farrar et al. 2001) and the blowfly *Lucilia sericata* (O'Callaghan et al. 1996). A high *per os* dose of *S. marcescens* HR-3 is required to kill the grassland locust *Myrmeleotettix palpalis* (Jin et al. 2005), and in such instances, the active microbial compound has been identified as a zinc-dependent metal protease, which has a high median lethal dose of 12.1 μg per locust (Tao et al. 2006). In addition to insect activity, some strains are also nematode active, such as *S. marcescens* DB11, which is lethal to the roundworm *C. elegans* (Mallo et al. 2002). Some strains of *S. marcescens* are highly pathogenic to many insects when injected into the haemocoel, with as few as five bacterial cells required to kill the boll weevil *Anthonomus grandis* (Slatten and Larson 1967).

Of particular note to this chapter is another member of the genus *Serratia*, *S. entomophila*, which causes host-specific disease of the New Zealand grass grub *C. zealandica* (Grimont et al. 1988). To date, only three other geographically distinct isolates of *S. entomophila* have been reported: a non-pathogenic strain isolated from a French water well, the Mexican isolate Mor.4.1, reported to control the white grub *Phyllophaga blanchardi* (Nuñez-Valdez et al. 2008) and the Indian *S. entomophila* strain AB2, isolated from *Heliothis armigera* larvae (Chattopadhyay et al. 2011). High doses of *S. entomophila* have been reported to control *H. armigera*, *S. littoralis* and *P. xylostella* larvae (Chattopadhyay et al. 2012).

8.6 Case Studies: A Chronic and Hypervirulent Pathogen

8.6.1 *The Host-Specific* Serratia entomophila–Costelytra zealandica *Chronic Relationship*

In this section, we bring together knowledge from other systems to place an ecological perspective on the disease of *C. zealandica* grass grub larvae as caused by strains of the bacteria *S. entomophila* and *S. proteamaculans*. The relationship between *S. entomophila* and *C. zealandica* is thought to have coevolved after the separation of New Zealand from Gondwanaland, some 85 million years ago. This host–pathogen relationship is now of significant agricultural interest where the endemic grass grub pest impacts the productivity of recently introduced exotic grasses.

The grass grub *C. zealandica* is a ubiquitous endemic pest in New Zealand, feeding on the roots of pasture plants such as grasses and clover and also the roots of a wide range of commercial plants including cereal crops, raspberries and strawberries. The larvae typically reach damaging levels 2–4 years from sowing causing damaged patches in the pasture that increase in size from year to year (Kain and Atkinson 1970). In some cases, in excess of 400 larvae per m^2 have been recorded, resulting in lost production and preventing pasture regeneration (Stucki et al. 1984).

Two pathogenic bacteria were isolated independently from amber-coloured *C. zealandica* larvae and were later identified as *S. entomophila* and *S. proteamaculans* (Grimont et al. 1988). Both pathogenic and non-pathogenic isolates of *S. entomophila* have since been described (Glare et al. 1993; Dodd et al. 2006), and these bacteria are typically found at levels of 1×10^3–10^5 bacteria per gram of pasture soil (O'Callaghan et al. 1999). Assessment of the *Serratia* strains from soils relatively devoid of grass grub larvae has found that the majority of isolates are non-pathogenic (O'Callaghan et al. 1999). In the absence of pathogens or predation, grass grub populations will increase from year to year, until eventually disease enters the system. With disease levels in excess of 20 %, there is a greater than 90 % probability of disease decline (Jackson et al. 1999).

An overview of the *S. entomophila*–*C. zealandica* disease process is outlined in Fig. 8.3. Grass grub larvae cease feeding within 2–5 days of ingesting pathogenic strains of either *S. entomophila* or *S. proteamaculans*. The rapid onset of cessation of feeding activity by the larvae prior to death prevents further damage to the pasture. The larval gut, which is normally dark in colour, clears, and the larvae take on a characteristic amber colouration (Jackson et al. 1993; Fig. 8.3e). Levels of the major gut digestive enzymes trypsin and chymotrypsin decrease sharply in tandem with gut clearance (Jackson 1995). In agreement with this observation, transcriptomic analysis has revealed that the expression of serine protease enzymes is downregulated at this time (Marshall et al. 2008). Proteomic analysis revealed the rates of subcellular structural proteins tubulin and actin synthesis to be significantly increased in the larval midgut, which may interfere with the release of digestive proteases (Gatehouse et al. 2008). Histological studies have revealed that there is also a visible reduction in the number of fat cells (nutrient reserves) as the disease progresses. In contrast to many other entomopathogenic bacteria, no destruction or blebbing of the midgut epithelium has been observed within the alimentary tract (Jackson et al. 1993). Within 6 days of ingestion, a maximum population density of 1×10^6 cells is found in the larval gut, the majority of which reside in the hindgut (Jackson et al. 2001). Aside from food particles, no specific sites of colonisation have been identified (Hurst and Jackson 2002). The larvae may remain in this diseased state for a prolonged period (1–3 months) before bacteria eventually invade the haemocoel, resulting in rapid death of the insect by septicaemia (Jackson et al. 1993, 2001). Relative to the approximate lifespan of the larvae (~7–8 months), the disease period represents a large amount of time and is therefore termed a chronic infection.

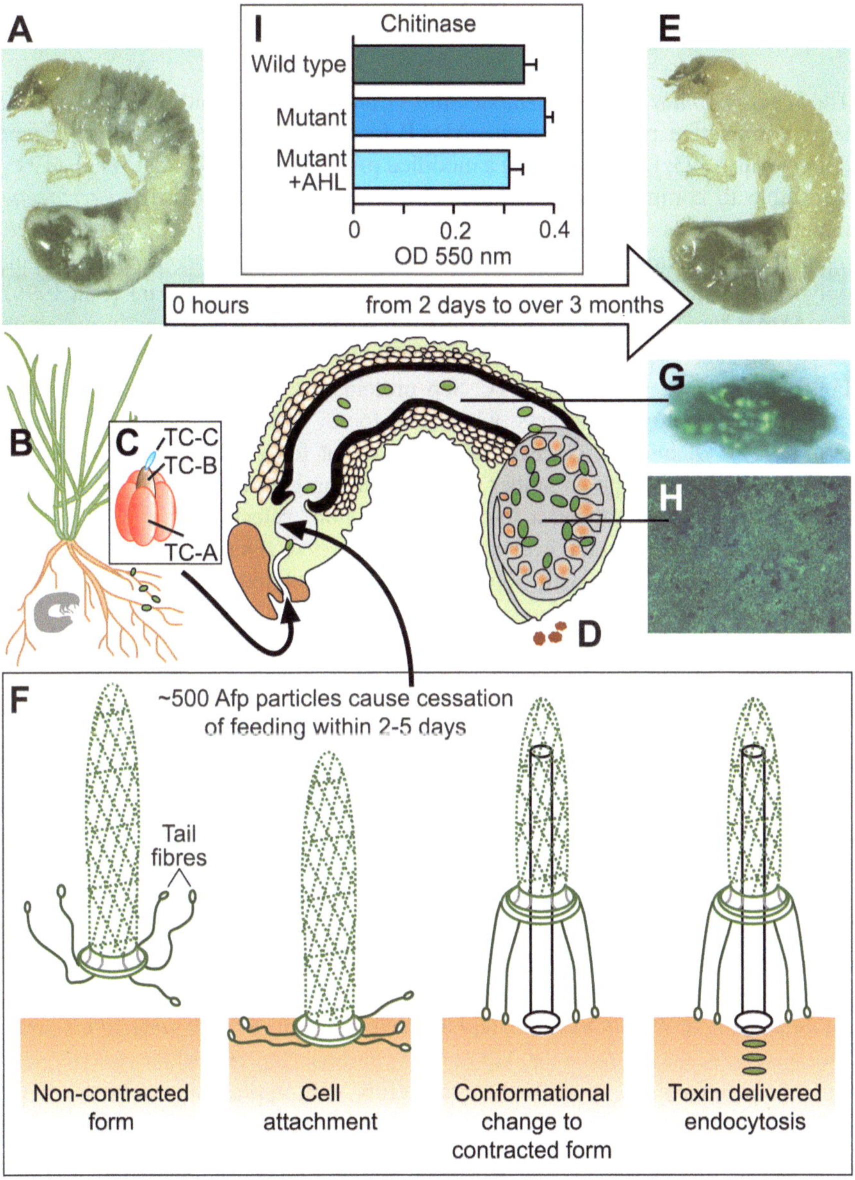

Fig. 8.3 Overview of the *Serratia entomophila–Costelytra zealandica* system. (**a**) Healthy *C. zealandica* larvae ingest the roots (**b**) of pasture grass colonised with *S. entomophila*. The *Serratia entomophila* pathogenicity (Sep) toxin complex (TC) (**c**) is released by ingested bacteria. This causes gut clearance, expulsion of frass pellets (**d**) and the larvae to become amber in colour (**e**). Within 2–5 days, the larvae cease feeding, a result of as few as 500 anti-feeding prophage (Afp) particles. The proposed model for the action of the Afp is depicted in (**f**), where the extended virus-like Afp particle latches onto a currently undetermined receptor on either the peritrophic membrane or the intestinal cells using its tail fibres. Once bound (cell attachment), the Afp particle changes conformation, and the eukaryotic target cell takes up the toxin possibly through

8.6.1.1 *Serratia entomophila* and *Serratia proteamaculans* Virulence Determinants

The *S. entomophila* and *S. proteamaculans* virulence determinants are encoded on a large conjugative plasmid termed pADAP for amber disease-associated plasmid (Glare et al. 1993; Fig. 8.4). Using a modified pADAP variant, Grkovic et al. (1995) were able to conjugate pADAP into other members of the *Enterobacteriaceae*, including strains of *S. marcescens*, *Serratia liquefaciens*, *Enterobacter agglomerans*, a *Klebsiella* species and *E. coli*. However, in *E. coli*, pADAP could only replicate as a plasmid-fused pADAP::pRP4-4 cointegrate (Glare et al. 1996).

pADAP encodes two virulence-associated regions: the Afp and an insect-active TC, designated *sepABC* for S. entomophila pathogenicity (Hurst et al. 2000; Fig. 8.3c). The *S. entomophila* TC Sep proteins are host specific and cause no observable histological affects when fed to *C. zealandica* larvae (Marshall et al. 2012). Of interest, *C. zealandica* larvae that fed small amounts of purified Sep proteins do undergo an amber phase, including gut clearance, but then revert back to a healthy pathotype, a so-called "reversion" effect. Following this reversion, the intestinal cells can regenerate and the insect recovers. Larvae that received a high *per os* challenge with Sep proteins still exhibited slight feeding activity (Hurst et al. 2007b).

The second pADAP virulence determinant is the previously mentioned Afp (Fig. 8.3f). The ability of purified Afp particles to cause cessation of feeding in *C. zealandica* larvae (Hurst et al. 2007a) demonstrated that the Afp can function independent of the bacterial cell. While high concentrations of Afp cause mortality within 11 days of application (Hurst et al. 2004), as few as 500 Afp particles are required to elicit cessation of feeding activity in 50 % of the larval population 3 days post *per os* challenge (Rybakova et al. 2013). This suggests that in the wild-type *S. entomophila*-grass grub system, a finite number of Afp particles are produced, causing cessation of feeding activity as a sublethal effect.

The ability of toxins such as Afp to stop insect feeding reduces the amount of incoming organic matter entering into the insect's gut, thus slowing the rate of gut metabolic processes. Reduced passage of food through the gut in turn facilitates the establishment of *S. entomophila* within the insect gut. *S. proteamaculans* strain 143, which does not contain the Afp but does harbour Sep TC orthologues, exhibits variable pathogenicity towards the grass grub, with approximately 65 % of the infected insects succumbing to amber disease, while the other 35 % remain healthy

---

cell-mediated endocytosis. The disease process can last for up to 3 months. There are no apparent sites of *S. entomophila* colonisation on the surface of gut cells, with bacteria only adhering to particulate matter of the gut (**g**). Within 6 days of ingestion, high levels of bacteria are observed in the fermentation chamber of the insect (**h**). Under conditions of high cell density, as is observed in the insect hindgut, levels of the *S. entomophila*-derived chitinases are reduced (**i**), possibly prolonging the time of infection. (**i**) Measurements of chitinase activity. WT denotes wild-type *S. entomophila*; mutant denotes quorum-sensing mutant; +AHL denotes the mutant complemented by addition of an exogenous supply of the quorum-sensing signal molecule N-Acyl homoserine lactone (AHL)

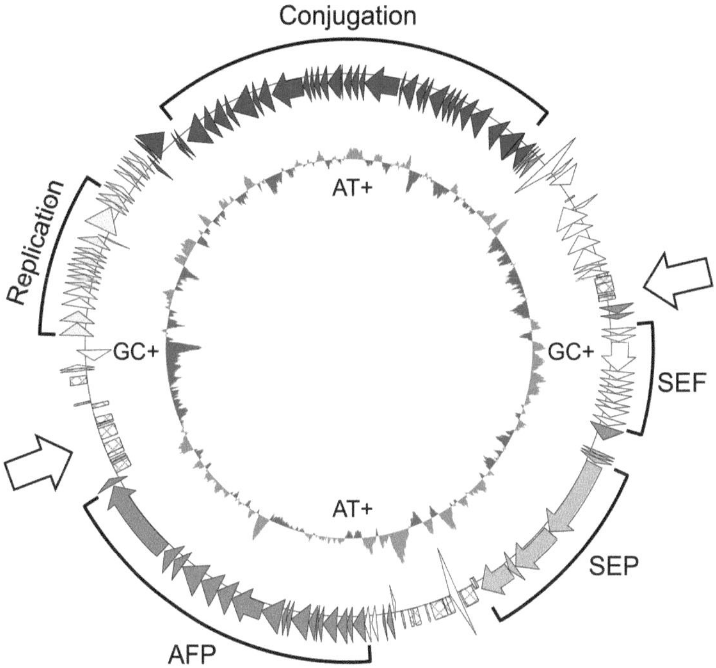

Fig. 8.4 Schematic of the *Serratia entomophila* plasmid pADAP. Based on the DNA G+C content, pADAP can be divided into two broad regions: the relatively AT-rich plasmid backbone (AT+), encoding the areas required for plasmid replication and conjugation (DNA transfer), and a GC-rich region (indicated by large inward-pointing arrows) of HGT-acquired DNA including the anti-feeding prophage (Afp) gene cluster and the *Serratia entomophila* pathogenicity (Sep) determinants. A further HGT-acquired region is denoted Sef for *Serratia entomophila* fimbrial gene cluster

(Glare et al. 1993; Hurst et al. 2011a). It would seem that in this instance, the pathogen was unable to establish itself in the subset of healthy larvae.

Both the Sep and Afp toxins need to be released from the bacterial cell to cause an effect. In *S. entomophila*, this process is achieved through bacterial cell lysis, killing the bacterial cell in an altruistic process. In line with this scenario, an intact lysis cassette, encoding an endopeptidase, a holin and a muramidase protein, is located upstream of the *afp* gene operon. These proteins co-ordinately interact to form perforations in the bacterial cell wall, leading to lysis of the bacterial cell (Catalão et al. 2013; Young 2014; Hurst et al. 2004). Lysis of the bacterium has been indirectly linked to the release of large proteins from *X. nematophilus* (Brillard et al. 2003) and *P. luminescens* (Waterfield et al. 2001). Based on the present understanding of the regulation of bacteriophage lytic systems, it would be safe to assume that populations of *S. entomophila* cells actively yielding Afp are undergoing cell lysis. Giddens et al. (2000) showed that expression of the anti-feeding gene *anfA1* in *S. entomophila* requires, and is regulated by, RpoS. Hurst et al. (2007a) subsequently determined that the *anfA1* expression product, AnfA1, induced the production of Afp. The global stationary phase regulator RpoS (RNA polymerase sigma S subunit) is a global regulator that triggers the expression of a subset of bacterial genes under conditions of stress or during the latter stages of cell growth—stationary phase. In *S. entomophila*, the expression of *anfA1* at stationary phase is lower in cells with a mutant *RpoS*. AnfA1 is a member of the RfaH family of transcriptional anti-terminators (Hurst et al. 2004) that bind to a specific DNA sequence (5′-GGCGGTAGNNT-3′), called the *ops* (operon polarity suppressor) element, located upstream or internal to the operon with which they are associated. These anti-terminators enable the reading through of terminator structures, specifically hairpin loops, thus preventing operon polarity (Bailey et al. 2003). Accordingly, an *ops* element is located immediately upstream of *afp1*, while a second *ops* element is positioned internal to the putative Sef class I fimbriae cluster (Hurst et al. 2007a, Fig. 8.4). Hence, through the induction of *anfA1*, both the *afp* and class I *sef* fimbriae are likely to be expressed, which would then aid in adhesion of the bacterium to a substrate or surface where the Afp might be effective. The induction of the Afp through the stationary phase stress response activator RpoS may mean that, within the insect, only a subset of mature bacterial cells are likely to produce Afp at any given point in time. This hypothesis furthermore suggests that the bacteria have evolved to prolong the period of infection by downregulating the disease process.

Assessment of a *S. entomophila* population density-dependant quorum-sensing mutant for chitinase production revealed that chitinase production was higher in the quorum-sensing mutant (Fig. 8.3i). Hence, under conditions of high cell density such as can occur in the *C. zealandica* hindgut, chitinase production is likely to be decreased. This limits the potential degradation of the insect's chitin-based peritrophic membrane or skeleton, thus prolonging the time required before the bacteria can invade the haemocoelic cavity.

8.6.1.2 Defining the Origin of pADAP and Relating This to Disease Epidemiology

Based on the completed DNA sequence of pADAP, it has been suggested that the *afp* and *sep* gene clusters have been independently acquired by pADAP (Hurst et al. 2011a). A plot of the frequency of G and C nucleotides in the pADAP plasmid revealed that the pADAP plasmid backbone, comprising the regions involved in plasmid replication, partition and conjugation, is more AT-rich relative to the GC-rich *afp*, *sep* and *sef* gene clusters (Fig. 8.4). This may indicate that these GC-rich gene clusters have been acquired by the pADAP backbone. In relation to the evolution of the TC *sep* and *afp* gene clusters, further assessment of the DNA GC skew revealed differences in the variable carboxyl termini of TC-C and the Afp genes *afp17* and *afp18* relative to their respective gene-encoding regions, indicative of acquisition of toxin components from another system by HGT (Hurst et al. 2000, 2004).

Being a plasmid, pADAP is capable of both vertical and horizontal genetic transmission. Assessment of the DNA restriction enzyme profiles of both the plasmids and genomes of virulent *S. entomophila* isolates showed that they are all quite similar (with only a single *S. entomophila* pADAP variant identified), relative to the dissimilarity observed for *S. proteamaculans* isolates (Dodd et al. 2006; Dodd 2003). It is likely that *S. entomophila* has recently acquired the plasmid and then spread clonally or that the *S. entomophila* genome has experienced a recent evolutionary bottleneck (Maynard-Smith 1990). It is therefore of interest that under laboratory conditions, the pADAP plasmid can be conjugated to several other members of the *Enterobacteriaceae* while in the field virulence-encoding variants of pADAP have only been found in strains of *S. entomophila* and *S. proteamaculans*. This suggests that the strain backgrounds of both *S. entomophila* and *S. proteamaculans* have properties that are suited to virulence and their maintenance within the environment. The occurrence of avirulent strains of both *S. entomophila* and *S. proteamaculans* suggests that, under some circumstances, the ability to confer virulence is not an overriding selection trait and reflects the availability of an insect host and that when present will ensure the maintenance of the plasmid-based genes in the ecosystem population. This scenario is in agreement with the epidemiology of the disease, where *C. zealandica* larval populations build up over a 5-year period before undergoing a natural decline as a result of disease. During this time, there is a concurrent increase in *S. entomophila* numbers in the field (O'Callaghan et al. 1999). This process reflects a classic cyclical "predator–prey" relationship.

Throughout the course of infection, the physical movement of grass grub larvae through the soil profile may inadvertently allow the spread of the bacterium, either through host-to-host contact or clearance of the gut. The chronic nature of infection gives the bacterium greater opportunity to infect other host insects. Further, colonisation and subsequent decay of the insect may occur at a time that coincides with the natural end of the univoltine *C. zealandica* larval stage. The cadaver may then "over season" the bacterium until the next generation of larvae emerge.

Linking this cyclic process to epidemiological theory, there is a critical minimum population size in which a pathogen can indefinitely persist. As the host population declines over several years of successional pathogen build up, the frequency of *S. entomophila* encountering its host is significantly reduced, leading to a decline in infected carryover cadavers. In relation to the grass grub system, the adult *C. zealandica* beetle can fly into areas of healthy pasture where disease may have previously occurred and pathogen numbers are at a nominal population density. Following egg deposition, a resurgence of healthy larvae enter the system, from where a pre-existing low-density disease will increase over time, with the cycle then repeating itself.

In general, the infection of *C. zealandica* by *S. entomophila* mirrors that of obligate Gram-positive pathogen *Paenibacillus popilliae*. *P. popilliae* is a haemocoelic pathogen, in contrast to the intestinal-based nature of *S. entomophila*, and yet in both instances the bacterial infection does not trigger any apparent host responses, such as melanisation or recruitment of haemocytes. However, in both systems, events are triggered that lead to the depletion of the larval fat bodies, from where the insect starves to death. In both of these scenarios, it appears that the purpose of pathogenicity is to provide an environment suitable for replication of the pathogen. Based on the current information on amber disease, it is proposed that *S. entomophila* is on its own path to form an association with *C. zealandica*.

8.6.2 *Yersinia entomophaga: A Broad-Range Hyper Pathogen*

In contrast to the chronic and host-specific nature of *S. entomophila*. The bacterium *Y. entomophaga*, isolated from a dead *C. zealandica* grass grub larva, is *per os* pathogenic to a wide range of coleopteran and lepidopteran species, including species not endemic to the country of origin, such as the migratory locust *Locusta migratoria* (Hurst et al. 2011b). The 6-day *per os* median lethal dose is approximately 3×10^4 *Y. entomophaga* bacterial cells per *C. zealandica* larva. During the infection process, the larvae vomit, expel frass pellets and change in appearance from a healthy grey to amber and then moribund brown in colour. This disease process is accompanied by reduced feeding activity. Bacteria enter the haemocoelic cavity within 48 h of ingestion, following which insect death is rapid (Hurst et al. 2014). The main *Y. entomophaga* virulence determinant is an insect-active TC derivative called the Yen-TC. The components of the Yen-TC are encoded on a ~32-kb PAI designated PAI_{Ye96}. The absence of any mobility determinants suggests that PAI_{Ye96} may be stably inherited and only capable of vertical transmission (Hurst et al. 2011c). Histological studies of the effects using either the purified Yen-TC toxin or the bacterium on *C. zealandica* larvae revealed that the larval midgut deteriorates over a 96-h period (Marshall et al. 2012). In the initial stages of the disease, *Y. entomophaga* can be observed on the peritrophic membrane, at the junction between the hindgut and the fermentation sack. This region is thought to be where the Yen-TC-associated chitinases degrade the peritrophic membrane allowing the bacterium to gain ingress to the haemocoelic cavity (Hurst et al. 2014).

Interestingly, the production of Yen-TC is temperature dependent, with large amounts released at 25 °C but not at 37 °C (Hurst et al. 2011c), a facet that affects the virulence of the bacterium. Challenge of *G. mellonella* larvae *per os* with *Y. entomophaga* revealed that the bacterium was non-virulent when the larvae were maintained at 37 °C but were virulent if the larvae were maintained at 25 °C. Injection of either *Y. entomophaga* or a derivative in which the Yen-TC gene cluster has been deleted revealed a 4-day median lethal dose of approximately three bacterial cells per larva, irrespective of temperature (25 or 37 °C). This number of bacteria is similar to the number of *P. luminescens* required to cause lethality by intra-haemocoelic injection and indicates that *Y. entomophaga* may have evolved to survive in the insect haemocoelic cavity (Hurst et al. 2015). The ability of *Y. entomophaga* and nematode-vectored pathogens such as *P. luminescens* to infect multiple species provides greater opportunity to infect other organisms, thereby allowing their maintenance in the ecosystem.

Interestingly, *Y. entomophaga* has yet to be re-isolated from the environment. The rapidity of killing by *Y. entomophaga* may limit its ability to spread in the environment, although this will be somewhat counteracted by the broad host range of *Y. entomophaga*. The inability so far to re-isolate *Y. entomophaga* from the environment might reflect small, highly localised disease outbreaks that go unnoted. Alternately, the low number of *Y. entomophaga* cells required to kill a host following intra-haemocoelic injection may indicate that the bacterium is either associated or vectored by another organism.

8.7 Summary

In this chapter, we have attempted to give a broad overview (as summarised in Fig. 8.5) of the wide range of variables that influence the potential efficacy and evolution of a pathogen. Often overlooked are the less effective pathogens that can only cause disease in a subset of the host population, where host susceptibility may depend on host factors such as age or health. This may allow greater maintenance of the less effective pathogen in the ecosystem, from where the surviving host population can proliferate, giving greater opportunity for later infection by the pathogen.

In other circumstances, infection may not be detrimental, particularly if the pathogen initiates a rather benign or chronic infection. In these instances, the host may instead act as a reservoir, allowing the maintenance of the pathogen until a more suitable host comes into proximity. Similarly, where disease does not affect the entire host population, the ability of a pathogen that is specific for one organism to persist in a different host may inadvertently increase the potential of the pathogen to interact with other microflora or new potential target species, increasing the opportunity for processes such as HGT. This may also increase the opportunity to form an association with a host.

In summary, we could assume that it is advantageous for a pathogen to overwhelm a host to essentially become a monoculture. This large pathogen population has increased opportunity to harbour progeny that have undergone natural mutations through processes such as point mutation or gross genomic rearrangements mediated by DNA recombination and/or HGT. It is plausible that within the decaying cadaver, the remaining resident host symbionts or other microbiota may offer a further source of novel DNA. These acquired genes or genetic alterations may further enhance the survival capabilities of the pathogen at a later time point or, alternately, lead to demise of the pathogen through a process termed short-sighted evolution (Fig. 8.5).

In recent times, the transfer of infected organisms or soil by human movement has become more frequent, dispersing microorganisms further afield. In such instances, potential hosts may be more susceptible to infection because they will not have evolved the appropriate pathogen countermeasures. Further to this, a change in land-usage and cropping systems can dramatically alter the resident

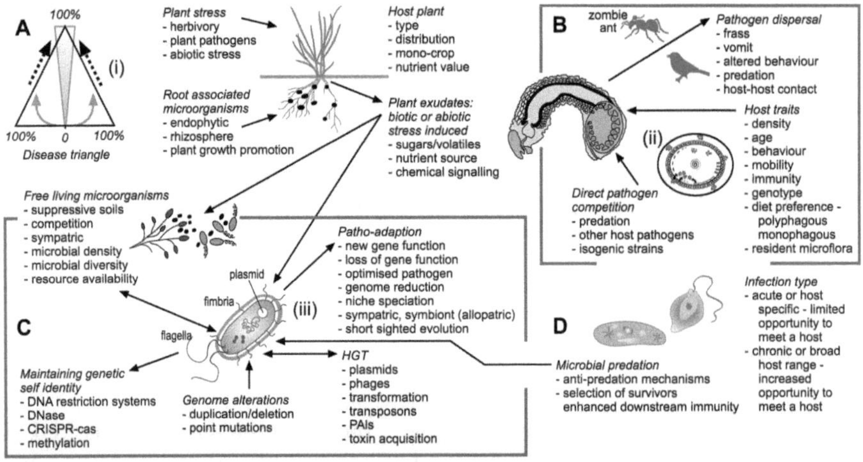

Fig. 8.5 (**a**) Abiotic and biotic variables that affect pathogenicity, including environmental influences on host and pathogen; (**i**) disease triangle (refer to Fig. 8.1). (**b**) Modes of transmission and host responses; (**ii**) cross section of insect gut (refer to Fig. 8.2). (**c**) Direct influences on the pathogen at any point in time; (**iii**) schematic of bacteria showing flagella, fimbriae and a plasmid. (**d**) Overarching influences on the long-term survival of the pathogen

soil biota and the nutrient value of the host's food, perhaps allowing changes in natural pathogen or host numbers.

Future research is needed to assess the lifestyle of pathogens, including growth or number of pathogen generations within the host. At an ecosystem level, a gross phylogenetic and transcriptomic assessment of all trophic levels, combined with an understanding of the surrounding environmental biotic and abiotic conditions (as summarised in Fig. 8.5), will allow a temporal snapshot of microbial communities and macroflora and fauna. Correlating this information with measurements of host abundance, host health and the relative abundance of pathogenic and non-pathogenic isogenic microbial strains will enable us to define why outbreaks of disease occur and determine the possible cues or drivers of pathogen outbreaks and evolution, respectively. This information may in turn facilitate changes in farming methods or assist in the development of improved field-based insect control measures.

Acknowledgements Tamsin R Sheen, for proofreading and helpful comments. Pauline Hunt, for the production of graphics. Christon J Hurst for his invitation to write the chapter and his efforts and input through the course of writing. Financial support for writing this chapter was provided by the Ministry of Business, Innovation and Employment, New Zealand grant: Next generation bio pesticides (C10X1310).

References

Aktories K, Schwan C, Papatheodorou P, Lang AE (2012) Bidirectional attack on the actin cytoskeleton. Bacterial protein toxins causing polymerization or depolymerization of actin. Toxicon 60:572–581

Alaux C, Ducloz F, Crauser D et al (2010) Diet effects on honeybee immunocompetence. Biol Lett 6:562–565

Albert-Weissenberger C, Cazalet C, Buchrieser C (2007) Legionella pneumophila—a human pathogen that co-evolved with fresh water protozoa. Cell Mol Life Sci 64:432–448

Alvarez-Martinez CE, Christie PJ (2009) Biological diversity of prokaryotic type IV secretion systems. Microbiol Mol Biol Rev 73:775–808

Aminov RI (2011) Horizontal gene exchange in environmental microbiota. Front Microbiol 2:158

Aperis G, Fuchs BB, Anderson CA et al (2007) *Galleria mellonella* as a model host to study infection by the *Francisella tularensis* live vaccine strain. Microbes Infect 9:729–734

Bailey MJ, Hughes C, Koronakis V (2003) RfaH and the ops element, components of a novel system controlling bacterial transcription elongation. Mol Microbiol 26:845–851

Ball OJ-P, Coudron TA, Tapper BA et al (2006) Importance of host plant species, Neotyphodium endophyte isolate, and alkaloids on feeding by *Spodoptera frugiperda* (Lepidoptera : Noctuidae) larvae. J Econ Entomol 99:1462–1473

Bapteste E, O'Malley MA, Beiko RG et al (2009) Prokaryotic evolution and the tree of life are two different things. Biol Direct 4:34. doi:10.1186/1745-6150-4-34

Behie SW, Bidochka MJ (2014) Ubiquity of insect-derived nitrogen transfer to plants by endophytic insect-pathogenic fungi: an additional branch of the soil nitrogen cycle. Appl Environ Microbiol 80:1553–1560

Bergin D, Reeves EP, Renwick J et al (2005) Superoxide production in *Galleria mellonella* hemocytes: identification of proteins homologous to the NADPH oxidase complex of human neutrophils. Infect Immun 73:4161–4170

Bidla G, Lindgren M, Theopold U et al (2005) Hemolymph coagulation and phenoloxidase in *Drosophila* larvae. Dev Comp Immunol 29:669–679

Blackburn M, Golubeva E, Bowen D et al (1998) A novel insecticidal toxin from *Photorhabdus luminescens*, toxin complex a (Tca), and Its histopathological effects on the midgut of *Manduca sexta*. Appl Environ Microbiol 64:3036–3041

Blackburn MB, Farrar RR, Novak NG et al (2006) Remarkable susceptibility of the diamondback moth (*Plutella xylostella*) to ingestion of Pir toxins from *Photorhabdus luminescens*. Entomologia Exper et Applica 121:31–37

Blackburn MB, Martin PAW, Kuhar D et al (2011) The occurrence of *Photorhabdus*-like toxin complexes in *Bacillus thuringiensis*. PLoS One. doi:10.1371/journal.pone.0018122

Blanford S, Thomas MB (1999) Host thermal biology: the key to understanding host–pathogen interactions and microbial pest control? Agric For Entomol 1:195–202

Bönemann G, Pietrosiuk A, Mogk A (2010) Tubules and donuts: a type VI secretion story. Mol Microbiol 76:815–821

Bordenstein SR, Reznikoff WS (2005) Mobile DNA in obligate intracellular bacteria. Nat Rev Microbiol 3:688–699

Borlee BR, Geske GD, Robinson CJ et al (2008) Quorum-sensing signals in the microbial community of the cabbage white butterfly larval midgut. ISME J 2:1101–1111

Bornberg-Bauer E, Albà MM (2013) Dynamics and adaptive benefits of modular protein evolution. Curr Opin Struct Biol 23:459–466

Boucias DG, Bradford DL, Barfield CS (1984) Susceptibility of the velvetbean caterpillar and soybean looper (Lepidoptera: Noctuidae) to Nomuraea rileyi: effects of pathotype, dosage, temperature, and host age. J Econ Entomol 77:247–253

Bowen D, Ensign JC (1998) Purification and characterisation of a high-molecular-weight insecticidal protein complex produced by the entomopathogenic bacterium *Photorhabdus luminescens*. Appl Environ Microbiol 64:3029–3035

Bowen D, Rocheleau TA, Blackburn M et al (1998) Insecticidal toxins from the bacterium *Photorhabdus luminescens*. Science 280:2129–2132
Bowman JP (2007) Bioactive compound synthetic capacity and ecological significance of marine bacterial genus *Pseudoalteromonas*. Mar Drugs 5:220–241
Bresolin G, Morgan JA, Ilgen D et al (2006) Low temperature-induced insecticidal activity of *Yersinia enterocolitica*. Mol Microbiol 59:503–512
Brillard J, Boyer-Giglio MH, Boemare N et al (2003) Holin locus characterisation from lysogenic *Xenorhabdus nematophila* and its involvement in *Escherichia coli* SheA haemolytic phenotype. FEMS Microbiol Lett 218:107–113
Brunder W, Karch H (2000) Genome plasticity in Enterobacteriaceae. Int J Med Microbiol 290:153–165
Buell CR, Joardar V, Lindeberg M et al (2003) The complete genome sequence of the *Arabidopsis* and tomato pathogen *Pseudomonas syringae* pv. tomato DC3000. Proc Natl Acad Sci U S A 100:10181–10186
Busby JN, Panjiker S, Landsberg MJ et al (2013) The BC component of ABC toxins is an RHS-repeat-containing protein encapsulation device. Nature 501:547–550
Büttner D, He SY (2009) Type III secretion in plant pathogenic bacteria. Plant Physiol 150:1656–1664
Cabral CM, Cherqui A, Pereira A et al (2004) Metalloproteases secreted by the entomopathogenic bacterium *Photorhabdus* sp. strain Az29. Appl Environ Microbiol 70:3831–3838
Casadevall A, Pirofski L-A (1999) Host-pathogen interactions: redefining the basic concepts of virulence and pathogenicity. Infect Immun 67:3703–3713
Casadevall A, Pirofski L-A (2001) Host-pathogen interactions: the attributes of virulence. J Infect Dis 184:337–344
Catalão MJ, Gil F, Moniz-Pereira J et al (2013) Diversity in bacterial lysis systems: bacteriophages show the way. FEMS Microbiol Rev 37:554–571
Cerenius L, Söderhäll K (2004) The prophenoloxidase-activating system in invertebrates. Immunol Rev 198:116–126
Chapman RF (2012) In: Simpson SJ, Douglas AE (eds) The insects: structure and function, 5th ed. Cambridge University Press, Cambridge
Chattopadhyay P, Gorthi S, Chatterjee S et al (2011) Characterization of bacterial isolates as natural biocontrolling agents of bollworm from an epizootic pest (*Heliothis armigera*). Pest Technol 5:81–85
Chattopadhyay P, Chatterjee S, Gorthi S et al (2012) Exploring agricultural potentiality of *Serratia entomophila* AB2: dual property of biopesticide and biofertilizer. Br Biotechnol J 2:1–12
Chouikha I, Hinnebusch BJ (2012) Yersinia-flea interactions and the evolution of the arthropod-borne transmission route of plague. Curr Opin Microbiol 15:239–246
Clarke DJ, Dowds BCQ (1991) Pathogenicity of *Xenorhabdus luminescens*. Biochem Soc Trans 20:65S
Coombes BK (2009) Type III secretion systems in symbiotic adaptation of pathogenic and non-pathogenic bacteria. Trends Microbiol 17:89–94
Daborn PJ, Waterfield NR, Silva CP et al (2002) A single *Photorhabdus* gene, makes caterpillars floppy (*mcf*), allows *Escherichia coli* to persist within and kill insects. Proc Natl Acad Sci U S A 88:10742–10747
Dale C, Moran NA (2006) Molecular interactions between bacterial symbionts and their hosts. Cell 126:453–465
Darmon E, Leach DR (2014) Bacterial genome instability. Microbiol Mol Biol Rev 78:1–39
de Bentzmann S, Giraud C, Bernard CS et al (2012) Unique biofilm signature, drug susceptibility and decreased virulence in *Drosophila* through the *Pseudomonas aeruginosa* two-component system PprAB. PLoS Pathog. doi:10.1371/journal.ppat.1003052
Derzelle S, Turlin E, Duchaud E et al (2004) The PhoP-PhoQ two-component regulatory system of *Photorhabdus luminescens* is essential for virulence in insects. J Bacteriol 186:1270–1279

Diene SM, Merhej V, Henry M et al (2013) The rhizome of the multidrug-resistant *Enterobacter aerogenes* genome reveals how new "killer bugs" are created because of a sympatric lifestyle. Mol Biol Evol 30:369–383

Dillon RJ, Vennard CT, Buckling A et al (2005) Diversity of locust gut bacteria protects against pathogen invasion. Ecol Lett 8:1291–1298

Dimijian GG (2000) Evolving together: the biology of symbiosis, part 1. Proc (Bayl Univ Med Cent) 13:217–226

Dodd SJ (2003) Horizontal transfer of plasmid-borne insecticidal toxin genes of *Serratia* species. PhD dissertation, University of Otago

Dodd SJ, Hurst MRH, Glare TR et al (2006) Occurrence of Sep insecticidal toxin complex genes in *Serratia* species and *Yersinia frederiksenii*. Appl Environ Microbiol 72:6584–6592

Dowling AJ, Daborn PJ, Waterfield NR et al (2004) The insecticidal toxin Makes caterpillars floppy (Mcf) promotes apoptosis in mammalian cells. Cell Microbiol 6:345–353

Elser JJ, Fagan WF, Denno RF et al (2000) Nutritional constraints in terrestrial and freshwater food webs. Nature 408:578–580

Engel P, Moran NA (2013) The gut microbiota of insects—diversity in structure and function. FEMS Microbiol Rev 37:699–735

Engleberg NC, DiRita VJ, Dermody TS (eds) (2012) Schaechter's mechanisms of microbial disease, 5th edn. Lippincott Williams and Wilkins, Philadelphia, PA

Fagan WF, Seimann E, Mitter C et al (2002) Nitrogen in insects: implications for trophic complexity and species diversification. Am Nat 160:784–802

Farrar RR, Martin PAW, Ridgway RL (2001) A strain of *Serratia marcescens* (Enterobacteriaceae) with high virulence per *os* to larvae of a laboratory colony of the corn earworm (Lepidoptera: Noctuidae). J Entomol Sci 36:380–390

Fauvarque MO, Bergeret E, Chabert J et al (2002) Role and activation of type III secretion system genes in *Pseudomonas aeruginosa*-induced *Drosophila* killing. Microb Pathog 32:287–295. doi:10.1006/mpat.2002.0504

Fedhila S, Daou N, Lereclus D et al (2006) Identification of *Bacillus cereus* internalin and other candidate virulence genes specifically induced during oral infection in insects. Mol Microbiol 62:339–355

Fedhila S, Buisson C, Dussurget O et al (2010) Comparative analysis of the virulence of invertebrate and mammalian pathogenic bacteria in the oral insect infection model *Galleria mellonella*. J Invertebr Pathol 103:24–29

Ferron P (1977) Influence of relative humidity on the development of fungal infection caused by *Beauveria bassiana* (Fungi Imperfecti, Moniliales) in imagines of *Acanthoscelides obtectus* (*Col. : Bruchidae*). Entomophaga 22:393–396

ffrench-Constant R, Waterfield N (2005) An ABC guide to the bacterial toxin complexes. Adv Appl Microbiol 58C:169–183

ffrench-Constant RH, Dowling A, Waterfield N (2007) Insecticidal toxins from *Photorhabdus* bacteria and their potential use in agriculture. Toxicon 49:436–451

Frank AC, Alsmark CM, Thollesson M et al (2005) Functional divergence and horizontal transfer of type IV secretion systems. Mol Biol Evol 22:1325–1336

Friman VP, Hiltunen T, Jalasvuori M et al (2011) High temperature and bacteriophages can indirectly select for bacterial pathogenicity in environmental reservoirs. PLoS One 6:e17651. doi:10.1371/journal.pone.0017651

Frost LS, Leplae R, Summers AO et al (2005) Mobile genetic elements: the agents of open source evolution. Nat Rev Microbiol 3:722–732

Fuchs TM, Bresolin G, Marcinowski L et al (2008) Insecticidal genes of *Yersinia* spp.: taxonomical distribution, contribution to toxicity towards *Manduca sexta* and *Galleria mellonella*, and evolution. BMC Microbiol 8:214

Galdiero S, Galdiero M, Pedone C (2007) β-Barrel membrane bacterial proteins: structure, function, assembly and interaction with lipids. Curr Protein Pept Sci 8:63–82

Gardner SN, Thomas MB (2002) Costs and benefits of fighting infection in locusts. Evol Ecol Res 4:109–131
Gatehouse HS, Marshall SD, Simpson RM et al (2008) *Serratia entomophila* inoculation causes a defect in exocytosis in *Costelytra zealandica* larvae. Insect Mol Biol 17:375–385
Gatsogiannis C, Lang AE, Meusch D et al (2013) A syringe-like injection mechanism in *Photorhabdus luminescens* toxins. Nature 495:520–523
Georgiades K (2012) Genomics of epidemic pathogens. Clin Microbiol Infect 18:213–217
Georgiades K, Raoult D (2010) Defining pathogenic bacterial species in the genomic era. Front Microbiol 1:article 151. doi:10.3389/fmicb.2010.00151
Giddens SR, Tormo A, Mahanty HK (2000) Expression of the antifeeding gene anfA1 in *Serratia entomophila* requires rpoS. Appl Environ Microbiol 66:1711–1714
Gill JJ, Young R (2011) Therapeutic applications of phage biology: history, practice and recommendations. In: Miller AA, Miller PF (eds) Emerging trends in antibacterial discovery: answering the call to arms. Caister Academic, Norfolk, UK
Givaudan A, Lanois A (2000) *flhDC*, the Flagellar Master Operon of *Xenorhabdus nematophilus*: requirement for motility, lipolysis, extracellular hemolysis, and full virulence in insects. J Bacteriol 182:107–115
Glare TR, Corbett GE, Sadler AJ (1993) Association of a large plasmid with amber disease of the New Zealand grass grub, *Costelytra zealandica*, caused by *Serratia entomophila* and *Serratia proteamaculans*. J Invertebr Pathol 62:165–170
Glare TR, Hurst MR, Grkovic S (1996) Plasmid transfer among several members of the family Enterobacteriaceae increases the number of species capable of causing experimental amber disease in grass grub. FEMS Microbiol Lett 139:117–120
González-Santoyo I, Córdoba-Aguilar A (2012) Phenoloxidase: a key component of the insect immune system. Entomol Exp App 142:1–16
Gophna U, Ron EZ, Graur D (2003) Bacterial type III secretion systems are ancient and evolved by multiple horizontal-transfer events. Gene 312:151–163
Grimont PA, Grimont F (1978) The genus Serratia. Annu Rev Microbiol 32:221–248
Grimont PAD, Jackson TA, Ageron E et al (1988) *Serratia entomophila* sp. nov. associated with amber disease in the New Zealand grass grub *Costelytra zealandica*. Int J Syst Bacteriol 38:1–6
Grkovic S, Glare TR, Jackson TA et al (1995) Genes essential for amber disease in grass grubs are located on the large plasmid found in *Serratia entomophila* and *Serratia proteamaculans*. Appl Environ Microbiol 61:2218–2223
Groisman EA, Ochman H (1996) Pathogenicity Islands: bacterial evolution in quantum leaps. Cell 87:791–794
Hachani A, Allsopp LP, Oduko Y et al (2014) Effectors delivery systems for bacterial Type VI the VgrG proteins are "à la Carte". J Biol Chem 289:17872–17884
Hacker J, Carniel E (2001) Ecological fitness, genomic islands and bacterial pathogenicity. A Darwinian view of the evolution of microbes. EMBO Rep 2:376–381
Hacker J, Kaper JB (2000) Pathogenicity islands and the evolution of microbes. Annu Rev Microbiol 54:641–679
Hacker J, Bender L, Ott M et al (1990) Deletions of chromosomal regions coding for fimbriae and hemolysins occur *in vitro* and *in vivo* in various extraintestinal *Escherichia coli* isolates. Microb Pathog 8:213–225
Haine ER, Rolff J, Siva-Jothy MT (2007) Functional consequences of blood clotting in insects. Dev Comp Immunol 31:456–464
Hall-Stoodley L, Costerton JW, Stoodley P (2004) Bacterial biofilms: from the natural environment to infectious diseases. Nat Rev Microbiol 2:95–108
Heermann R, Fuchs TM (2008) Comparative analysis of the *Photorhabdus luminescens* and the *Yersinia enterocolitica* genomes: uncovering candidate genes involved in insect pathogenicity. BMC Genomics 9:40. doi:10.1186/1471-2164-9-40
Hegedus D, Erlandson M, Gillott C et al (2009) New insights into peritrophic matrix synthesis, architecture, and function. Annu Rev Entomol 54:285–302

Hejazi A, Falkiner FR (1997) *Serratia marcescens*. J Med Microbiol 46:903–912

Held KG, LaRock CN, David A et al (2007) Metalloprotease secreted by the insect pathogen *Photorhabdus luminescens* induces melanization. Appl Environ Microbiol 73:7622–7628

Hill CW, Sandt CH, Vlazny DA (1994) *Rhs* elements of *Escherichia coli*: a family of genetic composites each encoding a large mosaic protein. Mol Microbiol 12:865–871

Hinnebusch BJ, Rosso ML, Schwan TG et al (2002) High-frequency conjugative transfer of antibiotic resistance genes to *Yersinia pestis* in the flea midgut. Mol Microbiol 46:349–354

Hosseinidoust Z, van de Ven TGM, Tufenkji N (2013) Evolution of *Pseudomonas aeruginosa* virulence as a result of phage predation. Appl Environ Microbiol 79:6110–6116

Huberty AF, Denno RF (2006) Consequences of nitrogen and phosphorus limitation for the performance of two planthoppers with divergent life-history strategies. Oecologia 149:444–455

Hughes DP, Andersen S, Hywel-Jones NL et al (2011) Behavioral mechanisms and morphological symptoms of zombie ants dying from fungal infection. BMC Ecol 11:13. doi:10.1186/1472-6785-11-13

Hurst MR, Jackson TA (2002) Use of the green fluorescent protein to monitor the fate of *Serratia entomophila* causing amber disease in the New Zealand grass grub, *Costelytra zealandica*. J Microbiol Methods 50:1–8

Hurst MR, Glare TR, Jackson TA et al (2000) Plasmid-located pathogenicity determinants of *Serratia entomophila*, the causal agent of amber disease of grass grub, show similarity to the insecticidal toxins of *Photorhabdus luminescens*. J Bacteriol 182:5127–5138

Hurst MRH, Glare TR, Jackson TA (2004) Cloning *Serratia entomophila* anti-feeding genes—a putative defective prophage active against the grass grub *Costelytra zealandica*. J Bacteriol 186:5116–5128

Hurst MR, Beard SS, Jackson TA et al (2007a) Isolation and characterization of the *Serratia entomophila* antifeeding prophage. FEMS Microbiol Lett 270:42–48

Hurst MR, Jones SM, Tan B et al (2007b) Induced expression of the *Serratia entomophila* Sep proteins shows activity towards the larvae of the New Zealand grass grub *Costelytra zealandica*. FEMS Microbiol Lett 275:160–167

Hurst MR, Becher SA, O'Callaghan M (2011a) Nucleotide sequence of the *Serratia entomophila* plasmid pADAP and the *Serratia proteamaculans* pU143 plasmid virulence associated region. Plasmid 65:32–41

Hurst MR, Becher SA, Young SD et al (2011b) *Yersinia entomophaga* sp. nov., isolated from the New Zealand grass grub *Costelytra zealandica*. Int J Syst Evol Microbiol 61:844–849

Hurst MRH, Jones SA, Tan B et al (2011c) The main virulence determinant of *Yersinia entomophaga* MH96 is a broad host range insect active, Toxin Complex. J Bacteriol 193:1966–1980

Hurst MR, van Koten C, Jackson TA (2014) Pathology of *Yersinia entomophaga* MH96 towards *Costelytra zealandica* (Coleoptera; Scarabaeidae) larvae. J Invertebr Pathol 115:102–107

Hurst MRH, Beattie A, Jones S et al (2015) *Galleria mellonella* mortality as a result of *Yersinia entomophaga* infection is temperature-dependent. App Environ Microbiol 81:6404–6414

Inglis GD, Lawrence AM (2001) Effects of *Serratia marcescens* on the F_1 generation of laboratory-reared *Heliothis virescens* (Lepidoptera: Noctuidae). J Econ Entomol 94:362–366

Jackson TA (1995) Amber disease reduces trypsin activity in midgut of *Costelytra zealandica* (Coleoptera: Scarabaeidae) larvae. J Invertebr Pathol 65:68–69

Jackson TA, McNeill MR (1998) Premature death in parasitized *Listronotus bonariensis* adults can be caused by bacteria transmitted by the parasitoid *Microctonus hyperodae*. Biocontrol Sci Technol 8:389–396

Jackson TA, Huger AM, Glare TR (1993) Pathology of amber disease in the New Zealand grass grub *Costelytra zealandica* (Coleoptera: Scarabaeidae). J Invertebr Pathol 61:123–130

Jackson TA, Townsend RJ, Barlow ND (1999) Predicting grassgrub population change in Canterbury. In: Mathiessen JN (ed) Proceedings of the 7th Australasian conference on grassland invertebrate ecology, CSIRO Entomology, Perth, pp 21–26

Jackson TA, Boucias DG, Thaler JO (2001) Pathobiology of amber disease, caused by *Serratia* spp., in the New Zealand grass grub, *Costelytra zealandica*. J Invertebr Pathol 78:232–243
Jaronski ST (2010) Ecological factors in the inundative use of fungal entomopathogens. Biocontrol 55:159–185
Jarrett P, Stephenson M (1990) Plasmid transfer between strains of *Bacillus thuringiensis* infecting *Galleria mellonella* and *Spodoptera littoralis*. Appl Environ Microbiol 56:1608–1614
Jin H, Ge SR, Tao Y et al (2005) Identification of a pathogenic strain of locusts and its toxicity and pathology. Acta Microbiol Sin 45:172–176
Joshi MC, Sharma A, Kant S et al (2008) An insecticidal GroEL protein with chitin binding activity from *Xenorhabdus nematophila*. J Biol Chem 283:28287–28296
Joyce SA, Brachmann AO, Glazer I et al (2008) Bacterial biosynthesis of a multipotent stilbene. Angew Chem Int Ed Engl 47:1942–1945
Jubelin G, Pagès S, Lanois A et al (2011) Studies of the dynamic expression of the *Xenorhabdus* FliAZ regulon reveal atypical iron-dependent regulation of the flagellin and haemolysin genes during insect infection. Environ Microbiol 13:1271–1284
Jurat-Fuentes JL, Jackson TA (2012) Bacterial entomopathogens. In: Vega FE, Kaya HA (eds) Insect pathology, 2nd edn. Elsevier, San Diego, CA, pp 265–349
Kain WM, Atkinson DS (1970) A rational approach to grass grub control. In: Matthews LJ (ed) Proceedings of the 23rd NZ weed and pest control conference, Palmerston North, August, pp 180–183
Kaska M (1976) The toxicity of extracellular proteases of the bacterium *Serratia marcescens* for larvae of greater wax moth, *Galleria mellonella*. J Invertebr 27:385–386
Kawanishi CY, Splittstoesser CM, Tashiro H (1978) Infection of the European chafer, *Amphimallon majalis*, by *Bacillus popilliae*: ultrastructure. J Invertebr Pathol 31:91–102
Khandelwal P, Banerjee-Bhatnagar N (2003) Insecticidal activity associated with the outer membrane vesicles of *Xenorhabdus nematophilus*. Appl Environ Microbiol 69:2032–2037
Kinkel LL, Bakker MG, Schlatter DC (2011) A coevolutionary framework for managing disease-suppressive soils. Annu Rev Phytopathol 49:47–67
Koskiniemi S, Lamoureux JG, Nikolakakis KC et al (2013) Rhs proteins from diverse bacteria mediate intercellular competition. Proc Natl Acad Sci USA 110:7032–7037
Koskiniemi S, Garza-Sánchez F, Sandegren L et al (2014) Selection of orphan Rhs toxin expression in evolved *Salmonella enterica* serovar Typhimurium. PLoS Genet 10:e1004255
Kostakioti M, Hadjifrangiskou M, Hultgren SJ (2013) Bacterial biofilms: development, dispersal, and therapeutic strategies in the dawn of the post antibiotic era. Cold Spring Harb Perspect Med 3:a010306. doi:10.1101/cshperspect.a010306
Kulkarni HM, Jagannadham MV (2014) Biogenesis and multifaceted roles of outer membrane vesicles from Gram-negative bacteria. Microbiology. doi:10.1099/mic.0.079400-0
Kupferschmied P, Maurhofer M, Keel C (2013) Promise for plant pest control: root-associated pseudomonads with insecticidal activities. Front Plant Sci 4:287. doi:10.3389/fpls.2013.00287
Kuraishi T, Binggeli O, Opota O et al (2011) Genetic evidence for a protective role of the peritrophic matrix against intestinal bacterial infection in *Drosophila melanogaster*. Proc Natl Acad Sci USA 108:15966–15971
Labrie SJ, Samson JE, Moineau S (2010) Bacteriophage resistance mechanisms. Nat Rev Microbiol 8:317–327
Landsberg MJ, Jones SA, Rothnagel R et al (2011) 3D structure of the *Yersinia entomophaga* toxin complex and implications for insecticidal activity. Proc Natl Acad Sci USA 108:20544–20549
Lavine MD, Strand MR (2002) Insect hemocytes and their role in immunity. Insect Biochem Mol Biol 32:1295–1309
Lee SC, Stoilova-McPhie S, Baxter L et al (2007) Structural characterisation of the insecticidal toxin XptA1, reveals a 1.15 MDa tetramer with a cage-like structure. J Mol Biol 366:1558–1568
Lehane MJ (1997) Peritrophic matrix structure and function. Annu Rev Entomol 42:525–550

Leiman PG, Basler M, Ramagopal UA et al (2009) Type VI secretion apparatus and phage tail-associated protein complexes share a common evolutionary origin. Proc Natl Acad Sci U S A 106:4154–4159

Lemichez E, Flatau G, Bruzzone M et al (1997) Molecular localization of the *Escherichia coli* cytotoxic necrotizing factor CNF1 cell-binding and catalytic domains. Mol Microbiol 24:1061–1070

Lemke T, Stingl U, Egert M et al (2003) Physicochemical conditions and microbial activities in the highly alkaline gut of the humus-feeding larva of *Pachnoda ephippiata* (Coleoptera: Scarabaeidae). Appl Environ Microbiol 69:6650–6658

Levin BR, Bull JJ (1994) Short-sighted evolution and the virulence of pathogenic microorganisms. Trends Microbiol 2:76–81

Loper JE, Buyer JS (1991) Siderophores in microbial interactions on plant surfaces. Mol Plant Microbe Interact 4:5–13

Loper JE, Hassan KA, Mavrodi DV et al (2012) Comparative genomics of plant- associated *Pseudomonas* spp. : insights into diversity and inheritance of traits involved in multitrophic inter actions. PLoS Genet 8:e1002784. doi:10.1371/journal.pgen.1002784

Louis C, Jourdan M, Cabanac M (1986) Behavioral fever and therapy in a Rickettsia-infected Orthoptera. Am J Physiol 250:R991–R995

Lysenko O (1976) Chitinase of *Serratia marcescens* and its toxicity to insects. J Invertebr Pathol 27:385–386

Maezawa K, Shigenobu S, Taniguchi TH et al (2006) Hundreds of flagellar basal bodies cover the cell surface of the endosymbiotic bacterium *Buchnera aphidicola* sp. strain APS. J Bacteriol 188:6539–6543

Mahajan-Miklos S, Rahme LG, Ausubel FM (2000) Elucidating the molecular mechanisms of bacterial virulence using non-mammalian hosts. Mol Microbiol 37:981–988

Mallo GV, Kurz CL, Couillault C et al (2002) Inducible antibacterial defense system in *C. elegans*. Curr Biol 12:1209–1214

Marshall SD, Gatehouse LN, Becher SA et al (2008) Serine proteases identified from a *Costelytra zealandica* (White) (Coleoptera: Scarabaeidae) midgut EST library and their expression through insect development. Insect Mol Biol 17:247–259

Marshall SDG, Hares MC, Jones SA et al (2012) Histolopathological effects of the Yen-Tc toxin complex from *Yersina entomophaga* MH96 (Enterobacteriaceae) on the midgut of *Costelytra zealandica* (Coleoptera: Scarabaeidae) larvae. Appl Environ Microbiol 78:4835–4847

Martínez JL (2013) Bacterial pathogens: from natural ecosystems to human hosts. Environ Microbiol 15:325–333

Mascher T, Helmann JD, Unden G (2006) Stimulus perception in bacterial signal-transducing histidine kinases. Microbiol Mol Biol Rev 70:910–938

Matz C, Deines P, Jurgens K (2002) Phenotypic variation in *Pseudomonas* sp. CM10 determines microcolony formation and survival under protozoan grazing. FEMS Microbiol Ecol 39:57–65

Maurelli AT (2007) Black holes, antivirulence genes, and gene inactivation in the evolution of bacterial pathogens. FEMS Microbiol Lett 267:1–8

Maynard-Smith J (1990) Models of a dual inheritance system. J Theor Biol 143:41–53

Mayr E (1961) Cause and effect in biology. Science 134:1501–1506

McCann HC, Guttman DS (2008) Evolution of the type III secretion system and its effectors in plant–microbe interactions. New Phytol 177:33–47

McDaniel LD, Young E, Delaney J et al (2010) High frequency of horizontal gene transfer in the oceans. Science 330:50

McNew GL (1960) The nature, origin, and evolution of parasitism. In: Horsfall JC, Dimond AE (eds) Plant pathology: an advanced treatise. vol II, The pathogen. Academic, New York, pp p19–p69

Medini D, Donati C, Tettelin H et al (2005) The microbial pan-genome. Curr Opin Genet Dev 15:589–594

Medrano-Soto A, Moreno-Hagelsieb G, Vinuesa P et al (2004) Successful lateral transfer requires codon usage compatibility between foreign genes and recipient genomes. Mol Biol Evol 21:1884–1894

Mekalanos JJ (1992) Environmental signals controlling expression of virulence determinants in bacteria. J Bacteriol 174:1–7

Mell JC, Redfield RJ (2014) Natural competence and the evolution of DNA uptake specificity. J Bacteriol 196:1471–1483

Merhej V, Notredame C, Royer-Carenzi M et al (2011) The rhizome of life: the sympatric *Rickettsia felis* paradigm demonstrates the random transfer of DNA sequences. Mol Biol Evol 28:3213–3223

Meusch D, Gatsogiannis C, Efremov RG et al (2014) Mechanism of Tc toxin action revealed in molecular detail. Nature 508:61–65

Michel-Briand Y, Baysse C (2002) The pyocins of *Pseudomonas aeruginosa*. Biochimie 84:499–510

Miest TS, Bloch-Qazi M (2008) Sick of mating: sexual transmission of a pathogenic bacterium. Fly (Austin) 2:215–219

Moliner C, Fournier PE, Raoult D (2010) Genome analysis of microorganisms living in amoebae reveals a melting pot of evolution. FEMS Microbiol Rev 34:281–294

Molmeret M, Horn M, Wagner M et al (2005) Amoebae as training grounds for intracellular bacterial pathogens. Appl Environ Microbiol 71:20–28

Moran NA (2002) Microbial minimalism: genome reduction in bacterial pathogens. Cell 108:583–586

Morgan JA, Sergeant M, Ellis D et al (2001) Sequence analysis of insecticidal genes from *Xenorhabdus nematophilus* PMFI296. Appl Environ Microbiol 67:2062–2069

Moya A, Pereto J, Gil R et al (2008) Learning how to live together: genomic insights into prokaryote-animal symbioses. Nat Rev Genet 9:218–229

Mukherjee K, Domann E, Hain T (2011) Insect Biotechnology the greater wax moth *Galleria mellonella* as an alternative model host for human pathogens. Biol Inspired Syst 2:3–14

Mulcahy H, Sibley CD, Surette MG et al (2011) *Drosophila melanogaster* as an animal model for the study of *Pseudomonas aeruginosa* biofilm infections *in vivo*. PLoS Pathog 7:e1002299. doi:10.1371/journal.ppat.1002299

Musser RO, Kwon HS, Williams SA et al (2005) Evidence that caterpillar labial saliva suppresses infectivity of potential bacterial pathogens. Arch Insect Biochem Physiol 58:138–144

Mylonakis E, Casadevall A, Ausubel FM (2007) Exploiting amoeboid and non-vertebrate animal model systems to study the virulence of human pathogenic fungi. PLoS Pathog 3:e101. doi:10.1371/journal.ppat.0030101

Nichol HH, Law JH, Winzerling JJ (2002) Iron metabolism in insects. Annu Rev Entomol 47:535–559

Nielsen-LeRoux C, Gaudriault S, Ramarao N et al (2012) How the insect pathogen bacteria *Bacillus thuringiensis* and *Xenorhabdus/Photorhabdus* occupy their hosts. Curr Opin Microbiol 15:220–231

Nishiwaki H, Ito K, Shimomura M et al (2007) Insecticidal bacteria isolated from predatory larvae of the antlion species *Myrmeleon bore* (Neuroptera: Myrmeleontidae). J Invertebr Pathol 96:80–88

Nuñez-Valdez ME, Calderón MA, Aranda E et al (2008) Identification of a putative Mexican strain of *Serratia entomophila* pathogenic against root damaging larvae of Scarabaeidae (Coleoptera). Appl Environ Microbiol 74:802–810

O'Callaghan M, Garnham ML, Nelson TL et al (1996) The pathogenicity of *Serratia* strains to *Lucilia sericata* (Diptera: Calliphoridae). J Invertebr Pathol 68:22–27

O'Callaghan M, Young SD, Barlow ND et al (1999) The ecology of grass grub pathogenic *Serratia* spp. in New Zealand pastures. In: Mathiessen JN (ed) Proceedings of the 7th Australasian conference on grassland invertebrate ecology. CSIRO Entomology, Perth, pp 85–91

Ogata H, La Scola B, Audic S et al (2006) Genome sequence of *Rickettsia bellii* illuminates the role of amoebae in gene exchanges between intracellular pathogens. PLoS Genet 2:e76

Oliver KM, Moran NA, Hunter MS (2005) Variation in resistance to parasitism in aphids is due to symbionts not host genotype. Proc Natl Acad Sci U S A 102:12795–12800

Oliver KM, Moran NA, Hunte MS (2006) Costs and benefits of a superinfection of facultative symbionts in aphids. Proc Biol Sci 273:1273–1280

Oliver KM, Degnan PH, Burke GR et al (2010) Facultative symbionts in aphids and the horizontal transfer of ecologically important traits. Annu Rev Entomol 55:247–266

Opota O, Vallet-Gély I, Vincentelli R et al (2011) Monalysin, anovel β-pore-forming toxin from the Drosophila pathogen Pseudomonas entomophila, contributes to host intestinal damage and lethality. PLoS Pathog 7:e1002259. doi:10.1371/journal.ppat.1002259

Owuama CI (2001) Entomopathogenic symbiotic bacteria, *Xenorhabdus* and *Photorhabdus* of nematodes. World J Microbiol Biotechnol 17:505–515

Pallen MJ, Wren BW (2007) Bacterial pathogenomics. Nature 449:835–842

Park D, Forst S (2006) Co-regulation of motility, exoenzyme and antibiotic production by the EnvZ-OmpR-FlhDC-FliA pathway in *Xenorhabdus nematophila*. Mol Microbiol 61:1397–1412

Paz-Y-Mino CG, Espinosa A (2010) Integrating horizontal gene transfer and common descent to depict evolution and contrast it with "common design". J Eukaryot Microbiol 57:11–18

Péchy-Tarr M, Bruck DJ, Maurhofer M et al (2008) Molecular analysis of a novel gene cluster encoding an insect toxin in plant-associated strains of *Pseudomonas fluorescens*. Environ Microbiol 10:2368–2386

Péchy-Tarr M, Borel N, Kupferschmied P et al (2013) Control and host-dependent activation of insect toxin expression in a root-associated biocontrol pseudomonad. Environ Microbiol 15:736–750

Pedrini N, Ortiz-Urquiza A, Huarte-Bonnet C et al (2013) Targeting of insect epicuticular lipids by the entomopathogenic fungus *Beauveria bassiana*: hydrocarbon oxidation within the context of a host-pathogen interaction. Front Microbiol 4:24. doi:10.3389/fmicb.2013.00024

Pedulla ML, Ford ME, Houtz JM et al (2003) Origins of highly mosaic mycobacteriophage genomes. Cell 113:171–182

Penz T, Horn M, Schmitz-Esser S (2010) The genome of the amoeba symbiont "*Candidatus* Amoebophilus asiaticus" encodes an Afp-like prophage possibly used for protein secretion. Virulence 1:541–545

Penz T, Schmitz-Esser S, Kelly SE et al (2012) Comparative genomics suggests an independent origin of cytoplasmic incompatibility in *Cardinium hertigii*. PLoS Genet 8(10):e1003012. doi:10.1371/journal.pgen.1003012

Persson OP, Pinhassi J, Riemann L et al (2009) High abundance of virulence gene homologues in marine bacteria. Environ Microbiol 11:1348–1357

Pingoud A, Fuxreiter M, Pingoud V et al (2005) Type II restriction endonucleases: structure and mechanism. Cell Mol Life Sci 62:685–707

Pinyon RA, Linedale EC, Webster MA et al (1996) Tn5-induced *Xenorhabdus bovienii* lecithinase mutants demonstrate reduced virulence for *Galleria mellonella* larvae. J App Bacteriol 80:411–417

Podgwaite JD, Cosenza BJ (1976) A strain of *Serratia marcescens* pathogenic for larvae of *Lymantria dispar*: infectivity and mechanisms of pathogenicity. J Invertebr Pathol 27:199–208

Poinar GO, Wassink HJM, Leegwater-van der Linden ME et al (1979) *Serratia marcescens* as a pathogen of tsetse flies. Acta Trop 36:223–227

Poole T, Crippen T (2009) Conjugative plasmid transfer between *Salmonella enterica* Newport and *Escherichia coli* within the gastrointestinal tract of the lesser mealworm beetle, *Alphitobius diaperinus* (Coleoptera: Tenebrionidae). Poult Sci 88:1553–1558

Popa O, Hazkani-Covo E, Landan G et al (2011) Directed networks reveal genomic barriers and DNA repair bypasses to lateral gene transfer among prokaryotes. Genome Res 21:599–609

Preston GM (2007) Metropolitan microbes: type III secretion in multihost symbionts. Cell Host Microbe 2:291–294

Rahme LG, Stevens EJ, Wolfort SF et al (1995) Common virulence factors for bacterial pathogenicity in plants and animals. Science 268:1899–1902

Rahme LG, Ausubel FM, Cao H et al (2000) Plants and animals share functionally common bacterial virulence factors. Proc Natl Acad Sci U S A 97:8815–8821

Rankin DJ, Rocha EP, Brown SP (2011) The evolutionary significance of costly punishment is still to be demonstrated. Heredity 106:1–10

Raoult D, Boyer M (2010) Amoebae as genitors and reservoirs of giant viruses. Intervirology 53:321–329

Raymond B, Wyres K, Sheppard S et al (2010) Environmental factors determining the epidemiology and population genetic structure of the *Bacillus cereus* group in the field. PLoS Pathog 6 (5):e1000905. doi:10.1371/jounrnal.ppat.1000905

Regev A, Keller M, Strizhov N et al (1996) Synergistic activity of a *Bacillus thuringiensis* delta-endotoxin and a bacterial endochitinase against *Spodoptera littoralis* larvae. Appl Environ Microbiol 62:3581–3586

Rodou A, Ankrah DO, Stathopoulos C (2010) Toxins and secretion systems of *Photorhabdus luminescens*. Toxins (Basel) 2:1250–1264

Rohmer L, Hocquet D, Miller SI (2011) Are pathogenic bacteria just looking for food? Metabolism and microbial pathogenesis. Trends Microbiol 19:341–348

Ruffner B, Péchy-Tarr M, Ryffel F et al (2013) Oral insecticidal activity of plant-associated pseudomonads. Environ Microbiol 15:751–763

Russell AB, Peterson SB, Mougous JD (2014) Type VI secretion system effectors: poisons with a purpose. Nat Rev Microbiol 12:137–148

Rybakova D, Radjainia M, Turner A et al (2013) Role of antifeeding prophage (Afp) protein Afp16 in terminating the length of the Afp tailocin and stabilizing its sheath. Mol Microbiol 89:702–714

Sarkar SF, Guttman DS (2004) Evolution of the core genome of *Pseudomonas syringae*, a highly clonal, endemic plant pathogen. Appl Environ Microbiol 70:1999–2012

Sarris PF, Ladoukakis ED, Panopoulos NJ et al (2014) A phage tail-derived element with wide distribution among both prokaryotic domains: a comparative genomic and phylogenetic study genome. Biol Evol 6:1739–1747

Schell MA, Lipscomb L, DeShazer D (2008) Comparative genomics and an insect model rapidly identify novel virulence genes of *Burkholderia mallei*. J Bacteriol 190:2306–2313

Schiavo G, van der Goot FG (2001) The bacterial toxin toolkit. Nat Rev Mol Cell Biol 2:530–537

Schikora M, Neupane B, Madhogaria S et al (2012) An image classification approach to analyze the suppression of plant immunity by the human pathogen *Salmonella* Typhimurium. BMC Bioinformatics 13:171. doi:10.1186/1471-2105-13-171

Scholthof K-B (2007) The disease triangle: pathogens, the environment and society. Nat Rev Microbiol 5:152–156

Senior NJ, Bagnall MC, Champion OL et al (2011) *Galleria mellonella* as an infection model for *Campylobacter jejuni* virulence. J Med Microbiol 60:661–669

Sergeant M, Jarrett P, Ousley M et al (2003) Interactions of insecticidal toxin gene products from *Xenorhabdus nematophilus* PMFI296. Appl Environ Microbiol 69:3344–3349

Sergeant M, Baxter L, Jarrett P et al (2006) Identification, typing, and insecticidal activity of *Xenorhabdus* isolates from entomopathogenic nematodes in United Kingdom soil and characterization of the *xpt* toxin loci. Appl Environ Microbiol 72:5895–5907

Sharp PM, Li WH (1987) The codon adaptation index-a measure of directional synonymous codon usage bias, and its potential applications. Nucleic Acids Res 15:1281–1295

Sharpe ES, Detroy RW (1979) Fat body depletion, a debilitating result of milky disease in Japanese beetle larvae. J Invertebr Pathol 34:92–94

Shen X, Hu H, Peng H et al (2013) Comparative genomic analysis of four representative plant growth-promoting rhizobacteria in *Pseudomonas*. BMC Genomics 14:271. doi:10.1186/1471-2164-14-271

Shikuma NJ, Pilhofer M, Weiss GL et al (2014) Marine tubeworm metamorphosis induced by arrays of bacterial phage tail-like structures. Science 343:529–533

Shneider MM, Buth SA, Ho BT et al (2013) PAAR-repeat proteins sharpen and diversify the type VI secretion system spike. Nature 500:350–353

Siefert JL (2009) Defining the mobilome. In: Gogarten MB, Gogarten JP, Olendzenski L (eds) Horizontal gene transfer: genomes in flux. Humana, New York, pp 13–27

Singh J, Banerjee N (2008) Transcriptional analysis and functional characterization of a gene pair encoding iron-regulated xenocin and immunity proteins of *Xenorhabdus nematophila*. J Bacteriol 190:3877–3885

Singh P, Park D, Forst S et al (2013) Xenocin export by the flagellar type III pathway in *Xenorhabdus nematophila*. J Bacteriol 195:1400–1410

Slatten BH, Larson AD (1967) Mechanism of pathogenicity of *Serratia marcescens* I. Virulence for the adult boll weevil. J Invertebr Pathol 9:78–81

Solano C, Echeverz M, Lasa I (2014) Biofilm dispersion and quorum sensing. Curr Opin Microbiol 18C:96–104

Somvanshi VS, Kaufmann-Daszczuk B, K-S K et al (2010) *Photorhabdus* phase variants express a novel fimbrial locus, *mad*, essential for symbiosis. Mol Microbiol 77:1021–1038

Somvanshi VS, Sloup RE, Crawford JM et al (2012) A single promoter inversion switches *Photorhabdus* between pathogenic and mutualistic states. Science 337:88–93

Song CJ, Seo S, Shrestha S et al (2011) Bacterial metabolites of an entomopathogenic bacterium, *Xenorhabdus nematophila*, inhibit a catalytic activity of phenoloxidase of the diamondback moth, *Plutella xylostella*. J Microbiol Biotechnol 21:317–322

Splittstoesser CM, Tashiro H, Lin SL et al (1973) Histopathology of the European chafer, *Amphimallon majalis*, infected with *Bacillus popilliae*. J Invertebr Pathol 22:161–167

Steinhaus EA (1941) A study of the bacteria associated with thirty species of insects. J Bacteriol 42:757–790

Steinhaus EA (1959) *Serratia marcescens* Bizio as an insect pathogen. Hilgardia 28:351–380

Stock AM, Robinson VL, Goudreau PN (2000) Two-component signal transduction. Annu Rev Biochem 69:183–215

Stucki G, Jackson TA, Noonan MJ (1984) Isolation and characterisation of *Serratia* strains pathogenic for larvae of the New Zealand grass grub *Costelytra zealandica*. N Z J Sci 27:255–260

Tao K, Long Z, Liu K et al (2006) Purification and properties of a novel insecticidal protein from the locust pathogen *Serratia marcescens* HR-3. Curr Microbiol 52:45–49

Terra WR (2001) The origin and functions of the insect peritrophic membrane and peritrophic gel. Arch Insect Biochem Physiol 47:47–61

Thamthiankul S, Suan-Ngay S, Tantimavanich S et al (2001) Chitinase from *Bacillus thuringiensis* subsp. *pakistani*. Appl Microbiol Biotechnol 56:395–401

Thomas CM, Nielsen KM (2005) Mechanisms of, and barriers to, horizontal gene transfer between bacteria. Nat Rev Microbiol 3:711–721

Thomas DJI, Morgan JAW, Whipps JM et al (2000) Plasmid transfer between the *Bacillus thuringiensis* subspecies *kurstaki* and *tenebrionis* in laboratory culture and soil and in lepidopteran and coleopteran larvae. Appl Environ Microbiol 66:118–124

Thomas DJI, Morgan JAW, Whipps JM et al (2001) Plasmid transfer between *Bacillus thuringiensis* subsp. *israelensis* strains in laboratory culture, river water, and dipteran larvae. Appl Environ Microbiol 67:330–338

Toft C, Andersson SGE (2010) Evolutionary microbial genomics: insights into bacterial host adaptation. Nat Rev Genet 11:465–475

Toft C, Fares MA (2008) The evolution of the flagellar assembly pathway in endosymbiotic bacterial genomes. Mol Biol Evol 25:2069–2076

Touchon M, Hoede C, Tenaillon O et al (2009) Organised genome dynamics in the *Escherichia coli* species results in highly diverse adaptive paths. PLoS Gent. doi:10.1371/journal.pgen.1000344

Tseng TT, Tyler BM, Setubal JC (2009) Protein secretion systems in bacterial-host associations, and their description in the gene ontology. BMC Microbiol 9(Suppl 1):S2. doi:10.1186/1471-2180-9-S1-S2

Vallet-Gely I, Lemaitre B, Boccard F (2008) Bacterial strategies to overcome insect defences. Nat Rev Microbiol 6:302–313

Vigneux FF, Zumbihl R, Jubelin G et al (2007) The *xaxAB* genes encoding a new apoptotic toxin from the insect pathogen *Xenorhabdus nematophila* are present in plant and human pathogens. J Biol Chem 282:9571–9580

Vodovar N, Vinals M, Liehl P et al (2005) *Drosophila* host defense after oral infection by an entomopathogenic *Pseudomonas* species. Proc Natl Acad Sci U S A 102:11414–11419

Vodovar N, Vallenet D, Cruveiller S et al (2006) Complete genome sequence of the entomopathogenic and metabolically versatile soil bacterium *Pseudomonas entomophila*. Nat Biotechnol 24:673–679

Watanabe K, Sato M (1998) Plasmid-mediated gene transfer between insect-resident bacteria, *Enterobacter cloacae*, and plant-epiphytic bacteria, *Erwinia herbicola*, in guts of silkworm larvae. Curr Microbiol 37:352–355

Waterfield NR, Bowen DJ, Fetherston JD et al (2001) The tc genes of *Photorhabdus*: a growing family. Trends Microbiol 9:185–191

Waterfield NR, Daborn PJ, ffrench-Constant RH (2002) Genomic islands in *Photorhabdus*. Trends Microbiol 10:541–545

Waterfield NR, Wren BW, ffrench-Constant RH (2004) Invertebrates as a source of emerging human pathogens. Nat Rev Microbiol 2:833–841

Waterfield N, Hares M, Dowling A et al (2005a) Potentiation and cellular phenotypes of the insecticidal toxin complexes of *Photorhabdus* bacteria. Cell Microbiol 7:373–382

Waterfield N, Kamita SG, Hammock BD et al (2005b) The *Photorhabdus* Pir toxins are similar to a developmentally regulated insect protein but show no juvenile hormone esterase activity. FEMS Microbiol Lett 245:47–52

Waterfield NR, Ciche T, Clarke D (2009) *Photorhabdus* and a host of hosts. Annu Rev Microbiol 63:557–574

Weinbauer MG (2004) Ecology of prokaryotic viruses. FEMS Microbiol Rev 28:127–181

Westra ER, Swarts DC, Staals RH et al (2012) The CRISPRs, they are a-changin: how prokaryotes generate adaptive immunity. Annu Rev Genet 46:311–339

Westra ER, Buckling A, Fineran PC (2014) CRISPR–Cas systems: beyond adaptive immunity. Nat Rev Microbiol 12:317–326

Wimley WC (2003) The versatile beta-barrel membrane protein. Curr Opin Struct Biol 13:404–411

Wiwat C, Thaithanum S, Pantuwatana S et al (2000) Toxicity of chitinase-producing *Bacillus thuringiensis* ssp. *kurstaki* HD-1 (G) toward *Plutella xylostella*. J Invertebr Pathol 76:270–277

Wommack KE, Colwell RR (2000) Virioplankton: viruses in aquatic ecosystems. Microbiol Mol Biol Rev 64:69–114

Yang G, Dowling AJ, Gerike U et al (2006) *Photorhabdus* virulence cassettes confer injectable insecticidal activity against the wax moth. J Bacteriol 188:2254–2261

Young R (2014) Phage lysis: three steps, three choices, one outcome. J Microbiol 52:243–258

Zhang D, de Souza RF, Anantharaman V et al (2012) Polymorphic toxin systems: comprehensive characterization of trafficking modes, processing, mechanisms of action, immunity and ecology using comparative genomics. Biol Direct 7:18. doi:10.1186/1745-6150-7-18

Chapter 9
Opportunistic Infections of Avians

Stephen M. Chege

Abstract Opportunistic infection is an infection by a microorganism that normally does not cause disease but becomes pathogenic when the body's immune system is impaired and unable to fight off infection. A compromised immune system presents an "opportunity" for the pathogens (bacteria, virus, fungus, or protozoa) to infect. Immunosuppression can be caused by malnutrition, fatigue, coinfections, recurrent infections, immunosuppressing agents for organ transplant recipients, genetic predisposition, skin damage, and long-term antibiotic treatment, among others.

This chapter describes opportunistic mycotic, bacterial, protozoal, and viral infections that do not infect the avian species under normal circumstances but do so when the body's immune system is compromised or there is some predisposing factor.

9.1 Mycotic Infections

Mycotic infections are relatively common in avian species. Many fungal agents exist in the environment as soilborne saprophytes. Most birds are exposed to them in their normal habitat or aviary environment without effect. Nutritional disorders in parrots, stress of captivity in raptors, and incubation-related disorders in hatchling galliformes, along with other causes of impaired immune function and environmental factors conducive to fungal proliferation, can cause disease to occur. Respiratory tract aspergillosis and alimentary tract infections due to *Candida* and other yeasts are the most frequent forms of fungal disease observed. The early diagnosis and successful management of systemic fungal disease present a diagnostic and management challenge to the avian practitioner. Advances in diagnostic methods, improved knowledge of therapeutic agents, and better management practices have reduced the morbidity and mortality associated with these agents.

S.M. Chege (✉)
Al Ain Zoo, P.O Box 1204, Al Ain, United Arab Emirates
e-mail: stevmah@yahoo.com; Stephen.chege@alainzoo.ae

© Springer International Publishing Switzerland 2016

C.J. Hurst (ed.), *The Rasputin Effect: When Commensals and Symbionts Become Parasitic*, Advances in Environmental Microbiology 3,
DOI 10.1007/978-3-319-28170-4_9

9.1.1 Aspergillosis

Aspergillosis is a noncontagious, opportunistic infection referring to any disease condition caused by members of the fungal genus *Aspergillus* (Zafra et al. 2008; Chege et al. 2013). The organism is an opportunistic, angioinvasive fungus that may act as an allergen, colonizer, or invasive pathogen. It can produce both acute and chronic disease varying in spectrum from local involvement to systemic dissemination. It is the most frequent cause of respiratory disease and the most commonly diagnosed fungal disease in pet birds (Lumeij et al. 1995). It is also considered the most common, non-traumatically induced medical problem in free-ranging birds of prey. The disease is known to occur in a wide variety of captive and free-living birds.

9.1.1.1 Epidemiology

Aspergillus spp. are ubiquitous fungi commonly found in the environment, soil and feed grains. They are distributed worldwide and proliferate in environments with high humidity and warm (>25 °C) temperatures (Jordan 1990). Moldy litter, grain, and bedding material contaminated with feces are common media for fungal growth. *Aspergillus fumigatus* is the most commonly isolated species from birds with aspergillosis, followed by *A. flavus* and *A. niger* (Carrasco et al. 1993; Hoppes et al. 2000). *Aspergillus clavatus*, *A. glaucus*, *A. nidulans*, *A. oryzae*, *A. terreus*, *A. ustus*, and *A. versicolor* are among the other species less commonly isolated.

An increased concentration of spores in the environment may predispose a bird to aspergillosis. A warm environment, humidity, poor ventilation (Phalen 2000; Tell 2005), poor sanitation (Oglesbee 1997), and the long-term storage of feed (Khosravi et al. 2008) may increase the amount of spores in the air.

9.1.1.2 Disease Predisposition

Immunosuppression is the major factor predisposing birds to the development of opportunistic *Aspergillus* infections. *Aspergillus* spores are widespread in the environment, and many birds may carry them in their lungs and air sacs until immunosuppression or stress triggers clinical disease. Stress alone (a strong immune suppressor) or other factors related to confinement, poor husbandry practice, malnutrition, preexisting disease, and the prolonged use of antibiotics and steroids increase the predilection to disease (Redig 2000). Overpopulated, poorly ventilated, and dusty aviary environments lead to pulmonary and air sac disease. Research has confirmed a causal relationship between high concentrations of *Aspergillus* spp. spores in the environment and aspergillosis. Damp feed or bedding in warm, humid environments and poor ventilation allow for a high concentration of *Aspergillus* spores to develop (Phalen 2000). Inhalation of large numbers of spores may occur. Birds exposed to the organism in quantities sufficient to establish a primary infection have developed acute disease. There is often a correlation between poor husbandry and a high concentration of spores. Eucalyptus leaves, which have been promoted as a "natural" insect repellent, are often heavily contaminated with this fungus. Acute, severe, untreatable aspergillosis has been associated with their use (Fudge and Reavill 1993). Aspergillosis is a common sequela to other respiratory tract diseases. A predominantly seed diet, with subsequent malnutrition including vitamin A deficiency, can lead to squamous metaplasia of the oral and respiratory epithelium and the establishment of fungal growth (McMillian and Petrak 1989). Aspergillosis is the most commonly occurring respiratory disease in captive wild birds (Redig 2000). Of the psittacine species, the African grey parrot (*Psittacus erithacus*) and Pionus parrots (*Pionus* spp.) are reported to have an increased susceptibility to the development of disease (Bauck 1994). Localized infection of the nasal passages is observed in Amazon parrots, possibly due to the higher incidence of hypovitaminosis A in this species (Bauck 1994). Raptorial species at particularly high risk of developing aspergillosis include goshawks (*Accipiter gentilis*), rough-legged hawks (*Buteo lagopus*), immature red-tailed hawks (*Buteo jamaicensis*), golden eagles (*Aquila chrysaetos*), and snowy owls (*Nyctea scandiaca*) (Forbes 1991; Redig 1993, 2000). Gyrfalcons (*Falco rusticolus*) are believed to be especially susceptible to *Aspergillus* infection (Forbes 1991; Redig 1993, 2000). Among waterfowl, a higher incidence of aspergillosis is seen in swans (*Cygnus* spp.). Captive penguins (Sphenisciformes) also are extremely susceptible to developing the disease. In these species, the increased incidence of aspergillosis may reflect environmental and husbandry deficiencies occurring in captivity in addition to increased species susceptibility.

9.1.2 Candidiasis

Candidiasis in birds, also known as thrush, moniliasis, or sour crop, refers to infections by yeasts of the genus *Candida*. The most commonly implicated species is *C. albicans*, although *C. parapsilosis*, *C. krusei*, and *C. tropicalis* also may cause disease (Samour and Naldo 2002). *Candida albicans* is opportunistic yeast and not regarded as a primary pathogen. Small numbers of the non-budding organism are commonly found in the digestive tract of normal birds and considered normal flora in healthy pigeons (Rupiper 1998). Host defense mechanisms and bacterial flora keep numbers of the organism controlled. *Candida* spp. can proliferate and cause disease when digestive tract flora is severely suppressed. In most cases, the infection is endogenous in origin, occurring secondarily to stress, immunosuppression, inadequate nutrition, poor sanitation, debilitation, or in birds that have been extensively treated with antibiotics (Deem 1999; Jones 2006).

9.1.2.1 Transmission and Pathogenesis

This disease is most often seen in psittacine neonates and cockatiels. Yeasts gain entry into the host by the oral route (Rupiper 1998). The organisms can be transmitted from the parent bird to chicks during regurgitative feeding. The infection also may be spread throughout the nursery population by the use of contaminated fomites and feeding utensils.

9.1.2.2 Clinical Signs and Gross Lesions

Affected birds are depressed and may exhibit delayed crop emptying, regurgitation, crop stasis, inappetence, and poor digestion of food. Droppings are often abnormal, appearing brownish in color and watery. Affected chicks do not grow or gain weight well and appear stunted.

Lesions vary in severity. They consist of thickening of the digestive tract mucosa with increased mucous and pseudomembranous patches. The choanae may become abscessed with formation of a diphtheritic membrane in the oropharynx as mycelial growth develops (Rupiper 1998). Whitish plaques may be evident under the tongue, in the mouth, and most frequently in the crop.

9.1.2.3 Diagnosis

Diagnosis is made by identifying the organism on wet or Gram-stained smears from lesions in the oral cavity, crop, or cloaca. Endoscopic examination of affected membranes of the oropharynx, crop, esophagus, proventriculus, and ventriculus may reveal white-gray to gray-green, thickened and diphtheritic membranes. A characteristic "Turkish towel" thickening of the crop lining is evident in advanced cases (Dahlhausen 2006).

9.1.2.4 Treatment, Control, and Prevention

Correction of the diet and husbandry are necessary for successful treatment of candidiasis. Nystatin is the first drug of choice for yeast infections confined to the alimentary tract. The recommended dose of 290,000 units/kg PO q8–12 h is safe and effective for use in psittacine neonates (Carpenter et al. 2001). For flock treatment, nystatin can be added to the drinking water at 100,000 IU/L (Rupiper 1998). Cages, equipment, and other materials in contact with infected birds should be disinfected because of the broad host range of species that can become infected.

9.1.3 Cryptococcosis

Cryptococcosis is most commonly caused by infection with *Cryptococcus neoformans* var. *neoformans*, an encapsulated saprophytic fungus with worldwide distribution. *Cryptococcus neoformans* var. *gattii* is more geographically restricted because of an ecological association with the river red gum (*Eucalyptus camaldulensis*) and other eucalyptus trees (Ellis and Pfeiffer 1990). The organism is commonly found in soils contaminated with bird droppings.

While cryptococcosis is a rare disease of birds, disseminated infection has been reported in the green-winged macaw (*Ara chloroptera*), Moluccan cockatoo (*Cacatua moluccensis*), thick-billed parrot (*Rhynchopsitta pachyrhyncha*), and North Island brown kiwi (*Apteryx australis mantelli*) (Hill *et al.* 1995; Curtis-Velasco 2000). Infections may involve the respiratory tract, digestive tract, and central nervous system, producing necrotic granulomatous lesions and a characteristic thick, pale, gelatinous exudate. Upper respiratory tract involvement can produce facial granulomas that distort the rhamphotheca (Curtis-Velasco 2000). A chronic rhinosinusitis resembling a neoplasm of the rhamphotheca was described in a Major Mitchell's cockatoo (*Cacatua leadbeateri*) and was due to *C. neoformans* var. *gattii* (Raidal and Butler 2001). An encephalitis or meningitis also may occur, causing blindness or paralysis in affected birds (Cooper and Harrison 1999) (Fig. 9.1).

Diagnosis of cryptococcosis is based on cytology and histopathology in combination with culture rather than culture of nasochoanal swabs or washes alone (Raidal and Butler 2001).

9.1.3.1 Zoonotic Risk

Most human infections occur through contact with contaminated exudates, fecal material, and nonclinically infected or diseased birds (Nosanchuck et al. 2000).

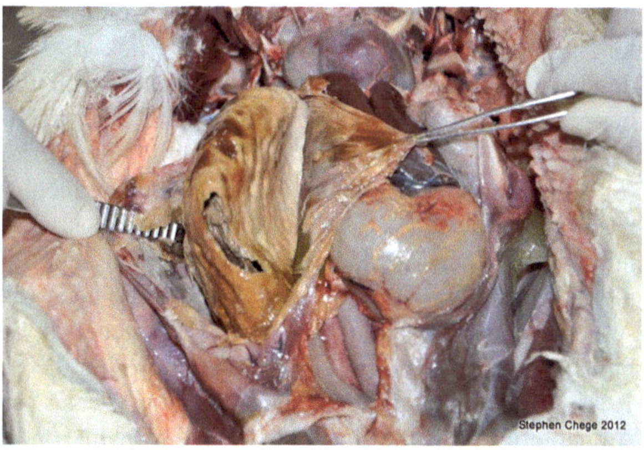

Fig. 9.1 Granulomas in caudal air sacs of a aspergillosis-infected Cape vulture (*Gyps coprotheres*)

While human infection with *C. neoformans* var. *neoformans* is well recognized in immunosuppressed patients, infection with *C. neoformans* var. *gattii* is commonly associated with otherwise healthy and immunocompetent individuals (Speed and Dunt 1995).

9.1.4 Histoplasmosis

Histoplasmosis is an infectious but not contagious mycotic disease that has been reported in poultry and zoo specimens. The soilborne organism *Histoplasma capsulatum* has worldwide distribution and is endemic in the eastern and central USA. It is commonly associated with fecal material from pigeons and gallinaceous birds and has the potential to grow within dirt substrates of enclosed aviaries (Bauck 1994). *Histoplasma* infections in birds produce disease signs similar to those seen with *Cryptococcus* spp. infections. An initial pneumonia can progress to disseminated disease with the formation of necrotic granulomas. Histoplasmosis was identified as the cause of an osteomyelitis and mineralized soft tissue granuloma of the shoulder and antebrachium in a Moluccan cockatoo (Vaughn 1996). The infection should be considered part of the differential diagnosis of granulomatous respiratory disease in avian patients. Diagnosis is based on culture of the organism and histopathologic examination of tissue samples.

9.1.5 Mucormycosis

The order *Mucorales* includes a number of saprophytic fungi that have been implicated as possible avian pathogens. They have been implicated as an etiologic agent of meningoencephalitis in birds (Bennett 1994). Hyphal invasion of cerebral blood vessels and dissemination of an *Absidia* sp. in the cerebrum were identified as the cause of progressive neurologic defects culminating in seizures in a chattering lory (*Lorius garrulus*) (Orcutt and Bartick 1994). Other clinical syndromes described include air sacculitis in a pigeon (*Columba* sp.), pneumonia in a rock hopper penguin (*Eudyptes crestatus*) and a group of rock ptarmigan (*Lagopus mutus*), and an osteolytic mass involving the ribs and air sacs of a penguin (Sphenisciformes) (Panigrahy et al. 1979). The feeding of damp, germinated seed has been implicated in disseminated mucormycosis causing alimentary granulomas in a group of canaries (*Serinus canarius*) and nephritis in an African grey parrot, glossitis in an African grey parrot, myocarditis in an Australian parakeet (*Psittacula* sp.), and nasal infection in waterfowl (Mitchell et al. 1986). Histopathology of biopsy specimens is more reliable in confirming the diagnosis (Orcutt and Bartick 1994). No effective treatment of mucormycosis in birds has been reported. Amphotericin B is the single most reliable agent used in humans. Other antifungal medications including nystatin 5-fluorocytosine, clotrimazole, and miconazole are reported to be inconsistent in vivo activity against the *Mucorales*.

9.1.6 Avian Gastric Yeast

Macrorhabdus ornithogaster is an anamorphic ascomycetous yeast that is the only known member of its genus (Tomaszewski et al. 2003). This colonizes the narrow junction (isthmus) of the glandular stomach (proventriculus) and grinding stomach (ventriculus) of birds and has not been identified elsewhere in the body or in the environment (Phalen 2005). Infection has been described in a range of species of birds including chickens, turkeys, ostriches, several species of parrots, and both captive-bred and wild finches (Wieliczko and Kuczkowski 2000; Gerlach 2001; Phalen 2005). Most *M. ornithogaster* infections cause little detectable disease. However, *M. ornithogaster* has been associated with a chronic wasting disease in budgerigars, canaries, and finches and an acute hemorrhagic gastritis in budgerigars and parrotlets (Phalen 2005).

9.2 Bacterial Infections

9.2.1 Streptococcosis

Streptococcosis in avian species is worldwide in distribution, occurring as both acute septicemic and chronic infections with mortality ranging from 0.5 to 50 %. Infection is considered secondary, because *Streptococci* may form part of the normal intestinal and mucosal flora of most avian species, including wild birds (Baele et al. 2002).

9.2.1.1 Etiology

The genus *Streptococcus* is composed of gram-positive, spherical bacteria occurring singly, in pairs, or short chains, which are nonmotile, non-spore-forming, facultative anaerobes. *Streptococcus* spp., isolated from avian species and associated with disease, includes *S. zooepidemicus* (occasionally referred to as *S. gallinarum*) from Lancefield antigenic serogroup type C, *S. bovis*, and *S. dysgalactiae*. A new species, *S. pleomorphus*, an obligate anaerobe in normal cecal contents of chickens, turkeys, and ducks, has also been described. Its possible role in disease for these species is undetermined (Abdul-Aziz 1994).

9.2.1.2 Epidemiology and Transmission

Streptococci are ubiquitous in nature and commonly found in various poultry environments. *Streptococcus dysgalactiae* has been cultured from broilers with cellulitis, a condition observed on the skin and subcutaneous tissue at processing (Vaillancourt et al. 1992). *Streptococcus* spp. has been isolated from lesions of osteomyelitis in turkeys, along with *E. coli* and *Staphylococcus* spp. (Clark et al. 1991).

Transmission of *Streptococci* occurs most commonly via oral and aerosol routes. Transmission can occur, however, through skin injuries, especially in caged layers. Aerosol transmission of *S. zooepidemicus* results in acute septicemia in chickens. Incubation periods range from 1 day to several weeks, with 5–21 days being most common. Endocarditis occurs when septicemic streptococcal infection progresses to a subacute or chronic stage.

9.2.1.3 Diagnosis

Isolation and identification of causative agent are easily done on blood agar. Demonstration of bacteria typical of *Streptococci* in blood films or impression smears of affected heart valves or lesions from birds with typical signs and lesions will provide a presumptive diagnosis of streptococcosis. Isolation of *S. zooepidemicus*, or any other Lancefield serogroup C *Streptococci* from typical lesions in poultry with appropriate clinical signs, will confirm streptococcosis. Rapid detection by latex agglutination test has been employed for the identification of antigenic serogroup C *Streptococci* in animals (Inzana and Irtani 1989).

9.2.1.4 Prevention and Treatment

Prevention and control require reducing stress and preventing immunodepressive diseases and conditions. Proper cleaning and disinfection can reduce environmental streptococcal resident flora to minimize external exposure. Treatment includes the use of antibiotics such as penicillin, erythromycin, novobiocin, oxytetracycline, chlortetracycline, and tetracycline in acute and subacute infections (Thayer et al. 2012).

9.2.2 Staphylococcosis

Staphylococcosis is a bacterial disease that can affect a wide range of avian species, including poultry, and is seen worldwide. *Staphylococcus aureus* is most commonly isolated from staphylococcosis cases, but species such as *S. hyicus* have also been reported as the causative agent of osteomyelitis in turkey poults. Omphalitis in young chicks caused by *Staphylococcus* species has also been reported (Andreasen 2013).

9.2.2.1 Transmission, Epidemiology, and Pathogenesis

Staphylococcus is part of the normal skin and mucosal flora; hence many infections are the result of a wound, mucosal damage, or both. Infection can also occur in the hatchery as a result of contamination of an open navel. Birds that are immunocompromised are also more prone to staphylococcal infections. Once in the host, *S. aureus* invades the metaphyseal area of the nearest joint, which leads to osteomyelitis and localization within that joint. Alternatively, the bacteria can invade the bloodstream and lead to a systemic infection in multiple organs.

S. aureus is found in water, dust, and air and primarily colonizes the mucosa of the nasopharynx and skin of humans and animals (Songer and Post 2005). The bacterium is considered to be a normal flora, isolated from the skin and feathers as well as in the respiratory and intestinal tracts (Skeeles 1997; Casey et al. 2007). However, some of the common forms of *S. aureus*-associated poultry infections include tenosynovitis (Butterworth 1999), omphalitis (Hill et al. 1989), femoral head necrosis, infected hock and stifle joints secondary to coccidiosis or vaccine reactions (McNamee and Smyth 2000), and "bumblefoot" (Skeeles 1997). Coagulase-negative *Staphylococci* are usually the most common inhabitants of healthy eyes of mammalian species and exotic birds (Dupont et al. 1994). Microorganisms inhabiting the external surface of the eye play a major role in its defense mechanisms. These microorganisms can protect the eye by taking nutrients from invading organisms or by secreting substances with antimicrobial properties (Eichenbaum et al. 1987).

Imbalances can occur in the normal microbial flora by the overuse of antibiotics or when the immune defense mechanisms of the host are compromised (Gerding 1990) and an overgrowth of gram-negative organisms or fungi can occur, as well as a selection of antibiotic-resistant species. *Staphylococcus aureus* and the coagulase-negative *Staphylococci* which are normal flora can occasionally cause ocular infections.

Staphylococcus aureus is important in relation to poultry and meat hygiene because of its ability to produce enterotoxins, which may cause food poisoning in humans. About 30 % of all outbreaks of food-borne diseases have been reported to be associated with poultry, and more than 25 % of poultry disease outbreaks have been attributed to *S. aureus* globally (Nortermans et al. 1982). Joint infections (swollen hocks, arthritis) and plantar abscess (bumblefoot) are associated with *S. aureus* (Saeed et al. 2000; John 2006). Infectious pododermatitis (bumblefoot), a common disorder of birds can be either unilateral or bilateral and is characterized by lameness, inflammation, and swelling of the footpad due to localized bacterial infection. Sequelae of infection can include chronic pododermatitis, septicemia, or amyloidosis. It can occur due to injury, infection, inappropriate substrate, obesity, or unilateral limb problems (trauma, arthritis) that result in excess and abnormal weight bearing on the contralateral foot. Treatment includes correction of the primary problem, local and systemic antibiotic and symptomatic treatment, and in more advanced cases, surgery.

9.2.2.2 Prevention

Hatchery sanitation and good egg management practices are important to reduce navel infections and omphalitis. Good litter management is important in controlling footpad injuries to prevent bumblefoot. Birds in captivity are prone to pododermatitis (bumblefoot) lesions due to sedentary habits, changes in normal activity patterns, prolonged time on hard and abrasive surfaces, and less time for exercises. Environmental enrichment allowing the use of creative and ingenious techniques that aim at keeping the captive animals occupied by increasing the range and the diversity of their behavioral opportunities always respecting the ethological needs of the species has helped in the prevention of bumblefoot (Reisfeld et al. 2013).

9.2.2.3 Resistance

Staphylococcus aureus has shown resistance to antimicrobials such as ampicillin and erythromycin (Pesavento et al. 2007; Otalu et al. 2011). The study by Nemati et al. (2008) detected for the first time methicillin-resistant *Staphylococcus aureus* (MRSA) strains in healthy poultry. This finding is of major public health concern since it is the drug of choice for human staphylococcal infections.

9.2.3 Mycobacterium

Environmental opportunistic mycobacteria are those that are recovered from natural and human-influenced environments and can infect and cause disease in humans, animals, and birds. The environmental opportunistic mycobacteria are normal inhabitants of natural waters, drinking waters, and soils.

Mycobacteria are acid-fast *Bacilli*, acidophilic, small, and slightly curved. They are aerobic, immobile, and non-sporulated bacteria. Their cell wall is lipid rich, forming smooth to wrinkled colonies when cultured in solid medium; some are pigmented (Carter and Wise 2004). Their high concentration in lipid cells is responsible for their resistance to immune system defense mechanisms and disinfectants (Brooks et al. 1996; Carter and Wise 2004). *Mycobacterium tuberculosis* complex species are recognized as being the cause of tuberculosis in mammals, although *M. tuberculosis* and *M. bovis* have been reported in birds and other species

Table 9.1 Mycobacterial species and the avian hosts affected by them

Avian host	Mycobacterial species
Eurasian jackdaw (*Corvus monedula*)	*M. fortuitum*
Common buzzard (*Buteo buteo*)	*M. terrae*
Common gull (*Larus canus*)	*M. nonchromogenicum*
Common pheasant (*Phasianus colchicus*)	*M. flavescens*
Other species	Mycobacterium avium complex and *M. genavense*

(Carter and Wise 2004). Another important group in human and veterinary medicine consists of non-tuberculous mycobacteria (NTM), these being opportunistic saprophytes causing mycobacteriosis in animals. Mycobacteria species and the wild bird hosts affected by them are shown in Table 9.1 (Carter and Wise 2004; Barragan and Brieva 2005). The mycobacterium avium complex (MAC) consists of opportunistic pathogens capable of causing disease in animals and humans (Inderlied et al. 1993). The complex has 28 serotypes from *M. avium* and *M. intracellulare* species (Inderlied et al. 1993; Hoenerhoff et al. 2004). Serotype 1 is most frequently reported in North American birds, while serotype 2 is more commonly reported in Europe. Serotype 3 has been sporadically isolated in European birds and serotypes 2 and 3 are more highly pathogenic than serotype 1 (Soler et al. 2009).

9.2.3.1 Epidemiology of Avian Mycobacteriosis

In birds three types of lesions are caused: the tubercular, focal, or multifocal form, the disseminated or diffuse form, and the paratuberculosis form in the gastrointestinal tract (Miller and Fowler 2012).

9.2.3.2 Species Affected and Prevalence

Avian mycobacteriosis is a ubiquitous disease in domestic, captive, and wild birds, and all birds appear to be susceptible with no reports of totally resistant species exist (Miller and Fowler 2012). Hejlicek and Treml in 1995 listed highly susceptible species as domestic fowl, sparrows, pheasants, partridges, and laughing gulls. Moderately susceptible species were turkeys and guinea fowl. Moderately resistant birds were reported to be geese and ducks. More highly mycobacteria-resistant species were thought to be pigeons, turtle doves, and rooks. Mycobacterial infection incidence in birds remains unknown and varies depending on the species, age, housing conditions, and whether birds are living in captivity or the wild. The incidence detected during avian necropsy has ranged from 0.5 to 14 % (Van DerHeyden 1997; Leite et al. 1998; Gerhold and Fischer 2005). In wild birds prevalence ranging from 1 to 39 % has been reported (Converse 2007; Witte et al. 2007). Mycobacterial infection usually affects older birds (more than 2 years old) and is more common in poultry and older aquatic species due to the microorganism's long incubation period, the cumulative risk of exposure, and decreased immune response (Soler et al. 2009).

9.2.3.3 Transmission

Contaminated environment and areas where birds congregate should be considered as potential danger spots, as the disease is mainly transmitted through inhalation and ingestion of contaminated food or water (Van DerHeyden 1997; Carter and Wise 2004).

9.2.3.4 Diagnosis

Diagnosis of mycobacteriosis is challenging. Clinical signs are variable and nonspecific and disease may go unnoticed thereby hampering diagnosis in live birds. Affected animals generally exhibit signs of depression, weakness, anorexia, and fever (Tell et al. 2004). There may also be diarrhea, polyuria, arthritis, claudication, pathological fractures, ascites, and subcutaneous tumors (Van DerHeyden 1997). Diagnosis is based on clinical signs, leukogram, serology, culture, radiography, and ultrasonography. Culture of the microorganism is the gold standard but it is time consuming and hence of limited clinical usefulness (Van DerHeyden 1997; Tell et al. 2004). Newer diagnostic methods include deoxyribonucleic acid–ribonucleic acid (DNA–RNA) and polymerase chain reaction (PCR) tests for specific mycobacterial genes (Tell et al. 2004; Bougiouklis et al. 2005);

9.2.3.5 Treatment, Control, and Prevention

Because of the poor prognosis, cost, labor, prolonged treatment periods, risks of resistance, and zoonotic risks, there is no generally recommended treatment protocol for birds affected with mycobacteriosis, and euthanasia on infected individual birds, decreasing population density, and strict hygiene are commonly recommended (Miller and Fowler 2012).

9.2.3.6 Zoonotic Potential

The risk of non-tuberculous mycobacteria being transmitted to immunocompetent adult humans appears to be extremely small (Leite et al. 1998). However, there is a significant risk of transmission to children and immunocompromised individuals. The recent concern about diagnosing and treating mycobacterial infections has resulted from opportunistic mycobacterial infections occurring in AIDS patients. Increasing contact between humans and wildlife, either in captivity when visiting zoos or having wildlife as pets (which is unsuitable) or in the wild with the rise of ecotourism and increasing colonization of wilderness areas, has thereby led to increasing contact between wild animals and humans and the subsequent transmission and spread of pathogens (Allchurch 2002).

9.2.4 *Avian Bordetellosis*

Avian bordetellosis is a highly infectious, acute upper respiratory tract disease of turkeys characterized by high morbidity and usually low mortality. *Bordetella avium* causes bordetellosis in domesticated turkeys (Skeeles and Arp 1997) and is an opportunistic pathogen in chickens (Jackwood et al. 1995). Damage to the upper respiratory tract from prior exposure to an upper respiratory disease agent or vaccine such as infectious bronchitis virus or Newcastle disease virus, or from an environmental irritant such as ammonia, is necessary to induce signs in chickens.

9.2.4.1 Etiology and Pathogenesis

The causative agent is *Bordetella avium*, a gram-negative, non-fermentative, motile, aerobic bacillus (Kersters et al. 1984). The mechanism of pathogenesis involves the ability of *B. avium* to destroy ciliated epithelial cells in the trachea. Certain strains of the bacteria adhere to the ciliated pseudostratified columnar epithelium and produce toxins, some of which appear to be similar to those from other *Bordetella* spp. Toxins associated with pathogenic strains of *B. avium* include a heat-labile toxin, tracheal cytotoxin, dermonecrotic toxin, and osteotoxin. Damage to the upper respiratory tract can lead to secondary infections with *Escherichia coli* or other agents, which can significantly increase the severity of the disease (Jackwood 2013).

9.2.4.2 Epidemiology of Bordetellosis

Bordetella avium has been isolated from domesticated species in Germany, including Muscovy ducks, domesticated geese, a yellow-crested cockatoo (*Kakatoe galeria*), parrot finches (*Erythrura psittacea*), and partridges (*Perdix perdix*) (Hinz and Glunder 1985). *Bordetella avium* has also been implicated in a respiratory syndrome of cockatiels (*Nymphicus hollandicus*) and ostriches (*Struthio camelus*) (Clubb et al. 1994). Hopkins et al. (1990) reported that 42 of 44 wild turkeys in Arkansas were seropositive. Raffel et al. (2002) recorded a high seroprevalence in wild birds; Canada geese had high seroprevalence in contrast with pigeons and doves which had low prevalence.

Environmental transmission may be important to the spread of the disease. *Bordetella avium* is known to be transmitted by water or litter contamination and can remain virulent in litter for 1–6 months (Skeeles and Arp 1997). *Bordetella avium* seems to be long lived in water as reported in *B. bronchiseptica*, and hence water may act as environmental reservoir for this microorganism. Birds commonly found in and around freshwater, such as those in the orders Ciconiiformes, Anseriformes, Pelecaniformes, Charadriiformes, and Gruiformes, have a relatively high prevalence of *B. avium* exposure (Raffel et al. 2002).

9.2.4.3 Zoonotic Risk

Bordetella avium may be a rare opportunistic pathogen in people. In addition, a closely related organism, *B. hinzii*, also isolated from poultry, has been associated with septicemia and bacteremia in older or immunocompromised people.

9.2.5 Clostridial Infections

9.2.5.1 Necrotic Enteritis

Necrotic enteritis is caused by an enterotoxemia or toxins in the blood produced in the intestine resulting from infections with *Clostridium perfringens*, a spore-forming bacterium and a natural inhabitant of soil and the intestinal tract of many warm-blooded animals (Brynestad and Granum 2002).

Pathogenesis and Predisposing Factors

The enterotoxemia that results in clinical disease most often occurs either following an alteration in the intestinal microflora or from a condition that results in damage to the intestinal mucosa (e.g., coccidiosis, mycotoxicosis, salmonellosis, ascarid larvae) (Williams 2005). High dietary levels of animal by-products (e.g., fishmeal), wheat, barley, oats, or rye predispose birds to the disease (Timbermont et al. 2011). Anything that promotes excessive bacterial growth and toxin production or slows feed passage rate in the small intestine could promote the occurrence of necrotic enteritis (McDevitt et al. 2006).

Clinical Findings and Lesions

Most often the only sign of necrotic enteritis in a flock is a sudden increase in mortality without premonitory signs. However, birds with depression, ruffled feathers, and diarrhea may also be seen (Timbermont et al. 2011). The gross lesions are primarily found in the small intestine (jejunum), which may be ballooned, friable, and contain a foul-smelling, brown fluid. The mucosa is usually covered with a tan to yellow pseudomembrane often referred to as a "Turkish towel" in appearance (Olkowski et al. 2006). This pseudomembrane may extend throughout the small intestine or be only in a localized area. The disease usually persists in a flock for 5–10 days, and mortality is 2–50 % (Riddell and Kong 1992).

Diagnosis

A presumptive diagnosis is based on gross lesions and a gram-stained smear of a mucosal scraping that exhibits large, gram-positive rods. Histologic findings consist of coagulative necrosis of one-third to one-half the thickness of the intestinal mucosa and masses of short, thick bacterial rods in the fibrinonecrotic debris (Bildfell et al. 2001).

Prevention, Control, and Treatment

Avoiding drastic changes in feed and minimizing the level of proteins and indigestible starch in the diet can aid in the prevention of necrotic enteritis. Control of coccidiosis or any other factor that may lead to intestinal mucosa damage should be done. Administration of probiotics or competitive exclusion cultures has been used to both prevent and treat clinical necrotic enteritis (presumably by preventing the proliferation of *C. perfringens*).

Treatment for necrotic enteritis is most commonly administered in the drinking water, with bacitracin (200–400 mg/gal. for 5–7 days), penicillin (1,500,000 U/gal. for 5 days), and lincomycin (64 mg/gal. for 7 days) most often used.

9.2.5.2 Ulcerative Enteritis (Quail Disease)

Ulcerative enteritis (UE) is an acute bacterial infection in young chickens, turkeys, and upland game birds characterized by sudden onset and rapidly increasing mortality. The disease was first seen in enzootic proportions in quail and was, therefore, named quail disease.

Etiology

Ulcerative enteritis is caused by a species of *Clostridium* named *Clostridium colinum* (Berkhoff 1985).

Pathogenesis and Epidemiology

Ulcerative enteritis is worldwide in its distribution and affects a wide variety of avian species. Natural infections have been found in bobwhite quail (*Colinus virginianus*), California quail (*Lophortyx californica*), Gambel quail (*L. gambelii*), mountain quail (*Oreortyx picta*), scaled quail (*Callipepla squamata*), and sharp-tailed grouse (*Pedioecetes phasianellus*) (Morse 1907); domestic turkeys (*Meleagris gallopavo*) and chickens (*Gallus gallus*) (Durant and Doll 1941); European partridge (*Perdix perdix*) and wild turkeys (*M. gallopavo*) (Durant and Doll 1941); and chukar partridge (*Alectoris graeca*) (Richards and Hunt 1973).

Under natural conditions, ulcerative enteritis is transmitted through droppings; birds become infected by ingesting contaminated feed, water, or litter. The organism produces spores, resulting in permanent contamination of premises after an outbreak has occurred (Berkhoff 1985).

Diagnosis

Diagnosis of ulcerative enteritis can be made on the basis of gross postmortem lesions. The presence of typical intestinal ulcerations accompanied by necrosis of the liver and an enlarged, hemorrhagic spleen suffices for clinical diagnosis. A fluorescent antibody (FA) test has been developed and found to be highly specific (Berkhoff and Kanitz 1976). An agar gel immunodiffusion test has also been used for diagnosis of UE (Berkhoff 1975). *Clostridium colinum* can be seen in gram-stained smears of the liver and intestinal lesions. Among similar diseases that must be differentiated from ulcerative enteritis are coccidiosis, necrotic enteritis, and histomoniasis.

Prevention, Treatment, and Control

Bacitracin in the feed at 200 g/ton is used for prevention in quail. Streptomycin (0.006 %) in the feed is effective to treat the disease. Prevention must start with good management practices (e.g., avoiding the introduction of new birds into existing flocks). High population density is a predisposing factor. The use of cages is recommended in quail breeding. Sick and dead birds should be removed promptly. Total cleanup between flocks, pest control in and around the premises, and periodic treatment of watering systems with innocuous chemicals that dissolve mineral and or biofilm buildup are good preventive measures (Otalora 2013).

9.2.6 Pseudomonas *Species Infections*

Pseudomonas is an environment-associated infection and may cause a serious problem in poultry farms. *Pseudomonas aeruginosa* is a motile, gram-negative, non-spore-forming rod occurring singly or in short chains under microscopic examination. The organism is strictly aerobic and grows readily on common bacteriologic media, usually producing a water-soluble green pigment composed of fluorescent and pyocyanin with a characteristic fruity odor. The organism is ubiquitous, often associated with soil, water, and humid environments, and occurs as intestinal flora in both mammals and birds (Bailey and Scott 1970). Generally, it is considered to be an opportunistic organism that produces respiratory infections, septicemia, and other forms when introduced into tissues of susceptible birds (Kabede 2010). *Pseudomonas* infection frequently leads to a chronic wasting disease in domestic birds too (especially hens) and free-ranging birds worldwide. Young birds, as well as severely stressed or immunodeficient birds, are more susceptible to the infection. Concurrent infections with viruses, other bacteria (especially *Mycoplasmas*), and *Aspergillus* are common and may enhance the infection of *Pseudomonas* (Kabede 2010; Chege et al. 2013). *Pseudomonas* stomatitis has also been reported as a sequel to trichomoniasis in saker falcons (*Falco cherrug*) (Samour 2000).

9.2.7 Avian Colibacillosis

9.2.7.1 Etiology

Colibacillosis is caused by *Escherichia coli* (*E. coli*), a gram-negative, non-acid-fast, uniform staining, non-spore-forming bacillus that grows both aerobically and anaerobically and may be variable in size and shape (La Ragione and Woodward 2002). *Escherichia coli* is normally present in the birds' body as commensal bacteria (Sarker et al. 2012). It is a normal inhabitant of the intestinal tracts of animals and is harmless as long as it is kept in check by other intestinal bacteria (Barnes et al. 2003). *Escherichia coli* is a bacterium that is often considered an opportunistic pathogen because it infects whenever it has the opportunity. When an imbalance occurs in bacterial flora of the intestinal tract, *E. coli* may grow and cause an outbreak of colibacillosis.

9.2.7.2 Epidemiology

Escherichia coli causes a variety of disease manifestations in poultry including yolk sac infection, omphalitis, respiratory tract infection, swollen head syndrome, septicemia, polyserositis, coligranuloma, enteritis, cellulitis, and salpingitis (Kabir 2010).

Enteritis caused by *Escherichia coli* (colibacillosis) is an important disease in the poultry industry because of increased mortality and decreased performance (Durairaj and Clark 2007). Chickens of all ages are susceptible to colibacillosis, but usually young birds are considered more susceptible.

Escherichia coli has also been identified in the feces of free-ranging psittacine nestling populations in which 100 % of samples were positive for *Enterobacteriaceae* and 62.5 % for *E. coli* (Saidenberg 2008). Such a high frequency of gram-negative bacteria in the nestlings' feces of free-ranging psittacines suggests that nests may be important sources of infection due to the accumulation of organic matter and feces and the likelihood of prior use by other species that have distinct intestinal microbiota (Hidasi et al. 2013).

Escherichia coli has been reported as the most common bacterial agent isolated from free-living waterfowl (composing 78 % of the bacteria isolated in mallards), and water contamination from dense congregations of waterfowl species has been suspected to be a source for human exposure (Samadpoour et al. 2002).

9.2.7.3 Predisposing Factors

Escherichia coli causes disease when underlying factors allow for a large bacterial population to overwhelm the host's normal immunity, or the birds themselves are already weakened due to stress, poor nutrition, or poor husbandry (Gregory et al. 2003). The disease often occurs in illegally traded birds mainly captured from the wild are exposed to inadequate conditions of transportation, diet, and sanitation. They are often placed in small and overcrowded cages which facilitate physiological imbalances, spread of microorganisms, and development of clinical disease (Raso et al. 2006). In poultry, systemic infection often involves predisposing environmental factors or infectious causes. Thus, mycoplasmosis, infectious bronchitis, Newcastle disease, hemorrhagic enteritis, and turkey bordetellosis, nutritional deficiencies, or exposure to poor air quality and other environmental stresses may precede colibacillosis (Nolan 2013).

Escherichia coli is normally present in the birds and sometimes an associated change from health to disease can be triggered by numerous events as illustrated in Fig. 9.2. Immunosuppressive diseases such as infectious bursal disease, Marek's disease, and chicken anemia may increase susceptibility to *E. coli* infection. However, other countless events or diseases can also increase susceptibility. For instance, an *E. coli* infection may appear if birds do not have regular access to feed or if their litter is too wet or if they are exposed to another disease. Generally, anything that causes stress in the bird may provide *E. coli* with the opening it needs.

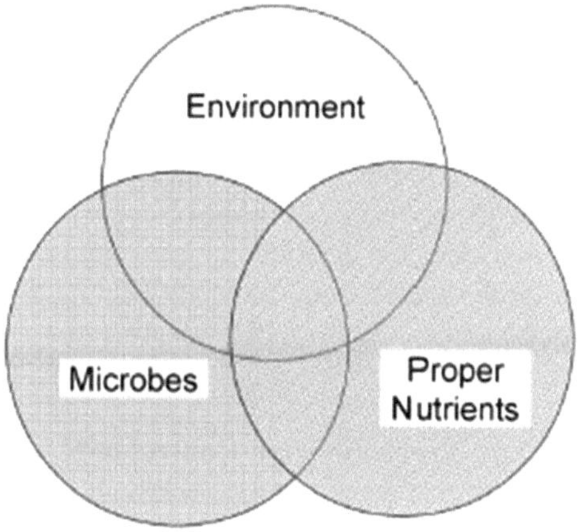

Fig. 9.2 Multiple factors that lead to disease (e.g., *Escherichia coli* infections) (adapted from McMullin 1998)

9.2.7.4 Antibiotic Resistance and Public Health Concern

Escherichia coli is not only associated with potentially pathogenic strains but also with the commensal strains which colonize the birds' microbiota and constitute a reservoir of resistance genes. These commensals may become a serious problem through the transfer of resistance genes and virulence factors among different bacterial strains in animal and human populations (Aminov et al. 2001).

At slaughter, resistant strains from the gut readily soil poultry carcasses and as a result poultry meats are often contaminated with multiresistant *E. coli* (Turtura et al. 1990), and likewise eggs become contaminated during laying (Lakhotia and Stephens 1973). Hence, resistant fecal *E. coli* from poultry can infect humans both directly and via food. These resistant bacteria may colonize the human intestinal tract and may also contribute resistance genes to human endogenous flora (Van den Bogaard et al. 2001).

9.2.7.5 Diagnosis

Colibacillosis is suspected based on the clinical features and the typical macroscopic lesions. The diagnosis is obtained by *E. coli* isolation from cardiac blood and affected tissues, like the liver, spleen, pericardium, or bone marrow. Experimentally it was shown that in acute cases, isolation is possible from 3 h to 3 days after infection; in subacute cases, isolation is only possible until 7 days after infection (Gomis et al. 1997). Contamination from the intestines is rarely a problem, if fresh material is used and standard bacteriological procedures are applied (Gross and Domermuth 1975). Selective media like McConkey, eosin-methylene blue, or drigalski agar are used for isolation. Further identification of the isolated colonies is based on biochemical reactions (indol production, fermentation of glucose with gas production, presence of β-galactosidase, absence of hydrogen sulfite production and urease, and the inability to use citrate as a carbon source) (Dho-Moulin and

Fairbrother 1999). O-serotyping is a frequently used typing method. An ELISA, based on sonicated *E. coli*, has been developed for detection of antibodies against two important pathogenic serotypes of *E. coli*: O78:K80 and O2:K1 (Leitner et al. 1990). Another ELISA was based on fimbrial antigen (Bell et al. 2002). Both have limited values because they can only detect homologous avian pathogenic *Escherichia coli* (APEC) types. All currently known virulence-associated factors, detected in strains isolated from colibacillosis lesions, can also be detected in fecal isolates from clinically healthy chickens. For this reason, none of these traits can be used for APEC identification.

9.2.7.6 Prevention and Control of *E. coli* Infections

Treatment strategies include attempts to control predisposing infections or environmental factors and early use of antibacterials indicated by susceptibility tests. Good housing hygiene and avoiding overcrowding help to prevent the disease. Most isolates are resistant to tetracyclines, streptomycin, and sulfa drugs, although therapeutic success can sometimes be achieved with tetracycline. Commercial bacterins (vaccines prepared from killed bacteria) administered to breeder hens or chicks have provided some protection against homologous *E. coli* serogroups (Nolan 2013).

9.2.8 Acinetobacter *Species Infections*

The genus *Acinetobacter* consists of strictly aerobic, nonmotile, catalase-positive, indole-negative, oxidase-negative, gram-negative, non-fermentative encapsulated *Coccobacilli* rods (Vallenet et al. 2008). They are widely distributed in the environment as well as in hospitals, where they can survive on moist or dry surfaces for long periods of time. *Acinetobacter* can also be found in soil and constitutes one of the predominant organisms in water (Baumann 1968).

9.2.8.1 Pathogenesis and Epidemiology

The pathogenic mechanisms of *Acinetobacter* spp. are little understood or studied (Peleg et al. 2009). *Acinetobacter baumannii* is the most commonly found *Acinetobacter* species in humans and has emerged as multidrug resistant (Schreckenberger et al. 2003). Moreover, it has evolved as one of the most important nosocomial pathogens in the past decade (Villers et al. 1998) particularly in immunosuppressed patients (Schreckenberger et al. 2003). *Acinetobacter baumannii* can affect different organs causing pneumonia, meningitis, septicemia, and urinary tract (Smith et al. 2007) and skin infections (Villers et al. 1998). *Acinetobacter baumannii* has been found in association with other infections such as aspergillosis (Tarello 2011) and in association with mycobacteriosis in falcons (Muller et al. 2010). *Acinetobacter baumannii* was isolated from feces randomly sampled from wild birds in a zoological collection in Japan; however, no pathological lesions were associated with it (Ahmed et al. 2007). *Acinetobacter lwoffii*, another ubiquitous and opportunistic bacterium, has been isolated by pulmonary and abdominal air sac swabs obtained from a lovebird (*Agapornis roseicollis*), which died of a severe respiratory disease (Robino et al. 2005). *Acinetobacter* spp. are normally found in healthy eyes but may occasionally cause ocular infections (Carter and Chengappa 1991).

9.2.8.2 Diagnosis of *Acinetobacter* Infections

Commercially available microbiology testing systems like mini-API and VITEK are not successful in the detection of *A. baumannii*, but its identification by internal transcribed spacer (ITS) sequencing is regarded as a very reliable method (Chang et al. 2005).

9.2.8.3 Prevention, Treatment, and Control

Wild birds have been shown to carry the disease agent asymptomatically; hence minimizing contact with contaminated feces may help in reducing the incidence in domestic and captive birds.

Different classes of antimicrobial agents that are considered as potentially effective for the treatment of *A. baumannii* infections include carbapenems, polymyxins, tetracyclines, glycylcyclines, aminoglycosides, and fluoroquinolones.

9.2.8.4 Antimicrobial Resistance

The major problem with *Acinetobacter* spp. is their resistance to antibiotics (Landman et al. 2002a, b). Savov et al. reported that these organisms are most commonly resistant to ampicillin, cephalothin, carbenicillin, gentamicin, amikacin, chloramphenicol, tetracycline, co-trimoxazole, ciprofloxacin, and cefoperazone. *Acinetobacter baumannii* has been developing resistance to all antibiotics used in treating infections. Currently, most *A. baumannii* strains are resistant to aminoglycosides, tetracyclines, cephalosporins, ampicillins, cefotaximes, chloramphenicols, gentamicins, and tobramycins (Prashanth and Badrinath 2005).

Antimicrobial resistance among *Acinetobacter* is either intrinsic or acquired via transformation. Several mechanisms of resistance have been reported including altered penicillin-binding proteins, low or decreased permeability of the outer membrane to antibiotics or increase in the active efflux of the antibiotics, target site mutations, and inactivation via modifying enzymes (Jain and Danziger 2004; Simor et al. 2002).

9.2.9 Enterococcus *Infections*

Enterococci are commensals of the human and animal gastrointestinal tract (Murray 1990). Disease outbreaks caused by *Enterococci* are therefore considered opportunistic, while predisposing factors (e.g., deficiencies, concurrent infections, immunosuppression, antibiotic treatment, vaccinations) are considered crucial for outcome of the pathology induced (Steentjes et al. 2002).

In poultry, *Enterococci* are frequently isolated from dead-in-shell and day-old chicks, often affecting the yolk sac (Cortes et al. 2004; Deeming 2005).

9.2.9.1 Epidemiology and Clinical Signs

Enterococcus faecalis has been associated most frequently with poultry diseases among which endocarditis, hepatic granulomas in turkeys, and arthritis and amyloidosis in both brown layers (Landman et al. 1994) and broiler breeders (Steentjes et al. 2002) have been most frequently reported. It has also been isolated from arthritic joints of ducks. Furthermore, *E. faecalis* has been associated with ascites in hens and pulmonary hypertension in broilers (Tankson et al. 2001, 2002). *Enterococcus faecalis* and *E. faecium* constitute the enterococcal species most commonly isolated from day-old chicks (Devriese et al. 2002). *Enterococci* can affect the quality of day-old chickens, having major impact on first-week mortality, and *E. coli* and *E. faecalis* have been shown to be the most significant bacterial pathogens associated with first-week mortality (Olsen et al. 2012). *Enterococcus durans* infection in young chickens has been associated with bacteremia and encephalomalacia (Cardona et al. 1993), whereas, *E. hirae* has been associated with brain lesions (focal necrosis) in young chicks (Randall et al. 1993).

9.2.10 Enterobacter *Infections*

Enterobacter is a normal inhabitant of the avian digestive tract (Binek et al. 2000). Similar to other gram-negative bacteria in the *Enterobacteriaceae* family, it can infect eggs and young birds causing embryo loss, omphalitis, yolk sac infections, and mortality in young birds (Salmon and Watts 2000). *Enterobacter* was isolated infrequently from turkeys with cellulitis.

9.2.11 Corynebacterium *and* Trueperella

Corynebacterium spp. have been recovered at different frequencies from the intestinal and cloacal flora of different wild birds (Bangert et al. 1988) and have occasionally been reported as causal agents of disease in birds (Fiennes 1982).

9.2.11.1 Epidemiology and Pathogenesis

A chronic, disseminated granulomatous disease of turkeys in Canada suspected to be actinomycosis has been observed sporadically since 1955 (Senior et al. 1962). Serious outbreaks of osteomyelitis involving the proximal tibia, thoracic vertebra, and proximal tibiotarsi caused by *Trueperella pyogenes* in commercial male turkey flocks resulted in considerable economic loss (Brinton et al. 1993).

Septicemia, visceral lesions, cutaneous abscesses, mortality of nearly 14 %, and a decrease in egg production of over 27 % have occurred in caged layers infected with *T. pyogenes*. Portal of entry was through skin lesions caused by poor caging (Corrales et al. 1988).

9.2.11.2 Diagnosis

Club-shaped, pleomorphic, gram-positive *Bacilli* in smears of lesions provide a rapid diagnosis (Brinton et al. 1993). Biochemical and serologic evaluation of isolates and agar gel precipitin test is highly effective at detecting antibodies (Barbour et al. 1991).

9.2.11.3 Prevention and Treatment

Treatment of an affected flock can be done with penicillin in the feed (100 g/ton).

9.2.12 Proteus *Infections*

Proteus is a genus in the family *Enterobacteriaceae* that inhabits the lower intestinal tract. The organism is capable of penetrating the eggshell, which is facilitated by fecal contamination. Experimental inoculation of fertile eggs resulted in 100 % embryonic mortality. *Proteus* occasionally causes embryonic death, yolk sac infections, and mortality in young chickens, turkeys, and ducks (Salmon and Watts 2000; Baruah et al. 2001). *Proteus* also can be a contaminant of artificially collected semen (Bale et al. 2000). Septicemia due to *Proteus* has occurred in quail (Mohamed 2004) and in broilers suspected of having immunologic deficiency

(Randall et al. 1984). *Proteus* has been recovered occasionally from a low percent of salpingitis and oophoritis lesions in layers (Batra et al. 1982) and has been associated with respiratory disease in chickens (Ye et al. 1995). An isolate from chickens with respiratory disease caused 50 % mortality in experimentally inoculated 4-week-old chickens. *Proteus mirabilis* was isolated from the lung, trachea, and kidney of chickens experiencing respiratory signs, diarrhea, paralysis, and high mortality. The disease was reproduced with isolates of the organism (Ye et al. 1995). *Proteus* was isolated infrequently from turkeys with cellulitis (Gomis et al. 2002) and white leghorn pullets with necrotic dermatitis that had seroconverted to reticuloendotheliosis virus (Howell et al. 1982). In waterfowl, *Proteus* can occasionally produce arthritis, salpingitis, airsacculitis, septicemia, and granulomatous inflammation of salt glands (Klopfleisch et al. 2005).

9.2.12.1 Diagnosis

Serological typing of *P. mirabilis* and *P. vulgaris* traditionally has been done using the slide agglutination test and the indirect hemagglutination test. Bacteriophage typing (Sekaninova et al. 1998) and modern molecular methods employing the polymerase chain reaction (PCR) to produce DNA fingerprints and other 16S ribosomal RNA gene (ribotyping) methods of strain differentiation have been applied to distinguish *P. mirabilis*, *P. vulgaris*, and *P. penneri* strains (Hoffmann et al. 1998).

9.2.12.2 Antimicrobial Resistance

Antimicrobial testing of 19 *Proteus* isolates from diseased or dead turkey poults between one and 35 days of age indicated sensitivity (MIC50 < 1 μg/ml) to enrofloxacin and ceftiofur (Salmon and Watts 2000).

9.2.13 Bacillus *Infections*

Bacillus spp. occasionally have been associated with embryo mortality and yolk sac infections in chickens (Venkanagouda et al. 1996), turkeys (Bruce and Drysdale 1983), ducks (Baruah et al. 2001), and ostriches (Deeming 1996). *Bacillus* spp. and *E. coli* were the most commonly cultured bacteria from reproductive disorders of hens (Goswami et al. 1988). *Bacillus cereus*, an organism that can cause food-borne illness in people, infected turkey hens following artificial insemination and was found in 25 % of their unhatched eggs. Certain strains of *Bacillus* interfere with intestinal colonization by enteric pathogens and have value as probiotics (La Ragione and Woodward 2003; Barbosa et al. 2005). Keratinases (subtilisins) produced by *B. licheniformis* have the ability to degrade feathers (Evans et al. 2000). *Bacillus cereus* was reported to cause infection and death in 12 captive psittacines (Godoy et al. 2012).

9.2.14 Klebsiella

Members of the genus *Klebsiella*, especially *K. pneumoniae* and *K. oxytoca*, are opportunistic pathogens associated with severe nosocomial infections such as septicemia, pneumonia, and urinary tract infections. *Klebsiella pneumoniae* has been taxonomically subdivided into three subspecies: *K. pneumoniae* subsp. *pneumonia*, *K. pneumoniae* subsp. *ozaenae*, and *K. pneumoniae* subsp. *rhinoscleromatis* (Brisse and Verhoef 2001). *Klebsiella pneumoniae* occur primarily as major components of the gut flora of domestic animals, making such animals a common medium of transmission for infections caused by these microorganisms and facilitating cross-contamination of food and drinking water sources (Gyles 2008). Concurrent infection of young poultry with *Klebsiella pneumoniae* increases the severity of respiratory disease. *Klebsiella* is an environmental contaminant that occasionally causes embryo mortality, yolk sac infections, and mortality in young chickens, turkeys, and ostriches (Venkanagouda et al. 1996; Salmon and Watts 2000).

9.2.14.1 Epidemiology and Coinfection

The organism has been associated with cutaneous, respiratory, ocular, systemic, and reproductive diseases of poultry. *Klebsiella* are among the aerobic bacteria isolated from turkeys with cellulitis (Gomis et al. 2002). Concurrent infection of young turkeys with *K. pneumoniae* increased the severity of respiratory disease resulting from *Bordetella avium* and *Chlamydophila psittaci* infections (Hinz et al. 1992). *Klebsiella* has been isolated from turkey flocks with adenoviral inclusion body tracheitis that experienced respiratory disease and increased mortality. *Klebsiella* and *Staphylococcus aureus* have been isolated from an outbreak of septicemic disease of 20-week-old layers experiencing increased mortality. Mortality and clinical disease occurred following oral inoculation of young chicks with three *Klebsiella* biotypes and chicks inoculated with *K. pneumoniae* had the highest mortality (Dessouky et al. 1982).

9.3 Protozoal Infections

9.3.1 Cryptosporidiosis

9.3.1.1 Etiology

Cryptosporidium which is ubiquitous within its geographic distribution is a small (4–6 mm) coccidian parasite that causes widespread disease in humans and many other vertebrates (Fayer 2004). In birds, cryptosporidiosis was first described in the ceca of chicken by Tyzzer in 1929. Three different species have been implicated in causing the disease in avian species: *Cryptosporidium meleagridis*, (Slavin 1955), *Cryptosporidium baileyi* (Sréter and Varga 2000; Fayer 2004), and *Cryptosporidium galli* (Xiao et al. 2004).These three *Cryptosporidium* spp. can each infect numerous bird species, but they differ in host range and predilection sites. Even though both *C. meleagridis* and *C. baileyi* are found in the small and large intestine and bursa, they differ significantly in oocyst size (Sréter and Varga 2000), and only *C. baileyi* is also found in the respiratory tissues such as the conjunctiva, sinus, and trachea. In contrast, *C. galli* infects only the proventriculus. Other strains of *Cryptosporidium* have been reported as possible pathogens in bobwhite quail (*Colinus virginianus*) and ostrich (*Struthio camelus*) (Penrith et al. 1994); isolation of *Cryptosporidium anserinum* has been reported from a domestic goose (*Anser domesticus*) and *Cryptosporidium tyzzeri* from chickens (*Gallus gallus domesticus*) (Proctor and Kemp 1974), but thus far, not all are considered valid microbial species designations.

9.3.1.2 Epidemiology

Cryptosporidiosis generally is a disease of the young and immunocompromised birds and is one of the most prevalent parasitic infections in domesticated, caged, and wild birds (Sréter and Varga 2000), and the parasite has been reported worldwide in more than 30 avian species belonging to orders Anseriformes, Charadriiformes, Columbiformes, Galliformes, Passeriformes, Psittaciformes, and Struthioniformes (Sréter and Varga 2000; Ng et al. 2006).

9.3.1.3 Pathogenesis and Transmission

The infection is transmitted by shedding of thick-walled, sporulated oocysts (containing four sporozoites), either via feces from infected hosts or directly coughing out of oocysts from the respiratory tract into the environment. Oocysts present in water, food, or contaminated materials can subsequently be ingested or directly inhaled, after which the sporozoites excyst and may penetrate epithelial cells of either the gastrointestinal or respiratory tract (Sréter and Varga 2000). The

Cryptosporidia subsequently undergo endogenous sporulation, leading to autoinfection in the parasitized host and shedding of new oocysts. As these thick-walled oocysts are relatively resistant to environmental influences, the infection can easily be transmitted to other hosts (Sréter and Varga 2000).

9.3.1.4 Clinical Signs

Cryptosporidiosis in birds can manifest in three clinical forms: respiratory disease, enteritis, and renal disease. Generally only one form of disease is present during an outbreak (Blagburn et al. 1991). *Cryptosporidium meleagridis* was originally described from the intestines of turkeys (*Meleagris gallopavo*) (Slavin 1955) and is thought to be more frequently associated with enteritis (Lindsay and Blagburn 1990). *Cryptosporidium baileyi* was first described from the bursa, cloaca, and respiratory tract of chickens (Goodwin et al. 1996; Murakami et al. 2002) and is more frequently associated with respiratory cryptosporidiosis, whereas *C. galli* has been described from the proventriculus only (Pavlásek 1999; Ryan et al. 2003).

9.3.1.5 Diagnosis

Staining techniques have proven to be useful tools in rapidly identifying the protozoan's endogenous stages in cytologic smears and feces, for which the Kinyoun modification of the Ziehl–Neelsen acid-fast staining is the most commonly used (Abbassi et al. 2000; Sréter and Varga 2000). The different endogenous developmental stages can also be detected in hematoxylin- and eosin-stained histologic samples, in which *Cryptosporidia* appear as 2.0–7.5 mm basophilic bodies. A combined polymerase chain reaction–restriction fragment length polymorphism (PCR-RFLP) assay has been developed recently for detecting cryptosporidial organisms (Wang et al. 2011). This test provides strong evidence for the presence of this particular agent and also differentiates between specific strains (*C. baileyi*, *Cryptosporidium muris*, and *Cryptosporidium parvum*) (Leng et al. 1996; Morgan et al. 2001).

9.3.1.6 Prevention and Treatment

To prevent the disease spread, identification of the contaminating source is important. Furthermore, proper hygiene measures and quarantine protocols should be implemented (Fujino et al. 2002).

At present no effective chemotherapy is available for the treatment of avian cryptosporidiosis. Experimental studies have shown that most commonly used anticoccidials when used either alone or in combination with the dihydroquinoline antioxidant duokvin do not prevent or reduce respiratory disease in chickens inoculated with *C. baileyi* oocysts (Varga et al. 1995). Oxytetracycline, amprolium,

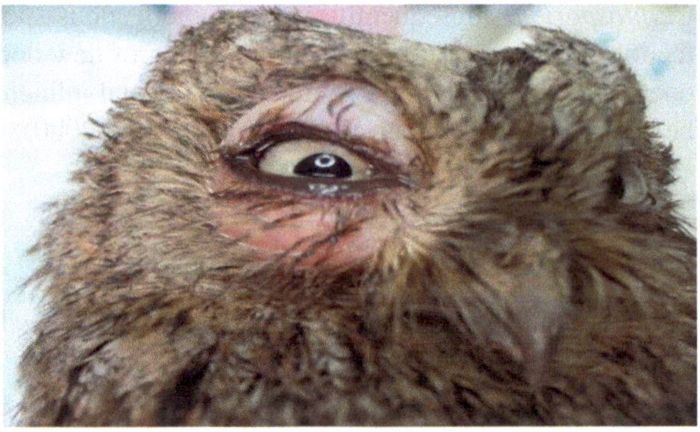

Fig. 9.3 Photograph of a young *Otus* owl (*O. scops*) showing eyelid edema due to cryptosporidiosis (source: Molina-López et al. 2010)

and chlortetracycline had no effect on respiratory cryptosporidiosis in turkeys and peafowl chicks (Glisson et al. 1984). Some authors have reported aminoglycosides, macrolide and ionophore antibiotics, halofuginone, nitazoxanide, and toltrazuril as effective therapeutic agents (Hornok et al. 1999; Sréter and Varga 2000). Paromomycin, a broad spectrum macrolide antibiotic that closely resembles the antibacterial action of neomycin, was the first drug that proved to be effective for treatment of (intestinal) cryptosporidiosis in humans, cattle, and poultry. More recently, azithromycin (40 mg/kg once daily for 15 days) was shown to be effective in *Cryptosporidium baileyi* affecting young *Otus* owls (*O. scops*) (Molina-López et al. 2010) (Fig. 9.3).

9.3.2 *Coccidiosis*

Coccidiosis is a disease that is caused by protozoan parasites of the genus *Eimeria*, developing within the intestine of most domestic and wild animals and birds. Species of *Eimeria* found in different bird species include *E. acervulina*, *E. brunetti*, *E. maxima*, *E. mitis*, *E. necatrix*, *E. praecox*, and *E. tenella* (recognized as infecting chickens); *Eimeria adenoeides*, *E. dispersa*, *E. gallopavonis*, *E. meleagrimitis*, *E. innocua*, *E. meleagridis*, and *E. subrotunda (in turkeys)*; *Eimeria phasiani*, *E. colchici*, *E. duodenalis*, *E. tetartooimia*, and *E. pacifica (in pheasants)*; *E. kofoidi* and possibly *E. legionensis (in chukar partridge)*; *E. lettyae (in bobwhite quail)*; and *Eimeria truncata (in geese)* (McDougald 2012).

9.3.2.1 Pathogenesis

Pathogenicity is influenced by host genetics, nutritional factors, concurrent diseases, and species of the coccidium. *Eimeria necatrix* and *E. tenella* are the most pathogenic in chickens because schizogony occurs in the lamina propria and crypts of Lieberkühn of the small intestine and ceca, respectively, and causes extensive hemorrhage. Most *Eimeria* species develop in epithelial cells lining the villi. Protective immunity usually develops in response to moderate and continuing infection. True age immunity does not occur, but older birds are usually more resistant than young birds because of earlier exposure to infection (McDougald 2012).

9.3.2.2 Diagnosis

Effective diagnosis for *Eimeria* parasites remains a challenge. A number of different methods have been employed to aid in the diagnosis and these include (1) oocysts per gram of feces and (2) lesion scoring based on macroscopic visible lesions caused by *Eimeria*. Other advanced methods include multilocus enzyme electrophoresis, Southern blot analysis, pulsed-field gel electrophoresis, and PCR techniques (Morris and Gasser 2006).

9.3.2.3 Prevention and Control of Coccidiosis

There are basically two means of preventing coccidiosis in poultry: chemoprophylaxis and vaccination (Chapman 2005).

9.3.3 Isosporiasis

Isosporiasis is a disease of passerine birds. Canaries, finches, sparrows, and species of the Sturnidae family (starlings, mynahs) are most often affected. Very rare cases of infection have also been reported in raptors. Poultry are not known to be affected. Infection is most often not pathogenic, and high parasitemia may be seen in young birds. Mortality can be high (as much as 80 %) and rapid in susceptible bird species or weakened birds, especially in fledglings.

9.3.3.1 Diagnosis

Diagnosis is difficult in chronically infected older birds. Very few parasites are present in blood and tissues, and oocysts tend to be shed only intermittently, although sometimes in high numbers. A polymerase chain reaction (PCR) test is available and can be performed on blood, tissues, or feces, although its sensitivity is poor in fecal samples.

9.3.3.2 Treatment, Prevention, and Control

Toltrazuril and sulfachloropyrazine have successfully reduced mortality and oocyst shedding. It is unlikely that these drugs clear the organism from the bird. Good management procedures, including isolation of age groups and scrupulous cleanliness (particularly daily cleaning before oocysts sporulate) help control the disease. Disinfectants have little effect on oocysts.

9.4 Viral Infections

There are some viral infections that may potentially be considered as opportunistic in nature as listed below.

9.4.1 Low Pathogenic Avian Influenza Virus

Low pathogenic avian influenza viruses (LPAIV) usually cause mild to moderate infections in various domestic and wild bird species (Alexander 2000). Bano et al. (2003) indicated that H9N2 subtype of AIV as a nonpathogenic virus can cause a severe infection under field conditions in the presence of opportunistic secondary pathogens. A coinfection of infectious bursal disease (IBD) with low pathogenic avian influenza subtype H9N2 resulted in disease due to the immunosuppression effects of IBD. Nili and Asasi (2003) suggested that concurrent infections with infectious bronchitis and secondary bacterial infection such as *Ornithobacterium rhinotracheale*, *Escherichia coli*, and *Mycoplasma gallisepticum* may be important enhancers of the signs indicating H9N2 infection in chickens. Previous infection of infectious bursal disease virus in chickens subsequently may render the birds more susceptible to avian influenza virus (AIV) infection, allowing for the potential introduction of AIVs in an otherwise resistant population.

9.4.2 Avian Circovirus Infection

Little is known about the pathogenesis of circovirus infection in animals. However, these viruses share many epizootiological and pathological similarities with other known opportunistic pathogens (i.e., young age of affected animals, particular tropism for lymphoid tissue and organs, related acquired immunosuppression and secondary infections) (Woods and Shivaprasad 1997). Avian circoviruses identified to date include chicken anemia virus (CAV), psittacine beak and feather disease (PBFD) virus, canary circovirus, raven circovirus, pigeon circovirus (PiCV), duck circovirus, ostrich circovirus, and goose circovirus (Todd et al. 1991). Infection with psittacine herpesvirus 3 and an *Aspergillus* spp. was identified in two eclectus parrots coinfected with the beak and feather disease virus.

9.4.3 Avian Herpesviruses

All avian herpesviruses that have been genetically characterized are members of the *Alphaherpesviruses*. It is likely that avian herpesviruses have coevolved with one or more species of birds and they generally do not cause disease. Infection in host-adapted species is likely to be subclinical and lifelong. Reactivation of the virus resulting into disease rarely occurs and when it does so, results from immunosuppression of the host and may be in association with a concurrent infection. Some of the diseases caused by avian herpesvirus include, duck virus enteritis (duck plague), columbid/falcon/owl herpesvirus (CoHV-1), Pacheco's disease (mucosal papillomas), psittacid herpesvirus-2 (PsHV-2), passerid herpesvirus (PaHV-1), psittacid herpesvirus-3 (PsHV-3), and vulturine herpesvirus (Phalen et al. 2011).

9.4.4 Avian Reovirus Infections

Avian reoviruses are ubiquitous among poultry flocks and although pathogenic strains do exist, 85–90 % of reoviruses isolated from poultry are nonpathogenic. Reoviruses are involved in a variety of disease conditions in domestic poultry of which the most important is viral arthritis/tenosynovitis in chickens, where the cause-and-effect relationship is well established (Rosenberger and Olson 1997). Other reoviral disease conditions include cloacal pasting and associated mortality, ulcerative enteritis, enteric disease, respiratory disease, inclusion body hepatitis, and the runting/malabsorption syndrome. Reovirus infections are mainly associated with young age, concurrent infectious agents, for example, *Mycoplasma synoviae* and *Staphylococcus aureus*, infectious bursal disease virus, and chicken anemia virus (Jones 2000).

References

Abbassi H, Wyers M, Cabaret J et al (2000) Rapid detection and quantification of *Cryptosporidium baileyi* oocysts in feces and organs of chickens using a microscopic slide flotation method. Parasitol Res 86:179–187

Abdul-Aziz TA (1994) Pathogenicity of *Enterococcus hirae* for chicken embryos and betamethasone-treated chicks. Res Vet Sci 56:397–398

Ahmed AM, Motoi Y, Sato M (2007) Zoo animals as reservoirs of gram-negative bacteria harboring integrons and antimicrobial resistance genes. Appl Environ Microbiol 73 (20):6686–6690

Alexander DJ (2000) A review of avian influenza in different bird species. Vet Microbiol 74:3–13

Allchurch A (2002) Zoological parks in endangered species recovery and conservation. In: Aguirre A, Ostfeld R, Tabor G, House C, Pearl M (eds) Conservation medicine, ecological health in practice. Oxford University Press, New York, p 278

Aminov RI, Garrigues-Jeanjean N, Mackie RI (2001) Molecular ecology of tetracycline resistance: development and validation of primers for detection of tetracycline resistance genes encoding ribosomal protection proteins. Appl Environ Microbiol 67:22–32

Andreasen CB (2013) Overview of Staphylococcosis in poultry. In: The Merck Veterinary manual. Available via http://www.merckmanuals.com/vet/poultry/staphylococcosis/overview_of_staphylococcosis_in_poultry.html. Accessed 16 Aug 2014

Baele M, Devriese LA, Butaye P et al (2002) Composition of enterococcal and streptococcal flora from pigeon intestines. J Appl Microbiol 92:348–351

Bailey R, Scott E (1970) Diagnostic microbiology, 3rd edn. Mosby, Saint Louis, MO

Bale JOO, Nwagu BI, Abubakar BY et al (2000) Semen bacterial flora of Rhode Island breeder cocks in Zaria, Kaduna State, Nigeria. Niger J Anim Prod 27:16–18

Bangert RL, Cho BR, Widders PR et al (1988) A survey of aerobic bacteria and fungi in the feces of healthy psittacine birds. Avian Dis 32:46–52

Bano S, Naeem K, Malik SA (2003) Evaluation of pathogenic potential of avian influenza virus serotype H9N2 in chickens. Avian Dis 47(3):817–822

Barbosa TM, Serra CR, La Ragione RM et al (2005) Screening for bacillus isolates in the broiler gastrointestinal tract. Appl Environ Microbiol 71:968–978

Barbour EK, Brinton MK, Caputa A et al (1991) Characteristics of *Actinomyces pyogenes* involved in lameness of male turkeys in north-central United States. Avian Dis 35:192–196

Barnes HJ, Vaillancourt JP, Gross WB (2003) Colibacillosis. In: Saif Y (ed) Diseases of poultry, 11th edn. Iowa State University Press, Ames, IA

Barragan K, Brieva C (2005) Tuberculosis mycobacteriosis en primates neotropicales en cautiverio: un enfoque desde la conservación. Rev Med Vet Zoot 52(2):128–143

Baruah KK, Sharma PK, Bora NN (2001) Fertility, hatchability and embryonic mortality in ducks. Indian Vet J 78:529–530

Batra GL, Balwant S, Grewal GS et al (1982) Aetiopathology of oophoritis and salpingitis in domestic fowl. Indian J Anim Sci 52:172–176

Bauck L (1994) Mycosis. In: Ritchie BW, Harrison GJ, Harrison LR (eds) Avian medicine: principles and application. Wingers, Lake Worth, FL, pp 997–1006

Baumann P (1968) Isolation of *Acinetobacte*r from soil and water. J Bacteriol 96(1):39–42

Bell CJ, Finlay DA, Clarke HJ et al (2002) Development of a sandwich ELISA and comparison with PCR for the detection of F11 and F165 fimbriated *Escherichia coli* isolates from septicemic disease in farm animals. Vet Microbiol 85:251–257

Bennett RA (1994) Neurology. In: Ritchie BW, Harrison GJ, Harrison LR (eds) Avian medicine: principles and application. Wingers, Lake Worth, FL, pp 723–747

Berkhoff GA (1975) Ulcerative enteritis-clostridial antigens. Am J Vet Res 36:583–585

Berkhoff HA (1985) *Clostridium colinum* sp. nov., nom. rev., the causative agent of ulcerative enteritis (quail disease) in quail, chickens, and pheasants. Int J Syst Bacteriol 35:155–159

Berkhoff GA, Kanitz CL (1976) Fluorescent antibody test in diagnosis of ulcerative enteritis. Avian Dis 20:525–533

Bildfell RJ, Eltzroth EK, Songer JG (2001) Enteritis as a cause of mortality in the western bluebird (*Sialia mexicana*). Avian Dis 45:760–763

Binek M, Borzemska W, Pisarski R et al (2000) Evaluation of the efficacy of feed providing on development of gastrointestinal microflora of newly hatched broiler chickens. Archiv fur Geflugelkunde 64:147–151

Blagburn BL, Lindsay DS, Hoerr FJ et al (1991) Pathobiology of cryptosporidiosis (*C. baileyi*) in broiler chickens. J Protozool 38:25S–27S

Bougiouklis P, Brellou G, Fragkiadaki E et al (2005) Outbreak of avian mycobacteriosis in a flock of two-year-old domestic pigeons (*Columba livia f. domestica*). Avian Dis 49:442–445

Brinton MK, Schellberg LC, Johnson JB, Frank RK et al (1993) Description of osteomyelitis lesions associated with *Actinomyces pyogenes* infection in the proximal tibia of adult male turkeys. Avian Dis 37:259–262

Brisse S, Verhoef J (2001) Phylogenetic diversity of *Klebsiella pneumoniae* and *Klebsiella oxytoca* clinical isolates revealed by randomly amplified polymorphic DNA, gyrA and parC genes sequencing and automated ribotyping. Int J Syst Evol Microbiol 51:915–924

Brooks G, Butel J, Ornston L (1996) Micobacterias. In: Brooks G, Butel J, Ornston L (eds) Microbiología Médica de Jawetz, Melnick y Adelberg, 15th edn. Editorial Manual Moderno, México, pp 323–334

Bruce J, Drysdale EM (1983) The bacterial flora of candling-reject and dead-in-shell turkey eggs. Br Poult Sci 24:391–395

Brynestad S, Granum PE (2002) *Clostridium perfringens* and food borne infections. Int J Food Microbiol 74:195–202

Butterworth A (1999) Infectious components of broiler lameness: a review. World Poult Sci J 56 (4):327–352

Cardona J, Bickford AA, Charlton BR et al (1993) *Enterococcus durans* infection in young chickens associated with bacteremia and encephalomalacia. Avian Dis 37:234–239

Carpenter JW, Mashima TY, Rupiper DJ (2001) Exotic animal formulary. WB Saunders, Philadelphia, PA

Carrasco L, Bautista MJ, Mulas JM (1993) Application of enzyme-immunochemistry for the diagnosis of aspergillosis, candidiasis, and zygomycosis in three lovebirds. Avian Dis 37:923–927

Carter GR, Chengappa MM (1991) Essentials of veterinary bacteriology and mycology, 4th edn. Lea & Febiger, Philadelphia, PA, pp 109–224

Carter G, Wise D (2004) Mycobacterium. In: Essentials of veterinary bacteriology and mycology, 6th edn. Iowa State Press, Ames, IA, pp 207–213

Casey AL, Lambert PA, Elliot TSJ (2007) Staphylococci. Int J Antimicrob Agents 29:23–32

Chang HC, Wei YF, Dijkshoorn L et al (2005) Species-level identification of isolates of the *Acinetobacter calcoaceticus-Acinetobacter baumannii* complex by sequence analysis of the 16S-23S rRNA gene spacer region. J Clin Microbiol 43(4):1632–1639

Chapman HD (2005) Perspectives for the control of coccidiosis in poultry by chemotherapy and vaccination. In: Proceedings of the 9th international coccidiosis conference, Foz do Iguassu

Chege S, Howlett J, Al Qassimi M et al (2013) Opportunistic infection of Aspergillus and bacteria in captive Cape vultures (*Gyps coprotheres*). Asian Pac J Trop Biomed 3(5):401–406

Clark SR, Barnes HJ, Bickford AA et al (1991) Relationship of osteomyelitis and associated soft-tissue lesions with green liver discoloration in tom turkeys. Avian Dis 35:139–146

Clubb SL, Homer BL, Pisani J et al (1994) Outbreaks of bordetellosis in psittacines and ostriches. In: Proceedings of the Association of Avian Veterinarians. American Association of Avian Veterinarians, Orlando, FL, pp 63–68

Converse K (2007) Avian tuberculosis. In: Thomas NJ, Hunter DB, Atkinson CT (eds) Infectious diseases of wild birds. Blackwell, Oxford, pp 289–302

Cooper JE, Harrison GJ (1999) Dermatology. In: Richie BW, Harrison GJ, Harrison LR (eds) Avian medicine: principles and application. HBD, Brentwood, TN, pp 997–1006

Corrales W, Vivo LM, Gutierrez E (1988) Cutaneous abscesses in a flock of caged layers. Report of an outbreak. Revista Avicultura 32:15–27

Cortes CR, Tellez Isaias G, Lopez Cuello C et al (2004) Bacterial isolation rate from fertile eggs, hatching eggs, and neonatal broilers with yolk sac infection. Rev Latinoam Microbiol 46:12–16

Curtis-Velasco M (2000) Candidiasis and cryptococcosis in birds. Sem Avian Exotic Pet Med 9:75–81
Dahlhausen RD (2006) Implications of mycoses in clinical disorders. In: Harrison GJ, Lightfoot T (eds) Clinical avian medicine. Spix, Palm Beach, FL, pp 700–701
Deem SL (1999) Infectious and parasitic diseases of raptors. Compend Contin Vet Educ Pract Vet 21:329–338
Deeming DC (1996) Microbial spoilage of ostrich (*Struthio camelus*) eggs. Br Poult Sci 37:689–693
Deeming DC (2005) Yolk sac, body dimensions and hatchling quality of ducklings, chicks and poults. Br Poult Sci 46:560–564
Dessouky MI, Moursy A, Niazi ZM et al (1982) Experimental *Klebsiella* infection in baby chicks. Archiv Geflugelkunde 46:145–150
Devriese LA, Crauwerts K, Hermans K et al (2002) *Enterococcus cecorum* septicemia as a cause of bone and joint lesions resulting in lameness in broiler chickens. Vlaams Diergeneeskundig Tijdschrift 71:219–221
Dho-Moulin M, Fairbrother JM (1999) Avian pathogenic *Escherichia coli* (APEC). Vet Res 30:299–316
Dupont C, Carrier M, Higgins R (1994) Bacterial and fungal flora in healthy eyes of birds of prey. Can Vet J 34:699–701
Durairaj V, Clark FD (2007) *Escherichia coli* an opportunist that causes enteritis. Avian Adv 9(3)
Durant AJ, Doll ER (1941) Ulcerative enteritis in quail. Mo Agric Exp Stn Res Bull 325:3–27
Eichenbaum JD, Lavach JD, Severin GA et al (1987) Immunology of the ocular surface. Compend Contin Educ Pract Vet 9:1101–1108
Ellis DH, Pfeiffer TJ (1990) Natural habitat of *Cryptococcus neoformans* var. *gattii*. J Clin Microbiol 228:1642–1644
Evans KL, Crowder J, Miller ES (2000) Subtilisins of *Bacillus* spp. hydrolyze keratin and allow growth on feathers. Can J Microbiol 46:1004–1011
Fayer R (2004) Cryptosporidium. A water-borne zoonotic parasite. Vet Parasitol 126:37–56
Fiennes RNTW (1982) Diseases of bacterial origin. In: Petrak ML (ed) Diseases of cage and aviary birds, 2nd edn. Lea & Febiger, Philadelphia, PA, pp 497–515
Forbes NA (1991) Aspergillosis in raptors. Vet Rec 128(11):263
Fudge AM, Reavill DR (1993) Bird litter and aspergillosis warning. J Avian Med Surg 7:15
Fujino T, Matsui T, Kobayashi F et al (2002) The effect of heating against Cryptosporidium oocysts. J Vet Med Sci 64:199–200
Gerding PA (1990) Microbiology of the canine and feline eye. Vet Clin North Am Small Anim Pract 20:615–625
Gerhold R, Fischer J (2005) Avian mycobacteriosis in a wild turkey. Avian Dis 49:164–166
Gerlach H (2001) Megabacteriosis. Sem Avian Exot Pet Med 10:12–19
Glisson JR, Brown TP, Brugh M et al (1984) Sinusitis in turkeys associated with respiratory cryptosporidiosis. Avian Dis 28:783–790
Godoy SN, Matushima ER, Chaves JQ et al (2012) *Bacillus cereus* infection outbreak in captive psittacines. Vet Microbiol 161(1–2):213–217
Gomis SM, Watts T, Riddell C et al (1997) Experimental reproduction of *Escherichia coli* cellulitis and septicemia in broiler chickens. Avian Dis 41:234–240
Gomis S, Amoako AK, Ngeleka AM et al (2002) Histopathologic and bacteriologic evaluations of cellulitis detected in legs and caudal abdominal regions of turkeys. Avian Dis 46:192–197
Goodwin MA, Brown J, Resurreccion RS et al (1996) Respiratory coccidiosis (*Cryptosporidium baileyi*) among northern Georgia broilers in one company. Avian Dis 40:572–575
Goswami S, Chaudhury B, Mukit A (1988) Reproductive disorders of domestic hen. Indian Vet J 65:747–749
Gregory RD, Noble D, Field R et al (2003) Using birds as indicators of biodiversity. Ornis Hungarica 13:11–24

Gross WB, Domermuth CH (1975) Colibacillosis. In: Hitchner SB, Domermuth CH, Purchase HG, Williams JE (eds) Isolation and identification of avian pathogens. Arnold, New York, pp 34–37

Gyles CL (2008) Antimicrobial resistance in selected bacteria from poultry. Anim Health Res Rev 9:149–158

Hejlicek K, Treml F (1995) Comparison of the pathogenesis and epizootiologic importance of avian mycobacteriosis in various types of domestic and free-living syntropic birds. Vet Med 40:187–194

Hidasi HW, Neto JH, Moraes DM et al (2013) Enterobacterial detection and *Escherichia Coli* antimicrobial resistance in parrots seized from the illegal wildlife trade. J Zoo Wildl Med 44 (1):1–7

Hill JE, Rowland GN, Glisson JR et al (1989) Comparative microscopic lesions in Reoviral and Staphylococcal tenosynovitis. Avian Dis 33:401–410

Hill FI, Woodgyer AJ, Lintott MA (1995) Cryptococcosis in a North Island brown kiwi (*Apteryx australis mantelli*) in New Zealand. J Med Vet Mycol 33:305–309

Hinz KH, Glunder G (1985) Zum Vorkommen von *Bordetella avium* sp. nov. und *Bordetella bronchiseptica* bei Voegeln. Berl Munch Tierarztl Wochenschr 98:369–373

Hinz KH, Ryll M, Heffels Redmann U, Poppel M (1992) Multicausal infectious respiratory disease of turkey poults. Dtsch Tierarztl Wochenschr 99:75–78

Hoenerhoff M, Kiupel M, Sikarskie J et al (2004) *Mycobacteriosis* in an American Bald Eagle (*Haliaeetus leucocephalus*). Avian Dis 48:437–441

Hoffmann G, Gajdos G, Czako M et al (1998) Diversity among clinical isolates of *Proteus penneri* detected by random amplified polymorphic DNA analysis. Zentralbl Bakteriol 288:351–360

Hopkins BA, Skeeles JK, Houghten GE et al (1990) A survey of infectious diseases in wild turkeys (*Meleagridis gallopavo silvestris*) from Arkansas. JWD 26:468–472

Hoppes S, Gurfield N, Flammer K et al (2000) Mycotic keratitis in a blue-fronted Amazon parrot (*Amazona aestiva*). J Avian Med Surg 14:185–189

Hornok S, Szell Z, Nieuwenhuijs J et al (1999) Immunogenicity of three oocyst extracts of *Cryptosporidium baileyi* in experimentally infected chickens. Parasitol Res 85:71–77

Howell LJ, Hunter R, Bagust TJ (1982) Necrotic dermatitis in chickens. N Z Vet J 30:87–88

Inderlied C, Kemper C, Bermudez L (1993) The *Mycobacterium avium* complex. Clin Microbiol Rev 6(3):266–310

Inzana TC, Irtani B (1989) Rapid detection of group C streptococci from animals by latex agglutination. J Clin Microbiol 27:309–312

Jackwood MW (2013) Overview of bordetellosis in poultry. In: Merck veterinary manual, available via http://www.merckmanuals.com/vet/poultry/bordetellosis/overview_of_bordetellosis_in_poultry.html. Accessed 17 Aug 2014

Jackwood MW, Mccarter SM, Brown TP (1995) *Bordetella avium*: an opportunistic pathogen in leghorn chickens. Avian Dis 39:360–367

Jain R, Danziger LH (2004) Multi-drug resistant *Acinetobacter* infections: an emerging challenge to clinicians. Ann Pharmacother 38:1449–1459

John HL (2006) Occurrence of methicillin-resistant *Staphylococcus aureus* strains from cattle and chickens, and analyses of their mec A mec R1 and mec 1 genes. Vet Microbiol 114:155–159

Jones RC (2000) Avian reovirus infections. Rev Sci Tech Off Int 19(2):614–625

Jones MP (2006) Selected infectious diseases of birds of prey. J Exot Pet Med 15:5–17

Jordan FTW (1990) Fungal diseases. In: Jordan FTW (ed) Poultry diseases, 3rd edn. Bailliere Tindall, Philadelphia, PA, pp 216–225

Kabede F (2010) Pseudomonas infection in chickens. J Vet Med Anim Health 2(4):55–58

Kabir SML (2010) Avian Colibacillosis and Salmonellosis: a closer look at epidemiology, pathogenesis, diagnosis, control and public health concerns. Int J Environ Res Public Health 7:89–114

Kersters K, Hinz KH, Hertle A et al (1984) *Bordetella avium* sp. nov. isolated from the respiratory tracts of turkeys and other birds. Int J Syst Bacteriol 34:56–70

Khosravi AR, Shokri H, Ziglari T et al (2008) Outbreak of severe disseminated aspergillosis in a flock of ostrich (*Struthio camelus*). Mycoses 51:557–559

Klopfleisch R, Muller C, Polster U et al (2005) Granulomatous inflammation of salt glands in ducklings (*Anas platyrhynchos*) associated with intralesional Gram-negative bacteria. Avian Pathol 34:233–237

La Ragione RM, Woodward MJ (2002) Virulence factors of *Escherichia coli* serotypes associated with avian colisepticemia. Res Vet Sci 73:27–35

La Ragione RM, Woodward MJ (2003) Competitive exclusion by *Bacillus subtilis* spores of *Salmonella enterica* serotype Enteritidis and *Clostridium perfringens* in young chickens. Vet Microbiol 94(3):245–256

Lakhotia RL, Stephens JF (1973) Drug resistance and R factors among enterobacteria isolated from eggs. Poult Sci 52:1955–1962

Landman WJM, Gruys E, Dwars RM (1994) A syndrome associated with growth depression and amyloid arthropathy in layers: a preliminary report. Avian Pathol 23:461–470

Landman DC, Georgescu C, Martin D et al (2002a) Polymyxins revisited. Clin Microbiol Rev 21:449–465

Landman D, Quale JM, Mayonga D et al (2002b) City ward clonal outbreak of multi-resistant *Acinetobacter baumannii* and *Pseudomonas aeruginosa* in Brooklyn, NY: the pre-antibiotic era has returned. Arch Int Med 162:1515–1520

Leite C, Souza C, Leite S (1998) Identification of mycobacteria by thin layer chromatographic analysis of mycolic acids and conventional biochemical method: four years of experience. Mem Inst Oswaldo Cruz 93(6):801–805

Leitner G, Melamed D, Drabkin N et al (1990) An enzyme-linked immunosorbent assay for detection of antibodies against *Escherichia coli*: association between indirect hemagglutination test and survival. Avian Dis 34:58–62

Leng X, Mosier DA, Oberst RD (1996) Differentiation of *Cryptosporidium parvum*, *C. muris* and *C. baileyi* by PCR-RFLP analysis of the 18S rRNA gene. Vet Parasitol 62:1–7

Lindsay DS, Blagburn BL (1990) Cryptosporidiosis in birds. In: Dubey JP, Speer CA, Fayer R (eds) Cryptosporidiosis in man and animals. CRC, Boca Raton, FL, pp 133–148

Lumeij JT, Gorgevska D, Woestenborghs R (1995) Plasma and tissue concentrations of itraconazole in racing pigeons (*Columba livia domestica*). J Avian Med Surg 9:32–35

McDevitt RM, Brooker JD, Acamovic T et al (2006) Necrotic enteritis; a continuing challenge for the poultry industry. World Poult Sci J 62:221–247

McDougald LR (2012) Overview of coccidiosis in poultry. In: Merck's veterinary manual, available via http://www.merckmanuals.com/vet/poultry/coccidiosis/overview_of_coccidiosis_in_poultry.html. Accessed 15 May 2014

McMillian M, Petrak M (1989) Retrospective study of aspergillosis in pet birds. J Avian Med Surg 3:211–215

McMullin P (1998) Wet litter in turkeys: diseases and interactions. Presentation at ADAS.NFU Turkey conference, http://www.poultryhealth.com/library/turkeys/adastu98.htm. Visited 8/10/07

McNamee PT, Smyth JA (2000) Bacterial chondronecrosis with osteomyelitis (femoral head necrosis) of broiler chickens: a review. Avian Pathol 29:253–270

Miller E, Fowler M (2012) Zoo and wild animal medicine current therapy, vol 7. W.B. Elsevier Saunders, St Louis, MO, pp 266–274

Mitchell G, Esnouf D, Pritchard R (1986) Mucormycosis in canaries (*Serinus canaries*) fed damp germinated seed. Vet Pathol 23:625–627

Mohamed HAAE (2004) *Proteus* spp. infection in quails in Assiut governorate. Assiut Vet Med J 50:196–204

Molina-López RA, Ramis A et al (2010) *Cryptosporidium baileyi* infection associated with an outbreak of ocular and respiratory disease in otus owls (*Otus scops*) in a rehabilitation centre. Avian Pathol 39(3):171–176

Morgan UM, Monis PT, Xiao L et al (2001) Molecular and phylogenetic characterization of Cryptosporidium from birds. Int J Parasitol 31:289–296

Morris GM, Gasser RB (2006) Biotechnological advances in the diagnosis of avian coccidiosis and the analysis of genetic variation in Eimeria. Biotechnol Adv 24:590–603

Morse GB (1907) Quail disease in the United States. United States Department of Agriculture, BAI Circ 109

Muller MG, George AR, Walochnik J (2010) *Acinetobacter baumannii* in localized cutaneous mycobacteriosis in Falcons. Vet Med Int 2010:321797

Murakami S, Miyama M, Ogawa A et al (2002) Occurrence of conjunctivitis, sinusitis and upper region tracheitis in Japanese quail (*Coturnix coturnix japonica*), possibly caused by *Mycoplasma gallisepticum* accompanied by *Cryptosporidium* sp. infection. Avian Pathol 31:363–370

Murray BE (1990) The life and times of the *Enterococcus*. Clin Microbiol Rev 3:46–65

Nemati M, Hermans K, Lipinska U et al (2008) Antimicrobial resistance of old and recent *Staphylococcus aureus* isolates from poultry: first detection of Livestock-Associated Methicillin-Resistant strain ST398. Antimicrob Agents Chemother 52(10):3817

Ng J, Pavlásek I, Ryan U (2006) Identification of novel *Cryptosporidium* genotypes from avian hosts. Appl Environ Microbiol 72(12):7548–7553

Nili H, Asasi K (2003) Avian influenza (H9N2) outbreak in Iran. Avian Dis 47:828–831

Nolan LK (2013) Overview of Colibacillosis in poultry. In: Merck veterinary manual, available via http://www.merckmanuals.com/vet/poultry/colibacillosis/overview_of_colibacillosis_in_poultry.html. Accessed 18 Aug 2014

Nortermans SJ, Dufrenni WJ, Leeuwen V (1982) Contamination of broiler chickens by *Staphylococcus aureus during* processing: incidence and origin. J Appl Bacteriol 55:275–28080

Nosanchuck JD, Shoham S, Bettina CF et al (2000) Evidence of zoonotic transmission of *Cryptococcus neoformans* from a pet cockatoo to an immunocompromised patient. Ann Intern Med 132:205–208

Oglesbee BL (1997) Mycotic diseases. In: Altman RB (ed) Avian medicine and surgery, 1st edn. W.B. Saunders, Philadelphia, PA, pp 323–361

Olkowski AA, Wojnarowicz C, Chinio-Trejo M et al (2006) Responses of broiler chickens orally challenged with *Clostridium perfringens* isolated from field cases of necrotic enteritis. Res Vet Sci 81:99–108

Olsen RH, Frantzen C, Christensen H et al (2012) An investigation on first-week mortality in layers. Avian Dis 56(1):51–57

Orcutt CJ, Bartick TE (1994) Mucormycotic meningoencephalitis and pneumonia in a chattering lory (Lorius Garrulus). J Avian Med Surg 8:85–89

Otalora RE (2013) Overview of ulcerative enteritis in poultry. In: Mercks veterinary manual, available via http://www.merckmanuals.com/vet/poultry/ulcerative_enteritis/overview_of_ulcerative_enteritis_in_poultry.html. Accessed 18 Aug 2014

Otalu OJ, Kabir J, Okolocha EC et al (2011) Multi-drug resistant coagulase positive *Staphylococcus aureus* from live and slaughtered chickens in Zaria, Nigeria. Int J Poult Sci 10(11):871–875

Panigrahy B, Naqi SA, Grumbles LC et al (1979) Candidiasis in cockatiel nestlings and mucormycosis in a pigeon. Avian Dis 23:757–760

Pavlásek I (1999) Cryptosporidia: biology, diagnosis, host spectrum, specificity, and the environment. Rem Klin Mikrobiol 3:290–301

Peleg AY, Jara S, Monga D et al (2009) *Galleria mellonella* as a model system to study *Acinetobacter baumannii* pathogenesis and therapeutics. Antimicrob Agents Chemother 53 (6):2605–2609

Penrith ML, Bezuidenhout AJ, Burger WP et al (1994) Evidence for cryptosporidial infection as a cause of prolapse of the phallus and cloaca in ostrich chicks (*Struthio camelus*). Onderstepoort J Vet Res 61:283–289

Pesavento G, Ducci B, Comodo N et al (2007) Antimicrobial resistance profile of *Staphylococcus aureus* isolated from raw meat: a research for methicillin resistant *Staphylococcus aureus* (MRSA). Food Control 18:196–200

Phalen DN (2000) Respiratory medicine of cage and aviary birds. Vet Clin North Am Exot Anim Pract 3:423–452

Phalen D (2005) Diagnosis and management of *Macrorhabdus ornithogaster* (formerly megabacteria). Vet Clin Exot Anim Pract 8:299–306

Phalen DN, Holz P, Rasmussen L et al (2011) Fatal Columbid herpesvirus-1 infection in three species of Australian birds of prey. Aust Vet J 89(5):193–196

Prashanth K, Badrinath S (2005) Epidemiological investigation of nosocomial *Acinetobacter* infections using arbitrarily primed PCR and pulse field gel electrophoresis. Indian J Med Res 122:408–418

Proctor SJ, Kemp RL (1974) *Cryptosporidium anserinum* sp. n. (Sporozoa) in a domestic goose *Anser anser* L., from Iowa. J Protozool 21:664–666

Raffel TR, Register KB, Marks SA et al (2002) Prevalence of *Bordetella avium* infection in selected wild and domesticated birds in the Eastern USA. JWD 38(1):40–46

Raidal SR, Butler R (2001) Chronic rhinosinusitis and rhamphotheca destruction in a Major Mitchell's cockatoo (*Cacatua leadbeateri*) due to *Cryptococcus neoformans* var. *gattii*. J Avian Med Surg 15:121–122

Randall CJ, Siller WG, Wallis AS et al (1984) Multiple infections in young broilers. Vet Rec 114:270–271

Randall CJ, Wood AM, MacKenzie G (1993) Encephalomalacia in first-week chicks. Vet Rec 132:419

Raso TF, Seixas GHF, Guedes NMR et al (2006) *Chlamydophila psittaci* in free-living blue-fronted Amazon parrots (Amazona aestiva) and hyacinth macaws (Anodorhynchus hyacinthinus) in the Pantanal of Mato Grosso do Sul, Brazil. Vet Microbiol 117:235–241

Redig PT (1993) Mycotic infections of birds of prey. In: Fowler ME (ed) Zoo and wild animal medicine, current therapy 3. WB Saunders, Philadelphia, PA, pp 178–181

Redig PT (2000) Aspergillosis. In: Samour J (ed) Avian medicine. Mosby, Philadelphia, PA, pp 275–287

Reisfeld L, Barbirato M, Ippolito L et al (2013) Reducing bumblefoot lesions in a group of captive Magellanic penguins (*Spheniscus magellanicus*) with the use of environmental enrichment. Pesq Vet Bras 33(6):791–795

Richards SM, Hunt BW (1973) Ulcerative enteritis in partridges. Vet Rec 111(25–26):591–592

Riddell C, Kong XM (1992) The influence of diet on necrotic enteritis in broiler chickens. Avian Dis 36:499–503

Robino P, Bert E, Tramuta C et al (2005) Isolation of *Acinetobacter lwoffii* from a lovebird (*Agapornis roseicollis*) with severe respiratory symptoms. Schweitz Arch Tierheilkd 147 (6):267–269

Rosenberger JK, Olson NO (1997) Viral arthritis. In: Calnek BW, Barnes HJ, Beard CW, McDougald LR, Saif YM (eds) Diseases of poultry, 10th edn. Mosby-Wolfe, London, pp 711–720

Rupiper DJ (1998) Diseases that affect race performance of homing pigeons. Part II: bacterial, fungal, and parasitic diseases. J Avian Med Surg 12:138–148

Ryan UM, Xiao L, Read C et al (2003) A redescription of *Cryptosporidium galli* Pavlasek, 1999 (Apicomplexa: Cryptosporidiidae) from birds. J Parasitol 89:809–813

Saeed AK, Mohammed SN, Ashraf AK et al (2000) Transfer of erythromycin resistance from poultry to human clinical strains of *Staphylococcus aureus*. J Clin Microbiol 38:1832–1838

Saidenberg ABS (2008) Deteccao de fatores de virulencia de Escherichia coli isoladas de psitacídeos com diferentes manifestacoes clínicas. M.S. Dissertation, Univ. de São Paulo, São Paulo, Brazil

Salmon SA, Watts JL (2000) Minimum inhibitory concentration determinations for various antimicrobial agents against 1570 bacterial isolates from turkey poults. Avian Dis 44(85):98

Samadpoour M, Stewart J, Steingart K et al (2002) Laboratory investigation of an *E. coli* O157:H7 outbreak associated with swimming in Battle Ground Lake, Vancouver, Washington. J Environ Health 64:16–20

Samour JH (2000) *Pseudomonas aeruginosa* stomatitis as a sequel to trichomoniasis in captive saker falcons (*Falco cherrug*). J Avian Med Surg 142:113–117

Samour JH, Naldo JL (2002) Diagnosis and therapeutic management of candidiasis in falcons in Saudi Arabia. J Avian Med Surg 16:129–132

Sarker MAH, Jahan M, Parvin MN et al (2012) Identification of bacterial flora isolated from apparently healthy water birds of Dhaka Zoo of Bangladesh. Bangladesh J Vet Med 10 (1&2):21–26

Schreckenberger PC, Daneshvar MI, Weyant RS et al (2003) *Acinetobacter*, *Achromobacter*, *Chryseobacterium*, *Moraxella*, and other nonfermentive gram-negative rods. In: Murray PR (ed) Manual of clinical microbiology, 8th edn. ASM, Washington, DC, pp 749–779

Sekaninova G, Rychlik I, Kolarova M et al (1998) A new bacteriophage typing scheme for *Proteus mirabilis* and *Proteus vulgaris* strains. 3: Analysis of lytic properties. Folia Microbiol 43:136–140

Senior VE, Lake R, OPratt C (1962) Suspected actinomycosis of turkeys. Can Vet J 3:120–125

Simor AE, Lee M, Vearncombe M, Jones-Paul L et al (2002) An outbreak due to multi-drug resistant *Acinetobacter baumannii* in a burn unit: risk factors for acquisition and management. Infect Control Hosp Epidemiol 23:261–267

Skeeles JK (1997) Staphylococcosis. In: Calnek BW (ed) Diseases of poultry, 10th edn. Iowa State University Press, Ames, IA, pp 247–253

Skeeles JK, Arp LH (1997) Bordetellosis (turkey coryza). In: Calnek BW, Barnes HJ, Beard CW, Reid WN, Yoder HW Jr (eds) Diseases of poultry. Iowa State University Press, Ames, IA, pp 277–288

Slavin D (1955) *Cryptosporidium meleagridis* (sp. nov.). J Comp Pathol 65:262–266

Smith MG, Gianoulis TA, Pukatzki S (2007) New insights into *Acinetobacter baumannii* pathogenesis revealed by high-density pyrosequencing and transposon *mutagenesis*. Genes Dev 21 (5):601–614

Soler D, Brieva C, Ribon W (2009) Mycobacteriosis in wild birds: the potential risk of disseminating a little-known infectious disease. Rev Salud Publica 11(1):134–144

Songer JG, Post KW (2005) Gram-positive aerobic Cocci. In: Glenn Songer J, Post KW (eds) Veterinary microbiology. Bacterial and fungal agents of animal disease. Illustrated edition. Elsevier, St. Louis, MO, pp 35–42

Speed B, Dunt D (1995) Clinical and host differences between infections with the two varieties of *Cryptococcus neoformans*. Clin Infect Dis 21:28–34

Sréter T, Varga I (2000) Cryptosporidiosis in birds—a review. Vet Parasitol 87:261–279

Steentjes A, Veldman KT, Mevius DJ et al (2002) Molecular epidemiology of unilateral amyloid arthropathy in broiler breeders associated with *Enterococcus faecalis*. Avian Pathol 31:31–39

Tankson JD, Thaxton JP, Vizzier-Thaxton Y (2001) Pulmonary hypertension syndrome in broilers caused by *Enterococcus faecalis*. Infect Immun 69:6318–6322

Tankson JD, Thaxton JP, Vizzier-Thaxton Y (2002) Morphological changes in heart and lungs of broilers experiencing pulmonary hypertension syndrome caused by *Enterococcus faecalis*. Poult Sci 81:365–370

Tarello W (2011) Aetiologic agents and diseases found associated with clinical aspergillosis in Falcons. Int J Microbiol. doi:10.1155/2011/176963

Tell LA (2005) Aspergillosis in mammals and birds: impact on veterinary medicine. Med Mycol 43(Suppl 1):S7–S73

Tell L, Ferrell S, Gibbons P (2004) Avian mycobacteriosis in free-living raptors in California: 6 Cases (1997-2001). J Avian Med Surg 18(1):30–40

Thayer SG, Waltman DW, Wages DP (2012) *Streptococcus* and *Enterococcus*. In: Saif YM, Fadly AM, Glisson JR, McDougald LR, Nolan LK, Swayne DE (eds) Diseases of poultry, 12th edn. Blackwell, Ames, IA, pp 900–903

Timbermont L, Haesebrouck F, Ducatelle R et al (2011) Necrotic enteritis in broilers: an updated review on pathogenesis. Avian Pathol 40(4):341–347

Todd D, Niagro FD, Ritchie BW et al (1991) Comparison of three animal viruses with circular single-stranded DNA genomes. Arch Virol 117:129–135

Tomaszewski EK, Logan KS, Snowden KF et al (2003) Phylogenetic analysis identifies the 'megabacterium' of birds as a novel anamorphic ascomycetous yeast, Macrorhabdus ornithogaster gen. nov., sp. nov. Int J Syst Evol Microbiol 53:1201–1205

Turtura GC, Massa S, Chazvinizadeh H (1990) Antibiotic resistance among coliform bacteria isolated from carcasses of commercially slaughtered chickens. Int J Food Microbiol 11:351–354

Tyzzer EE (1929) Coccidiosis in gallinaceous birds. Am J Hyg 10:269–383

Vaillancourt JP, Elfadil A, Bisaillon JR (1992) Cellulitis in the broiler fowl. Med Vet Que 22:168–172

Vallenet D, Nordmann P, Barbe V et al (2008) Comparative analysis of Acinetobacters: three genomes for three lifestyles. PLoS One 3:1805e

Van den Bogaard E, London N, Driessen C et al (2001) Antibiotic resistance of faecal *Escherichia coli* in poultry, poultry farmers and poultry slaughterers. J Antimicrob Chemother 47:763–771

Van DerHeyden N (1997) Clinical manifestations of mycobacteriosis in pet birds. Semin Avian Exot Pet Med 6(1):18–24

Varga I, Sreter T, Bekesi L (1995) Potentiation of ionophorous anticoccidials with duokvin—battery trials against *Cryptosporidium baileyi* in chickens. J Parasitol 81:777–780

Vaughn S (1996) What is your diagnosis? J Avian Med Surg 10:33–35

Venkanagouda G, Krishnappa J, Upadhye AS (1996) Bacterial etiology of early chick mortality. Indian Vet J 73:253–256

Villers D, Espaze E, Coste-Burel M (1998) Nosocomial *Acinetobacter baumannii* infections: microbiological and clinical epidemiology. Ann Intern Med 129(3):182–189

Wang R, Qi M, Jingjing Z et al (2011) Prevalence of *Cryptosporidium baileyi* in ostriches (*Struthio camelus*) in Zhengzhou, China. Vet Parasitol 175(1–2):151–154

Wieliczko A, Kuczkowski M (2000) Selected issues of infectious diseases in ostrich (*Struthio camelus*). Med Weter 56:23–28

Williams RB (2005) Intercurrent coccidiosis and necrotic enteritis of chickens: rational, integrated disease management by maintenance of gut integrity. Avian Pathol 34:159–180

Witte CL, Hungerford LL, Papendick R et al (2007) Factors associated with mycobacteriosis in captive birds. In: Proceedings of the American Association of Zoo Veterinarians, Yulee, FL. American Association of Zoo Veterinarians, p 73

Woods LW, Shivaprasad HL (1997) Pigeon circovirus infection. In: Calnek BW (ed) Diseases of poultry, 10th edn. Iowa State University Press, Ames, IA, pp 1050–1053

Xiao L, Fayer R, Ryan U et al (2004) Cryptosporidium taxonomy: recent advances and implications for public health. Clin Microbiol Rev 17:72–97

Ye S, Xu S, Huang Y et al (1995) Investigation of *Proteus* infections in chickens. Chin J Vet Sci Technol 25:14–15

Zafra R, Perez J, Perez-Ecija RA et al (2008) Concurrent aspergillosis and ascites with high mortality in a farm of growing broiler chickens. Avian Dis 52:711–713

Chapter 10
Opportunistic Algae, Fungi and Ichthyosporea Associated with Mammalian Livestock Disease

Christon J. Hurst

Abstract This chapter presents a current compendium of knowledge regarding the opportunistically pathogenic algae, fungi and ichthyosporea associated with those livestock mammals used as sources of meat and milk for consumption by humans. All of these mammalian species are terrestrial and the majority are bovids. However, as important as the bovids are to us, they do indeed represent only part of a diverse collection which includes rodents as small as the Edible dormouse and cervids as large as the moose or Eurasian elk. The opportunistic microorganisms mentioned in this chapter represent a critical concern with respect to human food supplies. This chapter lists the pathogens by genus along with their higher taxonomy, provides information regarding the ecology of these genera, and summarizes their associated disease information with respect to livestock mammals. The genus listings include the individual pathogen species by name and also identify those host mammals which are known to be affected by each pathogen species.

10.1 Introduction

Humans are by definition omnivores, which means that we can and indeed generally do eat as food almost anything whose ingestion is not either physically or chemically lethal. Animals certainly are an important source of dietary protein for most humans and as livestock those animals are most easily and inexpensively fed if either they are herbivores or can otherwise forage for themselves. The biggest factors in our choices of food animals are the affordability and availability of the animals and, of course, our level of hunger. A major challenge for us when keeping the animals as livestock is that we maintain their health, which includes a need to understand the pathogens which affect these animals.

C.J. Hurst (✉)
1814 Woodpine Lane, Cincinnati, OH 45255, USA

Universidad del Valle, Cali, Colombia
e-mail: christonjhurst@fuse.net

© Springer International Publishing Switzerland 2016

C.J. Hurst (ed.), *The Rasputin Effect: When Commensals and Symbionts Become Parasitic*, Advances in Environmental Microbiology 3,
DOI 10.1007/978-3-319-28170-4_10

While this chapter focuses on those mammalian species kept as livestock, avians are the subject of Chap. 9 of this book, and fish are the subject of Chaps. 4 and 5.

10.2 Livestock and Their Use

Livestock, in its broader definition, can include any type of animal that either is domesticated, semi-domesticated, or captured from the wild and then maintained under agricultural supervision. That supervision is termed animal husbandry and it has made a big change in the cultural evolution of our species by allowing a longer term control of meat resources and enabling dairy industry. Maintaining livestock freed us from continually hunting wild animals in order to meet our food requirements and subsequently enabled the development of larger human communities. The list of animals considered to be livestock includes not only mammals, but also amphibians, most notably frogs; those numerous avian groups which are categorized as poultry; crustaceans, most notably decapods such as crab, crayfish, lobster, prawn and shrimp; fish including eels; insects, most notably bees, crickets and silk worms; gastropod molluscs including abalone and edible terrestrial snails; bivalve molluscs including clams, mussels and oysters; and among the reptiles we keep as livestock not only snakes and turtles, but also various members of the order Crocodilia including alligators, caimans, and crocodiles.

While the definition of livestock is not limited either to mammals or the other groups of animals that are intended as food for humans, this chapter must have a more limited focus and therefore considers only those mammals maintained for the goal of consumption by humans. Some of the mammals listed in this chapter as livestock serve us for more than a single function. In addition to eating them as a source of meat, many of the larger mammals are of course considered to produce sufficient milk and be adequately passive in terms of their temperament that they are used as dairy animals. Those animals utilized in dairy industries generally are represented by the bovids, but the category of dairy animals also includes such cervids as moose and reindeer, camelids including camel and to a far lesser extent llama, and among the equines we use both the horse and donkey as sources of milk. Body fat from many of these animals is considered a food in its own right and also perceived as an important cooking ingredient. The larger mammals often have been used as a source of labor and transportation, either as beasts of burden for animal portage or harnessed for use as drayage animals. Indeed the very largest often are ridden.

There are, of course, many additional mammalian species kept as livestock and primarily of note are three species perceived as fur animals with their meat generally used to feed other animals. These three fur animals are the American mink (*Neovison vison*) and the two remaining species of chinchilla, Short-tailed chinchilla (*Chinchilla chinchilla*) and Long-tailed chinchilla (*Chinchilla lanigera*). Still other terrestrial mammals such as rats are hunted and eaten by humans but they generally are not kept as livestock.

Component parts of livestock that neither are eaten by humans nor fed as meat to other animals are utilized for a variety purposes with a few of those examples mentioned here. Body fat and butter fat from mammals has been burned as a source of lighting. We make not only health and beauty products, but also household furnishings and decorations from animal materials. Body fat serves as an ingredient for making soap. Hydrolysed keratin is a common cosmetic ingredient. Animal body fat, butter fat, and waxes such as lanolin are used for skin emollients. The hides of larger animals often are used as a source of leather, collagen, and derivatives of collagen including gelatin. Bones and connective tissue similarly are used as a source of collagen for making gelatin. Bones also are used as an ingredient in ceramic furnishings. Mammals which have long hair that easily can be spun and woven often become a source of fiber for clothing and other domestic items made of cloth. Mammals are, of course, not the only livestock animals that we view as raw materials. As an example, the shells of crustaceans and molluscs can be used as ingredients in dietary supplements for other animals. Birds often produce desirable feathers used for decorative purposes including clothing. Chemical industries have developed commercial applications for a broad range of additional products manufactured from shells and feathers.

10.2.1 The Mammals Kept as Livestock for Human Consumption

The mammalian species addressed in this chapter are those used as sources of either meat, milk, or both, for human consumption. These mammals are listed in Table 10.1 and taxonomically described in this section. All of the listed mammals are used as sources of meat. Those mammals which also are used as sources of milk for human ingestion are members of the genera *Alces*, *Bos*, *Bubalus*, *Camelus*, *Capra*, *Cervus*, *Ovis*, and *Rangifer*.

Examining the list of mammals maintained as livestock intended for human consumption may reveal some interesting surprises. What we either now raise to eat or have raised in the past varies tremendously from one part of the world to another and the dietary choices include some cultural preferences whose origins often were distinctly local in nature. Our preferential decisions include aspects that are religious, perhaps most notably being the considerations regarding consumption of cattle and pig. There also is the truth that different societies often view as pets those same animals which other people would consider to be traditional foods, and this difference frequently has become a point of contention when cultural groups meet and intermix. Local taste preferences include choices of such mammals as cat, dog, dormouse, horse and guinea pig. It is unfortunately often considered easier to negatively criticise someone else's cultural practices when we should instead present acceptance and offer understanding of the differences between the many personal choices which are made by societal groups. We must not criticize what is

Table 10.1 Mammals kept as livestock for human consumption

Infraclass	Order	Family	Subfamily	Genus	Common name
Eutheria	Artiodactyla	Bovidae	Bovinae	*Bison*	American bison
				Bos	Cattle
					Gayal
					Yak
				Bubalus	Carabao
					Water buffalo
			Caprinae	*Capra*	Goat
				Ovis	Sheep
		Camelidae	Camelinae	*Camelus*	Camel
				Lama	Llama
				Vicugna	Alpaca
		Cervidae	Odocoileinae	*Alces*	Moose
				Rangifer	Reindeer
			Cervinae	*Cervus*	Elk
					Red deer
				Dama	Fallow deer
		Suidae	Suinae	*Sus*	Pig
	Carnivora	Canidae	Caninae	*Canis*	Dog
		Felidae	Felinae	*Felis*	Cat
	Lagomorpha	Leporidae	Leporinae	*Oryctolagus*	Rabbit
	Perissodactyla	Equidae	Equinae	*Equus*	Donkey (ass)
					Horse
	Rodentia	Caviidae	Caviinae	*Cavia*	Guinea pig
		Hydrochaeridae	Hydrochoerinae	*Hydrochoerus*	Capybara
		Gliridae	Glirinae	*Glis*	Dormouse
		Myocastoridae	Not assigned	*Myocastor*	Coypu (Nutria)
Metatheria	Diprotodontia	Macropodidae	Not assigned	*Macropus*	Kangaroo

All of these listed mammals are raised as sources of meat for human consumption. Some members of the following genera also are used as a sources of milk for human consumption: *Alces*, *Bos*, *Bubalus*, *Camelus*, *Capra*, *Cervus*, *Equus*, *Ovis*, and *Rangifer*. The Water buffalo also is called the Asian buffalo and sometimes those two names are combined as Asian water buffalo. The African buffalo or Cape buffalo does not seem to ever have been domesticated owing to its unpredictable nature and for which reason it is considered dangerous to humans. The terms yak, goat, sheep and pig respectively refer to the Domestic yak, Domestic goat, Domestic sheep, and Domestic pig. My usage of Camel refers to both the Bactrian camel and the Dromedary camel. The species indicated in this table as dormouse commonly is known as the Edible dormouse or Fat dormouse, it is live-trapped and then raised domestically. My usage of the term kangaroo refers as a group to three species, the Red kangaroo, Eastern grey kangaroo, and Western grey kangaroo. Of those three kangaroo species it is the Red kangaroo which seems to be the most suggested for domestication as livestock. Kangaroo indeed are bred in captivity and raised on farms but to date their production for human consumption generally is considered unprofitable because of cost competition against hunted animals. I have kept use of the order name Artiodactyla although Cetartiodactyla seems now to be on the road to replacing that order name eaten in other, and in particularly less wealthy cultures, simply because those who are fortunate enough to have wealthier lives can afford to have multiple options and be finicky eaters. We also should not allow our personal preferences to influence control over what human groups either can or cannot choose to eat.

10.2.2 Terminology for the Mammalian Livestock Groups Included in This Chapter

The following briefly summarizes some of the terminology that I have used in this chapter when referring to mammalian livestock.

The Carabao often is called a water buffalo, but currently these two names are not considered synonymous. For this chapter the common name Carabao corresponds to the species (*Bubalus carabanensis*). The common name buffalo is used in this chapter as a reference to the Water buffalo (*Bubalus bubalis*) which also is called the Asian buffalo, and sometimes those names are combined in common practice to form the name Asian water buffalo. The African buffalo or Cape buffalo (*Syncerus caffer*) seems not to have been domesticated owing to its unpredictable nature, for which reason it is considered dangerous to humans. The term bison, when used in this chapter, indicates the American bison (*Bison bison*) unless specifically identified as a referral of information pertaining to the European bison (*Bison bonasus*). The terms yak, goat, sheep and pig respectively refer to the Domestic yak (*Bos grunniens*), Domestic goat (*Capra hircus*), Domestic sheep (*Ovis aries*), and Domestic pig (*Sus scrofa*). My usage of camel refers to both the Bactrian camel (*Camelus bactrianus*) and the Dromedary camel (*Camelus dromedarius*).

Not all of the mammals listed in Table 10.1 are considered domesticated, a few such as the dormouse and kangaroo are captured wild and then raised in captivity. The species indicated in Table 10.1 as dormouse is that commonly known as the Edible dormouse or Fat dormouse (*Glis glis*), it is live-trapped and then raised domestically. My usage of the term kangaroo refers as a group to three species, the Red Kangaroo (*Macropus rufus*), Eastern grey kangaroo (*Macropus giganteus*), and Western grey kangaroo (*Macropus fuliginosus*). Of those three kangaroo species it is the Red kangaroo which seems to be the most suggested for domestication as livestock. Kangaroo indeed are bred in captivity and raised on farms but to date their production for human consumption generally is considered unprofitable because of cost competition against hunted animals.

Bovid is a category that refers to members of the family Bovidae, and is represented in this chapter by both bovines and the caprids, with members of the latter group also identified by the term caprines.

Bovine is a term which refers to members of the subfamily Bovinae, is represented in this chapter by the following species: the American bison (*Bison bison*); Cattle (*Bos taurus*); Gayal (*Bos frontalis*); Yak (referring in this chapter to the Domesticated yak, *Bos grunniens*); Water Buffalo (*Bubalus bubalis*); and Carabao (*Bubalus carabanensis*). Neither of these bovines biological relatives the European bison (*Bison bonasus*) nor the African buffalo or Cape buffalo (*Syncerus caffer*), seem to have been either domesticated or otherwise considered as livestock and thus those two relatives are not included in this chapter. Interestingly, the American bison currently is raised agriculturally in Europe.

Canine is a term which represents those members of the subfamily Caninae that are grouped within the family Canidae. The category of canines is represented in this chapter by only one subspecies recognized as Dog or Domestic dog (*Canis lupus familiaris*).

Caprid and caprine are terms which refer to members of the subfamily Caprinae, and that group is represented in this chapter by the following species: goat, which for the purposes of this chapter represents only the Domestic goat (*Capra hircus*), and sheep, which herein refers to only the Domestic sheep (*Ovis aries*).

Camelid is a term which refers to members of the family Camelidae, and they are represented in this chapter by the following animals: Alpaca (*Vicugna pacos*); camel, referring in this chapter to both the Bactrian camel (*Camelus bactrianus*) and the Dromedary camel (*Camelus dromedarius*); and Llama (*Lama glama*).

Caviid is a term which refers to members of the family Caviidae, a group that is represented in this chapter by the Guinea pig (*Cavia porcellus*).

Cervid is a term which refers to members of the family Cervidae, and they are represented in this chapter by the following animals: Elk (*Cervus canadensis*); Fallow deer (*Dama dama*); Moose or Eurasian Elk (*Alces alces*); Red deer (*Cervus elaphus*); and Reindeer, also called Caribou (*Rangifer tarandus*). The information provided for pathogens of moose does not distinguish between species of *Alces*, all of which are termed to be moose.

Equine is a term that refers to members of the subfamily Equinae, which includes the genus *Equus*. That group is represented in this chapter by two species, those being the Donkey or Ass (*Equus asinus*) and the Horse or Domestic Horse (*Equus ferus caballus*).

Feline is a term which refers to members of the subfamily Felinae within the family Felidae, as represented in this chapter by only that species know as Cat or Domestic cat (*Felis catus*).

Glirid is a term which refers to members of the family Gliridae, a group that is represented in this chapter by the name dormouse, herein referring to only the Edible dormouse or Fat dormouse (*Glis glis*).

Leporid is a term which refers to members of the family Leporidae, and that group is represented in this chapter by the rabbit, also known as the European rabbit or Common rabbit (*Oryctolagus cuniculus*).

Macropod is a term which refers to members of the family Macropodidae, and three of its member species are represented in this chapter under the general name kangaroo; those are the Eastern grey kangaroo (*Macropus giganteus*), Western grey kangaroo (*Macropus fuliginosus*), and Red Kangaroo (*Macropus rufus*).

Rodent is a term that refers to members of the order Rodentia, which includes both the caviids and glirids as well as the hydrochaerids. That order also includes the Coypu or Nutria (*Myocastor coypus*).

Ruminant is a term which in this chapter refers only to those mammals belonging to the suborder Ruminantia, as represented by members of the families Bovidae and Cervidae. Camelids also are foregut fermenters but not classed as ruminants because of differences in stomach structure.

Suid is a term which refers to members of the family Suidae, and that group is represented in this chapter by only one species, the Pig or Domestic pig (*Sus scrofa*).

10.3 Opportunistic Pathogenicity

Opportunistic pathogens are those which typically cause disease subsequent to an opportunity. Some opportunistic pathogens are environmental residents that may live naturally in soil and water. But, if one of these opportunistic environmental organisms arrives at a location either on or within a potential host animal or plant which suits the metabolic capabilities and requirements of that environmental organism, then an infection can ensue and the host may become diseased. Many other opportunistic pathogens reside naturally either at various times or exclusively as symbionts or commensals of host plants and animals, and yet should the host become weakened in its capacity to mount a protective defense, that otherwise harmless symbiotic or commensal organism may reveal its pathogenic potential. The ability of a host to defend against the pathogenic potential inherent in a symbiotic or commensal organism often involves a component of site specificity. For example, a microorganism which normally is commensal in either the mouth or colon of an animal may cause disease if it reaches a cutaneous wound on that same animal. The term 'primary pathogen' often is used as a defining reference to those pathogenic organisms which can cause disease in otherwise healthy hosts. Correspondingly, opportunistic pathogens are frequently, although often confusingly, termed 'secondary pathogens' perhaps suggesting that opportunistic pathogens cannot be the sole attacker of a host. Opportunistic pathogens very often can act alone as the sole cause of pathogenicity related to an infection. Perhaps a clearer usage of the term secondary pathogen would involve a more restricted reference to uniquely those organisms whose pathogenicity can be expressed only as a compounding factor to an underlying infection that has been caused by some other pathogen.

There are numerous factors which affect opportunistic pathogenicity and although they strongly overlap, it is possible to group these as: host factors, microbial factors, ecological factors, and medical factors. Some of them are included in Table 10.2. A general introduction to disease transmission was published by Hurst and Murphy (1996).

10.3.1 Host Factors

There are several component host factors including immunosuppression and immune disfunction which can increase the susceptibility of mammals to pathogenic attack.

10.3.1.1 Immunosuppression

Immunosuppression can be present because of such conditions as concurrent infections, diabetes, dietary deficiencies, injury, medically induced suppression which generally is a side effect of treatment for some other illness, and stress whose components include not only climatic factors like seasonal weather but also the effects of transport and relocating animals including capture and release programs.

10.3.1.2 Immune Disfunction

Underlying immune disfunction generally is due to genetic deficiencies and often the impaired immune functions can be improved by selective breeding programs. Table 10.2 mentions the category of secretory system dysfunctions, representing either the absence, overproduction, or functional failure of metabolic components including those normally associated with mucosal surfaces of the respiratory and genitourinary tracts. An example of secretory failure is the chloride channel disfunction which causes cystic fibrosis and increases susceptibility to respiratory infections. Disfunction also can occur in the adaptive and innate immune systems, which in turn will compromise those systems and often results in the host more easily being targeted by potentially pathogenic organisms.

Adaptive Immunity

Adaptive immunity also is termed specific defense. In mammals, its main components are B cells and T cells. The B cells produce secretory proteins including antibodies as well as signaling proteins termed cytokines. The central roll of T cells is termed cell mediated immunity, and its component categories include Helper T cells which assist other lymphocytes, Cytotoxic T cells which destroy tumor cells and virus-infected cells, Memory T cells which allow the immune system to

Table 10.2 Factors affecting opportunistic pathogenicity

Host factors
Immunosuppression
Concurrent infection
Diabetes
Dietary deficiency
Injury
Medically induced effects
Stress
Underlying immune disfunction
Secretory systems
Adaptive immunity, also termed specific defense
B cells
T cells
Innate immunity, also termed nonspecific defense
Defensins
Extrachromosomal histone H2B
Lysozymes
Macrophage
Magainins
Pattern recognition receptors
Transferrins
Microbial factors
Zoonotic potential and host range
Ecological factors
Demography of an infectious disease
Presence of a susceptible animal host
Source of pathogenic organism
Environmental presence
Presence of another infected host serving as reservoir
Presence of suitable vector
Medical factors
Severity and duration of disease and symptoms
Acute versus chronic
Overt versus subacute
Appropriate diagnosis
Prevention and treatment
Antimicrobial compounds
Microbial treatments
Probiotics
Prebiotics
Synbiotics
Vaccines

respond more quickly to subsequent infections by similar pathogens, Suppressor T cells which help maintain immunological tolerance, and Natural killer T cells which are perceived as connecting the adaptive and innate immune systems. Some of the T cells also release cytokines.

Innate Immunity

Innate immunity also is termed nonspecific defense. There are numerous components which comprise the innate immune system and only some examples are mentioned here. Defensins are cationic proteins that typically bind to and assist in the destruction of microbes by forming pore-like defects. Defensins are present in some types of immune cells and many are secretory in nature. Extrachromosomal histone H2B represents a category of proteins that typically seem to remain in the cytoplasm where they aid in antiviral defense by attacking intracellular double stranded DNA. Lysozymes are proteins that damage bacterial cell walls by catalyzing hydrolysis, they are present in some types of immune cells and also are secretory. Macrophage are a category of phagocytes which act in conjunction with both adaptive and innate immune responses. The macrophage can engulf and then destroy microbes, and yet some pathogens have evolved a capability to survive inside the host animal by establishing residence within macrophage. Magainins are secretory peptides which act at least in part by permeabilizing cell membranes. Pattern recognition receptors also are proteins, of which some types are secretory, that act by recognizing molecules termed pathogen-associated molecular patterns. Transferrins are secretory plasma proteins that bind and control the level of available iron.

10.3.2 Microbial Factors

Among the microbial factors are zoonotic potential and host range. These are evolutionarily related to the success of past host–pathogen interactions and determined by metabolic requirements of the pathogen in conjunction with the pathogens ability to resist the hosts defenses. Zoonotic potential of environmental microbes can be influenced by ambient environmental conditions that affect genetic expression of the pathogens metabolism. An example of environmental preconditioning occurs when ambient environmental conditions such as elevated temperature activate the genetic expression of microbial metabolic traits that act as pathogenic traits, priming the capability of a microbe to initiate infection when it enters the warm body of a mammalian host.

10.3.3 Ecological Factors

Ecological factors can influence the demography of an infectious disease, often acting in conjunction to determine the likelihood of pathogen acquisition. From the perspective of a pathogen, demographic success requires a source of the pathogenic organism, simultaneous presence of a susceptible animal host, and some way in which the pathogen can encounter the host. The source of many pathogenic organisms arises from their possessing a natural environmental presence, such presence may range from ubiquitous to geographically limited and also can be seasonal. Other pathogenic species are acquired from infected hosts and importantly for these pathogens the likely simultaneous presence of both infected reservoir hosts and susceptible hosts can be geographically limited and also may be seasonal. Some pathogens require transferral by a suitable vector, but the presence of their potential vector species likewise can be geographically limited and also include a component of seasonality.

10.3.4 Medical Factors

There are a large number of medically related factors which can influence the incidence of opportunistic pathogenicity and outcome. Prevention of disease is the best goal. Among the disease prevention measures, those with proven success are good general animal husbandry practices, use of vaccines, and administration of prophylactic antimicrobials. Vaccines often protect the majority of animals but also will kill some animals and so their usage involves considering the net balance of risk. Prophylactic use of antimicrobials clearly can benefit animals but potentially also may have a negative affect by increasing the antibiotic resistance capabilities of microbes.

In addition to understanding that environmental conditions and host–microbe interactions can affect the occurrence of opportunistic infections, we must recognize the role of microbe–microbe interactions. It is for this reason that preventative treatments based upon the concept of assisting an animals protective commensal microbial populations have seen an increased level of interest in recent years. These treatments often are categorized as probiotics, prebiotics, and synbiotics. Probiotics are microbes administered to commensally compete against pathogens, but some of the microbes used as probiotics are themselves opportunistic pathogens and their administration can cause disease. Prebiotics are compounds administered with the intention of benefiting commensal microbes, with an example being nutritional supplements. Synbiotics is a concept that represents simultaneous administration of probiotic microbes and prebiotic compounds.

When disease does occur, obtaining an appropriate diagnosis in a timely fashion better allows for successful treatment of the affected animal. Disease may be acute, meaning that it has either a rapid onset or short duration, versus chronic, meaning that the disease has long term persistence. Overt disease is more likely to be recognized and adequately treated as compared with subacute disease. Following successful diagnosis, the issues of key importance in supporting the affected animals become reducing the duration of disease and severity of symptoms. Sometimes supportive care is sufficient, but when that accompanied by antimicrobial compounds becomes inadequate then further measures must be taken including the possibility of destroying the affected animals to protect healthy members of the host population.

10.4 Listing of Opportunistic Algae, Fungi and Ichthyosporea Affecting Livestock Mammals

Many of the listed algae, fungi, and Ichthyosporea naturally are environmental organisms and most typically they reside either in soil or water. Others are naturally associated with plants. Often these environmental organisms are saprophytically associated with decaying organic material and their association with animals typically arises from either a passive presence on the skin, accidental introduction into wounds, ingestion or inhalation. Some of the listed fungi either are opportunistic dermatophytes or commensal residents of the mucosal tissues, and many of these microbes may have either no natural environmental presence or only a brief presence in the natural environment as contaminants associated with carcasses and naturally shed materials including animal excretions.

Some pathogen species are very specific with respect to their host range and in particular that tendency exists with viruses, while other pathogens are generalists capable of infecting many different host animals. The algal, bacterial, fungal, metazoan and protozoan pathogens tend to be generalists in terms of their infective capability and you could presume that any association listed between a member of these pathogen groups with one member of a mammalian genus could suggest that the pathogen similarly has an ability to infect other members of that same mammalian genus. Some disease associations found to exist for the European bison but not for the American bison are listed in this chapter under the presumption that a pathogen which affects the European bison likely would also be capable of affecting the American bison. I have listed all of the potentially opportunistic associations that I could find by internet researching. The fact that a disease association between some particular pathogen species and host species cannot be found listed in this chapter should not result in the readers presumption that such an association does not exist.

Some mammalian groups seem to be represented in more detail among these listings. That fact represents differences in the amount of modern veterinary research which is conducted regarding the different pathogen and host relationships. The ability to conduct those studies depends upon availability of research funding, the relative value of the animal in question plus acceptable cost of treating opportunistic infections where the animal is raised, and indeed financial capacity for treating the animal. More research is done on larger animals and on those animals considered as pets because these categories tend to have more financial value per individual animal. Historically traditional veterinary studies would always have been a part of animal husbandry. Those historical studies identified diseases but generally not the affecting pathogens. I suggest the following references as good sources of additional information on these opportunistic pathogens: Kahn and Line (2010), Pathogen Safety Data Sheets and Risk Assessment (2015), Sumbali (2011), The Center for Food Security and Public Health (2015), and Vetbook (2015).

10.4.1 Algae

10.4.1.1 *Prototheca*

Phylum: Chlorophyta; Class: Trebouxiophyceae; Order: Chlorellales; Family: Chlorellaceae; Genus: *Prototheca*. The members of this genus are green algae that lack chlorophyll and have become parasitic. They exist naturally as saprophytes, relying upon saprophytic nutrition which is an extracellular chemoheterotrophic process that digests and decays organic material. The *Prototheca* are found in soil, animal feces, raw and treated sewage from both rural as well as municipal sources, and also in slime flux which is liquid produced by bacterial infection and fermentation of the sap in wounded trees. With regard to disease in mammals, members of this genus can affect the feet, cause facial dermatitis and mastitis. Cutaneous infections caused by *Prototheca* can become ulcerous. Infections caused by *Prototheca* also can be acquired through the nose causing rhinitis, and through the mouth causing enteritis. Protothecal enteritis is able to disseminate producing a systemic disease which is known to attack the eye resulting in blindness, the brain resulting in ataxia and seizures, the myocardium causing myocarditis, and the kidney. Those species belonging to this microbial genus which have been indicated as having a potentially opportunistic disease association with mammalian livestock are: *Prototheca wickerhamii* (cat, dog, cattle, horse, goat, cervids, guinea pig) and *Prototheca zopfii* (dog, cattle, horse, sheep, pig).

10.4.2 Fungi

10.4.2.1 *Alternaria*

Phylum: Ascomycota; Class: Dothideomycetes; Order: Pleosporales; Family: Pleosporaceae; Genus: *Alternaria*. The members of this genus are plant pathogens, and in particular *Alternaria alternata* is a noted cause of seed mold in carrot.

Members of the genus *Alternaria* attack the upper respiratory tract causing sinusitis and erosion of the nasal septum. They also infect the skin and nails. *Alternaria* have been found as causative agents in subcutaneous infections termed hyphomycosis both acting alone and as a mixed infection with the species *Phaeosclera dematioides* (see the listing for *Phaeosclera*). *Alternaria* additionally are a common cause of allergic rhinitis and asthma. Those species belonging to this microbial genus which have been indicated as having a potentially opportunistic disease association with mammalian livestock are: *Alternaria alternata* (cat, horse) and *Alternaria infectoria* (dog).

10.4.2.2 *Aphanoascus*

Phylum: Ascomycota; Class: Eurotiomycetes; Order: Onygenales; Family: Onygenaceae; Genus: *Aphanoascus*. The members of this genus typically are found in soil and feces. The species *Aphanoascus fulvescens* generally is considered to be a nonpathogenic commensal of the skin, but it has been determined to cause dermatophytosis which is a fungal infection of the skin (cat, horse).

10.4.2.3 *Apophysomyces*

Phylum: (not assigned); Class: (not assigned); Order: Mucorales; Family: Mucoraceae; Genus: *Apophysomyces*. The members of this genus typically are found in soil and decaying vegetation. The species *Apophysomyces elegans* causes ulcerative dermatitis (necrotizing fasciitis), which can become a systemic infection and possibly is associated with abortion (cattle).

10.4.2.4 *Aspergillus*

Phylum: Ascomycota; Class: Eurotiomycetes; Order: Eurotiales; Family: Aspergillaceae; Genus: *Aspergillus*. The members of this genus typically are found in mesophilic environments such as decaying vegetation including leaves and soil. The species *Aspergillus thermophilus* is a thermophilic fungus found in mushroom compost (typically contains wheat straw, dried blood, horse manure and ground chalk that have been composted together). Members of the genus *Aspergillus* cause subcutaneous mycotic nodules, produce both mastitis and foot infection, and can cause abscesses. This fungal genus additionally causes enteritis, disease in both the upper as well as lower respiratory tracts including pneumonia, ocular infection, cerebral infection, urinary tract infection, and notably causes infection of the gutteral pouch in equids. They also can produce systemic infections. Their effects upon the unborn offspring of placental mammals result both from placentitis as well as directly infecting the fetus, causing abortion during the second and third trimesters of pregnancy as noticed particularly in bovines. The species *Aspergillus flavus*, *Aspergillus fumigatus*, and *Aspergillus nidulans* are by themselves capable of causing abortion in cattle, and these same three *Aspergillus* species variously also have been found to coexist as part of mixed fungal infections with *Lichtheimia corymbifera*, *Rhizomucor pusillus*, and *Rhizopus oryzae* that caused abortion in cattle (see the listings for *Lichtheimia*, *Rhizomucor*, and *Rhizopus*). There have been indications of *Aspergillus* causing disease in kangaroo but in those cases the fungi were not identified to the level of species. Many members of the genus *Aspergillus* also produce neurotoxins and carcinogenic toxins which can complicate the concept of infections caused by those fungi. In particular, both the species *Aspergillus versicolor* which causes disseminated aspergillosis, and *Aspergillus nidulans* which apparently causes urinary tract infections, also produce the known carcinogenic toxin sterigmatocystin. Those species belonging to this microbial genus which have been indicated as having a potentially opportunistic disease association with mammalian livestock are: *Aspergillus* sp. (kangaroo), *Aspergillus amstelodami* (cattle), *Aspergillus deflectus* (dog), *Aspergillus felis* (cat), *Aspergillus flavus* (cattle, dog), *Aspergillus flavipes* (dog), *Aspergillus fumigatus* (guinea pig, American bison, sheep, goat, cattle, dog, cat, camel, capybara, rabbit, elk, horse), *Aspergillus nidulans* (cat, cattle), *Aspergillus niger* (alpaca, cat, cattle, cervids, dog, sheep, goat), *Aspergillus rugulosus* (cattle), *Aspergillus terreus* (dog, cattle), *Aspergillus versicolor* (dog), and *Aspergillus vitricola* (cat).

10.4.2.5 *Arthroderma*

Phylum: Ascomycota; Class: Eurotiomycetes; Order: Onygenales; Family: Arthrodermataceae; Genus: *Arthroderma*. The members of this genus typically are found in soil and skin. Members of the genus *Arthroderma* typically affect the skin, causing a condition known as dermatophytosis. Interestingly, the reservoir for *Arthroderma benhamiae* is the European hedgehog, *Erinaceus europaeus*. Those species belonging to this microbial genus which have been indicated as having a potentially opportunistic disease association with mammalian livestock are: *Arthroderma benhamiae* (dog), *Arthroderma obtusum* (cattle), *Arthroderma otae* (cat, dog, horse, guinea pig; presumably also affects kangaroo), and *Arthroderma simii* (cattle).

10.4.2.6 *Aureobasidium*

Phylum: Ascomycota; Class: Dothideomycetes; Order: Dothideales; Family: Aureobasidiaceae; Genus: *Aureobasidium*. The species *Aureobasidium pullulans* typically is found in soil, water, air, limestone, and as mildew on wood and finishes. The species *Aureobasidium pullulans* causes subcutaneous infections that include nodule formation (cat).

10.4.2.7 *Basidiobolus*

Phylum: Entomophthoromycota; Class: Basidiobolomycetes; Order: Basidiobolales; Family: Basidiobolaceae; Genus: *Basidiobolus*. The species *Basidiobolus ranarum* is a saprotroph typically associated with leaf litter. The species *Basidiobolus ranarum* causes subcutaneous infections including skin ulcers and mycotic nodules, and it also can produce systemic infections (horse, dog).

10.4.2.8 *Bipolaris*

Phylum: Ascomycota; Class: Dothideomycetes; Order: Pleosporales; Family: Pleosporaceae; Genus: *Bipolaris*. The members of this genus typically are plant pathogens found in subtropical and tropical regions, but they also can cause cutaneous infections of mammals and in particular *Bipolaris drechsleri* is known to affect livestock (cat).

10.4.2.9 *Blastomyces*

Phylum: Ascomycota; Class: Eurotiomycetes; Order: Onygenales; Family: Ajellomycetaceae; Genus: *Blastomyces*. The members of this genus typically are found in soil and wet decaying wood. The species *Blastomyces dermatitidis* causes mastitis, endophthalmitis, lymphadenitis, pneumonia, and systemic infections (cat, dog, horse).

10.4.2.10 *Byssochlamys*

Phylum: Ascomycota; Class: Eurotiomycetes; Order: Eurotiales; Family: Thermoascaceae; Genus: *Byssochlamys*. The members of this fungal genus typically are associated with food spoilage. The species *Byssochlamys spectabilis* more specifically is a thermophilic fungus found in soil, plants, and mushroom compost. The species *Byssochlamys spectabilis* infects the nails, causes cutaneous infections, wound infections, mastitis, sinusitis, endophthalmitis, otitis media, osteomyelitis, and pyelonephritis (horse, dog, goat). Additional information regarding the species *Byssochlamys spectabilis* can be found by searching for literature under its previous name, *Paecilomyces variotii*.

10.4.2.11 *Candida*

Phylum: Ascomycota; Class: Saccharomycetes; Order: Saccharomycetales; Family: Debaryomycetaceae; Genus: *Candida*. Most members of this genus live on plants and rotting vegetation, typically degrading xylose (wood degradation), cellobiose (cellulose degradation), starch, and the aliphatic hydrocarbons which are components of plant cuticles. Members of the genus *Candida* typically attack insects that feed on plants, but this genus also includes fungal species that variously are either commensals or symbionts of additional invertebrate as well as vertebrate animal hosts, including humans. Members of the genus *Candida* cause numerous diseases and that lengthy list includes dermatitis, mastitis, vaginitis, necrotizing placentitis resulting in abortion, septicemia (termed candidemia for this genus, meaning *Candida* in the blood), enteritis, gastritis, necrotizing stomatitis (necrotic periodontal disease), endophthalmitis, ascending urinary tract infections, meningitis, and arthritis. The *Candida* species also can attack the oropharynx, heart, brain, and lungs. Those species belonging to this microbial genus which have been indicated as having a potentially opportunistic disease association with mammalian livestock are: *Candida albicans* (broadly infective of mammals including cat, dog, pig, horse, rabbit, kangaroo, sheep, goat, guinea pig, water buffalo, alpaca), *Candida famata* (horse), *Candida glabrata* (cattle, dog), *Candida parapsilosis* (cattle), *Candida rugosa* (cattle), and *Candida tropicalis* (kangaroo, cattle).

10.4.2.12 *Chrysosporium*

Phylum: Ascomycota; Class: Eurotiomycetes; Order: Onygenales; Family: (not assigned); Genus: *Chrysosporium*. The members of this genus typically are found in soil and plant material where they effect cellulose hydrolysis.

The *Chrysosporium* affect reptiles in addition to mammals. Members of the genus *Chrysosporium* often are keratinolytic and in mammals they typically cause dermatitis and foot rot. Those species belonging to this microbial genus which have been indicated as having a potentially opportunistic disease association with mammalian livestock are: *Chrysosporium keratinophilum* (dog), *Chrysosporium parvum* (cat, cattle, horse), and *Chrysosporium tropicum* (dog).

10.4.2.13 *Cladophialophora*

Phylum: Ascomycota; Class: Eurotiomycetes; Order: Chaetothyriales; Family: Herpotrichiellaceae; Genus: *Cladophialophora*. The members of this genus typically are found in soil and decomposing plant materials including trees and timber, some species are more prevalent in subtropical and tropical forests. There are suggestions that the species *Cladophialophora bantiana* is the same as *Xylohypha bantiana*, but at this time both of these two species names are officially recognized (see the listing for *Xylohypha bantiana*). Members of the genus *Cladophialophora* typically cause chronic subcutaneous dermatitis. They also can cause meningitis and brain abscesses. Those species belonging to this microbial genus which have been indicated as having a potentially opportunistic disease association with mammalian livestock are: *Cladophialophora bantiana* (alpaca, cat, dog), *Cladophialophora carrionii* (cattle, horse), and *Cladophialophora emmonsii* (cat).

10.4.2.14 *Cladosporium*

Phylum: Ascomycota; Class: Dothideomycetes; Order: Capnodiales; Family: Cladosporiaceae; Genus: *Cladosporium*. The members of this genus typically are found on living, decaying and dead plant material, as well as on other decaying organic material. Some members of this genus parasitize other fungi. Members of the genus *Cladosporium* cause ulcerative dermatitis, respiratory infections, and meningitis. The species *Cladosporium cladosporioides* is known to opportunistically cause disease in livestock (dog, sheep, cat).

10.4.2.15 *Clavispora*

Phylum: Ascomycota; Class: Saccharomycetes; Order: Saccharomycetales; Family: Metschnikowiaceae; Genus: *Clavispora*. The species *Clavispora lusitaniae* is a plant pathogen typically found associated with cactus, fruit, necrotic plant tissues and materials of plant origin, birds, and manure. It also is opportunistic of invertebrate animals. In mammals, the species *Clavispora lusitaniae* causes invasive infections and generally is associated with immunocompromised humans. Additional information can be found by researching under the previous name of this species, which was *Candida lusitaniae*. Interestingly, this fungal species has been found as a natural inhabitant of the caecum in pig, where its potential role as an opportunistic pathogen remains to be explored (pig—disease association remains unclear).

10.4.2.16 *Coccidioides*

Phylum: Ascomycota; Class: Eurotiomycetes; Order: Onygenales; Family: (not assigned); Genus: *Coccidioides*. The members of this genus are soil microbes that also have evolved by interacting with their animal hosts. Members of the genus *Coccidioides* can cause ocular infections, lymphadenopathy which can disseminate, and pneumonia. Those species belonging to this microbial genus which have been indicated as having a potentially opportunistic disease association with mammalian livestock are: *Coccidioides immitis* (cat, cattle, dog, horse, camel, llama, rodents), and *Coccidioides posadasii* (dog, cat, cattle, sheep, horse, llama).

10.4.2.17 *Cokeromyces*

Phylum: (not assigned); Class: (not assigned); Order: Mucorales; Family: Mucoraceae; Genus: *Cokeromyces*. The species *Cokeromyces recurvatus* is found in decaying organic material, typically vegetation and feces. The species *Cokeromyces recurvatus* also is found normally in the gastrointestinal tract and cervix of many mammalian species. It causes enteritis including intestinal perforation, urogenital infections, chronic hemorrhagic cystitis, and can invade the pleural as well as peritoneal fluids in association with peritonitis (cat, dog).

10.4.2.18 *Colletotrichum*

Phylum: Ascomycota; Class: Sordariomycetes; Order: Glomerellales; Family: Glomerellaceae; Genus: *Colletotrichum*. The members of this genus typically are symbiotic to plants, exist as plant endophytes and are phytopathogens. However, they also can cause infections of many animals. *Colletotrichum* infections of mammals typically result in subcutaneous lesions including those involving the foot, and they also can cause eye infections. Among the species known to infect humans are *Colletotrichum dematium*, *Colletotrichum gloeosporioides* which notably also causes Umbel blight in carrot, and *Colletotrichum acutatum*. *Colletotrichum acutatum* also has been found to cause a disseminated infection in an Atlantic ridley sea turtle (*Lepidochelys kempii*). The subject of *Colletotrichum* in livestock mammals seems to have received extremely limited attention. An instance of subcutaneous infection which also affected the lung and kidney has been reported in cat but the fungal species was not determined (cat).

10.4.2.19 *Conidiobolus*

Phylum: Entomophthoromycota; Class: Entomophthoromycetes; Order: Entomophthorales; Family: Ancylistaceae; Genus: *Conidiobolus*. The members of this genus typically can be found as saprotrophs in leaf litter. Members of the genus *Conidiobolus* characteristically cause subcutaneous infections and also infect the nasal mucosa. They additionally can cause systemic infections and produce gastrointestinal ulcerations. Those species belonging to this microbial genus which have been indicated as having a potentially opportunistic disease association with mammalian livestock are: *Conidiobolus coronatus* (dog, horse, llama), *Conidiobolus incongruus* (sheep, red deer), and *Conidiobolus lamprauges* (horse).

10.4.2.20 *Coniochaeta*

Phylum: Ascomycota; Class: Sordariomycetes; Order: Coniochaetales; Family: Coniochaetaceae; Genus: *Coniochaeta*. The members of this genus are found in water and soil. They typically inhabit woody plants by living independently on the outside of lichens, growing as saprotrophs on leaves, and existing as root endophytes. The *Coniochaeta* also are associated with tree wounds or dead wood and penetrate to feed on the host plant resulting in an association with necrotic lesions. Members of the genus *Coniochaeta* cause subcutaneous infections in mammals that include nodule formation, have been found in the bone marrow as an association with osteomyelitis, and can cause abortion. Those species belonging to this microbial genus which have been indicated as having a potentially opportunistic disease association with mammalian livestock are: *Coniochaeta cateniformis* (dog, cat) and *Coniochaeta hoffmannii* (dog, cattle).

10.4.2.21 *Cryptococcus*

Phylum: Basidiomycota; Class: Tremellomycetes; Order: Filobasidiales; Family: (not assigned); Genus: *Cryptococcus*. The members of this genus primarily are found in bird guano and around eucalyptus trees. They also affect invertebrate animal hosts. In mammals, members of the genus *Cryptococcus* cause mastitis and subcutaneous infections which include nodule formation, both upper and lower respiratory infections, and they can become internally disseminated resulting in such complications as encephalitis and meningitis. Those species belonging to this microbial genus which have been indicated as having a potentially opportunistic disease association with mammalian livestock are: *Cryptococcus albidus* (horse, cat, dog), *Cryptococcus flavescens* (dog), *Cryptococcus gattii* (cat, dog, goat, llama, alpaca, donkey), *Cryptococcus laurentii* (dog, horse), *Cryptococcus magnus* (cat), and *Cryptococcus neoformans* (pig, goat, sheep, dog, cat, horse, guinea pig, cattle, water buffalo).

10.4.2.22 *Curvularia*

Phylum: Ascomycota; Class: Dothideomycetes; Order: Pleosporales; Family: Pleosporaceae; Genus: *Curvularia*. The members of this genus typically are found in soil and are opportunistic pathogens of plants. They also attack mammals causing subcutaneous infections, sinusitis, peritonitis, granulomatous encephalitis, and abortion. Various species that now belong to the genus *Curvularia* have had previous membership in the genera *Bipolaris* and *Drechslera*, and appropriate information on these fungi can be found by researching under those other genus names. Those species belonging to this microbial genus which have been indicated as having a potentially opportunistic disease association with mammalian livestock are: *Curvularia geniculata* (cattle), *Curvularia lunata* (cat), and *Curvularia spicifera* (cat, dog, horse, cattle).

10.4.2.23 *Cyniclomyces*

Phylum: Ascomycota; Class: Saccharomycetes; Order: Saccharomycetales; Family: Saccharomycetaceae; Genus: *Cyniclomyces*. The members of this genus typically inhabit the stomach and intestine, and they are found in deposited feces. The species *Cyniclomyces guttulatus* causes chronic diarrhoea, cholecystitis, and hepatitis (cat, dog, rabbit, guinea pig).

10.4.2.24 *Cyphellophora*

Phylum: Ascomycota; Class: Eurotiomycetes; Order: Chaetothyriales; Family: Cyphellophoraceae; Genus: *Cyphellophora*. The species *Cyphellophora suttonii* lives as a commensal on the skin and nails. It causes subcutaneous lesions including those of the ear (dog).

10.4.2.25 *Cystobasidium*

Phylum: Basidiomycota; Class: Cystobasidiomycetes; Order: Cystobasidiales; Family: Cystobasidiaceae; Genus: *Cystobasidium*. Members of this genus grow on other fungi, lichens, and wood. *Cystobasidium minutum* in particular lives on and causes the decay of wood, as well as living in dead wood, leaves, sticks, and other organic debris. The species *Cystobasidium minutum* also attacks mammals causing systemic infections including those which affect the kidney (sheep).

10.4.2.26 *Debaryomyces*

Phylum: Ascomycota; Class: Saccharomycetes; Order: Saccharomycetales; Family: Debaryomycetaceae; Genus: *Debaryomyces*. As a genus, the *Debaryomyces* are salt tolerant and typically marine, but they also can be found in cheeses and sausages, and are associated with invertebrate animal hosts. It has been suggested that the fungal species *Debaryomyces nepalensis* can behave as an opportunistic pathogen in human. Curiously, this species has been found in the oral cavity of an apparently healthy dog such that the fungus was presumed to have commensal status (dog—disease association remains unclear).

10.4.2.27 *Dipodascus*

Phylum: Ascomycota; Class: Saccharomycetes; Order: Saccharomycetales; Family: Dipodascaceae; Genus: *Dipodascus*. The species *Dipodascus capitatus* typically is found in decaying plant tissue including rotten fruit and leaves, and is associated with industrial food spoilage. The species *Dipodascus capitatus* also is known to cause mastitis and abortion (cattle, horse).

10.4.2.28 *Epidermophyton*

Phylum: Ascomycota; Class: Eurotiomycetes; Order: Onygenales; Family: Arthrodermataceae; Genus: *Epidermophyton*. The species *Epidermophyton floccosum* is a dermatophyte which naturally lives on keratin, a key structural protein on the outer skin, hair and nails. As a pathogen this species typically affects the hair and nails, although it also can cause invasive disease (cat, dog, goat, guinea pig).

10.4.2.29 *Encephalitozoon*

Phylum: Microsporidia; Class: (not assigned); Order: (not assigned); Family: Unikaryonidae; Genus: *Encephalitozoon*. Members of the genus *Encephalitozoon* are obligate intracellular parasites which have a typical association with the urinary bladder and intestinal epithelium. They are transiently found in animal urine and produce environmentally resistant spores that can be found both in animal urine and feces. Generally these listed members of the genus *Encephalitozoon* do not cause a clinically obvious disease although they opportunistically can cause respiratory tract illness and disseminated infections that produce meningitis, infect the brain, cause focal interstitial nephritis, and may additionally involve the heart, trachea, spleen, and lymph nodes. They possibly have been a contributive cause in cases of abortion. Those species belonging to this microbial genus which have been indicated as potentially opportunistic for mammalian livestock are: *Encephalitozoon cuniculi* (cattle, rabbit, guinea pig, goat, sheep, pig, horse, dog, cat); *Encephalitozoon hellem* (cat, dog), and *Encephalitozoon intestinalis* (cat, cattle, donkey, dog, pig, goat).

10.4.2.30 *Enterocytozoon*

Phylum: Microsporidia; Class: (not assigned); Order: (not assigned); Family: Enterocytozoonidae; Genus: *Enterocytozoon*. Members of the genus *Enterocytozoon* are obligate intracellular parasites which are associated with intestinal epithelial cells and typically do not cause obvious clinical disease. The species

Enterocytozoon bieneusi can cause enteritis by infecting epithelial cells of the small intestine. This fungal species also has been noted as contributing to viral induced diarrhea in bovines (cat, dog, pig, cattle).

10.4.2.31 *Exophiala*

Phylum: Ascomycota; Class: Eurotiomycetes; Order: Chaetothyriales; Family: Herpotrichiellaceae; Genus: *Exophiala*. The *Exophiala* commonly are found in soil and water as saprotrophs associated with plants and decaying wood. In mammals, members of the genus *Exophiala* notably produce abscesses. The members of this genus infect the nasal cavity and can invade the lung, cause subcutaneous infections which include nodule formation and tumor-like cysts that are termed mycetomas, and the subcutaneous infections can become progressively granulomatous involving not only the skin and subcutaneous tissues but additionally becoming invasive of underlying muscle and bone. Infections caused by this genus can become disseminated with systemic consequences and cause abortion. Additionally, although more rarely, the *Exophiala* can attack the cornea. *Exophiala hongkongensis* is a recently discovered microbe which causes an infection of the toenails termed dermatophytic onychomycosis in humans and may eventually be found to affect livestock mammals. *Exophiala hongkongensis* may be the same species as *Phialemoniopsis hongkongensis* although both of these names currently are recognized (see the listing for genus *Phialemoniopsis*). Those species belonging to this microbial genus which have been indicated as having a potentially opportunistic disease association with mammalian livestock are: *Exophiala attenuata* (cat), *Exophiala dermatitidis* (cat, dog, cattle), *Exophiala jeanselmei* (cat, cattle), and *Exophiala spinifera* (cat).

10.4.2.32 *Fonsecaea*

Phylum: Ascomycota; Class: Eurotiomycetes; Order: Chaetothyriales; Family: Herpotrichiellaceae; Genus: *Fonsecaea*. The members of this genus are saprotrophs typically found in soil and on living plants including trees. They have been isolated from rotting wood including tree stumps, woodpiles and fence posts. Members of the genus *Fonsecaea* are known to produce ulcerations localized to the skin and subcutaneous tissues, but they also can cause cerebral abscess. Those species belonging to this microbial genus which have been indicated as having a potentially opportunistic disease association with mammalian livestock are: *Fonsecaea multimorphosa* (cat) and *Fonsecaea pedrosoi* (cat).

10.4.2.33 *Fusarium*

Phylum: Ascomycota; Class: Sordariomycetes; Order: Hypocreales; Family: Nectriaceae; Genus: *Fusarium*. Members of the genus *Fusarium* are saprophytes typically isolated from plants and in soil where they decay organic matter. Members of the genus *Fusarium* cause dermatitis that can become granulomatous in nature and these infections may include skin ulcers as well as subcutaneous nodule formation. Fusarial infections may cause ulceration of the cornea. *Fusarium* infections also can produce invasive disease with pulmonary involvement and attack the brain resulting in meningoencephalitis. There has been a possible association of *Fusarium oxysporum* with neonatal anencephalia in human. Infections by this genus can be complicated by the production of fungal toxins that cause infertility. Those species belonging to this microbial genus which have been indicated as having a potentially opportunistic disease association with mammalian livestock are: *Fusarium oxysporum* (cats), *Fusarium proliferatum* (cat), *Fusarium solani* (dog, goat, sheep), and *Fusarium sporotrichioides* (dog).

10.4.2.34 *Galactomyces*

Phylum: Ascomycota; Class: Saccharomycetes; Order: Saccharomycetales; Family: Dipodascaceae; Genus: *Galactomyces*. The species *Galactomyces candidum* previously has been known by many names including *Geotrichum candidum*. It typically is found in plants, soil, water, sewage, cereals and dairy products, and is noted as a cause of sour rot in carrot. The species *Galactomyces candidum* causes mastitis, attacks the tonsils, and causes chronic infection of the lungs, mouth and intestine. These infections can become disseminated and pyogranulomatous resulting in pneumonia, hepatitis, and nephritis (dog, sheep, goat, pig, cattle, water buffalo).

10.4.2.35 *Histoplasma*

Phylum: Ascomycota; Class: Eurotiomycetes; Order: Onygenales; Family: Ajellomycetaceae; Genus: *Histoplasma*. The species *Histoplasma capsulatum* is a common soil organism which also is found in association with rotting excrement from bats and birds. Typically for immunocompetent hosts, the fungal species *Histoplasma capsulatum* which is very broadly infective of mammals will produce a transiently noticed although potentially never eliminated granulomatous pulmonary infection, and in that regard the pulmonary disease histoplasmosis which this fungus produces is similar to the illness tuberculosis which is caused by the bacterial species *Mycobacterium tuberculosis*. *Histoplasma capsulatum* also causes conjunctivitis and opthalmic infection, the latter of which can result in permanent blindness and can be an occupational disease for people who wash windows and window ledges. This fungus also can invade the skin, producing subcutaneous mycotic nodules, and will cause systemic granulomas. The fungal species *Histoplasma capsulatum* is all too well know in geographical areas such as the Ohio River Valley in the United States of America, which is where I live. Here, it has an association with the cloacal excrement of Feral pigeons (*Columba livia*) because the fungus grows in that excrement. In particular, my home city of Cincinnati, Ohio, once was referred to as being the Histoplasmosis Capital of North America with more than 86 % of its residents being serological positive for the organism. Histoplasmosis likely could have caused the blindness of the "Bird Woman" character who sells crumbs to feed pigeons in the play and movie "Mary Poppins", as based upon the books of Pamela Lyndon Travers. This fungal species has been identified as attacking many species of mammalian livestock (cat, dog, llama, horse, kangaroo, cattle, sheep).

10.4.2.36 *Hortaea*

Phylum: Ascomycota; Class: Dothideomycetes; Order: Dothideales; Family: not assigned; Genus: *Hortaea*. The species *Hortaea werneckii* is extremely halotolerant and is a dominant fungal species in hypersaline aquatic environments. The species *Hortaea werneckii* causes infections of the skin and footpad, often as a consequence of scratch wounds and punctures (cat, guinea pig).

10.4.2.37 *Kluyveromyces*

Phylum: Ascomycota; Class: Saccharomycetes; Order: Saccharomycetales; Family: Saccharomycetaceae; Genus: *Kluyveromyces*. The species *Kluyveromyces marxianus* commonly is found in milk which may be a part of its ecological habitat although it does additionally have invertebrate animal hosts. This species is used commercially in fermenting wine and is produced as both a nutritional component and binder for animal feed despite the fact of it being an opportunistic pathogen. The species *Kluyveromyces marxianus* causes mastitis (cattle).

10.4.2.38 *Lacazia*

Phylum: Ascomycota; Class: Eurotiomycetes; Order: Onygenales; Family: (not assigned); Genus: *Lacazia*. No information could be found about the ecology of *Lacazia loboi* except the indication that it is a geographically restricted environmental fungus. This fungal species causes mucocutaneous lesions typically of the nose and lips. It does as well cause cutaneous and subcutaneous infections including nodules following injuries from thorns or insect bites (cattle, sheep, goat, pig).

10.4.2.39 *Lichtheimia*

Phylum: (not assigned); Class: (not assigned); Order: Mucorales; Family: Lichtheimiaceae; Genus: *Lichtheimia*. The species *Lichtheimia corymbifera* is a thermophilic fungus that usually inhabits soil. It is associated with decomposing plant material including leaves, also with mushroom compost, and possibly contributes to the decay of grass and hay in agricultural settings. Infections by the species *Lichtheimia corymbifera* cause abortion and that disease outcome also has been noted for instances of mixed fungal infections that included this species and members of the genus *Aspergillus* (see the listing for *Aspergillus*). *Lichtheimia corymbifera* also can be found in the lung, liver, kidney and spleen of immunodeficient hosts simultaneous to infection with *Aspergillus fumigatus* (cattle, horse, alpaca, llama).

10.4.2.40 *Lomentospora*

Phylum: Ascomycota; Class: Sordariomycetes; Order: Microascales; Family: Microascaceae; Genus: *Lomentospora*. The species *Lomentospora prolificans* is a soil organism that has been found in the soil of both house plants and greenhouse plants. A general disease association exists between this fungus and subcutaneous lesions caused by either splinters or plant thorns. *Lomentospora prolificans* additionally causes arthritis and degenerative osteomyelitis in horse. This fungal species has been isolated from cat although apparently without a clear disease association. The previous species name for this fungus was *Scedosporium prolificans* (horse; also found in cat although a disease association remains unclear for cat).

10.4.2.41 *Madurella*

Phylum: Ascomycota; Class: Sordariomycetes; Order: Sordariales; Family: (not assigned); Genus: *Madurella*. The species *Madurella mycetomatis* is a soil fungus which seems to be limited to tropical and subtropical regions. As a cause of infection it produces cutaneous and subcutaneous lesions that can involve fascia, it also has been identified as causing abdominal mycetomas (dog).

10.4.2.42 *Malassezia*

Phylum: Basidiomycota; Class: Malasseziomycetes; Order: Malasseziales; Family: Malasseziaceae; Genus: *Malassezia*. The *Malassezia* live on the skin where they consume naturally excreted oils and fats. Members of this species typically cause the benign dermal condition called dandruff, but they can infect the skin surrounding nails and claws causing the condition termed paronychia. The *Malassezia* also cause otitis externa, otitis media, and otitis interna. Additional information can be found by researching members of this genus under the previously assigned genus name *Pityrosporum*. Those species belonging to this microbial genus which have been indicated as having a potentially opportunistic disease association with mammalian livestock are: *Malassezia caprae* (goat), *Malassezia cuniculi* (rabbit), *Malassezia equina* (horse), *Malassezia furfur* (cat, cattle, dog, goat, pig, guinea pig, horse), *Malassezia globosa* (cat, cattle, goat, horse, sheep), *Malassezia nana* (cat, cattle), *Malassezia obtusa* (cat, goat, horse), *Malassezia pachydermatis* (dog, cat, cattle, goat), *Malassezia restricta* (cat, goat, horse, sheep), *Malassezia slooffiae* (cat, cattle, horse), and *Malassezia sympodialis* (cat, cattle, goat, sheep).

10.4.2.43 *Metschnikowia*

Phylum: Ascomycota; Class: Saccharomycetes; Order: Saccharomycetales; Family: Metschnikowiaceae; Genus: *Metschnikowia*. The species *Metschnikowia pulcherrima* occurs naturally on plant buds, floral parts and fruits. The species *Metschnikowia pulcherrima* has invertebrate animal hosts in addition to causing dermal infections of mammals (dog).

10.4.2.44 *Meyerozyma*

Phylum: Ascomycota; Class: Saccharomycetes; Order: Saccharomycetales; Family: Debaryomycetaceae; Genus: *Meyerozyma*. The species *Meyerozyma guilliermondii* is a soil microorganism which causes mastitis and abortion (cattle, horse).

10.4.2.45 *Microsphaeropsis*

Phylum: Ascomycota; Class: (not assigned); Order: (not assigned); Family: (not assigned); Genus: *Microsphaeropsis*. The species *Microsphaeropsis arundinis* normally is an inhabitant of plants although it opportunistically causes dermal infections including ulcerations, typically in diabetic mammals. A suggested association was found between this fungal genus and horse, although the possible clinical nature of that interaction is uncertain. Those species belonging to this microbial genus which have been indicated as having a potentially opportunistic disease association with mammalian livestock are: *Microsphaeropsis* sp. (horse—disease association remains unclear) and *Microsphaeropsis arundinis* (cat, dog).

10.4.2.46 *Microsporum*

Phylum: Ascomycota; Class: Eurotiomycetes; Order: Onygenales; Family: Arthrodermataceae; Genus: *Microsporum*. The members of this genus naturally reside in soil wherein they presumably are involved with keratin degradation. Members of the genus *Microsporum* cause subcutaneous infections which can include nodule formation. Those species belonging to this microbial genus which have been indicated as having a potentially opportunistic disease association with mammalian livestock are: *Microsporum distortum* (dog), *Microsporum equinum* (horse), *Microsporum gypseum* (kangaroo, dog, cat, cattle, horse, sheep, goat, guinea pig, water buffalo), *Microsporum persicolor* (dog, kangaroo), *Microsporum praecox* (horse), and *Microsporum rivalieri* (horse).

10.4.2.47 *Moniliella*

Phylum: Basidiomycota; Class: Moniliellomycetes; Order: Moniliellales; Family: Moniliellaceae; Genus: *Moniliella*. The species *Moniliella suaveolens* normally is associated with oils and oil-based foods including press cake made from seeds, and it interestingly causes spoilage of margarine. It could be surmised that the natural ecology of *Moniliella suaveolens* might be related to plant seeds and plant oils. With regard to mammalian livestock, the species *Moniliella suaveolens* is known to cause subcutaneous disease (cat).

10.4.2.48 *Monocillium*

Phylum: Ascomycota; Class: (not assigned); Order: (not assigned); Family: (not assigned); Genus: *Monocillium*. The species *Monocillium indicum* normally is found in soil. As an opportunistic pathogen of mammals it causes respiratory infections, lymphadenopathy, and splenitis (dog).

10.4.2.49 *Mortierella*

Phylum: (not assigned); Class: (not assigned); Order: Mortierellales; Family: Mortierellaceae; Genus: *Mortierella*. Members of the genus *Mortierella* are soil fungi typically living on decaying leaves and other organic material, they are colonizers on the surface of roots, and they also live on fecal pellets and the exoskeletons of arthropods. Members of the genus *Mortierella* cause lung infections and abortion. Those species belonging to this microbial genus which have been indicated as having a potentially opportunistic disease association with mammalian livestock are: *Mortierella polycephala* (cattle) and *Mortierella wolfii* (cattle).

10.4.2.50 *Mucor*

Phylum: (not assigned); Class: (not assigned); Order: Mucorales; Family: Mucoraceae; Genus: *Mucor*. Members of the genus *Mucor* are saprophytes typically isolated from plants and found in soil where they decay organic matter. Members of the genus *Mucor* characteristically cause ulcerative dermatitis and nasal nodules. They also can produce systemic infections resulting in meningitis. Those species belonging to this microbial genus which have been indicated as having a potentially opportunistic disease association with mammalian livestock are: *Mucor amphibiorum* (cat, horse) and *Mucor circinelloides* (cattle, pig).

10.4.2.51 *Neoscytalidium*

Phylum: Ascomycota; Class: Dothideomycetes; Order: Botryosphaeriales; Family: Botryosphaeriaceae; Genus: *Neoscytalidium*. The *Neoscytalidium* typically are latent pathogens involved with dieback of woody plants. In mammals, members of the genus *Neoscytalidium* characteristically produce subcutaneous infections typically affecting the feet and nails. *Neoscytalidium* can, however, cause invasive and systemic disease in immunosuppressed animals. The fungal species *Neoscytalidium dimidiatum* is known to causes pulmonary infections in non-livestock mammals including Risso's dolphin (*Grampus griseus*). Those species belonging to this microbial genus which have been indicated as having a potentially opportunistic disease association with mammalian livestock are: *Neoscytalidium dimidiatum* (dog) and *Neoscytalidium hyalinum* (dog).

10.4.2.52 *Ochroconis*

Phylum: Ascomycota; Class: Dothideomycetes; Order: Venturiales; Family: Sympoventuriaceae; Genus: *Ochroconis*. The species *Ochroconis humicola* typically lives on the skin of animals including marine fish. As an opportunistic pathogen of mammals, this species causes subcutaneous infections and attacks the nasal tissue (cat).

10.4.2.53 *Oxyporus*

Phylum: Basidiomycota; Class: Agaricomycetes; Order: Polyporales; Family: Coriolaceae; Genus: *Oxyporus*. The species *Oxyporus corticola* is a plant pathogen which typically affects fruit of the genus *Prunus*. As an opportunistic pathogen of mammals, *Oxyporus corticola* causes skin infections as well as lymphadenopathy and osteomyelitis (dog).

10.4.2.54 *Paracoccidioides*

Phylum: Ascomycota; Class: Eurotiomycetes; Order: Onygenales; Family: (not assigned); Genus: *Paracoccidioides*. The species *Paracoccidioides brasiliensis* commonly is associated with soils in which coffee is cultivated and can grow in animal excrement. In mammals, the species *Paracoccidioides brasiliensis* causes disease of the respiratory and gastrointestinal tracts in addition to producing lymphadenopathy and organomegaly (dog).

10.4.2.55 *Penicillium*

Phylum: Ascomycota; Class: Eurotiomycetes; Order: Eurotiales; Family: Aspergillaceae; Genus: *Penicillium*. The members of this genus typically live in the soil where they process dead and decaying organic matter. Members of the genus *Penicillium* cause a canine pneumonia and also have been associated with bone lesions of a dog with osteomyelitis. The species *Penicillium digitatum* causes opportunistic pneumonia in humans and presumably likewise would be pathogenic for livestock mammals, although a causative association between that fungal species and livestock disease remains uncertain. Those species belonging to this microbial genus which have been indicated as having a potentially opportunistic disease association with mammalian livestock are: *Penicillium brevicompactum* (dog), *Penicillium canis* (dog), and *Penicillium digitatum* (association with animal disease remains uncertain).

10.4.2.56 *Phaeosclera*

Phylum: Ascomycota; Class: (not assigned); Order: (not assigned); Family: (not assigned); Genus: *Phaeosclera*. The species *Phaeosclera dematioides* has been found on rocks and been isolated from the pith of fire dependent pine trees (*Pinus contorta*). The species *Phaeosclera dematioides* can, when acting alone, attack the skin and mucous membranes. *Phaeosclera dematioides* also has been found to cause hyphomycosis as a mixed infection with members of the genus *Alternaria* (see the listing for *Alternaria*) (cattle).

10.4.2.57 *Phialemoniopsis*

Phylum: Ascomycota; Class: Sordariomycetes; Order: Xylariales; Family: (not assigned); Genus: *Phialemoniopsis*. The species *Phialemoniopsis curvata* is broadly present in the environment, being found in air, soil, industrial water and sewage. *Phialemoniopsis hongkongensis*, which causes subcutaneous nodules in human, presumably likewise would be pathogenic for livestock mammals although a causative association between this fungal species and livestock disease remains uncertain. *Phialemoniopsis hongkongensis* may be the same species as *Exophiala hongkongensis* although both of these names currently are recognized (see the listing for genus *Exophiala*). A natural ecology has not yet been defined for *Phialemoniopsis hongkongensis*. Members of the genus *Phialemoniopsis* mainly are known as opportunistic pathogens of immunocompromised people and rarely are seen to affect immunocompetent people. Among the diseases which this fungal genus causes in mammals are cutaneous infections of wounds and burns, and such infections can include the formation of subcutaneous nodules. *Phialemoniopsis* species also cause internal infections among which are peritonitis with fungus recoverable from the pleural fluid, osteomyelitis, and endovascular disease including endocarditis. The species *Phialemoniopsis curvata* has been identified as attacking mammalian livestock (dog).

10.4.2.58 *Phialemonium*

Phylum: Ascomycota; Class: Sordariomycetes; Order: Sordariales; Family: Cephalothecaceae; Genus: *Phialemonium*. The species *Phialemonium obovatum* normally resides on the skin. As an opportunistic pathogen, *Phialemonium obovatum* infects the skin in addition to causing granulomatous lung infections, endocarditis, and corneal keratitis (cat).

10.4.2.59 *Phialophora*

Phylum: Ascomycota; Class: Eurotiomycetes; Order: Chaetothyriales; Family: Herpotrichiellaceae; Genus: *Phialophora*. The species *Phialophora verrucosa* is isolated from soils and rotting wood including logs and bark. As a pathogen of mammals, *Phialophora verrucosa* causes long term infections of the skin and subcutaneous tissues including the formation of subcutaneous fungal nodules (cat, dog).

10.4.2.60 *Phialosimplex*

Phylum: Ascomycota; Class: Eurotiomycetes; Order: Eurotiales; Family: Aspergillaceae; Genus: *Phialosimplex*. The *Phialosimplex* presumably are by nature soil and subsurface organisms. Members of the genus *Phialosimplex* attack immunosuppresed animals. They can cause disease of the skin and invade subcutaneous tissues. This fungal genus also can cause fatally invasive infections including disseminated disease evidenced as myelitis (inflammation of the spinal cord), and can be found in the bone marrow. Those species belonging to this microbial genus which have been indicated as having a potentially opportunistic disease association with mammalian livestock are: *Phialosimplex caninus* (dog), *Phialosimplex chlamydosporus* (dog), and *Phialosimplex sclerotialis* (dog).

10.4.2.61 *Pichia*

Phylum: Ascomycota; Class: Saccharomycetes; Order: Saccharomycetales; Family: Pichiaceae; Genus: *Pichia*. The members of this fungal species are found in soil as well as on fruits, and in natural fermentations. The *Pichia* are responsible for spoilage of pickled vegetables such as kimchi. They affect both invertebrate and vertebrate animal hosts. The species *Pichia kudriavzevii* also is used in fermenting wine despite the fact of it being an opportunistic pathogen. As an opportunistic pathogen of mammals, the species *Pichia kudriavzevii* causes mastitis (cattle, dog).

10.4.2.62 *Pneumocystis*

Phylum: Ascomycota; Class: Pneumocystidomycetes; Order: Pneumocystidales; Family: Pneumocystidaceae; Genus: *Pneumocystis*. The species *Pneumocystis carinii* does not seem to have an ecology other than its association with a host animal. Asymptomatic infections caused by this species apparently are very common, but it also attacks the lower respiratory tract causing pneumonia. In humans, this same fungal species often is called *Pneumocystis jirovecii* (very broadly infective of mammals including cat, cattle, sheep, pig, dog, donkey, horse, kangaroo).

10.4.2.63 *Pseudogymnoascus*

Phylum: Ascomycota; Class: Leotiomycetes; Order: (not assigned); Family: Pseudeurotiaceae; Genus: *Pseudogymnoascus*. The species *Pseudogymnoascus pannorum* is a soil organism which degrades keratin based substrates. It also causes superficial infections of the skin and nails in numerous mammalian species. Additional information can be found by researching under its previous names *Chrysosporium pannorum* and *Geomyces pannorum* (dog, camel, cervids). Another member of this genus, *Pseudogymnoascus destructans*, has become infamous by invading the skin of bats and causing ulcers in a syndrome termed white-nose disease.

10.4.2.64 *Purpureocillium*

Phylum: Ascomycota; Class: Sordariomycetes; Order: Hypocreales; Family: Ophiocordycipitaceae; Genus: *Purpureocillium*. The species *Purpureocillium lilacinum* is a saprophyte that has been isolated from soil in numerous environmental zones, as well as from estuarine sediments, sewage sludge, insects, root nematodes and nematode eggs. In mammals, *Purpureocillium lilacinum* causes nodular cutaneous granulomas. It also infects other organs including eye, sinus, and lung (cat).

10.4.2.65 *Pythium*

Phylum: (not assigned); Class: (not assigned); Order: Pythiales; Family: Pythiaceae; Genus: *Pythium*. Members of the genus *Pythium* typically are plant pathogens. The species *Pythium insidiosum* is a water mold found in subtropical and tropical areas and it seemingly cannot survive in freezing water. Environmental sources of *Pythium insidiosum* include irrigation water and rice paddy fields. Infection by this organism is acquired when mammals immerse themselves in standing surface water that contains the microorganism. This fungus typically invades existing skin wounds to cause cutaneous and subcutaneous lesions that characteristically appear on the head and lower legs. These lesions are ulcerative, fistulative, granulomatous and necrotic, and the lesions enlarge by subcutaneous extension. The subcutaneous skin lesions also can spread to the underlying bones. This fungal species additionally can, although more rarely, attack the nasopharynx as well as causing naso-orbital abscesses. It can attack the gastrointestinal tract including the esophagus producing gastrointestinal tract granulomas, and these gastrointestinal consequences particularly occur in dog. The infections caused by this organism are capable of becoming disseminated and chronic systemic infections caused by *Pythium insidiosum* can move into the lung. Additional information can be found by researching this fungal species under its previous name, *Hyphomyces destruens* (dog, horse, cat, cattle).

10.4.2.66 *Rasamsonia*

Phylum: Ascomycota; Class: Eurotiomycetes; Order: Eurotiales; Family: Trichocomaceae; Genus: *Rasamsonia*. The species *Rasamsonia argillacea* is found in soil and air. This species attacks the skin and can cause a disseminated systemic infection. The previous names of *Rasamsonia argillacea* include *Geosmithia argillacea*, and it often is listed as being an anamorph of *Talaromyces eburneus* (dog).

10.4.2.67 *Rhinocladiella*

Phylum: Ascomycota; Class: Eurotiomycetes; Order: Chaetothyriales; Family: Herpotrichiellaceae; Genus: *Rhinocladiella*. The species *Rhinocladiella mackenziei* is found naturally on the skin. The fact that human disease caused by this fungus seems geographically limited, generally to the middle eastern region and parts of south Asia, suggests the possibility that this fungal species may have a broader natural ecology but I could find no information about its natural ecology. In mammals, *Rhinocladiella mackenziei* causes a long-term fungal infection of the skin and subcutaneous tissue. This fungal species also causes respiratory distress that can be accompanied by constant coughing, severe swelling of the nose that impairs breathing, and deep infections of the nasal cavity extending to the point of impairing vision. In a disseminated form, there may be severe systemic inflammation and fungal lesions can be found throughout the body including the kidney and brain. Brain infections caused by this organism can include the formation of brain abscesses and attack the cerebellum. The additional symptoms of anorexia, lethargy, hypothermia, and hair loss which are noticed for infections caused by this fungal species do sadly seem almost inconsequential in comparison with the often fatal systemic effects (cat, dog, bovines, equines). The previous name of this species was *Ramichloridium mackenziei*.

10.4.2.68 *Rhizomucor*

Phylum: (not assigned); Class: (not assigned); Order: Mucorales; Family: Lichtheimiaceae; Genus: *Rhizomucor*. The species *Rhizomucor pusillus* is a thermophilic fungus most commonly found in compost piles and also in mushroom compost. In mammals, this species produces subcutaneous infections and by itself also can cause abortion. *Rhizomucor pusillus* additionally has been found to cause abortion as a member of mixed infections that included members of the genus *Aspergillus* (see the listing for *Aspergillus*) (cat, cattle).

10.4.2.69 *Rhizopus*

Phylum: (not assigned); Class: (not assigned); Order: Mucorales; Family: Rhizopodaceae; Genus: *Rhizopus*. The species *Rhizopus microsporus* is a plant pathogen known to infect maize, sunflower, and rice. *Rhizopus microsporus* interestingly contains as a bacterial endosymbiont *Burkholderia rhizoxinica*, which produces the antitumor drug rhizoxin. *Rhizopus oryzae* can degrade both plant and fungal polysaccharides. It lives worldwide in dead organic matter plus it infects and causes a soft rot of carrot, pineapple and mango. The commercial uses of *Rhizopus oryzae* include fermentation of rice for saki, and as part of a fungal consortium *Rhizopus oryzae* also can ferment rice straw to produce ethanol. As opportunistic pathogens of mammals, members of the genus *Rhizopus* cause both cutaneous as well as subcutaneous infections, and they can produce especially severe rhinocerebral infections when there is underlying diabetes. *Rhizopus microsporus* has a gastrointestinal presence and commonly is found in gastric ulcers of pig although a causal relationship with those ulcers remains uncertain. Both of the *Rhizopus* species listed here can independently cause abortion. *Rhizopus oryzae* additionally has been found to cause abortion with members of the genus

Aspergillus in mixed fungal infections (see the listing for *Aspergillus*). Note: the indicated capability of this fungal genus to cause infection of rabbit was achieved under experimental immunosuppression and thus far has not been noted to occur naturally, although that natural occurrence must be presumed possible. Those species belonging to this microbial genus which have been indicated as having a potentially opportunistic disease association with mammalian livestock are: *Rhizopus microsporus* (cattle, pig) and *Rhizopus oryzae* (cattle, dog, rabbit).

10.4.2.70 *Rhodotorula*

Phylum: Basidiomycota; Class: Microbotryomycetes; Order: Sporidiobolales; Family: (not assigned); Genus: *Rhodotorula*. Members of the genus *Rhodotorula* commonly are found as environmental inhabitants and can be isolated from soil, water, and air. They also are found in milk and fruit juices. These members of the genus *Rhodotorula* typically produce superficial infections of the skin and secondarily infect skin lesions in addition to causing mastitis and otitis externa. *Rhodotorula* can as well cause epididymitis and produce a septicemia which associatively includes infection of various internal organs among which are the lung and spleen. They very notably cause liver abscesses and endocarditis. Those species belonging to this microbial genus which have been indicated as having a potentially opportunistic disease association with mammalian livestock are: *Rhodotorula glutinis* (guinea pig, dog) and *Rhodotorula mucilaginosa* (cattle, sheep).

10.4.2.71 *Saksenaea*

Phylum: (not assigned); Class: (not assigned); Order: Mucorales; Family: Saksenaeaceae; Genus: *Saksenaea*. The species *Saksenaea vasiformis* is found in tropical and subtropical soils. It causes subcutaneous infections which present as necrotizing fasciitis and can become severely invasive (cattle).

10.4.2.72 *Sarocladium*

Phylum: Ascomycota; Class: Sordariomycetes; Order: Hypocreales; Family: (not assigned); Genus: *Sarocladium*. The species *Sarocladium kiliense* is a ubiquitous cellulolytic soil saprophyte and also found in the gut of the termite *Reticulitermes santonensis*. In mammals, the species *Sarocladium kiliense* causes keratoconjunctivitis and abortion (cattle, dog).

10.4.2.73 *Scedosporium*

Phylum: Ascomycota; Class: Sordariomycetes; Order: Microascales; Family: Microascaceae; Genus: *Scedosporium*. The members of this genus are soil fungi. The illnesses associated with members of this genus are infections of the skin and mucous membranes, nasal granuloma, osteomyelitis, discospondylitis (infection of the vertebral disks), and abortion. They also have been noted as potentially causing pulmonary infections. Those species belonging to this microbial genus which have been indicated as having a potentially opportunistic disease association with mammalian livestock are: *Scedosporium apiospermum* (dog) and *Scedosporium boydii* (cat, dog, cattle).

10.4.2.74 *Schizophyllum*

Phylum: Basidiomycota; Class: Agaricomycetes; Order: Agaricales; Family: Schizophyllaceae; Genus: *Schizophyllum*. The species *Schizophyllum commune* commonly is found growing in and on rotting wood and is a mushroom eaten by human populations. It infects the skin and also causes osteomyelitis (dog).

10.4.2.75 *Scopulariopsis*

Phylum: Ascomycota; Class: Sordariomycetes; Order: Microascales; Family: (not assigned); Genus: *Scopulariopsis*. The species *Scopulariopsis brevicaulis* occurs in soil and decaying organic matter. In mammals, *Scopulariopsis brevicaulis* causes keratitis, nodular skin lesions (mycetomas), invasive sinusitis, endophthalmitis, and plumonary infections. The infections can disseminate to produce systemic disease including endocarditis and brain abscess. The previous name of this species was *Microascus brevicaulis*. It has been identified as affecting the European bison (*Bison bonasus*), and so presumably also would infect the American bison (cat, cattle, goat, horse, likely also affects bison).

10.4.2.76 *Setosphaeria*

Phylum: Ascomycota; Class: Dothideomycetes; Order: Pleosporales; Family: Pleosporaceae; Genus: *Setosphaeria*. The species *Setosphaeria rostrata* is found in the mobile surface layer of Saharan desert soil. In mammals, this species causes cutaneous granulomas and corneal ulcers that can fulminantly disseminate resulting in meningitis. Additional information on this fungal species can be found by researching under its previous names *Exserohilum rostratum* and *Drechslera rostrata* (cattle).

10.4.2.77 *Sporobolomyces*

Phylum: Basidiomycota; Class: Microbotryomycetes; Order: Sporidiobolales; Family: (not assigned); Genus: *Sporobolomyces*. The species *Sporobolomyces roseus* frequently is associated with plants and also has invertebrate animal hosts. In mammals, *Sporobolomyces roseus* causes meningoencephalitis (dog).

10.4.2.78 *Sporothrix*

Phylum: Ascomycota; Class: Sordariomycetes; Order: Ophiostomatales; Family: Ophiostomataceae; Genus: *Sporothrix*. The species *Sporothrix schenckii* is present in soil, and can as well be found living upon and decomposing such plant materials as peat moss. It also infects invertebrate animal hosts. In mammals, *Sporothrix schenckii* causes subcutaneous infections that include the formation of fungal nodules and ulcerations. It can target the genital area including the penis, and cause nodules of the oral cavity, upper and lower respiratory tracts, eyes and conjunctiva. These infections can become lymphocutaneous by following the lymphatic system through the body. Immunosuppression can lead to these infections becoming multifocal and disseminated to produce systemic disease including meningitis (broadly infective of birds and mammals including cat, dog, horse, cattle, camel, goat, sheep, pig, rodents).

10.4.2.79 *Stachybotrys*

Phylum: Ascomycota; Class: Sordariomycetes; Order: Hypocreales; Family: (not assigned); Genus: *Stachybotrys*. The species *Stachybotrys chartarum* is found in soil and rotting grain. Its growth requires a high moisture content and one of its more common habitats is cellulose-rich water damaged construction materials. Its original isolation was from the wall of a house. The species *Stachybotrys chartarum* primarily is known as a cause of toxicity when either inhaled or ingested, but it also has been indicated to have an association with mastitis and diarrhea in cattle which might be either infectious or allergic in nature (cattle).

10.4.2.80 *Staphylotrichum*

Phylum: Ascomycota; Class: (not assigned); Order: (not assigned); Family: (not assigned); Genus: *Staphylotrichum*. The species *Staphylotrichum coccosporum* is a soil organism which causes subcutaneous infections (cat).

10.4.2.81 *Talaromyces*

Phylum: Ascomycota; Class: Eurotiomycetes; Order: Eurotiales; Family: Trichocomaceae; Genus: *Talaromyces*. The members of this genus typically are found in soil. Members of the genus *Talaromyces* cause disseminated infections that can affect the lung and also cause abortion. Those species belonging to this microbial genus which have been indicated as having a potentially opportunistic disease association with mammalian livestock are: *Talaromyces flavus* (cattle), *Talaromyces helicus* (dog); *Talaromyces marneffei* (dog), and *Talaromyces purpureogenus* (dog).

10.4.2.82 *Thermoascus*

Phylum: Ascomycota; Class: Eurotiomycetes; Order: Eurotiales; Family: Thermoascaceae; Genus: *Thermoascus*. The species *Thermoascus thermophilus* is a thermophilic soil fungus that also has been found in mushroom compost. In mammals, the species *Thermoascus thermophilus* causes abortion (cattle).

10.4.2.83 *Trematosphaeria*

Phylum: Ascomycota; Class: Dothideomycetes; Order: Pleosporales; Family: Trematosphaeriaceae; Genus: *Trematosphaeria*. The species *Trematosphaeria grisea* is a soil fungus which seems to be limited to tropical and subtropical regions. The previous name of this species was *Madurella grisea*. In mammals, *Trematosphaeria grisea* produces mycetomas (cat).

10.4.2.84 *Trichophyton*

Phylum: Ascomycota; Class: Eurotiomycetes; Order: Onygenales; Family: Arthrodermataceae; Genus: *Trichophyton*. The members of this genus are soil organisms but many of them more naturally seem associated with the skin and keratin structures produced by animals. Members of the genus *Trichophyton* typically cause dermatitis and rhinitis. Those species belonging to this microbial genus which have been indicated as having a potentially opportunistic disease association with mammalian livestock are: *Trichophyton equinum* (horse); *Trichophyton mentagrophytes* (rabbit, kangaroo, cat, cattle, guinea pig, coypu, horse, dog, sheep, goat), *Trichophyton rubrum* (dog, horse, cattle), and *Trichophyton verrucosum* (cattle, sheep, goat, kangaroo, horse).

10.4.2.85 *Trichosporon*

Phylum: Basidiomycota; Class: Tremellomycetes; Order: Tremellales; Family: (not assigned); Genus: *Trichosporon*. Members of this genus typically are isolated from soil but many also occur as a constituent of the natural skin microbiota, they also affect invertebrate animal hosts. Members of the genus *Trichosporon* typically cause cutaneous infections and mastitis. They also can produce disseminated infections resulting in meningoencephalitis. Those species belonging to this microbial genus which have been indicated as having a potentially opportunistic disease association with mammalian livestock are: *Trichosporon asahii* (cattle), *Trichosporon beigelii* (cat, horse), *Trichosporon cutaneum* (dog, cattle), and *Trichosporon montevideense* (dog).

10.4.2.86 *Verruconis*

Phylum: Ascomycota; Class: Dothideomycetes; Order: Venturiales; Family: Sympoventuriaceae; Genus: *Verruconis*. The species *Verruconis gallopava* is thermophilic and occurs in such natural environments as hot springs and thermal soils, as well as poultry litter and other self-heating organic waste. In mammals, *Verruconis gallopava* produces necrotic lesions of the skin, pulmonary symptoms similar to allergic bronchitis, and fatal systemic infections that include such results as encephalitis (cat, dog).

10.4.2.87 *Westerdykella*

Phylum: Ascomycota; Class: Dothideomycetes; Order: Pleosporales; Family: Sporormiaceae; Genus: *Westerdykella*. Members of the genus *Westerdykella* typically are found in a variety of natural environments including soil, and in general the Sporormiaceae are found in dung and rotting plant material. The species *Westerdykella reniformis* typically attacks the skin. Infections caused by this fungus can disseminate to the kidney and vertebral disks (dog).

10.4.2.88 *Xylohypha*

Phylum: Ascomycota; Class: Eurotiomycetes; Order: Chaetothyriales; Family: (not assigned); Genus: *Xylohypha*. The species *Xylohypha bantiana* typically is found in soil and rotting plant material. There are suggestions that the species *Xylohypha bantiana* is the same as *Cladophialophora bantiana*, but at this time both of these two species names are officially recognized (see the listing for *Cladophialophora bantiana*). The species *Xylohypha bantiana* is an opportunistic dermatophyte which can produce infections of the skin and disseminate to the internal organs. Systemic infections caused by this fungal species can invade the liver, spleen, kidney, adrenal glands, and produce cerebral abscess. This fungal species additionally has been found contributing as a secondary infection to the bacterium *Ehrlichia canis* (dog).

10.4.3 Ichthyosporea

10.4.3.1 *Rhinosporidium*

Phylum: (not assigned); Class: Ichthyosporea; Order: Dermocystida; Family: (not assigned); Genus: *Rhinosporidium*. The species *Rhinosporidium seeberi* has been found naturally residing in a surface water impoundment reservoir. It should be noted that officially the Ichthyosporea are not considered to be fungi. In mammals, *Rhinosporidium seeberi* causes rhinitis and attacks the nasal passages producing fungal nodules that resemble nasopharyngeal polyps. It also can cause conjunctivitis and produce systemic infections (cat, cattle, dog, horse).

References

Hurst CJ, Murphy PA (1996) The transmission and prevention of infectious disease. In: Hurst CJ (ed) Modeling disease transmission and its prevention by disinfection. Cambridge University Press, Cambridge, pp 3–54

Kahn CM, Line S (eds) (2010) The Merck veterinary manual, 10th edn. Wiley, New York

Pathogen Safety Data Sheets and Risk Assessment (2015) Public Health Agency of Canada, Ottawa. http://www.phac-aspc.gc.ca/lab-bio/res/psds-ftss/index-eng.php. Accessed 9 May 2015

Sumbali G (2011) The fungi, 2nd edn. Alpha Science, Oxford

The Center for Food Security and Public Health (2015) Iowa State University, Ames. http://www.cfsph.iastate.edu/. Accessed 9 May 2015

Vetbook (2015) Centre for Veterinary Education, The University of Sidney, Sidney. http://www.cve.edu.au/vetbook. Accessed 9 May 2015

Chapter 11
Opportunisitic Pathogens of Humans

Kristin M. Burkholder and Mary X.D. O'Riordan

Abstract Opportunistic pathogens (OP) pose a serious threat to human health, and the frequency of opportunistic infections (OI) is increasing worldwide. Rising rates of human OI are attributed to the emergence of antimicrobial-resistant or more virulent microbes, as well as advances in medical technology that have led to the extended lifespan of individuals suffering from chronic diseases and living in an immunocompromised state. This chapter highlights the populations at greatest risk for OI, as well as the major bacterial and fungal OPs that threaten human health and burden modern healthcare systems. We emphasize here characteristics of infection, epidemiology, microbial pathogenesis, and antimicrobial resistance.

11.1 Introduction

An OP is commonly defined as an organism which causes overt infection in immunocompromised hosts, but which does not cause clinical disease in healthy individuals (Casadevall and Pirofski 1999; Falkow 1997). These pathogens have a broad range of natural habitats, including soil, water, interior environments such as those found in clinical settings, and even the human body. Therefore, infections may arise from introduction of the pathogen into host tissues from outside sources, or may originate endogenously, from the patient's own normal microbiota. Individuals at greatest risk for OI are the immunocompromised and the very young and elderly, although some maternal pathogens also pose a special threat to pregnant women and their fetuses. The increased incidence of reported OIs is likely due to

K.M. Burkholder (✉)
Department of Biology, University of New England, Biddeford, ME, USA
e-mail: kburkholder@une.edu

M.X.D. O'Riordan
Department of Microbiology and Immunology, University of Michigan Medical School, Ann Arbor, MI, USA
e-mail: oriordan@une.edu

© Springer International Publishing Switzerland 2016
C.J. Hurst (ed.), *The Rasputin Effect: When Commensals and Symbionts Become Parasitic*, Advances in Environmental Microbiology 3,
DOI 10.1007/978-3-319-28170-4_11

many factors, including improved laboratory expertise in pathogen detection (Caston-Osorio et al. 2008) as well as genetic and physiological changes within the pathogens themselves. The spectrum of OPs that infect humans is changing, such that new and more virulent pathogens are emerging, and many opportunists are evolving elaborate mechanisms of drug resistance. In addition, advances in medical therapy for patients suffering from cancer, receiving organ transplantation, or afflicted with other chronic illnesses have led to a greater population of people living in an immunocompromised state. This places on the healthcare system the great burden of preventing and treating OIs in highly susceptible patients, in an era of unprecedented antimicrobial drug resistance.

In the following review, we discuss the major bacterial and fungal OPs that currently plague humans. Emphasis is placed on the frequency and types of diseases caused by each pathogen, as well as modes of disease transmission, pathogenic mechanisms, and concerns related to antibiotic resistance. For additional information regarding opportunistic protozoal and viral human infections, the reader is directed to several excellent reviews (Visvesvara et al. 2007; Webb et al. 2012; Andreani et al. 2012; Fishman 2013).

11.1.1 Opportunistic Infections in Individuals with Human Immunodeficiency Virus and Acquired Immune Deficiency Syndrome

Since the 1980s, human immunodeficiency virus and acquired immune deficiency syndrome (HIV-AIDS) has become part of the global landscape and has greatly shaped the scope of OI facing healthcare systems worldwide. The AIDS pandemic was first recognized in 1981, when the U.S. Centers for Disease Control and Prevention (CDC) reported five cases of pneumonia, caused by the opportunistic fungus *Pneumocystis jirovecii*, in gay men living in California (Centers for Disease Control 1981). Although the disease was initially believed to be restricted to homosexual males, additional cases were soon reported among non-homosexual intravenous drug users in the USA, Europe, Haiti, and African countries (Merson et al. 2008). In 1983, a retrovirus, which would later be named HIV, was isolated from an AIDS patient in France, and soon after the U.S. Food and Drug Administration (FDA) approved a commercial test for the virus (Barre-Sinoussi et al. 1983). By 1985, 71 countries had reported over 17,000 cases of AIDS to the World Health Organization (WHO). To date, every country in the world has acknowledged the presence of HIV-positive individuals in its population, and it is estimated that 35 million people are infected with HIV worldwide (UNAIDS 2013). Although new infections occur globally, approximately 95 % of new infections occur in low- to middle-income countries, particularly in sub-Saharan Africa (UNAIDS 2012).

The emergence of an HIV-positive immunodeficient population led to a dramatic spike in incidence of OI. However, the adoption of highly active antiretroviral therapy (HAART) in the 1990s led to a delay in the onset of AIDS in many HIV-infected individuals, and subsequently, the incidence of OI and death in AIDS patients has declined (Forrest et al. 1998; Centers for Disease Control and Prevention 1997; Hogg et al. 1998; Brooks et al. 2009). However, despite these advances, HIV infection is still a major risk factor for development of OI (Bonnet et al. 2005; Puhan et al. 2010; Palella et al. 2006). Because AIDS patients are subjected to frequent antibiotic prophylaxis and treatment, these individuals are also a potentially important reservoir for antibiotic-resistant microbes in hospitals and communities.

11.1.2 Opportunistic Infections in Non-HIV Infected Immunocompromised Individuals

Medical advances have led to improved treatment and longer life expectancy for patients with chronic diseases such as cancer, asthma, rheumatoid arthritis, cystic fibrosis, chronic obstructive pulmonary disease, and for organ transplant recipients (Sepkowitz 2002). As treatments for these conditions often involve prolonged immunosuppressive therapy, the end result is a larger population of HIV-free immunocompromised individuals who are at risk of developing OIs. These patients suffer frequent disease caused by all major classes of microbial OP, but specific causes of immune dysregulation can be more associated with particular OPs, such as *Pseudomonas aeruginosa*, which is a major cause of morbidity and mortality in patients with cystic fibrosis (Kontoyiannis et al. 2010; Fishman 2014).

11.1.3 Opportunistic Infections in Healthy Individuals

Although OI are typically defined as those that occur in immunocompromised populations, it is possible for healthy people to acquire OI, a phenomenon which blurs the lines of traditional distinctions between opportunistic and primary pathogens. In an immunocompetent person, OI might arise if the pathogen gains access to host tissues that it would normally not encounter, either via trauma or insertion of an indwelling medical device during hospitalization. Thus, if a normally intact host barrier is breached, an OP could establish infection in an otherwise healthy individual. The incidence of OI in healthy adults has increased with the emergence of opportunists with hypervirulent phenotypes, both within hospitals and in the community. For example, community-acquired methicillin-resistant *Staphylococcus aureus* (CA-MRSA) is a pathogen known to infect healthy humans with no predisposing risk factors for staphylococcal infection. Some particularly virulent strains of CA-MRSA can spread rapidly through communities, sports teams, schools, and prisons, infecting healthy and immunocompromised individuals alike, with potentially devastating outcomes (McDougal et al. 2003; Tattevin et al. 2012).

11.1.4 Role of Microbial Biofilms in OI

In nature, pathogens commonly exist as part of complex biofilms rather than as free-swimming planktonic cells, and therefore biofilms serve as important reservoirs for opportunists. A biofilm is a complex community of microbes attached to a surface through the use of microbial adhesion factors and extracellular polymers. Biofilms can form on animate and inanimate substrates and therefore are utilized by microbes to colonize and persist on surfaces both within and outside of the host. These structures are particularly common in clinical environments, where they easily contaminate medical devices such as indwelling catheters, mechanical ventilators, dialysis tubing, and dental equipment (Donlan 2002; Donlan and Costerton 2002; Anaissie et al. 2002). Cells within mature biofilms exhibit altered phenotypes compared to the corresponding planktonic cells. Biofilm-resident microbes often become increasingly resistant to disinfectants and antimicrobial drugs; therefore, biofilms are major reservoirs for antibiotic-resistant OP within the clinical setting (Lindsay and von Holy 2006). However, the mechanisms associated with antibiotic resistance in biofilm-resident bacteria can be distinct from those characterized in planktonic bacteria.

Biofilm formation follows a series of stages, including surface conditioning, microbial adhesion and colonization of the surface, and eventual dispersal of cells to seed new sites (Watnick and Kolter 2000). During the earliest stage of biofilm formation, surfaces are conditioned by the adsorption of organic and inorganic molecules that aid in subsequent binding of microbial cells. This stage can occur quickly on indwelling medical devices such as catheters, where host proteins and cellular debris coat the plastic surface and provide a nutritive and adhesive substrate for microbial colonization (Donlan 2002). Soon after, microbial cells bind to the surface via adhesion factors and secreted extracellular matrix polymers, largely comprised of exopolysaccharides. The biofilm community expands and forms a complex structure via replication and addition of new species of microbes. Cells of the mature biofilm are subject to increased level of genetic mutation and activation of quorum sensing systems, which induce antimicrobial tolerance or resistance mechanisms and hyperproduction of biofilm polymers (Hoiby et al. 2010). Eventually, small clusters of biofilm cells will detach and transfer to seed new surfaces (Lindsay and von Holy 2006). If biofilms are present on indwelling medical devices, during the final dispersal stage, antimicrobial-resistant pathogens may migrate and colonize host tissues, leading to disseminated disease.

11.2 Gram-Positive Opportunistic Bacteria

11.2.1 Staphylococcus *Species*

The Staphylococci are non-sporeforming, Gram-positive cocci that are frequent colonizers of mammals, particularly of skin surfaces. Of the 36 known Staphylococcus species, many of which are part of the normal human microbiota, *Staphylococcus aureus* and *Staphylococcus epidermidis* are particularly versatile as pathogens and are the species most often associated with human OI. Most Staphylococcal infections originate as skin and soft tissue infections (SSTIs), but contamination of the bloodstream or respiratory mucosa with Staphylococci can lead to serious and fatal diseases (Boucher and Corey 2008; Otto 2009).

11.2.2 Staphylococcus aureus

Of all the Staphylococci, *S. aureus* is most commonly implicated in human disease. *Staphylococcus aureus* causes a wide range of infections, including SSTIs, pneumonia, endocarditis, osteomyelitis, septicemia, and even self-limiting gastroenteritis due to the ability of the bacterium to contaminate food products and produce enterotoxins (Mistry 2013; Boucher and Corey 2008). Immunocompromised individuals and hospitalized patients are most commonly afflicted by invasive *S. aureus* infections, although as discussed below, community-acquired *S. aureus* strains have recently emerged. *Staphylococcus aureus* is highly prevalent in humans; it is estimated that up to one-half of the population is asymptomatically colonized by *S. aureus*, either transiently or persistently (Frank et al. 2010), and these individuals can serve as carriers of the microbe, providing a reservoir within the community. In these carriers, *S. aureus* can reside in many parts of the body, but is most frequently found in the anterior nares and, to a lesser degree, on the skin. The high prevalence of *S. aureus* in the human population likely contributes to the abundance of the pathogen in the hospital environment and its success both as a nosocomial and community-acquired pathogen. Indeed, *S. aureus* is the most commonly isolated pathogen from hospitalized patients in North America (Vincent et al. 2009). The ease with which *S. aureus* is transmitted also aids in its spread within hospitals and communities; the bacteria are easily transmitted through casual contact with infected and transient carriers, as well as through indirect contact with contaminated materials or medical devices. In addition, inevitable breaches of skin or mucosal surfaces provide opportunity for infection by *S. aureus* that is already present on the skin or other body sites.

11.2.2.1 *Staphylococcus aureus* Pathogenesis, Immune Evasion, and Biofilm Formation

Initial colonization of intact epithelial surfaces, such as the moist epithelium of the anterior nares, is initially dependent on the activity of *S. aureus* cell wall teichoic acid (WTA) and MSCRAMMs (microbial surface components recognizing adhesive matrix molecules), which bind to receptor molecules located within the host tissue's extracellular matrix (Weidenmaier et al. 2008). For example, *S. aureus* clumping factor B (ClfB) is an MSCRAMM that is highly expressed during nasal colonization and binds to host cytokeratins on the nasal epithelium (Wertheim et al. 2008; Walsh et al. 2004). Other factors, such as collagen-binding protein (Cbp) and fibronectin-binding protein (FnBP) permit colonization of exposed collagen, fibrinogen, and elastin, as could be found in abrasions, prosthetic devices, or even on heart valves (Foster and Hook 1998; Wann et al. 2000; Roche et al. 2004). Although simple colonization of intact epithelial surfaces does not induce a strong host immune response or result in damage to host tissues, colonization is a strong risk factor for Staphylococcal infection (von Eiff et al. 2001), and infections that do arise often initiate at sites where the skin barrier is breached, either through an abrasion or insertion of a medical device.

Upon breaching initial host barriers, *S. aureus* induces a strong inflammatory response, characterized by robust neutrophil recruitment, followed by infiltration of macrophages and fibroblasts. In some cases, the host response is sufficient to contain the infection, whereas in other cases the bacteria may spread to the bloodstream. Not surprisingly, *S. aureus* possesses an armament of virulence factors that enable it to colonize a variety of host tissues while evading destruction by the host inflammatory response. For example, *S. aureus* relies on multiple virulence factors to evade destruction by host phagocytes. The bacteria resist phagocytosis by expressing anti-phagocytic surface molecules such as protein A, capsular polysaccharides, clumping factor A, and a variety of complement inhibitors, which interfere with opsonization by host antibodies and complement proteins (Foster 2005, 2009; Veldkamp and van Strijp 2009). *Staphylococcus aureus* also produces leukotoxins, such as phenol-soluble modulins (PSM) and Panton–Valentine leukocidin (PVL) which specifically lyse neutrophils (Loffler et al. 2010; Surewaard et al. 2013). Staphylococcal exoenzymes such as nucleases and adenosine synthase contribute to degradation of neutrophil toxic products (Thammavongsa et al. 2013), while collagenases, coagulases, and staphylokinases promote tissue destruction and invasion into deeper extracellular spaces (Shaw et al. 2004).

Although *S. aureus* has historically been characterized as an extracellular pathogen, recent reports demonstrate that *S. aureus* pathogenesis involves colonization of both extracellular and intracellular host environments (Fraunholz and Sinha 2012). While extracellular *S. aureus* can replicate quickly and cause extensive damage to host tissues via exoenzyme and toxin production (Shaw et al. 2004), the population of *S. aureus* that enter and survive within host cells may use the intracellular niche as a means to persist and even disseminate in the face of the host immune response (Loffler et al. 2013; Thwaites and Gant 2011).

The noteworthy ability of *S. aureus* to form biofilms on implanted medical devices as well as native host tissues (Scherr et al. 2014) is at least in part attributed to bacterial MSCRAMMS and teichoic acid which mediate attachment onto abiotic and biotic surfaces that have been preconditioned with host matrix proteins (Navarre and Schneewind 1999). *Staphylococcus aureus* also secretes polysaccharide intercellular adhesion (PIA) which aids in bacterial attachment to other microbial cells and to surfaces (Mack et al. 1996). These PIA molecules constitute a large portion of the "slime" often associated with biofilms, and the genes encoding PIA are contained within the inducible *ica* operon, which is positively regulated by quorum sensing molecules. The formation of biofilms during infection are especially dangerous because Staphylococcal biofilms are recalcitrant to many antibiotics and can also alter the host immune response such that bacterial persistence is favored (Scherr et al. 2014).

11.2.2.2 Antibiotic Resistance in *S. aureus*

The emergence of methicillin-resistant *S. aureus* (MRSA) poses a significant threat to public health and is a great challenge to modern medicine, with MRSA being responsible for an increasing proportion of *S. aureus* infections. In the USA alone, approximately 14 million people seek medical treatment for *S. aureus*-associated SSTIs each year (Hersh et al. 2008), of which over 50 % are caused by methicillin-resistant *S. aureus* (MRSA) (Moran et al. 2006; Van de Velde et al. 2008). Like drug-susceptible *S. aureus*, MRSA infections can also persist and cause systemic disease, and invasive MRSA infections are estimated to cause over 18,000 deaths per year in the USA (Klevens et al. 2007). The high hospitalization and mortality rates associated with MRSA are attributed to increasing drug resistance; MRSA are resistant to most beta-lactam drugs, as well as to aminoglycosides, fluoroquinolones, erythromycin, clindamycin, and rifampin. Emergence of strains resistant to vancomycin, which is typically the drug of last resort for MRSA infections, means that few treatment options remain (Weigel et al. 2003).

11.2.2.3 Emergence of Community-Acquired MRSA

Methicillin-resistant *S. aureus* was once recognized primarily as a hospital-acquired (nosocomial) pathogen that gained access to immunocompromised or physiologically weakened hosts via indwelling medical devices. However, new strains of CA-MRSA have surfaced that cause infection in healthy individuals with no predisposing risk factors for staphylococcal infection (Herold et al. 1998). The infection rates for CA-MRSA have risen such that these strains are now the leading cause of SSTIs in patients admitted to US emergency rooms (Moran et al. 2006; Popovich et al. 2008; Talan et al. 2011). Of particular concern is the evolution of the highly virulent CA-MRSA clone USA300, which causes aggressive and persistent SSTIs that can spread systemically (McDougal et al. 2003; Tattevin et al. 2012). The USA300 strain has been isolated from outbreaks plaguing prisons, hospitals, sports teams, and small communities (McCaskill et al. 2007), and it is well-documented that patients with USA300 infections are more likely to suffer life-threatening complications than are patients with non-USA300 MRSA infections (Kreisel et al. 2011).

The success of USA300 as a human pathogen is attributed to unique virulence traits and its remarkable adaptation to the host environment. In comparison to other MRSA strains, USA300 overexpresses the global virulence gene regulator *agr*, as well as Agr-regulated toxins such as PVL, Hla, and PSMs (Li et al. 2010; Kobayashi et al. 2011). In addition, in contrast to other MRSA strains, USA300 strains harbor the arginine catabolism mobile element (ACME), which was horizontally acquired from *S. epidermidis* and which makes USA300 uniquely suited for survival in the harsh skin environment (Planet et al. 2013; Thurlow et al. 2013). The ACME encodes a constitutive arginine-deiminase system (Arc), which promotes bacterial ammonia production and aids in bacterial survival in the low pH environment of the skin. Further, while most strains of *S. aureus* are susceptible to destruction by host polyamines produced during the post-inflammatory stages of wound infection, ACME-encoded SpeG enables USA300 to degrade host spermine to avoid polyamine toxicity (Thurlow et al. 2013; Joshi et al. 2011), which greatly enhances the ability of USA300 to cause persistent wound infections.

11.2.3 Staphylococcus epidermidis

Previously considered an innocuous commensal microbe of human skin, *Staphylococcus epidermidis* is now recognized as a leading agent of nosocomial and medical device-associated infections (Van Mellaert et al. 2012). The high incidence of hospital-acquired *S. epidermidis* infections is due to the fact that the bacterium is carried on the skin of most humans, and therefore medical devices can be contaminated by *S. epidermidis* from the patient's own microbiota or from the skin of a healthcare worker (Uckay et al. 2009; Gomes et al. 2011). *S. epidermidis* is the most common cause of nosocomial bloodstream infections, cardiovascular infections, and infections of the eye, ear, nose, and throat. Although *S. epidermidis* infections are rarely fatal, their frequency, along with the fact that they are difficult to treat, presents a serious burden for the healthcare industry (Otto 2009). For example, over 20 % of bloodstream infections associated with central intravenous catheters are caused by *S. epidermidis*, and the economic burden of such infections is around US $2 billion per year in the USA (National Nosocomial Infections Surveillance System 2004).

11.2.3.1 *Staphylococcus epidermidis* Pathogenesis

Unlike its virulent relative *S. aureus*, *S. epidermidis* seemingly lacks the virulence factors necessary to directly invade and damage host tissues. Instead, *S. epidermidis* pathogenesis relies almost entirely on the bacterium's ability to form adherent biofilms on the abiotic surface of medical devices (Gomes et al. 2011; Uckay et al. 2009). As is the case with *S. aureus*, *S. epidermidis* biofilms are difficult to eradicate, as the bacteria within the biofilms become resistant to common disinfectants, antimicrobial drugs, and host defenses (Hanke and Kielian 2012; O'Gara and Humphreys 2001). Biofilm formation on synthetic surfaces is also dependent upon surface-associated MSCRAMMs, teichoic acids, and the exopolysaccharide PIA (Otto 2012; Gross et al. 2001; Holland et al. 2011). Although *S. epidermidis* lacks most exotoxins of *S. aureus*, *S. epidermidis* does produce at least five types of PSMs (PSMα, PSMβ, PSMδ, PSMε, and PSM-mec). In particular, PSMα and PSMδ damage human neutrophils and erythrocytes, but are produced at relatively low levels compared to similar PSMs in *S. aureus* (Cheung et al. 2010). The low-level expression of PSMs in *S. epidermidis* correlates with the bacterial lifestyle as a minimally virulent commensal and OP.

11.2.3.2 Antibiotic Resistance in *S. epidermidis*

Like *S. aureus*, *S. epidermidis* are becoming increasingly antibiotic resistant. Drug tolerance can result from existence within a biofilm, as the thick polysaccharide matrix minimizes bacterial exposure to certain drugs (Stewart 2002). In addition, numerous strains of *S. epidermidis* have acquired drug resistance genes from other bacteria within the community. Up to 90 % of clinical *S. epidermidis* isolates contain the methicillin-resistance gene *mecA* and exhibit moderate to high levels of methicillin resistance (Diekema et al. 2001). Prevalence of *S. epidermidis* strains tolerant to macrolide and aminoglycoside antibiotics as well as tetracycline, chloramphenicol, and clindamycin is also increasing, and drug resistance is found most frequently in clinical isolates (Rogers et al. 2009). However, unlike *S. aureus*, there is little evidence for spread of vancomycin-resistant strains of *S. epidermidis* (Otto 2012).

11.2.4 Streptococcus *Species*

The genus *Streptococcus* includes important commensal and pathogenic bacteria that asymptomatically colonize or cause infections of the upper respiratory tract and the intestines of man and animals. Streptococci are nonmotile, non-sporeforming, facultatively anaerobic Gram-positive bacteria. Nearly forty species of Streptococci exist, and most species are exclusively associated with a living host. Virtually all the commensal Streptococci can also cause OI, particularly if they gain access to the bloodstream from their site of colonization in the host oral cavity or intestinal tract. The species most often associated with OI are *S. pneumonia*, *S. agalactiae*, *S. sanguinis*, and *S. mutans*. Although *S. pyogenes* is also a human pathogen of great importance, we will not focus on its role in infection as the bacterium can cause a variety of diseases in healthy humans and therefore does not fit the classic definition of an opportunist. *S. pyogenes* disease states and pathogenesis are thoroughly reviewed elsewhere (Cole et al. 2011).

11.2.5 Streptococcus pneumoniae

Streptococcus pneumoniae, commonly referred to as pneumococcus due to its round morphology, is a human nasopharyngeal commensal which asymptomatically colonizes 10–40 % of healthy individuals (Marks et al. 2013). In the young, elderly, and immunocompromised, pneumococcus can cause a variety of diseases such as pneumonia, meningitis, and sepsis. In the USA, *S. pneumoniae* causes approximately 40,000 fatalities per year. In particular, pneumococcal meningitis can have a very high case fatality rate of 20 % in developed countries and up to 50 % in developing countries (Lynch and Zhanel 2010). Pneumococcal meningitis survivors may also suffer from serious sequelae such as hearing loss or neurological or neuropsychological impairment (Jit 2010).

11.2.5.1 *Streptococcus pneumoniae* Pathogenesis

The bacterium's thick exopolysaccharide capsule is its most critical virulence factor. The capsule mediates both adhesion to epithelial surfaces and is also antiphagocytic, as it inhibits complement- and antibody-mediated opsonophagocytosis (Hyams et al. 2010). In fact, *S. pneumoniae* mutants deficient in the gene encoding capsular polysaccharides are rapidly cleared by host phagocytes and are severely attenuated for virulence (Magee and Yother 2001). The *S. pneumoniae* capsule has also played a key role in our understanding of basic biological concepts (Watson and Musher 1999). For example, in 1928, the British scientist Frederich Griffith reported genetic exchange between bacteria when nonencapsulated; avirulent *S. pneumoniae* mutants were transformed to an encapsulated, virulent phenotype following co-infection of mice with capsule-deficient and wild-type strains (Griffith 1928). Later, this "transforming principle" was attributed to the ability of a bacterial cell to incorporate environmental DNA originating from neighboring lysed cells and led to the discovery of an important mode of horizontal gene transfer.

Streptococcus pneumoniae also relies on non-capsular adhesion factors to bind to host epithelial surfaces during infection. The pathogenicity island-encoded (PAI) serine-rich repeat protein PsrP contributes to colonization and virulence in lower respiratory tract infections by binding to Keratin 10 on the lung epithelium (Shivshankar et al. 2009; Obert et al. 2006). In addition, several PAI-encoded pili facilitate binding to host cells. Previously characterized simply as colonization factors, pneumococcal pili are now recognized as virulence factors that enable the bacteria to survive in the lung where they mediate attachment to lung epithelial cells, particularly during early stages of colonization (Pancotto et al. 2013). In addition, piliated strains of *S. pneumoniae* induce a stronger TNFα-mediated host inflammatory response and subsequently cause greater tissue damage and bacterial invasion in infected lungs (Barocchi et al. 2006).

Streptococcus pneumoniae uses several secreted factors to evade destruction by the host immune system. The bacterial Iga1 protease inactivates human IgA1 (Kilian et al. 1980), an abundant antibody in the epithelial mucosa of the upper respiratory tract (Fagarasan and Honjo 2004). This protease contributes to evasion of phagocytosis by preventing IgA-mediated opsonization of bacterial cells. In addition, some strains of *S. pneumoniae* produce an intracellular membrane-damaging toxin called pneumolysin, which is released from the bacterial cells via autolysis (Hirst et al. 2008; Canvin et al. 1995). Pneumolysin inhibits neutrophil chemotaxis (Paton and Ferrante 1983) and alters the neutrophil oxidative burst in a manner that damages the neutrophils rather than the bacteria (Martner et al. 2008).

11.2.5.2 Pneumococcal Vaccines and Vaccine Resistance

In the USA, anti-pneumococcal vaccines are recommended for children as well as individuals over age 65 (Assaad et al. 2012). Vaccines are often based on *S. pneumoniae* polysaccharide capsular antigens (Zhang et al. 2002). A seven-valent pneumococcal conjugate vaccine (PCV7), which contains polysaccharide antigens from seven *S. pneumoniae* serotypes, was introduced in 2000 in the USA, and it has been successful in reducing invasive pneumococcal disease, especially in children (Dagan 2008). However, the initial effectiveness of PCV7 is now waning, as non-vaccine strains of *S. pneumoniae* are emerging as a major cause of illness. The emergence of these non-vaccine strains has led to the subsequent replacement of PCV7 by 10-valent (PCV-10) and 13-valent (PCV13) vaccines (Myint et al. 2013). The higher-valency vaccines include polysaccharide antigens from emerging strains of *S. pneumoniae* and offer hope for continued protection of at-risk populations from pneumococcal disease.

11.2.5.3 Antibiotic Resistance in *S. pneumoniae*

The prevalence of antibiotic-resistant pneumococcus has increased worldwide. The adoption of the PCV7 vaccine led to decreased antibiotic resistance in the PCV7 serotypes due to reduction in invasive infections requiring drug treatment (Kaplan et al. 2004). However, PCV7 introduction led to increased incidence of infection by, and antibiotic resistance in, non-PCV7 serotypes. The most important serotypes that currently cause pediatric pneumococcal infection are serotypes 19A, 15A, and 35B. In particular, the rise in 19A infections has been accompanied by increased rates of penicillin-resistance and multidrug-resistance in that serotype (Dagan and Klugman 2008).

11.2.6 Streptococcus agalactiae

Streptococcus agalactiae, or Group B Streptococcus (GBS), is an OP that primarily causes disease in pregnant women and their newborns and in immunocompromised or physiologically weakened adults. GBS can be carried asymptomatically in the lower gastrointestinal tract or genital tract of pregnant women, where it may be transmitted to infants during delivery. Infant infection often occurs via aspiration of GBS from the mother's birth canal, resulting in contamination of the oral cavity and respiratory tract and potential systemic invasion through the neonatal lung (Mitchell 2003). Infection is associated with severe diseases such as pneumonia, meningitis, and septicemia in newborns and mothers (Nandyal 2008; Mullaney 2001). Infants can suffer from early-onset or late-onset disease caused from GBS. Symptoms of early-onset infection appear within hours of delivery, and the disease manifests as pneumonia that can rapidly progress to bacteremia and sepsis. Late-onset infant infection occurs weeks to months after birth, and usually presents as meningitis, resulting from bacteria crossing the respiratory epithelial barrier to enter the bloodstream and migrate to the central nervous system (Melin 2011). In immunocompromised adults, such as those with diabetes or cancer, GBS can cause pneumonia, endocarditis, as well as infections of soft tissue, bones, and joints. In very rare circumstances, some particularly virulent strains of GBS can cause necrotizing fasciitis or toxic shock syndrome (Mitchell 2003). Maternal screening for GBS and antibiotic treatment has been an effective method for decreasing neonatal infections.

11.2.6.1 Group B Streptococcus Pathogenesis

The virulence associated with *S. agalactiae* is attributed to factors that enable the bacteria to colonize epithelia of the genital and intestinal tracts, evade host defenses, and damage host tissues. Colonization is facilitated by the bacterial laminin-binding protein (Lmb), which binds to the host extracellular matrix protein laminin (Spellerberg et al. 1999). Like *S. pneumoniae*, *S. agalactiae* also produces a polysaccharide capsule that promotes colonization as well as immune evasion. In particular, the GBS capsule inhibits opsonization by complement proteins, thereby preventing activation of the complement pathway (Lindahl et al. 2005). In addition, the streptococcal fibrinogen-binding protein FbsA protects against opsonophagocytosis by inducing fibrinogen-dependent aggregation of platelets around the bacterial cells, essentially shielding the bacteria from host phagocytes. Expression of FbsA is also implicated as an important virulence factor in infective endocarditis, as it promotes adhesion to host tissues and thrombus formation (Pietrocola et al. 2005).

11.2.7 Oral Streptococci

Streptococci are dominant members of the oropharyngeal resident microbiota in humans of all ages. Oral streptococci are generally referred to as viridans streptococci, a serologically based classification based on bacterial reaction with Lancefield group antigens. These commensal streptococci are classified into five distinct groups: (1) sanguinis group, containing *S. sanguinis* and *S. gordonii*; (2) mutans group, which includes *S. mutans* and *S. sobrinus*; (3) salivarius group, which contains *S. salivarius*; (4) anginosus group, which contains *S. anginosus* and *S. intermedius*; and (5) mitis group, including *S. mitis* and *S. oralis*. Some of these commensal streptococci are purported to have a positive impact on oral health. In particular, *S. salivarius* exhibits probiotic behaviors by producing bacteriocins to inhibit colonization of oropharyngeal pathogens (Wescombe et al. 2009) and by modulating the epithelial inflammatory response (Cosseau et al. 2008). However, despite the beneficial impact of commensals like *S. salivarius*, other oral streptococci such as *S. sanguinis* and *S. mutans* are important and common OP of humans.

11.2.8 Streptococcus sanguinis *Infections and Pathogenesis*

Streptococcus sanguinis is the most common cause of subacute infective endocarditis (Mylonakis and Calderwood 2001; Moreillon and Que 2004) and is also associated with septic arthritis and periodontal disease. As one of the major bacterial species in dental plaque, *S. sanguinis* is an early colonizer of the tooth surface, on which it produces biofilms. Whether or how *S. sanguinis* contributes to pathogenesis of periodontitis is unclear; some strains are reported to interact synergistically with another periodontal pathogen, *Porphyromonas gingivalis*, to cause gingival infections (Stinson et al. 1991), while other strains may actually reduce incidence of periodontal disease (Stingu et al. 2008). However, in patients with weakened gingiva due to gingivitis or periodontal disease, small tears in the gum tissue provide a route for bloodstream contamination by *S. sanguinis*. In these patients, *S. sanguinis* can cause native valve endocarditis, a disease in which the bacteria colonize and form biofilms on heart valves. *Streptococcus sanguinis* endocarditis most often afflicts elderly patients with a prior history of cardiac illness. In these patients, *S. sanguinis* biofilms impede normal cardiac function and can also disperse from the initial site of colonization to form vascular emboli (Mylonakis and Calderwood 2001). Although rare, *S. sanguinis* can also cause septic arthritis, a condition in which the bacteria directly invade and cause inflammation in joint tissue, particularly in patients with history of cardiovascular disease (Mandac et al. 2010; Edson et al. 2002; Papaioannides et al. 2006).

Streptococcus sanguinis pathogenesis relies on factors that enable the bacteria to colonize while evading immune defenses of the host oral mucosa and circulatory system. Several unique virulence genes, *purB*, *purl*, *thrB*, and *pyrE*, contribute to *S. sanguinis* biofilm formation in vitro. However, these genes are dispensable for induction of infective endocarditis in vivo (Ge et al. 2008), suggesting that *S. sanguinis* biofilm formation in the host is a complex process that remains to be fully elucidated. In vivo infection does involve *S. sanguinis* adhesion to host fibronectin, as well as bacterially-induced platelet aggregation. The consolidation of platelets around bacterial cells aids in biofilm development as well as evasion of host phagocytes, and requires bacterial interaction with the platelet receptors glycoprotein (GP) IIb/IIIa, GPIbα, FcγRIIa, and also complement receptors (Cox et al. 2011; Kerrigan et al. 2002). In addition, the *S. sanguinis* cell-surface ectonucleotidase, Nt5e, was implicated as a potential immunosuppressive agent that potentiates biofilm formation in vivo (Fan et al. 2012).

11.2.9 Clostridium difficile

Clostridium difficile is a Gram-positive, obligate anaerobic, spore-forming bacterium that is widespread in the environment and is also carried asymptomatically in the human intestine. The bacterial species *C. difficile* was believed to be a non-pathogenic human commensal until 1978, when the bacterium was identified as the source of enterotoxin in the stools of patients with pseudomembranous colitis (Bartlett et al. 1978). Now, *C. difficile* is recognized as the leading cause of nosocomial diarrhea in industrialized nations. In certain regions of the USA, *C. difficile* is the leading cause of healthcare-associated infections, and the incidence of hospital patients discharged with *C. difficile* infections more than doubled between 2000 and 2009 (Lessa et al. 2012).

Disruption of the normal gut microbiota has long been known to play a central role in the pathogenesis of *C. difficile* infection. Recent findings suggest that *C. difficile* proliferates in the intestinal tract following antibiotic therapy by exploiting metabolites that are more abundant in the intestine after antibiotic administration (Theriot et al. 2014). Therefore, both antimicrobial therapy and underlying illnesses that alter intestinal microbial populations or host immunity are key risk factors for development of *C. difficile* enterocolitis. *Clostridium difficile* most commonly infects individuals following treatment with antibiotics such as cephalosporins, clindamycin, penicillins, and fluoroquinolones (Warren and Guerrant 2011; Young and Schmidt 2004). Populations with highest rates of *C. difficile* infections include the elderly, individuals with HIV/AIDS, and cancer patients undergoing chemotherapy (Collini et al. 2013; Keller and Surawicz 2014; Taslim 2009). In particular, *C. difficile* infection disproportionately affects adults over age 65, likely due to their increased susceptibility to other diseases that result in longer hospital stays with antimicrobial therapy. In 2008, *C. difficile* was ranked as the 18th leading cause of death in people aged 65 and older, and approximately 92 % of deaths from *C. difficile* occurred in this age group (Minino et al. 2011). In addition, *C. difficile* is one of the most commonly isolated pathogens in HIV patients with diarrheal disease. Increased risk of *C. difficile* infections in HIV-positive individuals is attributed to their higher frequency of hospitalization and antimicrobial drug use, but also likely involves HIV-induced alterations in the gut microbiota and compromised immune response (Collini et al. 2013).

11.2.9.1 *Clostridium difficile* Infections and Pathogenesis

Clostridium difficile is acquired via the ingestion of bacterial spores which germinate in the intestine. In the hospital environment, the spores are difficult to eliminate and are easily transmitted from patient to patient or on the hands of healthcare workers (Walker et al. 2012). In general, *C. difficile* is a poor competitor against other members of the gut microbiota and therefore does not proliferate or produce symptomatic disease in healthy individuals. However, in individuals with an altered microbiome, *C. difficile* will replicate and produce exotoxins that destroy intestinal epithelial cells and induce sloughing of the intestinal lining, which leads to classic symptoms of diarrhea, pseudomembranous stools, painful abdominal cramping, and dehydration. Patients with severe infections can experience complications such as toxic megacolon, in which colonic inflammation triggers muscle and neural injury and results in intestinal distension and poor gut motility (Earhart 2008). A rarer complication of *C. difficile* infection is development of fulminant colitis, in which the large bowel can become perforated, leading to peritonitis and high rate of mortality (Koss et al. 2006).

The main *C. difficile* virulence factors are its exotoxins, *C. difficile* toxins A (TcdA) and B (TcdB) and the A-B type toxin *C. difficile* toxin (CDT). Both TcdA and TcdB are cytotoxic to intestinal epithelial cells, because they disrupt the actin cytoskeleton and impair tight junction integrity (Thelestam and Chaves-Olarte 2000). After the two toxins TcdA and TcdB enter host cells, these toxins glycosylate and inactivate Ras and Rho GTPases, inhibiting their normal regulation of cytoskeletal dynamics (Jank et al. 2007; Stubbs et al. 2000). This decreases transepithelial resistance of the gut barrier and results in destruction of the epithelium, promoting fluid accumulation within the intestinal lumen. The toxin CDT also contributes to cytoskeletal dysfunction by ADP ribosylation, altering the normal function of host actin molecules (Stubbs et al. 2000). However, the role of CDT in disease remains unclear, as CDT-deficient mutants lose their cytotoxic effect in some infection models but not others (Rupnik et al. 2009).

11.2.9.2 Emergence of Highly Virulent *C. difficile* NAPI/BI/027

Increased rates of *C. difficile* infection are at least partly attributed to the emergence of the highly virulent NAPI/BI/027 strain in the USA and Europe (Rupnik et al. 2009). Several characteristics are thought to contribute to hypervirulence of NAPI/BI/027. This strain possesses polymorphisms in *tcdC*, a negative regulator of toxin production, which results in hyperproduction of cytotoxins (Warny et al. 2005). In addition, NAPI/BI/027 produces a TcdB toxin with greater binding affinity for host receptors compared to other *C. difficile* strains (Stabler et al. 2008). The NAPI/BI/027 strain also exhibits high-level resistance to fluoroquinolone antibiotics, leading to its selection over other strains in medical settings where fluoroquinolone therapy is used. Although the majority of *C. difficile* strains cause hospital-associated infections, NAPI/BI/027 has been isolated from patients with diarrheal illness who have not undergone antibiotic therapy and who have no predisposing risk factors for clostridial infection (Rupnik et al. 2009). These incidents suggest that the future of *C. difficile* control will focus not only on hospital-acquired but also on community-acquired infections.

11.2.9.3 Novel Treatments for *C. difficile* Infection

Treatment for *C. difficile* enterocolitis involves both prudent use of antimicrobials and, more recently, attempts to restore the gut microbiome to create a microenvironment not suitable for *C. difficile* growth. Initial therapy usually entails an alteration in antibiotic use in an effort to find an antibiotic that will alleviate rather than exacerbate the infection. Currently, vancomycin and metronidazole are most commonly used to treat *C. difficile* infections. While most strains of *C. difficile* remain susceptible, there is a trend towards increased resistance to these drugs and therefore much current research focuses on non-antimicrobial strategies to treat *C. difficile* infections (Rea et al. 2013). For example, a few studies indicate that frequent consumption of probiotic bacteria, such as *Lactobacillus* and *Streptococcus* species, can reduce incidence of antibiotic-associated diarrheal disease in hospitalized patients (Hickson 2011; Gao et al. 2010). Perhaps the therapeutic strategy that has gained the most recent attention is fecal transplantation, in which stool from an uninfected donor is transplanted into the colon of a person with *C. difficile* infection. This treatment concept is not entirely new, and its development was first published by Eisenman and his colleagues in 1958 (Eiseman et al. 1958). Fecal transplantation has potential for success because it directly addresses the underlying pathophysiology of *C. difficile* enterocolitis (Burke and Lamont 2013). Recent reports show that the success rate for fecal transplantation in individuals with recurrent *C. difficile* infections is over 90 % (Koenigsknecht and Young 2013). However, fecal transplantation is less effective in individuals colonized with the emerging NAPI/BI/027 strain (Mattila et al. 2012), suggesting that additional exploration is needed to find an appropriate therapeutic for the more virulent *C. difficile* infections.

11.2.10 Listeria monocytogenes

Listeria monocytogenes is a Gram-positive, facultative intracellular pathogen that causes the systemic foodborne disease listeriosis. The majority of human listeriosis cases occur in the young, elderly, immunocompromised, and pregnant women (Bortolussi 2008). In nonpregnant individuals, listeriosis primarily manifests as meningitis, encephalitis, or septicemia. In these patients, the incubation period is approximately 3 weeks, and symptoms include fever, intense headache, nausea, and neurological impairment (Peeters et al. 1989). Pregnant women are most susceptible to infection during the third trimester of pregnancy, when maternal T-cell immunity is most impaired (Pamer 2004). Although the pregnant woman usually remains asymptomatic or experiences mild flu-like symptoms, listeriosis poses a severe threat to the fetus and can result in spontaneous abortion, stillbirth, or birth of a child with sepsis or meningitis (Allerberger and Wagner 2010). Listeriosis infections cause nearly 1500 hospitalizations per year in the USA and results in over 250 deaths. Although *Listeria monocytogenes* is not the most frequent cause of foodborne infections due to extensive food safety surveillance, it is one of the most deadly foodborne pathogens, with an outbreak-associated mortality rate of approximately 20 % (Scallan et al. 2011).

11.2.10.1 *Listeria monocytogenes* Contamination of Food

Listeria monocytogenes is ubiquitous in the environment and can be isolated from soil, water, and the intestinal tract of food-producing animals. Therefore, the bacterium can contaminate raw food ingredients, which provide a route for the pathogen to enter the food processing environment. *Listeria monocytogenes* is especially hardy and can survive and replicate at high salt concentrations, a wide pH range, and at refrigeration temperatures. These characteristics allow the bacteria to persist in the food processing plant environment, where *Listeria monocytogenes* forms biofilms that are especially resistant to many disinfectants (Folsom and Frank 2006; Chavant et al. 2004; Saa Ibusquiza et al. 2011; Pan et al. 2006). *L. monocytogenes* has been isolated from a variety of raw and processed food products, including produce, dairy products, meat and egg products, seafood, and ready-to-eat foods (Farber and Peterkin 1991). Recent deadly listeriosis outbreaks were associated with consumption of contaminated cantaloupe in several US states (Laksanalamai et al. 2012), and deli meat in Canada (Gilmour et al. 2010).

11.2.10.2 *Listeria monocytogenes* Pathogenesis

Infections caused by *L. monocytogenes* are thought to initiate in the distal small intestine, where the bacterium uses multiple virulence factors to interact with the intestinal epithelium. Initial adhesion to host cells involves adhesion factors such as autolysin amidase (Ami) (Milohanic et al. 2001), fibronectin-binding protein (FbpA) (Osanai et al. 2013), cysteine transport-associated protein (CTAP) (Xayarath et al. 2009) and *Listeria* adhesion protein (LAP) (Wampler et al. 2004; Burkholder and Bhunia 2010). Internalin J (InlJ), a member of the large family of internalin proteins, binds to MUC2, a major component of intestinal mucus (Linden et al. 2008), and also promotes adhesion to epithelial cells of non-intestinal origin (Sabet et al. 2008). Following initial adhesion, the bacteria breach the intestinal barrier through the concerted action of multiple virulence factors. Direct invasion of epithelial cells is largely driven by InlA and InlB; InlA stimulates entry into intestinal epithelial cells, while InlB mediates invasion into different non-phagocytic cell types (Dramsi et al. 1995; Mengaud et al. 1996). The virulence invasion protein (Vip) also directs epithelial invasion by interacting with host receptor GP96 (Cabanes et al. 2005). In addition to using invasins to enter intestinal epithelial cells, *Listeria* also secretes the LAP protein, which has been shown to weaken tight junctions in intestinal epithelial cell lines, and promotes LAP-mediated bacterial translocation through the epithelial barrier (Kim and Bhunia 2013; Burkholder and Bhunia 2010).

Central to *Listeria* pathogenesis is the ability of the bacterium to survive and replicate to high numbers inside of nonphagocytic and phagocytic host cells and spread from cell to cell. In particular, intraphagocyte survival is critical for dissemination of the bacteria from the initial site of intestinal infection to deeper tissues such as the central nervous system (CNS) (Drevets et al. 2001). An arsenal of virulence factors contribute to the *L. monocytogenes* intracellular lifestyle. Soon after phagocytosis, the bacterium secretes listeriolysin O (LLO), a pore-forming toxin, which perforates the phagosomal membrane and facilitates bacterial escape from the potentially degradative phagosome into the more permissive environment of the host cytoplasm (Gedde et al. 2000). Once in the cytoplasm, the listerial surface protein ActA nucleates host actin into a tail-like structure, which confers to the bacterium actin-based motility and the ability to spread to adjacent cells (Tilney and Portnoy 1989; Brundage et al. 1993; Welch et al. 1998). Cytoplasmic infection is also characterized by rapid bacterial replication, which is aided by adaptations that allow for metabolic efficiency within the intracellular environment. For example, *L. monocytogenes* uses the hexose phosphate transporter Hpt to make use of glucose-1-phosphate present in the host cytosol. *Listeria monocytogenes* also relies on the bacterial lipoate ligase LplA1 to add a critical lipoyl modification to pyruvate dehydrogenase in the cytosol, where host lipoyl substrates are limiting (Keeney et al. 2007; O'Riordan et al. 2003; Keeney et al. 2009).

In addition to being an important opportunist of humans, *L. monocytogenes* is a powerful tool for cell biologists and immunologists. Early studies of

L. monocytogenes actin-based motility informed current understanding of the role of critical host proteins in regulating actin dynamics. *Listeria monocytogenes* is also valuable for studying both innate and adaptive immunity, as both arms of the immune system contribute to recognition and clearance of the pathogen. New studies which investigate the use of *L. monocytogenes* as a vaccine for intracellular delivery of antigens promise to keep *L. monocytogenes* at the forefront of research into the host response to infection (Paterson et al. 2010).

11.3 Gram-Negative Opportunistic Bacteria

11.3.1 Pseudomonas aeruginosa

Pseudomonas aeruginosa is an aerobic, Gram-negative bacterium that exists in diverse environments, including water, soil, plants, and in animal and human hosts. In humans, *P. aeruginosa* is an opportunist that causes both community- and hospital-acquired infections. Community-acquired infections most often occur in the immunocompromised and in diabetics and include acute infections of the eye, such as otitis externa or ulcerative keratitis, as well as SSTIs. Nosocomial infections can be acute or chronic, and include those of the urinary tract, surgical site incisions, the bloodstream, and the lung (Driscoll et al. 2007), and often occur in association with epithelial damage induced by indwelling medical devices such as catheters or ventilators. *P. aeruginosa* lung infections are a particular concern in patients with preestablished ventilator-associated pneumonia (VAP), chronic obstructive pulmonary disease (COPD), and cystic fibrosis (CF), and these infections have mortality rates approaching 70–80 % (Chastre and Fagon 2002). Infection in these patients is common, as *P. aeruginosa* has been identified as a leading cause of hospital-acquired pneumonia, second only to *Staphylococcus aureus* (Rotstein et al. 2008; Kollef et al. 2005).

11.3.1.1 Pathogenesis of *P. aeruginosa*

Pseudomonas aeruginosa is one of the best-studied human pathogens, and numerous studies have demonstrated that *P. aeruginosa* pathogenesis is achieved through the coordinated action of bacterial surface-associated and secreted virulence factors. Together, these factors drive host colonization, manipulation of host physiology, and evasion of host defenses. The bacterium depends on five secretory systems to deliver bacterial effectors directly into host cells or to the extracellular milieau; *P. aeruginosa* possesses a type I secretion system (TISS), TIISS, TIIISS, TVSS, and TVISS. Others have thoroughly reviewed *P. aeruginosa* virulence factors and their regulation, as well as the structure and function of the *Pseudomonas* secretory apparatuses (Bleves et al. 2010; Filloux 2011); here we highlight the major virulence factors that contribute to acute and chronic infection.

Initial stages of *P. aeruginosa* colonization require bacterial flagella, pili, and secreted exopolysaccharides. Type IV pili promote motility and also bind to the host glycolipid asialo-ganglioside MI (αG_{M1}) (Comolli et al. 1999). Interestingly, αG_{M1} is maximally expressed by injured tissues during epithelial repair processes, which may explain why *P. aeruginosa* preferentially colonizes injured epithelial tissue (de Bentzmann et al. 1996). Following initial attachment, *P. aeruginosa* secretes alginate, a viscous exopolysaccharide whose production is regulated by bacterial quorum sensing systems. Alginate is crucial for biofilm development and biofilm-mediated adhesion to host and inanimate surfaces (Driscoll et al. 2007). Alginate also impairs key aspects of the host innate immune response, by scavenging free radicals produced by macrophages, inhibiting neutrophil chemotaxis, inhibiting IFNγ-mediated macrophage phagocytosis, and preventing complement activation (Leid et al. 2005; Simpson et al. 1989; Pedersen et al. 1990).

Pseudomonas aeruginosa dramatically alters the function of host cells, leading to loss of epithelial integrity and impaired function of innate immune cells. For instance, the TIISS-dependent effectors elastase, phospholipases, and exotoxin AB all contribute to epithelial damage. The LasB elastase degrades elastin, a major component of lung tissue, and therefore contributes to lung deterioration in vivo (Williams et al. 1992). Phospholipase C (PlcC), PlcH, and PlcN target lipids present in eukaryotic cell membranes and surfactants, compromising integrity of the epithelial mucosa (Lopez et al. 2011; Wargo et al. 2011). Exotoxin AB inactivates host elongation factor 2, disrupting protein synthesis and triggering host cell death. The TIIISS directly secretes effectors such as ExoS, ExoT, and ExoU into host cells, using a syringe-like structure. The actions of ExoS and ExoT perturb the cytoskeletal dynamics of phagocytes to interfere with phagocytosis and to trigger host cell apoptosis (Sun et al. 2012; Barbieri and Sun 2004). The effector ExoU is a cytotoxin that triggers rapid necrotic cell death of phagocytes (Diaz and Hauser 2010). The function of ExoS and ExoY also enables the bacteria to establish a novel niche within the plasma membrane of epithelial cells, called bleb niches, within which the bacteria replicate and even swim (Angus et al. 2010; Hritonenko et al. 2011). Bleb niches can detach from live cells and travel significant distances with bacteria still swimming inside them, which suggests that they may be important for infection dissemination (Angus et al. 2008).

11.3.1.2 Chronic *P. aeruginosa* Infection in CF Patients

Cystic fibrosis patients suffer from compromised lung integrity due to a mutation in the gene encoding the CF transmembrane conductance regulator (CFTR). The key importance of CFTR is its function as a chloride ion channel, and CFTR mutations result in misfolding, improper localization, or complete lack of the protein. Deficiency of CFTR impairs chloride secretion and also enhances sodium import into the cell by the epithelium sodium channel (ENaC) (Berdiev et al. 2009). The influx of sodium into CFTR-deficient epithelial cells is followed by enhanced water movement into cells. The altered ion and water transport leads to decreased airway surface liquid (ASL) (Smith et al. 1996), the liquid layer which maintains mucus hydration in the respiratory epithelium and provides a substrate for ciliary movement. Altogether, CFTR deficiency results in dehydration and thickening of the mucus layer. Cystic fibrosis patients have great difficulty expelling mucus secretions and are plagued by frequent and chronic lung infections by multiple pathogens, including *P. aeruginosa* and *Burkholderia cepacia* (Govan and Deretic 1996).

Recent reports suggest that *P. aeruginosa* is adept at establishing chronic infections in both patients with and without CF. During acute and chronic infection, the bacterium produces and secretes the CFTR inhibitory factor (Cif), an epoxide hydrolase which reduces the presence of apical membrane CFTR by interfering with normal CFTR recycling (Bomberger et al. 2011; Swiatecka-Urban et al. 2006). *Pseudomonas aeruginosa* also secretes alkaline protease (AprA), which proteolytically activates ENaC (Butterworth et al. 2012). The combined action of Cif and AprA leads to enhanced influx of sodium and water into host cells, and represents a bacterial mechanism to decrease the ASL and dehydrate epithelial mucus. Therefore, Cif and AprA can promote infection in patients without CF by creating a CF-like microenvironment within the lung. In CF patients with incomplete CFTR deficiencies, Cif and AprA can exacerbate poor lung conditions. In addition, these virulence factors may limit the efficacy of CF drugs designed to target ENaC and CFTR, which should be considered for future development of anti-CF drugs (Ballok and O'Toole 2013).

11.3.1.3 Antibiotic Resistance in *P. aeruginosa*

Treatment of *P. aeruginosa* infections is complicated by increasing drug resistance, as with many other opportunistic pathogens. In fact, *P. aeruginosa* drug resistance has escalated such that the bacterium is now classified as an ESKAPE pathogen, a group of microbes (including *Enterococcus faecium*, *Staphylococcus aureus*, *Klebsiella pneumoniae*, *Acinetobacter baumannii*, *Pseudomonas aeruginosa*, and *Enterobacter* spp.) known for multidrug-resistant phenotypes (Pendleton et al. 2013). *Pseudomonas aeruginosa* is particularly adept at developing resistance during the course of antibiotic treatment. Multidrug-resistant *P. aeruginosa* (MDRPA) strains now exist that are resistant to drugs from the β-lactam, aminoglycoside, and fluoroquinolone classes (Obritsch et al. 2005). Emergence of MDRPA during antibiotic therapy has been reported in 27–72 % of patients who were originally infected with antibiotic-susceptible strains (Zhao and Hu 2010), and infections caused by MDRPA have greater morbidity and mortality rates than those caused by susceptible strains (Driscoll et al. 2007).

11.3.2 Klebsiella pneumoniae

Klebsiella pneumoniae is a member of the *Enterobacteriaceae*, a large family of Gram-negative, facultative anaerobic bacteria that are normal inhabitants of the intestinal tract of humans and animals (Guentzel 1996). The genus *Klebsiella* contains seven species, and *K. pneumoniae* is the species that causes opportunistic, often nosocomial, infections in immunocompromised individuals. *Klebsiella pneumoniae* is frequently associated with pneumonia and urinary tract infections, but in serious cases, *K. pneumoniae* infections can lead to bacteremia and even osteomyelitis (Podschun and Ullmann 1998). Humans are the primary reservoir for *K. pneumoniae*, and the prevalence of *K. pneumoniae* infections is due in part to high carrier rates. In the community, it is estimated that *K. pneumoniae* is carried in the intestines of up to 38 % and in the nasopharynx of 6 % of healthy people. The proportion of carriers is markedly higher in hospitalized patients, where carriage rates reach 77 % in the intestine and 19 % in the nasopharynx (Podschun and Ullmann 1998).

11.3.2.1 Antibiotic Resistance in *K. pneumoniae*

Klebsiella pneumoniae infections have gained much attention over the last decade due to the emergence of drug-resistant strains, which have led to the classification of *K. pneumoniae* as an ESKAPE pathogen. Most drug-resistant strains produce *Klebsiella pneumoniae* carbapenamase (KPC), a β-lactamase that confers resistance to all β-lactam antibiotics (Queenan and Bush 2007). The first KPC-positive *K. pneumoniae* isolate was obtained from a patient in a North Carolina hospital in 1996, and since then the strain has spread nationwide and is considered endemic in some US states (Landman et al. 2007). Although KPC is found primarily in *Klebsiella pneumoniae*, the gene encoding KPC is contained within a mobile genetic element and can be transferred between bacterial species (Naas et al. 2008). Indeed, KPC has been isolated from other species of *Enterobacteriaceae* in the USA, Europe, Asia, the Middle East, and South America (Kitchel et al. 2009). In addition, *K. pneumoniae* strains have recently been isolated which contain the New Delhi metallo-β-lactamase (NDM1), a mobile genetic element encoding multiple drug resistance genes. Some NDM1-positive bacteria are resistant to all antibiotics except the polymixins, drugs often used as a "last resort" due to their potential cytotoxic effects in humans (Li et al. 2005). The initial isolation of NDM1-positive *K. pneumoniae* occurred in 2007 from a patient who had been infected in India (Yong et al. 2009), but now NDM1-positive *Enterobacteriaceae* are present worldwide, including many Asian countries, Europe, and the USA (Moellering 2010). Throughout the world, nosocomial outbreaks of KPC- and NDM1-positive bacteria are on the rise, and are accompanied by very high mortality rates due to limited treatment options. For example, in 2011, an outbreak of KPC-positive *K. pneumoniae* at the National Institutes of Health hospital in

Bethesda, MD affected 18 patients and caused 11 deaths (Snitkin et al. 2012). Undoubtedly, the occurrence and spread of KPC- and NDM1-positive Gram-negative pathogens pose a serious challenge to human healthcare systems worldwide.

11.3.2.2 *Klebsiella pneumoniae* Pathogenesis

The most important *K. pneumoniae* virulence factor is the thick polysaccharide capsule, which the bacterium uses to colonize host surfaces such as the lung epithelium (Cortes et al. 2002; Simoons-Smit et al. 1986). The capsule protects the bacterium against opsonophagocytosis and provides the adhesive matrix for biofilm formation. In addition, most *K. pneumoniae* strains express Type I and Type III fimbriae, which aid in adhesion to epithelium and to medical devices, respectively (Struve et al. 2008; Murphy and Clegg 2012; Murphy et al. 2013).

11.3.2.3 Hypervirulent *K. pneumoniae*

Although most infections are caused by "classic" opportunistic *K. pneumoniae* as described above, a new hypervirulent strain of the bacterium has emerged and is causing sporadic infection outbreaks, even in healthy individuals with no predisposing risk factors for disease (Shon et al. 2013). These patients frequently presented with pyogenic liver abscesses, and in a number of patients, the infection disseminated to distant tissues (Shon et al. 2013). This strain is distinguished from classic *K. pneumoniae* in its ability to infect healthy hosts, as well as its propensity to cause atypical infections, such as meningitis and endopthalmitis (Lee et al. 2006). The hypervirulent strain exhibits a hypermucoviscous phenotype, due to production and secretion of a polysaccharide web, which exhibits amplified characteristics of a capsule and provides great protection against opsonophagocytosis (Fang et al. 2004). To date, there is no evidence of KPC or NDM1-mediated drug resistance in the hypervirulent strain (Li et al. 2014). However, with the worldwide spread of hypervirulent *K. pneumoniae*, it is to be expected that the hypervirulent and KPC or NDM1 strains will soon coexist within communities, and there is concern that hypervirulent *K. pneumoniae* will acquire multidrug-resistant capabilities.

11.3.3 Haemophilus influenzae

Haemophilus influenzae is a Gram-negative coccobacillus which primarily exists as a commensal microbe of the human nasopharyngeal mucosa. In many people, colonization occurs within the first year of life (Aniansson et al. 1992), and it is estimated that by age six over 50 % of healthy children are *H. influenzae* carriers (Mukundan et al. 2007). Although *H. influenzae* is often carried asymptomatically, the bacterium can cause a variety of OIs, primarily in children or individuals with compromised lung function from preexisting diseases such as chronic obstructive pulmonary disease (COPD). *Haemophilus influenzae* is associated with a range of diseases, including localized infections of the upper respiratory tract such as sinusitis, otitis media, bronchitis, and otitis, as well as more serious invasive diseases such as pneumonia, bacteremia, meningitis, and septic arthritis (Watt et al. 2009). The species *H. influenzae* is categorized into two main groups: typeable (encapsulated) and nontypeable (nonencapsulated) strains. Typeable *H. influenzae* are certainly the most virulent and historically have been most frequently associated with invasive forms of disease (Moxon and Kroll 1988). However, since the introduction of polysaccharide–protein conjugate vaccines in the late 1980s, invasive *H. influenzae* infections caused by typeable strains have largely been eliminated in the developed world (Agrawal and Murphy 2011). Current focus is on the emergence of non-typeable *H. influenzae* as a major cause of childhood mucosal infections, such as otitis media and sinusitis.

11.3.3.1 *Haemophilus influenzae* Pathogenesis and Evasion of Host Defenses

Haemophilus influenzae does not produce toxins or exoenzymes that directly damage the host, and therefore pathogenicity is mediated by virulence factors that promote colonization of host tissues and evasion of host defenses. Typeable *H. influenzae*, of which the most medically relevant strain is *H. influenzae* type b, rely on the capsule for adhesion and protection from the host. For nontypeable *H. influenzae* (NTHi), which lack a capsule, several surface structures drive interaction with host receptors. For example, NTHi express multiple fimbriae, such as the P2 and P5 fimbriae that mediate binding to the intercellular adhesion molecule-1 (ICAM-1) on host epithelial cells (Avadhanula et al. 2006). In addition, the bacterial adhesin Haemophilus adhesion protein (Hap) enables NTHi to form microcolonies via cell-to-cell interactions and also promotes adhesion to extracellular matrix molecules such as laminin, fibronectin, and collagen IV (Fink et al. 2003; Fink et al. 2002).

The pathogenesis of *H. influenzae* also relies on its ability to counter antibody- and complement-mediated host defenses. *Haemophilus influenzae* produces IgA proteases to protect against IgA opsonization; nearly all strains of *H. influenzae* produce a type 1 IgA protease (IgA-a), while strains associated with COPD also tend to encode a separate type 2 IgA protease (IgA-b) (Murphy et al. 2011). *Haemophilus influenzae* also evades destruction by the host complement system by either blocking complement activation on the bacterial surface or by acquiring soluble host factors that inactivate complement proteins (Hallstrom et al. 2009; Hallstrom et al. 2010).

11.3.3.2 *Haemophilus influenzae* Vaccines

Vaccines against *H. influenzae* type b were first introduced in the late 1980s. Since then, the epidemiology of *H. influenzae* infections has dramatically changed such that invasive type b infections have been nearly eradicated in industrialized countries. However, in recent decades the incidence of NTHi infections has risen worldwide. While NTHi strains cause mostly localized infections in healthy children, they can lead to invasive forms of disease in malnourished or immunocompromised children (Chisti et al. 2009). Future work may be aimed at developing vaccines against NTHi strains as a means of protecting vulnerable populations against potentially invasive disease.

11.3.4 Acinetobacter baumannii

Acinetobacter baumannii is a Gram-negative, nonmotile bacillus that is ubiquitous in nature and can be associated with water, soil, animals, and humans. Although incidence of disease caused by *A. baumannii* is increasing worldwide, *A. baumannii* infections are most common in members of the armed forces who have been deployed to conflict zones in the Middle East. Because *A. baumannii* is physiologically robust, it thrives in dry sandy climates such as Middle Eastern deserts and is a common contaminant of combat wounds and military field hospitals (Korzeniewski and Bochniak 2011). Although *A. baumannii* infections originated in conflict zones, the pathogen is now an important cause of disease in USA and European hospitals, likely introduced into the hospital environment by combat soldiers who, following exposure to the bacterium overseas, return home to recuperate (Tong 1972; Howard et al. 2012). In the USA and Europe, *A. baumannii* primarily infects immunocompromised individuals, particularly patients experiencing long-term hospitalization. The pathogen has recently gained attention and notoriety due to its increasing drug resistance; the majority of strains isolated from clinical infections exhibit resistance to most first-line antibiotics (Fournier et al. 2006). As such, *A. baumannii* is now recognized by the World Health Organization as an ESKAPE pathogen (Boucher et al. 2009).

The most common diseases caused by *A. baumannii* are skin and soft tissue infections (SSTIs), nosocomial and community-acquired pneumonia, bacteremia, and meningitis. The severity and outcome of infection depend on infection site, immune status of the host, and the relative drug resistance of the *A. baumannii* strain. *A. baumannii*-induced SSTIs are most often associated with trauma, burns, or surgical site infections, where wounds are contaminated by bacteria present within the outdoor or hospital environment (Howard et al. 2012). Severe SSTIs can lead to necrotizing infection, in which the soft tissue is degraded, potentially providing a route for the bacteria to enter the bloodstream and cause bacteremia or septicemia (Sebeny et al. 2008). The frequency of *A. baumannii*-induced SSTIs is unclear, but estimated to be low in comparison to other diseases caused by the pathogen (Moet et al. 2007). In addition, *A. baumannii* SSTIs are often polymicrobial, as the bacterium frequently co-colonizes wounds with pathogens such as *Enterococcus faecium*, *Citrobacter freundii*, and *Klebsiella pneumoniae* (Guerrero et al. 2010; Charnot-Katsikas et al. 2009).

Acinetobacter baumannii can cause pneumonia in both hospital and community settings. Nosocomial pneumonia occurs most frequently as VAP in patients receiving mechanical ventilation for an unrelated illness or those with prior use of antibiotics. Quite often, VAP results from *A. baumannii* biofilms present on the abiotic surface of catheters and ventilator tubing (Fournier et al. 2006). *Acinetobacter baumannii*-induced VAP is particularly serious, especially among patients housed within intensive care units (ICU), and has an estimated mortality rate of 30–60 % (Chaari et al. 2013; Chastre and Fagon 2002; Tablan et al. 2004). However, pneumonia caused by *A. baumannii* is not restricted to the hospital environment; reports indicate that community-acquired (CA) strains are emerging, particularly in Australia and Asia (Anstey et al. 2002). The CA infections are characterized by severe and sudden onset coupled with secondary bacteremia or septicemia. These infections appear to coincide with *A. baumannii* throat colonization and are associated with heavy alcohol consumption. The mortality rate of CA *A. baumannii* infections is especially high, as approximately 40–60 % of infections are fatal (Leung et al. 2006).

11.3.4.1 *Acinetobacter baumannii* Pathogenesis

The success of *A. baumannii* as a human pathogen, particularly in the hospital environment, is due to its ability to colonize and form biofilms on abiotic and biotic surfaces, to evade host defenses and physiologically alter host cells. *Acinetobacter baumannii* forms biofilms on devices such as indwelling catheters, respiratory equipment, artificial heart valves, and other instruments (Donlan and Costerton 2002). As with other pathogens, biofilm-resident *A. baumannii* exhibit increased tolerance to disinfectants and antimicrobial drugs (Gurung et al. 2013), making biofilm production an important factor for survival in the hospital environment. The pili and exopolysaccharides of *A. baumannii* pili also promote biofilm formation by aiding in bacterial adhesion to surfaces and other cells. The best-characterized *A. baumannii* virulence factor is the bacterial outer membrane protein, OmpA, which plays critical roles in biofilm production on plastic and in various stages of in vivo infection. During biofilm formation, OmpA enhances bacterial cell aggregation, and OmpA-deficient mutants exhibit simpler biofilms with fewer cell aggregates (Gaddy et al. 2009). In addition, OmpA promotes adhesion to and invasion of epithelial cells and may contribute to systemic dissemination of the bacteria during in vivo infection (Choi et al. 2008). The OmpA protein also aids in bacterial resistance to host innate defenses by reducing complement-mediated bacterial lysis and opsonophagocytosis (Kim et al. 2009). The multiple functions of OmpA in *A. baumannii* virulence make this factor a potential target for development of novel anti-infective therapeutics.

11.3.4.2 Antimicrobial Resistance in *A. baumannii*

In the last decade, *A. baumannii* has gained much attention, and ESKAPE classification, for its increasing drug resistance and emergence of some strains that are resistant to all available antibiotics. *Acinetobacter baumannii* exhibits intrinsic resistance to commonly used drugs such as aminopenicillins, cephalosporins, and chloramphenicol, due to the low permeability of the outer membrane and constitutive expression of efflux pumps (Neonakis et al. 2011). In addition, *A. baumannii* strains have acquired tolerance to a broad-spectrum of β-lactams (a class of antibiotic compounds whose molecular structure incorporates the beta lactam ring) including carbapenems, as well as aminoglycosides, fluoroquinolones, and tetracyclines (Dijkshoorn et al. 2007).

11.3.5 Stenotrophomonas maltophilia

Stenotrophomonas maltophilia is an aerobic, Gram-negative, motile bacillus which is associated with aquatic environments, soil, and plants (Looney et al. 2009). The nomenclature of this species has changed over several decades, as it was formerly called *Bacterium bookeri*, *Pseudomonas maltophilia* (Hugh and Ryschenkow 1961), and *Xanthomonas maltophilia* prior to its current identification as *S. maltophilia*. The bacterium is an OP of humans and is primarily associated with nosocomial infections. However, community-acquired infections have been reported, and typically afflict individuals with a comorbidity, such as malignancy or HIV infection (Falagas et al. 2009). In general, *S. maltophilia* most commonly infects immunocompromised individuals, patients exposed to broad-spectrum antibiotics, or those with chronic pulmonary diseases or lung cancer (Waters et al. 2012; Safdar and Rolston 2007). In these populations, *S. maltophilia* infects the respiratory tract, bloodstream, urinary tract, bones, joints, and soft tissues. Pneumonia is the most common manifestation of *S. maltophilia* infection in the immunocompromised, and reported mortality rates in patients with pneumonia range from 23 to 77 %. The greatest mortality rates are associated with cancer patients and those with secondary bacteremia (Looney et al. 2009). *Stenotrophomonas maltophilia* has also been categorized as an ESKAPE pathogen due to its high level of drug resistance (Boucher et al. 2009).

11.3.5.1 Nosocomial *S. maltophilia* Infections

Stenotrophomonas maltophilia is a rising threat to healthcare facilities worldwide. Surveys from multiple continents report increased incidence of *S. maltophilia* infections in hospitalized patients (Looney et al. 2009). For example, in the UK the annual incidence of bloodborne *S. maltophilia* infections increased by 93 % between 2000 and 2006 (Looney et al. 2009), and a Taiwanese hospital reported an 83 % increase in *S. maltophilia* infections between 1999 and 2004 (Tan et al. 2008). In addition, a US multihospital surveillance study of *S. maltophilia* infections from 1993 to 2004 reported that *S. maltophilia* was among the eleven most frequently isolated pathogens from ICU patients (Lockhart et al. 2007). In clinical settings, *S. maltophilia* has been isolated from water faucets, drains, and medical devices such as catheters, endotracheal tubes, and tubing associated with dental equipment. Of particular concern is the ability of this bacterium to adhere to and form biofilms on plastics; for this reason, infections occur most often in patients with indwelling medical devices such as endotracheal tubes and catheters.

Stenotrophomonas maltophilia poses a special threat to patients with underlying pulmonary disease, such as cystic fibrosis (CF), chronic obstructive pulmonary disease (COPD), or lung cancer (Nseir et al. 2006). Although *S. maltophilia* is not considered as a "classic" CF pathogen, as it is not recovered from CF patients with the same frequency as other pathogens like *P. aeruginosa* (Hansen 2012), CF patients are more susceptible to chronic lung infection with *S. maltophilia* than are non-CF patients, and infection enhances risk of pulmonary exacerbations requiring hospitalization and antibiotic therapy (Waters et al. 2012). In patients with obstructive lung cancer, or in cancer patients requiring mechanical ventilation, *S. maltophilia* is a growing concern. In these patients, *S. maltophilia* lung infection can cause localized necrosis and pleural hemorrhage and is associated with greater than 50 % mortality (Fujita et al. 1996).

11.3.5.2 *Stenotrophomonas maltophilia* Pathogenesis

The results of *S. maltophilia* whole-genome sequencing indicate that this bacterium possesses few identifiable virulence factors. However, its capacity for adhesion and biofilm formation on abiotic and biotic surfaces is mediated by the positively charged bacterial surface and its flagella (Brooke 2012). In particular, *S. maltophilia* flagella have been found to promote adhesion of the microbe to mouse tracheal mucus (Zgair and Chhibber 2011) and epithelial cell lines in vitro (Pompilio et al. 2010). Bacterial lipopolysaccharide (LPS) also promotes adhesion and aids in bacterial resistance to the host complement system *Stenotrophomonas maltophilia* produces exoenzymes, including proteases, lipases, DNases, RNases, and fibrolysins that may contribute to localized tissue damage and evasion of host innate defenses, particularly during lung infection (Brooke 2012).

11.3.5.3 Antimicrobial Resistance and Therapeutic Options

Stenotrophomonas maltophilia exhibits high-level intrinsic and acquired resistance to a wide variety of disinfectants and antibiotics, including β-lactams, aminoglycosides, quinolones, and tetracyclines. Intrinsic resistance is due to the low permeability of the outer membrane, as well as to chromosomally encoded multidrug efflux pumps, β-lactamases, and aminoglycoside-modifying enzymes (Nicodemo and Paez 2007; Zhang et al. 2000). Most acquired resistance comes from genetic determinants contained within integrons, transposons, or plasmids, which are acquired by *S. maltophilia* through horizontal gene transfer (Looney et al. 2009). Not surprisingly, there is great interest in pursuing antibiotic alternatives for treatment of *S. maltophilia* infections. Such development may take an ecologically based approach, by exploiting natural antimicrobial- or phage-induced mechanisms of killing *S. maltophilia*. In addition, much greater emphasis is likely to be placed on preventative actions, in an effort to minimize *S. maltophilia* spread from hospital sources to patients with weakened defenses (Brooke 2014).

11.4 Opportunistic Fungi

11.4.1 Candida *Species*

Candida species are pleomorphic yeasts that naturally inhabit the human oral, gastrointestinal, and vaginal mucosa. Although *Candida* species are commensal microbes of most healthy humans, they can cause overt disease (candidiasis), most often in immunocompromised adults and infants. The genus *Candida* contains around 154 species, of which over 17 can cause disease in humans. Over 90 % of candidiasis infections are caused by *C. albicans*, *C. glabrata*, *C. parapsilosis*, *C. tropicalis*, and *C. krusei* (Pfaller et al. 2007, 2012). Superficial candidiases most often include infections of the oral and vaginal mucosa, caused by fungal overgrowth resulting from immune suppression or decreased bacterial populations following antibiotic therapy. Invasive diseases are most common in hospitalized patients with indwelling medical devices such as catheters or endotracheal tubes and include urinary tract infections, pneumonia, and bloodstream infections (Sardi et al. 2013). Although candidiasis is most often caused by indigenous *Candida*, infections can be transmitted via the hands of healthcare workers or via contaminated healthcare materials and indwelling devices. Therefore, individuals most at risk of developing candidiasis are the immunocompromised, or those subjected to long-term hospitalization with indwelling medical devices, exposure to broad-spectrum antimicrobials, corticosteroids, or cytotoxic chemotherapy (Pfaller and Diekema 2007; Pappas et al. 2003). The frequency of candidiasis is increasing globally, but the greatest increases have occurred in North America. As a result, *Candida* species are the leading cause of invasive fungal infections in hospitalized patients (Sardi et al. 2013) and are the third to fourth most common nosocomial infection in the USA (Pfaller et al. 2012). In addition, *Candida* species are the dominant fungal pathogen in AIDS patients.

Of all *Candida* species, *C. albicans* is the predominant cause of invasive infections and is associated with highest mortality rates (Hope et al. 2013; Pfaller et al. 2012). However, non-albicans species do contribute to a significant proportion of fungal infections, particularly in hospitalized patients. In a study of 6845 patients with invasive fungal infections from 25 US hospitals, blood samples from 3648 patients (53.3 %) tested positive for some species of *Candida*, with 18 species detected in total (Vincent et al. 2009). In that study, *C. albicans* caused the majority of infections (42.1 %), followed by *C. glabrata* (26.7 %), *C. parapsilosis* (15.9 %), *C. tropicalis* (8.7 %), *C. krusei* (3.4 %), and *C. dubliniensis* (<3 %). During the course of that same study by Vincent et al. (2009), it was demonstrated that different *Candida* species may show a propensity for specific ecological niches within the body of an infected patient. *Candida albicans* caused the majority of infections in neonates and surgical patients. In contrast, *C. glabrata* tended to cause disease in older patients (with a mean age of 58 years), the majority of whom had previously undergone antifungal therapy. *C. parapsilosis* was most common in postsurgical patients, while C. *tropicalis* and *C. krusei* were most frequently isolated from patients with hematologic malignancies (Vincent et al. 2009). In addition, reports of antifungal resistance in multiple *Candida* species suggest that infections caused by *albicans* and non-*albicans* species will continue to pose a real threat to susceptible individuals for the foreseeable future (Mishra et al. 2007; Prasad and Kapoor 2005).

11.4.2 Candida albicans

As the most common agent of candidiasis, *C. albicans* causes superficial mucosal infections, urinary tract infections, pneumonia, as well as infections of the bloodstream (otherwise known as candidemia) and deeper tissues. Superficial infections typically affect the vaginal and oral mucosa, and are commonly known as thrush, characterized by the appearance of white fungal colonies overlying inflamed mucosal membranes (Kim and Sudbery 2011). Vulvo-vaginal candidiasis (VVC), or vaginal thrush, is the most common form of candidiasis; VVC is estimated to affect up to 75 % of women at least once in their lifetime (Sobel 1997). In contrast to VVC, which routinely affects even healthy adult women, oropharyngeal candidiasis (OPC), or oral thrush, is associated primarily with patients who are immunocompromised with hyposalivation, diagnosed with diabetes mellitus, or subjected to prolonged antibiotic therapy or immunosuppressive drugs. In fact, OPC is so common among individuals with AIDS that its occurrence is recognized as a marker for development of AIDS in HIV-positive patients (Cassone and Cauda 2012). Presence of OPC is also associated with oral cancers as well as the use of dentures in elderly patients (Kim and Sudbery 2011).

Invasive *C. albicans* infections usually manifest first as candidemia, with subsequent dissemination to deeper tissues. Since neutrophils mediate strong anti-*Candida* immunity, candidemia can develop from a superficial infection in neutropenic individuals, such as those subjected to immunosuppressive therapy or with certain types of hematalogic malignancy. In the hospital environment, candidemia is frequently the result of a breach in natural barriers, as occurs via surgical incision or catheter use. Both candidemia and disseminated candidiasis are serious infections that symptomatically resemble systemic bacteremia and which require rapid treatment. However, because of the similarities in symptoms between invasive bacterial and *Candida* infections, diagnosis may occur too late for proper antifungal treatment (Blot and Vandewoude 2004). For this reason, as well as the immunocompromised condition of many *C. albicans* patients, mortality rates for invasive *C. albicans* infections are estimated to be as high as 30–50 % (Kim and Sudbery 2011).

11.4.2.1 *Candida albicans* Morphogenesis

As a Pleiomorphic fungus, *C. albicans* has the ability to reversibly transition between yeast and filamentous hyphal forms. This morphological conversion is a key aspect of *C. albicans* virulence, as the switch from yeast to hyphae is associated with increased morbidity (Saville et al. 2003), potentially due to greater invasiveness of the filamentous fungi. Studies employing yeast-locked *C. albicans* mutants, which are unable to convert to the filamentous morphology, are noninvasive in animal models of infection and are subject to destruction by host immune defenses both in vitro and in vivo (McKenzie et al. 2010; Jayatilake et al. 2006). During nonpathogenic colonization, the pathogen exists primarily in yeast form, and transition from the commensal to pathogenic lifestyle correlates with a switch to the filamentous morphology (Gow et al. 2012). Perhaps not surprisingly, this morphogenesis involves complex and highly coordinated signaling events leading to activation of transcriptional regulators and is influenced by environmental factors such as presence of human serum, pH, nutrients, temperature, and oxygen concentration (Sudbery 2011). Such environmental regulation enables *C. albicans* to fine-tune its morphological response to different microenvironments within the host. Importantly, the host immune system discriminates between the colonizing yeast and invasive filamentous forms of *C. albicans*, such that it tolerates nonpathogenic colonization by the yeast form, but mounts a potent immune response against the invasive filamentous fungi (Gow et al. 2012).

11.4.2.2 *Candida albicans* Biofilm Formation

Candida albicans forms biofilms on host mucosa as well as abiotic surfaces of medical devices. During biofilm maturation, *C. albicans* cells switch from yeast to filamentous form, which corresponds with secretion of polysaccharides and exhibition of drug resistance (Nobile et al. 2012; Ramage et al. 2005; Blankenship and Mitchell 2006). Because *C. albicans* forms robust biofilms on medical tubing such as intravascular catheters, these implanted devices serve as reservoirs for the pathogen and aid in development of candidemia and dissemination of the infection to deep organs. Indeed, *C. albicans* is a leading cause of intravascular catheter-related infections and has the highest crude mortality rate of any pathogen associated with intravascular catheter-associated disease (Crump and Collignon 2000).

11.4.2.3 Drug Resistance in *C. albicans* Biofilms

Biofilms formed by *C. albicans* survive exposure to high concentrations of antifungal drugs such as fluconazole and polyenes like Amphotericin B and Nystatin (Ramage et al. 2001; Chandra et al. 2001). This drug resistance is attributed primarily to enhanced synthesis of β-1,3-glucan by the *C. albicans* Fks1p synthase. Cells growing in biofilms have thicker cell walls, resulting from greater cell wall deposition of β-1,3-glucan, and also secrete significantly higher concentrations of the glucan than do planktonic cells. β-1,3-glucan binds to antifungal drugs, essentially sequestering the drug and preventing its diffusion throughout the biofilm matrix (Nett et al. 2010a, b). Recent findings suggest that β-1,3-glucan synthesis may be regulated by the molecular chaperone Hsp90, as impairment of Hsp90 function reduced glucan synthesis and subsequent biofilm growth, and increased susceptibility of biofilm-associated cells to the antifungal azoles (Robbins et al. 2011).

11.4.2.4 *Candida albicans* Pathogenesis

Pathogenicity of *C. albicans* is mediated via biofilm formation, adhesins, and secretion of exoenzymes that damage host tissues and promote fungal invasion. The adhesins of *C. albicans* bind to host extracellular matrix proteins such as fibrinogen and fibronectin (Tronchin et al. 1991). *Candida albicans* also produces an array of hydrolytic proteases, phospholipases, and hemolysins that aid in adherence and tissue invasion and destruction. In particular, *C. albicans* produces ten aspartic proteases, SAP1–SAP10, and SAPs 1–6 are secreted virulence factors that facilitate adhesion, tissue damage, and immunomodulation of the host. Contrastingly, SAP9 and SAP10 are surface-associated proteases that aid in maintenance of cell wall integrity and interaction of *C. albicans* with host epithelial cells and neutrophils (Naglik et al. 2003; Schild et al. 2011). In addition, seven phospholipases (PLA, PLB1, PLB2, PLC1, PLC2, PLC3, and PLD1) have been identified which are implicated in vivo infection (Sardi et al. 2013). *Candida albicans* also produces α- and β-hemolysins, which aid in iron acquisition from host tissues during infection (Luo et al. 2001; Almeida et al. 2009). Several reports indicate a possible requirement for *C. albicans* to possess actively functional hydrolytic and hemolytic enzymes in order for the full expression of virulence; as production of proteases, lipases, and hemolysins is associated with clinical signs of severe candidiasis (Ribeiro et al. 2004; Bramono et al. 2006; Mane et al. 2012).

11.4.3 Pneumocystis jirovecii *(Formerly* Pneumocystis carinii*)*

Pneumocystis jirovecii is an ascomycetous yeast-like fungus which causes mild to no infection in healthy humans, but fulminant pneumonia in the immunocompromised patient. Pneumocystis pneumonia (PcP) is a serious and highly fatal disease; the mortality rate approaches 100 % in untreated infections and 10–20 % with treatment (Gigliotti and Wright 2012). Originally characterized as a zoonosis, members of the genus *Pneumocystis* are now recognized as pathogens that infect a variety of mammals, including humans, but which are not transmitted between mammalian species. Available data indicate that individual mammalian species harbor specifically corresponding *Pneumocystis* species or strains, and human infections are caused by *P. jirovecii* (Gigliotti and Wright 2012; Gigliotti et al. 1993). Typically, *P. jirovecii* presents itself in humans either as an asymptomatic colonization or is associated with mild infections. These outcomes are especially common in infants, and the microbe is easily transmitted via an airborne route from infant to infant or from infant to adult (Gigliotti and Wright 2012).

11.4.3.1 Populations at Risk for PcP

Individuals with greatest risk of PcP infection are those who are immunocompromised, and patients with HIV-AIDS make up the greatest number of PcP infections in the USA and Europe. During the early years of the US AIDS epidemic, PcP accounted for two-thirds of AIDS-defining illnesses. Since the advent of antiretroviral therapy and anti-Pneumocystis prophylaxis in the 1990s, the rate of PcP infections in HIV-positive patients has decreased somewhat, although the pathogen remains a major cause of OIs in AIDS patients in the developed world (Miller et al. 2013). In addition to its role as a dominant pathogen in AIDS patients, *P. jirovecii* is an emerging threat in non-HIV-infected individuals undergoing immunosuppressive treatment for conditions such as malignancy, organ transplantation, or rheumatoid arthritis (Tasaka and Tokuda 2012; Mori and Sugimoto 2012). The clinical manifestation of PcP infection differs between HIV-infected and noninfected PcP patients. In individuals with HIV, *P. jirovecii* respiratory infections gradually progress to pneumonia over a course of 2 weeks to 2 months, whereas in non-HIV-infected immunocompromised patients, *P. jirovecii* infections exhibit rapid onset and progress to fulminant pneumonia within a week.

11.4.3.2 Pathogenesis of PcP

Pneumocystis jirovecii cannot be cultivated under standard laboratory conditions, and therefore the study of its pathogenic mechanisms has been limited. However, the adoption of molecular techniques and genomic analysis for *P. jirovecii* research has contributed to our current understanding of how this pathogen interacts with host tissues. This microbe resides nearly exclusively within the lung alveoli, where the fungus attaches to extracellular matrix molecules of the alveolar epithelium. Specifically, *P. jirovecii* adhesins such as PCINT1 mediate adhesion to host fibronectin (Kottom et al. 2008). Adhesion triggers morphological conversion of the yeastlike cells to cyst form. The cyst wall is composed of β-1,3-glucan and chitin polymers (Thomas and Limper 2007). These cell wall carbohydrates can elicit a strong inflammatory response in the host, especially in the absence of HIV infections, and this inflammatory response is associated with rapid onset of fulminant pneumonia (Carmona et al. 2006; Lebron et al. 2003). Severe PcP is characterized by infiltration of neutrophils and CD8+ T cells, which damage the alveolar epithelium and impair gas exchange, which can lead to respiratory failure and death (Limper et al. 1989; Thomas and Limper 2007).

11.4.4 Cryptococcus neoformans

Cryptococcus neoformans is a pathogenic yeast that causes the life-threatening disease cryptococcosis in immunocompromised patients, but little to no disease in healthy individuals. Infection results from inhalation of dried yeast cells or spores from environmental sources (Velagapudi et al. 2009; Sabiiti and May 2012). Individuals at greatest risk for cryptococcal infection are those living in areas with concentrated avian populations, such as poultry farms or in urban areas with resident pigeon flocks, as *C. neoformans* is a frequent colonizer of the avian intestinal tract (Gonzalez-Hein et al. 2010; Levitz 1991).

Cryptococcus neoformans has been associated with infections of the skin, bone, urinary tract, spleen, lymph nodes, lungs, and CNS (Severo et al. 2009; Hernandez 1989; Perfect and Casadevall 2002). However, infection most frequently manifests as meningitis, which has a mortality rate of up to 65 % (Park et al. 2009). Approximately, one million cases of cryptococcal meningitis (CM) occur globally each year. The greatest density of cases is in Subsaharan Africa, although important outbreaks of *C. neoformans* and the related species *C. gattii* have occurred in Europe, Britain, and the USA (Park et al. 2009; Brandt et al. 2001; MacDougall et al. 2007). Cryptococcal infections are especially serious in immunocompromised individuals and are associated with over 60 % mortality within 3 months of infection. Cryptococcosis is most common in patients with HIV-AIDS, who are particularly susceptible to CM (Park et al. 2009). However, incidence of cryptococcosis in non-HIV-infected individuals, such as transplant patients undergoing immunosuppressive therapy, is increasing in the medically developed world (Singh et al. 2008, 2009; Sun et al. 2009).

11.4.4.1 *Cryptococcus neoformans* Pathogenesis

Following inhalation, *C. neoformans* infectious particles transit to the lung alveoli (Velagapudi et al. 2009; Sabiiti and May 2012). The alveolar epithelium is coated with a surfactant, SP-D, which promotes adhesion of cryptococcal cells via the fungal phospholipase (Plb1) (Ganendren et al. 2006). Survival of this fungus can occur both extracellularly or may follow being phagocytized by alveolar macrophages. At this relatively early stage of the encounter, the progress of infection depends upon the immune condition of the host. In immunocompetent individuals, the infection is restricted to the lung and is eventually cleared by the host immune response. However, in immunodeficient hosts, *C. neoformans* can disseminate from the lung to deeper tissues such as the central nervous system (CNS), where infection leads to meningoencephalitis. Survival and spread within the infected host requires that *C. neoformans* to simultaneously avoid destruction by host phagocytes while using them as a niche for systemic dissemination.

Cryptococcus neoformans possesses virulence factors that enable it to resist destruction within the phagocyte. For example, fungal melanin acts as an antioxidant that protects *C. neoformans* against oxidative killing during the phagocyte respiratory burst. Secreted capsular polysaccharides also protect the fungus from oxidative stress and induce damage to the phagosomal membrane (Zaragoza et al. 2008; Tucker and Casadevall 2002). *Cryptococcus neoformans* has elaborate mechanisms for obtaining essential nutrients within the host cell; the fungus uses a copper sensing and transport system (CFU1/CTR4) to acquire copper from the intracellular environment. Copper acquisition is essential for cryptococcal survival and virulence, as it is a key component of capsule polysaccharides and is required for systemic spread to the CNS (Waterman et al. 2007; Chun and Madhani 2010). Like many microbes, *C. neoformans* also requires free iron during infection, which is typically in short supply within the host. However, *C. neoformans* acquires iron from the host by producing its own high affinity iron transporters (the permease/ferroxidase system) as well as the siderophore Sit1 and iron uptake systems (Jung et al. 2009). Together, these virulence traits enable *C. neoformans* to survive intracellularly, which contributes to its use of phagocytes as vehicles for fungal persistence and spread. In fact, some reports point to phagocytes as one potential route for dissemination of the fungus to the CNS during CM (Kim 2008; Charlier et al. 2009).

11.4.5 Aspergillus fumigatus

Aspergillus is an opportunistic mold that causes both allergic and invasive disease, called aspergillosis. The genus *Aspergillus* contains 175 species, but only a few of these species are pathogenic to humans. Of these, *A. fumigatus* is associated with the most frequent and serious forms of aspergillosis. *Aspergillus fumigatus* is ubiquitous in nature and is found in soil, water, food, and decaying vegetation. The mold conidia (spores) are easily aerosolized and therefore the major route of transmission is through the air (Denning 1991). Although exposure to *A. fumigatus* is extremely common, infection almost exclusively occurs in patients who are immunocompromised or who have compromised lung function, such as individuals with HIV-AIDS, hematological malignancies, COPD, organ transplant recipients, or hospitalized patients with extended stays (Meersseman et al. 2007). In the immunocompromised, *A. fumigatus* can cause invasive pulmonary aspergillosis or systemic infections that affect deeper tissues such as the bones (Denning 1998). Aspergillosis is a major concern in hospitals, where airborne fungal spores can proliferate in hospital ventilation systems and in mechanical ventilators used for the treatment of individual patients. Although limitations in clinical diagnostic tools complicate efforts to determine precise rates of incidence of invasive aspergillosis, it is documented that these infections are especially severe in immunocompromised individuals and are associated with a mortality rate of up to 65 % (Richardson and Lass-Florl 2008).

11.4.5.1 *Aspergillus fumigatus* Pathogenesis

Inhalation of airborne conidia results in deposition of the spores in the bronchioles or alveoli. In immunocompetent individuals, the spores are phagocytized and destroyed by alveolar macrophages and neutrophils. In immunocompromised patients, the spores may survive and even germinate in the lung tissue. Adhesion of *A. fumigatus* to the airway epithelium is aided by the presence of sialic acid residues on the conidia (Wasylnka et al. 2001) and by adhesion of the conidia to host extracellular matrix molecules like fibrinogen, laminin, and fibronectin (Annaix et al. 1992; Bouchara et al. 1988; Bromley and Donaldson 1996). *Aspergillus fumigatus* also secretes specific proteases that induce actin cytoskeletal rearrangements, leading to epithelial cell detachment and loss of focal contacts between epithelial cells (Kogan et al. 2004; Robinson et al. 1990; Tomee et al. 1997). Secondary metabolites of *A. fumigatus*, such as gliotoxin, fumagilin, helvolic acid, and verruculogen, also modify transepithelial resistance and polarization in human respiratory epithelial cells (Amitani et al. 1995a, b; Botterel et al. 2002; Cody et al. 1997; Khoufache et al. 2007). Together, these secreted fungal products likely promote the invasion of fungal hyphae into lung tissue.

Aspergillus fumigatus evades destruction by host defenses through use of a surface-masking β(1,3)-glucan, which in turn delays macrophage activation and minimizes the early immune response. Its conidia also produce melanin and PksP that offer additional protection against phagocytic intracellular defenses. Melanin minimizes effects of oxidative stress, while PksP delays fusion of phagosomes with lysosomes (Jahn et al. 1997; Langfelder et al. 1998). In addition, *A. fumigatus* evades complement-mediated killing by binding onto two of the host complement regulators, factor H, and plasminogen. Both of these regulators, when bound to the conidial surface, cleave specific complement proteins and prevent activation of the complement cascade (Behnsen et al. 2008).

11.4.5.2 Drug Resistance in *A. fumigatus*

Antifungal azoles have been used for decades to successfully treat *A. fumigatus* infections, but their efficacy is waning as evidenced by the fact that azole-resistant *A. fumigatus* strains constantly seem to emerge across the globe (Lelievre et al. 2013). Azole-resistant invasive aspergillosis infections are associated with greater mortality (up to 88 %) than infections caused by susceptible strains (53 %) (van der Linden et al. 2011; Baddley et al. 2009). The increasing problem of resistance has led to great interest in potential development of novel, nonselective means to treat opportunistic fungal infections.

11.5 Challenges to Be Faced in the Future Study of Opportunistic Pathogens

The definition of a pathogen as opportunistic depends primarily on the state of the host. Many of the microbes described here in this chapter live on healthy humans as commensals or are ubiquitous in the environment, and therefore it would be extremely difficult to eradicate them in order to prevent exposure to immunocompromised or otherwise susceptible individuals. Limiting exposure in the hospital setting has led to some success, for example, by treating *S. aureus* carriers with topical antibiotic ointment to prevent infection of surgical sites (Bode et al. 2010). However, it seems clear that in the future, antibiotic therapy must be administered with caution, as the rapid emergence of antibiotic-resistant opportunistic pathogens is heralding an imminent public health crisis. Molecular epidemiological approaches coupled with current genomic sequencing technologies may allow for more rapid pathogen identification and clarification of the specific mechanisms that promote pathogenesis and antibiotic resistance. A broader approach to developing prophylaxis and treatment of opportunistic pathogens, which takes advantage of our growing knowledge of the healthy microbiota and other biological anti-infectives, will be important to limiting these devastating infections.

References

Agrawal A, Murphy TF (2011) Haemophilus influenzae infections in the H. influenzae type b conjugate vaccine era. J Clin Microbiol 49(11):3728–3732. doi:10.1128/JCM.05476-11

Allerberger F, Wagner M (2010) Listeriosis: a resurgent foodborne infection. Clin Microbiol Infect 16(1):16–23. doi:10.1111/j.1469-0691.2009.03109.x

Almeida RS, Wilson D, Hube B (2009) Candida albicans iron acquisition within the host. FEMS Yeast Res 9(7):1000–1012. doi:10.1111/j.1567-1364.2009.00570.x

Amitani R, Murayama T, Nawada R, Lee WJ, Niimi A, Suzuki K, Tanaka E, Kuze F (1995a) Aspergillus culture filtrates and sputum sols from patients with pulmonary aspergillosis cause damage to human respiratory ciliated epithelium in vitro. Eur Respir J 8(10):1681–1687

Amitani R, Taylor G, Elezis EN, Llewellyn-Jones C, Mitchell J, Kuze F, Cole PJ, Wilson R (1995b) Purification and characterization of factors produced by Aspergillus fumigatus which affect human ciliated respiratory epithelium. Infect Immun 63(9):3266–3271

Anaissie EJ, Penzak SR, Dignani MC (2002) The hospital water supply as a source of nosocomial infections: a plea for action. Arch Intern Med 162(13):1483–1492

Andreani G, Lodge R, Richard D, Tremblay MJ (2012) Mechanisms of interaction between protozoan parasites and HIV. Curr Opin HIV AIDS 7(3):276–282. doi:10.1097/COH.0b013e32835211e9

Angus AA, Lee AA, Augustin DK, Lee EJ, Evans DJ, Fleiszig SM (2008) Pseudomonas aeruginosa induces membrane blebs in epithelial cells, which are utilized as a niche for intracellular replication and motility. Infect Immun 76(5):1992–2001. doi:10.1128/IAI.01221-07

Angus AA, Evans DJ, Barbieri JT, Fleiszig SM (2010) The ADP-ribosylation domain of Pseudomonas aeruginosa ExoS is required for membrane bleb niche formation and bacterial survival within epithelial cells. Infect Immun 78(11):4500–4510. doi:10.1128/IAI.00417-10

Aniansson G, Alm B, Andersson B, Larsson P, Nylen O, Peterson H, Rigner P, Svanborg M, Svanborg C (1992) Nasopharyngeal colonization during the first year of life. J Infect Dis 165 (Suppl 1):S38–S42

Annaix V, Bouchara JP, Larcher G, Chabasse D, Tronchin G (1992) Specific binding of human fibrinogen fragment D to Aspergillus fumigatus conidia. Infect Immun 60(5):1747–1755

Anstey NM, Currie BJ, Hassell M, Palmer D, Dwyer B, Seifert H (2002) Community-acquired bacteremic Acinetobacter pneumonia in tropical Australia is caused by diverse strains of Acinetobacter baumannii, with carriage in the throat in at-risk groups. J Clin Microbiol 40 (2):685–686

Assaad U, El-Masri I, Porhomayon J, El-Solh AA (2012) Pneumonia immunization in older adults: review of vaccine effectiveness and strategies. Clin Interv Aging 7:453–461. doi:10.2147/CIA.S29675

Avadhanula V, Rodriguez CA, Ulett GC, Bakaletz LO, Adderson EE (2006) Nontypeable Haemophilus influenzae adheres to intercellular adhesion molecule 1 (ICAM-1) on respiratory epithelial cells and upregulates ICAM-1 expression. Infect Immun 74(2):830–838. doi:10.1128/IAI.74.2.830-838.2006

Baddley JW, Marr KA, Andes DR, Walsh TJ, Kauffman CA, Kontoyiannis DP, Ito JI, Balajee SA, Pappas PG, Moser SA (2009) Patterns of susceptibility of Aspergillus isolates recovered from patients enrolled in the Transplant-Associated Infection Surveillance Network. J Clin Microbiol 47(10):3271–3275. doi:10.1128/JCM.00854-09

Ballok AE, O'Toole GA (2013) Pouring salt on a wound: Pseudomonas aeruginosa virulence factors alter Na+ and Cl- flux in the lung. J Bacteriol 195(18):4013–4019. doi:10.1128/JB.00339-13

Barbieri JT, Sun J (2004) Pseudomonas aeruginosa ExoS and ExoT. Rev Physiol Biochem Pharmacol 152:79–92. doi:10.1007/s10254-004-0031-7

Barocchi MA, Ries J, Zogaj X, Hemsley C, Albiger B, Kanth A, Dahlberg S, Fernebro J, Moschioni M, Masignani V, Hultenby K, Taddei AR, Beiter K, Wartha F, von Euler A,

Covacci A, Holden DW, Normark S, Rappuoli R, Henriques-Normark B (2006) A pneumococcal pilus influences virulence and host inflammatory responses. Proc Natl Acad Sci U S A 103(8):2857–2862. doi:10.1073/pnas.0511017103

Barre-Sinoussi F, Chermann JC, Rey F, Nugeyre MT, Chamaret S, Gruest J, Dauguet C, Axler-Blin C, Vezinet-Brun F, Rouzioux C, Rozenbaum W, Montagnier L (1983) Isolation of a T-lymphotropic retrovirus from a patient at risk for acquired immune deficiency syndrome (AIDS). Science 220(4599):868–871

Bartlett JG, Chang TW, Gurwith M, Gorbach SL, Onderdonk AB (1978) Antibiotic-associated pseudomembranous colitis due to toxin-producing clostridia. N Engl J Med 298(10):531–534. doi:10.1056/NEJM197803092981003

Behnsen J, Hartmann A, Schmaler J, Gehrke A, Brakhage AA, Zipfel PF (2008) The opportunistic human pathogenic fungus Aspergillus fumigatus evades the host complement system. Infect Immun 76(2):820–827. doi:10.1128/IAI.01037-07

Berdiev BK, Qadri YJ, Benos DJ (2009) Assessment of the CFTR and ENaC association. Mol Biosyst 5(2):123–127. doi:10.1039/b810471a

Blankenship JR, Mitchell AP (2006) How to build a biofilm: a fungal perspective. Curr Opin Microbiol 9(6):588–594. doi:10.1016/j.mib.2006.10.003

Bleves S, Viarre V, Salacha R, Michel GP, Filloux A, Voulhoux R (2010) Protein secretion systems in Pseudomonas aeruginosa: a wealth of pathogenic weapons. Int J Med Microbiol 300 (8):534–543. doi:10.1016/j.ijmm.2010.08.005

Blot S, Vandewoude K (2004) Early detection of systemic infections. Acta Clin Belg 59(1):20–23

Bode LG, Kluytmans JA, Wertheim HF, Bogaers D, Vandenbroucke-Grauls CM, Roosendaal R, Troelstra A, Box AT, Voss A, van der Tweel I, van Belkum A, Verbrugh HA, Vos MC (2010) Preventing surgical-site infections in nasal carriers of Staphylococcus aureus. N Engl J Med 362(1):9–17. doi:10.1056/NEJMoa0808939

Bomberger JM, Ye S, Maceachran DP, Koeppen K, Barnaby RL, O'Toole GA, Stanton BA (2011) A Pseudomonas aeruginosa toxin that hijacks the host ubiquitin proteolytic system. PLoS Pathog 7(3):e1001325. doi:10.1371/journal.ppat.1001325

Bonnet F, Lewden C, May T, Heripret L, Jougla E, Bevilacqua S, Costagliola D, Salmon D, Chene G, Morlat P, Mortalité 2000 Study Group (2005) Opportunistic infections as causes of death in HIV-infected patients in the HAART era in France. Scand J Infect Dis 37 (6-7):482–487

Bortolussi R (2008) Listeriosis: a primer. CMAJ 179(8):795–797. doi:10.1503/cmaj.081377

Botterel F, Cordonnier C, Barbier V, Wingerstmann L, Liance M, Coste A, Escudier E, Bretagne S (2002) Aspergillus fumigatus causes in vitro electrophysiological and morphological modifications in human nasal epithelial cells. Histol Histopathol 17(4):1095–1101

Bouchara JP, Bouali A, Tronchin G, Robert R, Chabasse D, Senet JM (1988) Binding of fibrinogen to the pathogenic Aspergillus species. J Med Vet Mycol 26(6):327–334

Boucher HW, Corey GR (2008) Epidemiology of methicillin-resistant Staphylococcus aureus. Clin Infect Dis 46(Suppl 5):S344–S349. doi:10.1086/533590

Boucher HW, Talbot GH, Bradley JS, Edwards JE, Gilbert D, Rice LB, Scheld M, Spellberg B, Bartlett J (2009) Bad bugs, no drugs: no ESKAPE! An update from the Infectious Diseases Society of America. Clin Infect Dis 48(1):1–12. doi:10.1086/595011

Bramono K, Yamazaki M, Tsuboi R, Ogawa H (2006) Comparison of proteinase, lipase and alpha-glucosidase activities from the clinical isolates of Candida species. Jpn J Infect Dis 59 (2):73–76

Brandt ME, Pfaller MA, Hajjeh RA, Hamill RJ, Pappas PG, Reingold AL, Rimland D, Warnock DW, Cryptococcal Disease Active Surveillance Group (2001) Trends in antifungal drug susceptibility of Cryptococcus neoformans isolates in the United States: 1992 to 1994 and 1996 to 1998. Antimicrob Agents Chemother 45(11):3065–3069. doi:10.1128/AAC.45.11.3065-3069.2001

Bromley IM, Donaldson K (1996) Binding of Aspergillus fumigatus spores to lung epithelial cells and basement membrane proteins: relevance to the asthmatic lung. Thorax 51(12):1203–1209

Brooke JS (2012) Stenotrophomonas maltophilia: an emerging global opportunistic pathogen. Clin Microbiol Rev 25(1):2–41. doi:10.1128/CMR.00019-11
Brooke JS (2014) New strategies against Stenotrophomonas maltophilia: a serious worldwide intrinsically drug-resistant opportunistic pathogen. Expert Rev Anti Infect Ther 12(1):1–4. doi:10.1586/14787210.2014.864553
Brooks JT, Kaplan JE, Holmes KK, Benson C, Pau A, Masur H (2009) HIV-associated opportunistic infections—going, going, but not gone: the continued need for prevention and treatment guidelines. Clin Infect Dis 48(5):609–611. doi:10.1086/596756
Brundage RA, Smith GA, Camilli A, Theriot JA, Portnoy DA (1993) Expression and phosphorylation of the Listeria monocytogenes ActA protein in mammalian cells. Proc Natl Acad Sci U S A 90(24):11890–11894
Burke KE, Lamont JT (2013) Fecal transplantation for recurrent Clostridium difficile infection in older adults: a review. J Am Geriatr Soc 61(8):1394–1398. doi:10.1111/jgs.12378
Burkholder KM, Bhunia AK (2010) Listeria monocytogenes uses Listeria adhesion protein (LAP) to promote bacterial transepithelial translocation and induces expression of LAP receptor Hsp60. Infect Immun 78(12):5062–5073. doi:10.1128/IAI.00516-10
Butterworth MB, Zhang L, Heidrich EM, Myerburg MM, Thibodeau PH (2012) Activation of the epithelial sodium channel (ENaC) by the alkaline protease from Pseudomonas aeruginosa. J Biol Chem 287(39):32556–32565. doi:10.1074/jbc.M112.369520
Cabanes D, Sousa S, Cebria A, Lecuit M, Garcia-del Portillo F, Cossart P (2005) Gp96 is a receptor for a novel Listeria monocytogenes virulence factor, Vip, a surface protein. EMBO J 24(15):2827–2838. doi:10.1038/sj.emboj.7600750
Canvin JR, Marvin AP, Sivakumaran M, Paton JC, Boulnois GJ, Andrew PW, Mitchell TJ (1995) The role of pneumolysin and autolysin in the pathology of pneumonia and septicemia in mice infected with a type 2 pneumococcus. J Infect Dis 172(1):119–123
Carmona EM, Vassallo R, Vuk-Pavlovic Z, Standing JE, Kottom TJ, Limper AH (2006) Pneumocystis cell wall beta-glucans induce dendritic cell costimulatory molecule expression and inflammatory activation through a Fas-Fas ligand mechanism. J Immunol 177(1):459–467
Casadevall A, Pirofski LA (1999) Host-pathogen interactions: redefining the basic concepts of virulence and pathogenicity. Infect Immun 67(8):3703–3713
Cassone A, Cauda R (2012) Candida and candidiasis in HIV-infected patients: where commensalism, opportunistic behavior and frank pathogenicity lose their borders. AIDS 26 (12):1457–1472. doi:10.1097/QAD.0b013e3283536ba8
Caston-Osorio JJ, Rivero A, Torre-Cisneros J (2008) Epidemiology of invasive fungal infection. Int J Antimicrob Agents 32(Suppl 2):S103–S109. doi:10.1016/S0924-8579(08)70009-8
Centers for Disease Control (1981) Pneumocystis pneumonia—Los Angeles. MMWR Morb Mortal Wkly Rep 30(21):250–252
Centers for Disease Control and Prevention (1997) Update: trends in AIDS incidence, deaths, and prevalence—United States, 1996. MMWR Morb Mortal Wkly Rep 46(8):165–173
Chaari A, Mnif B, Bahloul M, Mahjoubi F, Chtara K, Turki O, Gharbi N, Chelly H, Hammami A, Bouaziz M (2013) Acinetobacter baumannii ventilator-associated pneumonia: epidemiology, clinical characteristics, and prognosis factors. Int J Infect Dis 17(12):e1225–e1228. doi:10.1016/j.ijid.2013.07.014
Chandra J, Kuhn DM, Mukherjee PK, Hoyer LL, McCormick T, Ghannoum MA (2001) Biofilm formation by the fungal pathogen Candida albicans: development, architecture, and drug resistance. J Bacteriol 183(18):5385–5394
Charlier C, Nielsen K, Daou S, Brigitte M, Chretien F, Dromer F (2009) Evidence of a role for monocytes in dissemination and brain invasion by Cryptococcus neoformans. Infect Immun 77 (1):120–127. doi:10.1128/IAI.01065-08
Charnot-Katsikas A, Dorafshar AH, Aycock JK, David MZ, Weber SG, Frank KM (2009) Two cases of necrotizing fasciitis due to Acinetobacter baumannii. J Clin Microbiol 47(1):258–263. doi:10.1128/JCM.01250-08

Chastre J, Fagon JY (2002) Ventilator-associated pneumonia. Am J Respir Crit Care Med 165 (7):867–903. doi:10.1164/ajrccm.165.7.2105078
Chavant P, Gaillard-Martinie B, Hebraud M (2004) Antimicrobial effects of sanitizers against planktonic and sessile Listeria monocytogenes cells according to the growth phase. FEMS Microbiol Lett 236(2):241–248. doi:10.1016/j.femsle.2004.05.040
Cheung GY, Rigby K, Wang R, Queck SY, Braughton KR, Whitney AR, Teintze M, DeLeo FR, Otto M (2010) Staphylococcus epidermidis strategies to avoid killing by human neutrophils. PLoS Pathog 6(10):e1001133. doi:10.1371/journal.ppat.1001133
Chisti MJ, Tebruegge M, La Vincente S, Graham SM, Duke T (2009) Pneumonia in severely malnourished children in developing countries—mortality risk, aetiology and validity of WHO clinical signs: a systematic review. Trop Med Int Health 14(10):1173–1189. doi:10.1111/j.1365-3156.2009.02364.x
Choi CH, Lee JS, Lee YC, Park TI, Lee JC (2008) Acinetobacter baumannii invades epithelial cells and outer membrane protein A mediates interactions with epithelial cells. BMC Microbiol 8:216. doi:10.1186/1471-2180-8-216
Chun CD, Madhani HD (2010) Ctr2 links copper homeostasis to polysaccharide capsule formation and phagocytosis inhibition in the human fungal pathogen Cryptococcus neoformans. PLoS One 5 (9). doi:10.1371/journal.pone.0012503
Cody DT 2nd, McCaffrey TV, Roberts G, Kern EB (1997) Effects of Aspergillus fumigatus and Alternaria alternata on human ciliated epithelium in vitro. Laryngoscope 107(11 Pt 1):1511–1514
Cole JN, Barnett TC, Nizet V, Walker MJ (2011) Molecular insight into invasive group A streptococcal disease. Nat Rev Microbiol 9(10):724–736. doi:10.1038/nrmicro2648
Collini PJ, Kuijper E, Dockrell DH (2013) Clostridium difficile infection in patients with HIV/AIDS. Curr HIV/AIDS Rep 10(3):273–282. doi:10.1007/s11904-013-0162-z
Comolli JC, Waite LL, Mostov KE, Engel JN (1999) Pili binding to asialo-GM1 on epithelial cells can mediate cytotoxicity or bacterial internalization by Pseudomonas aeruginosa. Infect Immun 67(7):3207–3214
Cortes G, Borrell N, de Astorza B, Gomez C, Sauleda J, Alberti S (2002) Molecular analysis of the contribution of the capsular polysaccharide and the lipopolysaccharide O side chain to the virulence of Klebsiella pneumoniae in a murine model of pneumonia. Infect Immun 70 (5):2583–2590
Cosseau C, Devine DA, Dullaghan E, Gardy JL, Chikatamarla A, Gellatly S, Yu LL, Pistolic J, Falsafi R, Tagg J, Hancock RE (2008) The commensal Streptococcus salivarius K12 downregulates the innate immune responses of human epithelial cells and promotes host-microbe homeostasis. Infect Immun 76(9):4163–4175. doi:10.1128/IAI.00188-08
Cox D, Kerrigan SW, Watson SP (2011) Platelets and the innate immune system: mechanisms of bacterial-induced platelet activation. J Thromb Haemost 9(6):1097–1107. doi:10.1111/j.1538-7836.2011.04264.x
Crump JA, Collignon PJ (2000) Intravascular catheter-associated infections. Eur J Clin Microbiol Infect Dis 19(1):1–8
Dagan R (2008) Use of pneumococcal conjugate vaccine to decrease rates of bacterial meningitis. Clin Infect Dis 46(11):1673–1676. doi:10.1086/587898
Dagan R, Klugman KP (2008) Impact of conjugate pneumococcal vaccines on antibiotic resistance. Lancet Infect Dis 8(12):785–795. doi:10.1016/S1473-3099(08)70281-0
de Bentzmann S, Roger P, Puchelle E (1996) Pseudomonas aeruginosa adherence to remodelling respiratory epithelium. Eur Respir J 9(10):2145–2150
Denning DW (1991) Epidemiology and pathogenesis of systemic fungal infections in the immunocompromised host. J Antimicrob Chemother 28(Suppl B):1–16
Denning DW (1998) Invasive aspergillosis. Clin Infect Dis 26(4):781–803, quiz 804–805
Diaz MH, Hauser AR (2010) Pseudomonas aeruginosa cytotoxin ExoU is injected into phagocytic cells during acute pneumonia. Infect Immun 78(4):1447–1456. doi:10.1128/IAI.01134-09

Diekema DJ, Pfaller MA, Schmitz FJ, Smayevsky J, Bell J, Jones RN, Beach M, Group SP (2001) Survey of infections due to Staphylococcus species: frequency of occurrence and antimicrobial susceptibility of isolates collected in the United States, Canada, Latin America, Europe, and the Western Pacific region for the SENTRY Antimicrobial Surveillance Program, 1997-1999. Clin Infect Dis 32(Suppl 2):S114–S132. doi:10.1086/320184

Dijkshoorn L, Nemec A, Seifert H (2007) An increasing threat in hospitals: multidrug-resistant Acinetobacter baumannii. Nat Rev Microbiol 5(12):939–951. doi:10.1038/nrmicro1789

Donlan RM (2002) Biofilms: microbial life on surfaces. Emerg Infect Dis 8(9):881–890. doi:10.3201/eid0809.020063

Donlan RM, Costerton JW (2002) Biofilms: survival mechanisms of clinically relevant microorganisms. Clin Microbiol Rev 15(2):167–193

Dramsi S, Biswas I, Maguin E, Braun L, Mastroeni P, Cossart P (1995) Entry of Listeria monocytogenes into hepatocytes requires expression of inlB, a surface protein of the internalin multigene family. Mol Microbiol 16(2):251–261

Drevets DA, Jelinek TA, Freitag NE (2001) Listeria monocytogenes-infected phagocytes can initiate central nervous system infection in mice. Infect Immun 69(3):1344–1350. doi:10.1128/IAI.69.3.1344-1350.2001

Driscoll JA, Brody SL, Kollef MH (2007) The epidemiology, pathogenesis and treatment of Pseudomonas aeruginosa infections. Drugs 67(3):351–368

Earhart MM (2008) The identification and treatment of toxic megacolon secondary to pseudomembranous colitis. Dimens Crit Care Nurs 27(6):249–254. doi:10.1097/01.DCC.0000338869.70035.2b

Edson RS, Osmon DR, Berry DJ (2002) Septic arthritis due to Streptococcus sanguis. Mayo Clin Proc 77(7):709–710. doi:10.4065/77.7.709

Eiseman B, Silen W, Bascom GS, Kauvar AJ (1958) Fecal enema as an adjunct in the treatment of pseudomembranous enterocolitis. Surgery 44(5):854–859

Fagarasan S, Honjo T (2004) Regulation of IgA synthesis at mucosal surfaces. Curr Opin Immunol 16(3):277–283. doi:10.1016/j.coi.2004.03.005

Falagas ME, Kastoris AC, Vouloumanou EK, Dimopoulos G (2009) Community-acquired Stenotrophomonas maltophilia infections: a systematic review. Eur J Clin Microbiol Infect Dis 28(7):719–730. doi:10.1007/s10096-009-0709-5

Falkow S (1997) What is a pathogen? ASM News 63:359–365

Fan J, Zhang Y, Chuang-Smith ON, Frank KL, Guenther BD, Kern M, Schlievert PM, Herzberg MC (2012) Ecto-5′-nucleotidase: a candidate virulence factor in Streptococcus sanguinis experimental endocarditis. PLoS One 7(6):e38059. doi:10.1371/journal.pone.0038059

Fang CT, Chuang YP, Shun CT, Chang SC, Wang JT (2004) A novel virulence gene in Klebsiella pneumoniae strains causing primary liver abscess and septic metastatic complications. J Exp Med 199(5):697–705. doi:10.1084/jem.20030857

Farber JM, Peterkin PI (1991) Listeria monocytogenes, a food-borne pathogen. Microbiol Rev 55(3):476–511

Filloux A (2011) Protein secretion systems in Pseudomonas aeruginosa: an essay on diversity, evolution, and function. Front Microbiol 2:155. doi:10.3389/fmicb.2011.00155

Fink DL, Green BA, St Geme JW 3rd (2002) The Haemophilus influenzae Hap autotransporter binds to fibronectin, laminin, and collagen IV. Infect Immun 70(9):4902–4907

Fink DL, Buscher AZ, Green B, Fernsten P, St Geme JW 3rd (2003) The Haemophilus influenzae Hap autotransporter mediates microcolony formation and adherence to epithelial cells and extracellular matrix via binding regions in the C-terminal end of the passenger domain. Cell Microbiol 5(3):175–186

Fishman JA (2013) Overview: cytomegalovirus and the herpesviruses in transplantation. Am J Transplant 13(Suppl 3):1–8. doi:10.1111/ajt.12002, quiz 8

Fishman JA (2014) From the classic concepts to modern practice. Clin Microbiol Infect. doi:10.1111/1469-0691.12593

Folsom JP, Frank JF (2006) Chlorine resistance of Listeria monocytogenes biofilms and relationship to subtype, cell density, and planktonic cell chlorine resistance. J Food Prot 69 (6):1292–1296

Forrest DM, Seminari E, Hogg RS, Yip B, Raboud J, Lawson L, Phillips P, Schechter MT, O'Shaughnessy MV, Montaner JS (1998) The incidence and spectrum of AIDS-defining illnesses in persons treated with antiretroviral drugs. Clin Infect Dis 27(6):1379–1385

Foster TJ (2005) Immune evasion by staphylococci. Nat Rev Microbiol 3(12):948–958

Foster TJ (2009) Colonization and infection of the human host by staphylococci: adhesion, survival and immune evasion. Vet Dermatol 20(5–6):456–470. doi:10.1111/j.1365-3164.2009.00825.x

Foster TJ, Hook M (1998) Surface protein adhesins of Staphylococcus aureus. Trends Microbiol 6 (12):484–488

Fournier PE, Vallenet D, Barbe V, Audic S, Ogata H, Poirel L, Richet H, Robert C, Mangenot S, Abergel C, Nordmann P, Weissenbach J, Raoult D, Claverie JM (2006) Comparative genomics of multidrug resistance in Acinetobacter baumannii. PLoS Genet 2(1):e7. doi:10.1371/journal.pgen.0020007

Frank DN, Feazel LM, Bessesen MT, Price CS, Janoff EN, Pace NR (2010) The human nasal microbiota and Staphylococcus aureus carriage. PLoS One 5(5):e10598. doi:10.1371/journal.pone.0010598

Fraunholz M, Sinha B (2012) Intracellular Staphylococcus aureus: live-in and let die. Front Cell Infect Microbiol 2:43

Fujita J, Yamadori I, Xu G, Hojo S, Negayama K, Miyawaki H, Yamaji Y, Takahara J (1996) Clinical features of Stenotrophomonas maltophilia pneumonia in immunocompromised patients. Respir Med 90(1):35–38

Gaddy JA, Tomaras AP, Actis LA (2009) The Acinetobacter baumannii 19606 OmpA protein plays a role in biofilm formation on abiotic surfaces and in the interaction of this pathogen with eukaryotic cells. Infect Immun 77(8):3150–3160. doi:10.1128/IAI.00096-09

Ganendren R, Carter E, Sorrell T, Widmer F, Wright L (2006) Phospholipase B activity enhances adhesion of Cryptococcus neoformans to a human lung epithelial cell line. Microbes Infect 8 (4):1006–1015. doi:10.1016/j.micinf.2005.10.018

Gao XW, Mubasher M, Fang CY, Reifer C, Miller LE (2010) Dose-response efficacy of a proprietary probiotic formula of Lactobacillus acidophilus CL1285 and Lactobacillus casei LBC80R for antibiotic-associated diarrhea and Clostridium difficile-associated diarrhea prophylaxis in adult patients. Am J Gastroenterol 105(7):1636–1641. doi:10.1038/ajg.2010.11

Ge X, Kitten T, Chen Z, Lee SP, Munro CL, Xu P (2008) Identification of Streptococcus sanguinis genes required for biofilm formation and examination of their role in endocarditis virulence. Infect Immun 76(6):2551–2559. doi:10.1128/IAI.00338-08

Gedde MM, Higgins DE, Tilney LG, Portnoy DA (2000) Role of listeriolysin O in cell-to-cell spread of Listeria monocytogenes. Infect Immun 68(2):999–1003

Gigliotti F, Wright TW (2012) Pneumocystis: where does it live? PLoS Pathog 8(11):e1003025. doi:10.1371/journal.ppat.1003025

Gigliotti F, Haidaris PJ, Haidaris CG, Wright TW, Van der Meid KR (1993) Further evidence of host species-specific variation in antigens of Pneumocystis carinii using the polymerase chain reaction. J Infect Dis 168(1):191–194

Gilmour MW, Graham M, Van Domselaar G, Tyler S, Kent H, Trout-Yakel KM, Larios O, Allen V, Lee B, Nadon C (2010) High-throughput genome sequencing of two Listeria monocytogenes clinical isolates during a large foodborne outbreak. BMC Genomics 11:120. doi:10.1186/1471-2164-11-120

Gomes F, Teixeira P, Cerca N, Ceri H, Oliveira R (2011) Virulence gene expression by Staphylococcus epidermidis biofilm cells exposed to antibiotics. Microb Drug Resist 17(2):191–196. doi:10.1089/mdr.2010.0149

Gonzalez-Hein G, Gonzalez-Hein J, Diaz Jarabran MC (2010) Isolation of Cryptococcus neoformans in dry droppings of captive birds in Santiago, Chile. J Avian Med Surg 24 (3):227–236

Govan JR, Deretic V (1996) Microbial pathogenesis in cystic fibrosis: mucoid Pseudomonas aeruginosa and Burkholderia cepacia. Microbiol Rev 60(3):539–574

Gow NA, van de Veerdonk FL, Brown AJ, Netea MG (2012) Candida albicans morphogenesis and host defence: discriminating invasion from colonization. Nat Rev Microbiol 10(2):112–122. doi:10.1038/nrmicro2711

Griffith F (1928) The significance of pneumococcal types. J Hyg (Lond) 27(2):113–159

Gross M, Cramton SE, Gotz F, Peschel A (2001) Key role of teichoic acid net charge in Staphylococcus aureus colonization of artificial surfaces. Infect Immun 69(5):3423–3426. doi:10.1128/IAI.69.5.3423-3426.2001

Guentzel MN (1996) Escherichia, Klebsiella, Enterobacter, Serratia, Citrobacter, and Proteus. In: Baron S (ed) Medical microbiology, 4th edn. University of Texas Medical Branch at Galveston, Galveston, TX

Guerrero DM, Perez F, Conger NG, Solomkin JS, Adams MD, Rather PN, Bonomo RA (2010) Acinetobacter baumannii-associated skin and soft tissue infections: recognizing a broadening spectrum of disease. Surg Infect (Larchmt) 11(1):49–57. doi:10.1089/sur.2009.022

Gurung J, Khyriem AB, Banik A, Lyngdoh WV, Choudhury B, Bhattacharyya P (2013) Association of biofilm production with multidrug resistance among clinical isolates of Acinetobacter baumannii and Pseudomonas aeruginosa from intensive care unit. Indian J Crit Care Med 17 (4):214–218. doi:10.4103/0972-5229.118416

Hallstrom T, Blom AM, Zipfel PF, Riesbeck K (2009) Nontypeable Haemophilus influenzae protein E binds vitronectin and is important for serum resistance. J Immunol 183 (4):2593–2601. doi:10.4049/jimmunol.0803226

Hallstrom T, Resman F, Ristovski M, Riesbeck K (2010) Binding of complement regulators to invasive nontypeable Haemophilus influenzae isolates is not increased compared to nasopharyngeal isolates, but serum resistance is linked to disease severity. J Clin Microbiol 48 (3):921–927. doi:10.1128/JCM.01654-09

Hanke ML, Kielian T (2012) Deciphering mechanisms of staphylococcal biofilm evasion of host immunity. Front Cell Infect Microbiol 2:62. doi:10.3389/fcimb.2012.00062

Hansen CR (2012) Stenotrophomonas maltophilia: to be or not to be a cystic fibrosis pathogen. Curr Opin Pulm Med 18(6):628–631. doi:10.1097/MCP.0b013e328358d4f8

Hernandez AD (1989) Cutaneous cryptococcosis. Dermatol Clin 7(2):269–274

Herold BC, Immergluck LC, Maranan MC, Lauderdale DS, Gaskin RE, Boyle-Vavra S, Leitch CD, Daum RS (1998) Community-acquired methicillin-resistant Staphylococcus aureus in children with no identified predisposing risk. JAMA 279(8):593–598

Hersh AL, Chambers HF, Maselli JH, Gonzales R (2008) National trends in ambulatory visits and antibiotic prescribing for skin and soft-tissue infections. Arch Intern Med 168(14):1585

Hickson M (2011) Probiotics in the prevention of antibiotic-associated diarrhoea and Clostridium difficile infection. Therap Adv Gastroenterol 4(3):185–197. doi:10.1177/1756283X11399115

Hirst RA, Gosai B, Rutman A, Guerin CJ, Nicotera P, Andrew PW, O'Callaghan C (2008) Streptococcus pneumoniae deficient in pneumolysin or autolysin has reduced virulence in meningitis. J Infect Dis 197(5):744–751. doi:10.1086/527322

Hogg RS, Heath KV, Yip B, Craib KJ, O'Shaughnessy MV, Schechter MT, Montaner JS (1998) Improved survival among HIV-infected individuals following initiation of antiretroviral therapy. JAMA 279(6):450–454

Hoiby N, Bjarnsholt T, Givskov M, Molin S, Ciofu O (2010) Antibiotic resistance of bacterial biofilms. Int J Antimicrob Agents 35(4):322–332. doi:10.1016/j.ijantimicag.2009.12.011

Holland LM, Conlon B, O'Gara JP (2011) Mutation of tagO reveals an essential role for wall teichoic acids in Staphylococcus epidermidis biofilm development. Microbiology 157 (Pt 2):408–418. doi:10.1099/mic.0.042234-0

Hope W, Natarajan P, Goodwin L (2013) Invasive fungal infections. Clin Med 13(5):507–510. doi:10.7861/clinmedicine.13-5-507

Howard A, O'Donoghue M, Feeney A, Sleator RD (2012) Acinetobacter baumannii: an emerging opportunistic pathogen. Virulence 3(3):243–250. doi:10.4161/viru.19700

Hritonenko V, Mun JJ, Tam C, Simon NC, Barbieri JT, Evans DJ, Fleiszig SM (2011) Adenylate cyclase activity of Pseudomonas aeruginosa ExoY can mediate bleb-niche formation in epithelial cells and contributes to virulence. Microb Pathog 51(5):305–312. doi:10.1016/j.micpath.2011.08.001

Hugh R, Ryschenkow E (1961) Pseudomonas maltophilia, an alcaligenes-like species. J Gen Microbiol 26:123–132

Hyams C, Camberlein E, Cohen JM, Bax K, Brown JS (2010) The Streptococcus pneumoniae capsule inhibits complement activity and neutrophil phagocytosis by multiple mechanisms. Infect Immun 78(2):704–715. doi:10.1128/IAI.00881-09

Jahn B, Koch A, Schmidt A, Wanner G, Gehringer H, Bhakdi S, Brakhage AA (1997) Isolation and characterization of a pigmentless-conidium mutant of Aspergillus fumigatus with altered conidial surface and reduced virulence. Infect Immun 65(12):5110–5117

Jank T, Giesemann T, Aktories K (2007) Rho-glucosylating Clostridium difficile toxins A and B: new insights into structure and function. Glycobiology 17(4):15R–22R. doi:10.1093/glycob/cwm004

Jayatilake JA, Samaranayake YH, Cheung LK, Samaranayake LP (2006) Quantitative evaluation of tissue invasion by wild type, hyphal and SAP mutants of Candida albicans, and non-albicans Candida species in reconstituted human oral epithelium. J Oral Pathol Med 35(8):484–491. doi:10.1111/j.1600-0714.2006.00435.x

Jit M (2010) The risk of sequelae due to pneumococcal meningitis in high-income countries: a systematic review and meta-analysis. J Infect 61(2):114–124. doi:10.1016/j.jinf.2010.04.008

Joshi GS, Spontak JS, Klapper DG, Richardson AR (2011) Arginine catabolic mobile element encoded speG abrogates the unique hypersensitivity of Staphylococcus aureus to exogenous polyamines. Mol Microbiol 82(1):9–20. doi:10.1111/j.1365-2958.2011.07809.x

Jung WH, Hu G, Kuo W, Kronstad JW (2009) Role of ferroxidases in iron uptake and virulence of Cryptococcus neoformans. Eukaryot Cell 8(10):1511–1520. doi:10.1128/EC.00166-09

Kaplan SL, Mason EO Jr, Wald ER, Schutze GE, Bradley JS, Tan TQ, Hoffman JA, Givner LB, Yogev R, Barson WJ (2004) Decrease of invasive pneumococcal infections in children among 8 children's hospitals in the United States after the introduction of the 7-valent pneumococcal conjugate vaccine. Pediatrics 113(3 Pt 1):443–449

Keeney KM, Stuckey JA, O'Riordan MX (2007) LplA1-dependent utilization of host lipoyl peptides enables Listeria cytosolic growth and virulence. Mol Microbiol 66(3):758–770. doi:10.1111/j.1365-2958.2007.05956.x

Keeney K, Colosi L, Weber W, O'Riordan M (2009) Generation of branched-chain fatty acids through lipoate-dependent metabolism facilitates intracellular growth of Listeria monocytogenes. J Bacteriol 191(7):2187–2196. doi:10.1128/JB.01179-08

Keller JM, Surawicz CM (2014) Clostridium difficile infection in the elderly. Clin Geriatr Med 30 (1):79–93. doi:10.1016/j.cger.2013.10.008

Kerrigan SW, Douglas I, Wray A, Heath J, Byrne MF, Fitzgerald D, Cox D (2002) A role for glycoprotein Ib in Streptococcus sanguis-induced platelet aggregation. Blood 100(2):509–516

Khoufache K, Puel O, Loiseau N, Delaforge M, Rivollet D, Coste A, Cordonnier C, Escudier E, Botterel F, Bretagne S (2007) Verruculogen associated with Aspergillus fumigatus hyphae and conidia modifies the electrophysiological properties of human nasal epithelial cells. BMC Microbiol 7:5. doi:10.1186/1471-2180-7-5

Kilian M, Mestecky J, Kulhavy R, Tomana M, Butler WT (1980) IgA1 proteases from Haemophilus influenzae, Streptococcus pneumoniae, Neisseria meningitidis, and Streptococcus sanguis: comparative immunochemical studies. J Immunol 124(6):2596–2600

Kim KS (2008) Mechanisms of microbial traversal of the blood-brain barrier. Nat Rev Microbiol 6 (8):625–634. doi:10.1038/nrmicro1952

Kim H, Bhunia AK (2013) Secreted Listeria adhesion protein (Lap) influences Lap-mediated Listeria monocytogenes paracellular translocation through epithelial barrier. Gut Pathog 5 (1):16. doi:10.1186/1757-4749-5-16

Kim J, Sudbery P (2011) Candida albicans, a major human fungal pathogen. J Microbiol 49 (2):171–177. doi:10.1007/s12275-011-1064-7

Kim SW, Choi CH, Moon DC, Jin JS, Lee JH, Shin JH, Kim JM, Lee YC, Seol SY, Cho DT, Lee JC (2009) Serum resistance of Acinetobacter baumannii through the binding of factor H to outer membrane proteins. FEMS Microbiol Lett 301(2):224–231. doi:10.1111/j.1574-6968.2009.01820.x

Kitchel B, Rasheed JK, Patel JB, Srinivasan A, Navon-Venezia S, Carmeli Y, Brolund A, Giske CG (2009) Molecular epidemiology of KPC-producing Klebsiella pneumoniae isolates in the United States: clonal expansion of multilocus sequence type 258. Antimicrob Agents Chemother 53(8):3365–3370. doi:10.1128/AAC.00126-09

Klevens RM, Morrison MA, Nadle J, Petit S, Gershman K, Ray S, Harrison LH, Lynfield R, Dumyati G, Townes JM (2007) Invasive methicillin-resistant Staphylococcus aureus infections in the United States. JAMA 298(15):1763–1771

Kobayashi SD, Malachowa N, Whitney AR, Braughton KR, Gardner DJ, Long D, Bubeck Wardenburg J, Schneewind O, Otto M, Deleo FR (2011) Comparative analysis of USA300 virulence determinants in a rabbit model of skin and soft tissue infection. J Infect Dis 204 (6):937–941. doi:10.1093/infdis/jir441

Koenigsknecht MJ, Young VB (2013) Faecal microbiota transplantation for the treatment of recurrent Clostridium difficile infection: current promise and future needs. Curr Opin Gastroenterol 29(6):628–632. doi:10.1097/MOG.0b013e328365d326

Kogan TV, Jadoun J, Mittelman L, Hirschberg K, Osherov N (2004) Involvement of secreted Aspergillus fumigatus proteases in disruption of the actin fiber cytoskeleton and loss of focal adhesion sites in infected A549 lung pneumocytes. J Infect Dis 189(11):1965–1973. doi:10.1086/420850

Kollef MH, Shorr A, Tabak YP, Gupta V, Liu LZ, Johannes RS (2005) Epidemiology and outcomes of health-care-associated pneumonia: results from a large US database of culture-positive pneumonia. Chest 128(6):3854–3862. doi:10.1378/chest.128.6.3854

Kontoyiannis DP, Marr KA, Park BJ, Alexander BD, Anaissie EJ, Walsh TJ, Ito J, Andes DR, Baddley JW, Brown JM, Brumble LM, Freifeld AG, Hadley S, Herwaldt LA, Kauffman CA, Knapp K, Lyon GM, Morrison VA, Papanicolaou G, Patterson TF, Perl TM, Schuster MG, Walker R, Wannemuehler KA, Wingard JR, Chiller TM, Pappas PG (2010) Prospective surveillance for invasive fungal infections in hematopoietic stem cell transplant recipients, 2001-2006: overview of the Transplant-Associated Infection Surveillance Network (TRANSNET) Database. Clin Infect Dis 50(8):1091–1100. doi:10.1086/651263

Korzeniewski K, Bochniak A (2011) Medical support of military operations in Iraq and Afghanistan. Int Marit Health 62(1):71–76

Koss K, Clark MA, Sanders DS, Morton D, Keighley MR, Goh J (2006) The outcome of surgery in fulminant Clostridium difficile colitis. Colorectal Dis 8(2):149–154. doi:10.1111/j.1463-1318.2005.00876.x

Kottom TJ, Kennedy CC, Limper AH (2008) Pneumocystis PCINT1, a molecule with integrin-like features that mediates organism adhesion to fibronectin. Mol Microbiol 67(4):747–761. doi:10.1111/j.1365-2958.2007.06093.x

Kreisel KM, Stine OC, Johnson JK, Perencevich EN, Shardell MD, Lesse AJ, Gordin FM, Climo MW, Roghmann MC (2011) USA300 methicillin-resistant Staphylococcus aureus bacteremia and the risk of severe sepsis: is USA300 methicillin-resistant Staphylococcus aureus associated with more severe infections? Diagn Microbiol Infect Dis 70(3):285–290. doi:10.1016/j.diagmicrobio.2011.03.010

Laksanalamai P, Joseph LA, Silk BJ, Burall LS, L Tarr C, Gerner-Smidt P, Datta AR (2012) Genomic characterization of Listeria monocytogenes strains involved in a multistate listeriosis

outbreak associated with cantaloupe in US. PLoS One 7(7):e42448. doi:10.1371/journal.pone.0042448
Landman D, Bratu S, Kochar S, Panwar M, Trehan M, Doymaz M, Quale J (2007) Evolution of antimicrobial resistance among Pseudomonas aeruginosa, Acinetobacter baumannii and Klebsiella pneumoniae in Brooklyn, NY. J Antimicrob Chemother 60(1):78–82. doi:10.1093/jac/dkm129
Langfelder K, Jahn B, Gehringer H, Schmidt A, Wanner G, Brakhage AA (1998) Identification of a polyketide synthase gene (pksP) of Aspergillus fumigatus involved in conidial pigment biosynthesis and virulence. Med Microbiol Immunol 187(2):79–89
Lebron F, Vassallo R, Puri V, Limper AH (2003) Pneumocystis carinii cell wall beta-glucans initiate macrophage inflammatory responses through NF-kappaB activation. J Biol Chem 278 (27):25001–25008. doi:10.1074/jbc.M301426200
Lee HC, Chuang YC, Yu WL, Lee NY, Chang CM, Ko NY, Wang LR, Ko WC (2006) Clinical implications of hypermucoviscosity phenotype in Klebsiella pneumoniae isolates: association with invasive syndrome in patients with community-acquired bacteraemia. J Intern Med 259 (6):606–614. doi:10.1111/j.1365-2796.2006.01641.x
Leid JG, Willson CJ, Shirtliff ME, Hassett DJ, Parsek MR, Jeffers AK (2005) The exopolysaccharide alginate protects Pseudomonas aeruginosa biofilm bacteria from IFN-gamma-mediated macrophage killing. J Immunol 175(11):7512–7518
Lelievre L, Groh M, Angebault C, Maherault AC, Didier E, Bougnoux ME (2013) Azole resistant Aspergillus fumigatus: an emerging problem. Med Mal Infect 43(4):139–145. doi:10.1016/j.medmal.2013.02.010
Lessa FC, Gould CV, McDonald LC (2012) Current status of Clostridium difficile infection epidemiology. Clin Infect Dis 55(Suppl 2):S65–S70. doi:10.1093/cid/cis319
Leung WS, Chu CM, Tsang KY, Lo FH, Lo KF, Ho PL (2006) Fulminant community-acquired Acinetobacter baumannii pneumonia as a distinct clinical syndrome. Chest 129(1):102–109. doi:10.1378/chest.129.1.102
Levitz SM (1991) The ecology of Cryptococcus neoformans and the epidemiology of cryptococcosis. Rev Infect Dis 13(6):1163–1169
Li J, Nation RL, Milne RW, Turnidge JD, Coulthard K (2005) Evaluation of colistin as an agent against multi-resistant Gram-negative bacteria. Int J Antimicrob Agents 25(1):11–25. doi:10.1016/j.ijantimicag.2004.10.001
Li M, Cheung GY, Hu J, Wang D, Joo HS, Deleo FR, Otto M (2010) Comparative analysis of virulence and toxin expression of global community-associated methicillin-resistant Staphylococcus aureus strains. J Infect Dis 202(12):1866–1876. doi:10.1086/657419
Li W, Sun G, Yu Y, Li N, Chen M, Jin R, Jiao Y, Wu H (2014) Increasing occurrence of antimicrobial-resistant hypervirulent (hypermucoviscous) Klebsiella pneumoniae isolates in China. Clin Infect Dis 58(2):225–232. doi:10.1093/cid/cit675
Limper AH, Offord KP, Smith TF, Martin WJ 2nd (1989) Pneumocystis carinii pneumonia. Differences in lung parasite number and inflammation in patients with and without AIDS. Am Rev Respir Dis 140(5):1204–1209. doi:10.1164/ajrccm/140.5.1204
Lindahl G, Stalhammar-Carlemalm M, Areschoug T (2005) Surface proteins of Streptococcus agalactiae and related proteins in other bacterial pathogens. Clin Microbiol Rev 18 (1):102–127. doi:10.1128/CMR.18.1.102-127.2005
Linden SK, Bierne H, Sabet C, Png CW, Florin TH, McGuckin MA, Cossart P (2008) Listeria monocytogenes internalins bind to the human intestinal mucin MUC2. Arch Microbiol 190 (1):101–104. doi:10.1007/s00203-008-0358-6
Lindsay D, von Holy A (2006) Bacterial biofilms within the clinical setting: what healthcare professionals should know. J Hosp Infect 64(4):313–325. doi:10.1016/j.jhin.2006.06.028
Lockhart SR, Abramson MA, Beekmann SE, Gallagher G, Riedel S, Diekema DJ, Quinn JP, Doern GV (2007) Antimicrobial resistance among Gram-negative bacilli causing infections in intensive care unit patients in the United States between 1993 and 2004. J Clin Microbiol 45 (10):3352–3359. doi:10.1128/JCM.01284-07

Loffler B, Hussain M, Grundmeier M, Bruck M, Holzinger D, Varga G, Roth J, Kahl BC, Proctor RA, Peters G (2010) Staphylococcus aureus panton-valentine leukocidin is a very potent cytotoxic factor for human neutrophils. PLoS Pathog 6(1):e1000715. doi:10.1371/journal.ppat.1000715

Loffler B, Tuchscherr L, Niemann S, Peters G (2013) Staphylococcus aureus persistence in non-professional phagocytes. Int J Med Microbiol. doi:10.1016/j.ijmm.2013.11.011

Looney WJ, Narita M, Muhlemann K (2009) Stenotrophomonas maltophilia: an emerging opportunist human pathogen. Lancet Infect Dis 9(5):312–323. doi:10.1016/S1473-3099(09)70083-0

Lopez DJ, Collado MI, Ibarguren M, Vasil AI, Vasil ML, Goni FM, Alonso A (2011) Multiple phospholipid substrates of phospholipase C/sphingomyelinase HR2 from Pseudomonas aeruginosa. Chem Phys Lipids 164(1):78–82. doi:10.1016/j.chemphyslip.2010.11.001

Luo G, Samaranayake LP, Yau JY (2001) Candida species exhibit differential in vitro hemolytic activities. J Clin Microbiol 39(8):2971–2974. doi:10.1128/JCM.39.8.2971-2974.2001

Lynch JP 3rd, Zhanel GG (2010) Streptococcus pneumoniae: epidemiology and risk factors, evolution of antimicrobial resistance, and impact of vaccines. Curr Opin Pulm Med 16 (3):217–225. doi:10.1097/MCP.0b013e3283385653

MacDougall L, Kidd SE, Galanis E, Mak S, Leslie MJ, Cieslak PR, Kronstad JW, Morshed MG, Bartlett KH (2007) Spread of Cryptococcus gattii in British Columbia, Canada, and detection in the Pacific Northwest, USA. Emerg Infect Dis 13(1):42–50. doi:10.3201/eid1301.060827

Mack D, Fischer W, Krokotsch A, Leopold K, Hartmann R, Egge H, Laufs R (1996) The intercellular adhesin involved in biofilm accumulation of Staphylococcus epidermidis is a linear beta-1,6-linked glucosaminoglycan: purification and structural analysis. J Bacteriol 178 (1):175–183

Magee AD, Yother J (2001) Requirement for capsule in colonization by Streptococcus pneumoniae. Infect Immun 69(6):3755–3761. doi:10.1128/IAI.69.6.3755-3761.2001

Mandac I, Prkacin I, Matovinovic MS, Sustercic D (2010) Septic arthritis due to Streptococcus sanguis. Coll Antropol 34(2):661–664

Mane A, Gaikwad S, Bembalkar S, Risbud A (2012) Increased expression of virulence attributes in oral Candida albicans isolates from human immunodeficiency virus-positive individuals. J Med Microbiol 61(Pt 2):285–290. doi:10.1099/jmm.0.036269-0

Marks LR, Davidson BA, Knight PR, Hakansson AP (2013) Interkingdom signaling induces Streptococcus pneumoniae biofilm dispersion and transition from asymptomatic colonization to disease. MBio 4(4). doi:10.1128/mBio.00438-13

Martner A, Dahlgren C, Paton JC, Wold AE (2008) Pneumolysin released during Streptococcus pneumoniae autolysis is a potent activator of intracellular oxygen radical production in neutrophils. Infect Immun 76(9):4079–4087. doi:10.1128/IAI.01747-07

Mattila E, Uusitalo-Seppala R, Wuorela M, Lehtola L, Nurmi H, Ristikankare M, Moilanen V, Salminen K, Seppala M, Mattila PS, Anttila VJ, Arkkila P (2012) Fecal transplantation, through colonoscopy, is effective therapy for recurrent Clostridium difficile infection. Gastroenterology 142(3):490–496. doi:10.1053/j.gastro.2011.11.037

McCaskill ML, Mason EO Jr, Kaplan SL, Hammerman W, Lamberth LB, Hultén KG (2007) Increase of the USA300 clone among community-acquired methicillin-susceptible Staphylococcus aureus causing invasive infections. Pediatr Infect Dis J 26(12):1122–1127

McDougal LK, Steward CD, Killgore GE, Chaitram JM, McAllister SK, Tenover FC (2003) Pulsed-field gel electrophoresis typing of oxacillin-resistant Staphylococcus aureus isolates from the United States: establishing a national database. J Clin Microbiol 41(11):5113–5120

McKenzie CG, Koser U, Lewis LE, Bain JM, Mora-Montes HM, Barker RN, Gow NA, Erwig LP (2010) Contribution of Candida albicans cell wall components to recognition by and escape from murine macrophages. Infect Immun 78(4):1650–1658. doi:10.1128/IAI.00001-10

Meersseman W, Lagrou K, Maertens J, Van Wijngaerden E (2007) Invasive aspergillosis in the intensive care unit. Clin Infect Dis 45(2):205–216. doi:10.1086/518852

Melin P (2011) Neonatal group B streptococcal disease: from pathogenesis to preventive strategies. Clin Microbiol Infect 17(9):1294–1303. doi:10.1111/j.1469-0691.2011.03576.x

Mengaud J, Ohayon H, Gounon P, Mege RM, Cossart P (1996) E-cadherin is the receptor for internalin, a surface protein required for entry of L. monocytogenes into epithelial cells. Cell 84 (6):923–932
Merson MH, O'Malley J, Serwadda D, Apisuk C (2008) The history and challenge of HIV prevention. Lancet 372(9637):475–488. doi:10.1016/S0140-6736(08)60884-3
Miller RF, Huang L, Walzer PD (2013) Pneumocystis pneumonia associated with human immunodeficiency virus. Clin Chest Med 34(2):229–241. doi:10.1016/j.ccm.2013.02.001
Milohanic E, Jonquieres R, Cossart P, Berche P, Gaillard JL (2001) The autolysin Ami contributes to the adhesion of Listeria monocytogenes to eukaryotic cells via its cell wall anchor. Mol Microbiol 39(5):1212–1224
Minino AM, Murphy SL, Xu J, Kochanek KD (2011) Deaths: final data for 2008. Natl Vital Stat Rep 59(10):1–126
Mishra NN, Prasad T, Sharma N, Payasi A, Prasad R, Gupta DK, Singh R (2007) Pathogenicity and drug resistance in Candida albicans and other yeast species. A review. Acta Microbiol Immunol Hung 54(3):201–235. doi:10.1556/AMicr.54.2007.3.1
Mistry RD (2013) Skin and soft tissue infections. Pediatr Clin North Am 60(5):1063–1082. doi:10.1016/j.pcl.2013.06.011
Mitchell TJ (2003) The pathogenesis of streptococcal infections: from tooth decay to meningitis. Nat Rev Microbiol 1(3):219–230. doi:10.1038/nrmicro771
Moellering RC Jr (2010) NDM-1—a cause for worldwide concern. N Engl J Med 363 (25):2377–2379. doi:10.1056/NEJMp1011715
Moet GJ, Jones RN, Biedenbach DJ, Stilwell MG, Fritsche TR (2007) Contemporary causes of skin and soft tissue infections in North America, Latin America, and Europe: report from the SENTRY Antimicrobial Surveillance Program (1998-2004). Diagn Microbiol Infect Dis 57 (1):7–13. doi:10.1016/j.diagmicrobio.2006.05.009
Moran GJ, Krishnadasan A, Gorwitz RJ, Fosheim GE, McDougal LK, Carey RB, Talan DA (2006) Methicillin-resistant S. aureus infections among patients in the emergency department. N Engl J Med 355(7):666–674
Moreillon P, Que YA (2004) Infective endocarditis. Lancet 363(9403):139–149. doi:10.1016/S0140-6736(03)15266-X
Mori S, Sugimoto M (2012) Pneumocystis jirovecii infection: an emerging threat to patients with rheumatoid arthritis. Rheumatology (Oxford) 51(12):2120–2130. doi:10.1093/rheumatology/kes244
Moxon ER, Kroll JS (1988) Type b capsular polysaccharide as a virulence factor of Haemophilus influenzae. Vaccine 6(2):113–115
Mukundan D, Ecevit Z, Patel M, Marrs CF, Gilsdorf JR (2007) Pharyngeal colonization dynamics of Haemophilus influenzae and Haemophilus haemolyticus in healthy adult carriers. J Clin Microbiol 45(10):3207–3217. doi:10.1128/JCM.00492-07
Mullaney DM (2001) Group B streptococcal infections in newborns. J Obstet Gynecol Neonatal Nurs 30(6):649–658
Murphy CN, Clegg S (2012) Klebsiella pneumoniae and type 3 fimbriae: nosocomial infection, regulation and biofilm formation. Future Microbiol 7(8):991–1002. doi:10.2217/fmb.12.74
Murphy TF, Lesse AJ, Kirkham C, Zhong H, Sethi S, Munson RS Jr (2011) A clonal group of nontypeable Haemophilus influenzae with two IgA proteases is adapted to infection in chronic obstructive pulmonary disease. PLoS One 6(10):e25923. doi:10.1371/journal.pone.0025923
Murphy CN, Mortensen MS, Krogfelt KA, Clegg S (2013) Role of Klebsiella pneumoniae type 1 and type 3 fimbriae in colonizing silicone tubes implanted into the bladders of mice as a model of catheter-associated urinary tract infections. Infect Immun 81(8):3009–3017. doi:10.1128/IAI.00348-13
Myint TT, Madhava H, Balmer P, Christopoulou D, Attal S, Menegas D, Sprenger R, Bonnet E (2013) The impact of 7-valent pneumococcal conjugate vaccine on invasive pneumococcal disease: a literature review. Adv Ther 30(2):127–151. doi:10.1007/s12325-013-0007-6

Mylonakis E, Calderwood SB (2001) Infective endocarditis in adults. N Engl J Med 345 (18):1318–1330. doi:10.1056/NEJMra010082

Naas T, Cuzon G, Villegas MV, Lartigue MF, Quinn JP, Nordmann P (2008) Genetic structures at the origin of acquisition of the beta-lactamase bla KPC gene. Antimicrob Agents Chemother 52 (4):1257–1263. doi:10.1128/AAC.01451-07

Naglik JR, Challacombe SJ, Hube B (2003) Candida albicans secreted aspartyl proteinases in virulence and pathogenesis. Microbiol Mol Biol Rev 67(3):400–428, table of contents

Nandyal RR (2008) Update on group B streptococcal infections: perinatal and neonatal periods. J Perinat Neonatal Nurs 22(3):230–237. doi:10.1097/01.JPN.0000333925.30328.fd

National Nosocomial Infections Surveillance System (2004) National Nosocomial Infections Surveillance (NNIS) System Report, data summary from January 1992 through June 2004, issued October 2004. Am J Infect Control 32(8):470–485. doi:10.1016/S0196655304005425

Navarre WW, Schneewind O (1999) Surface proteins of gram-positive bacteria and mechanisms of their targeting to the cell wall envelope. Microbiol Mol Biol Rev 63(1):174–229

Neonakis IK, Spandidos DA, Petinaki E (2011) Confronting multidrug-resistant Acinetobacter baumannii: a review. Int J Antimicrob Agents 37(2):102–109. doi:10.1016/j.ijantimicag.2010.10.014

Nett JE, Crawford K, Marchillo K, Andes DR (2010a) Role of Fks1p and matrix glucan in Candida albicans biofilm resistance to an echinocandin, pyrimidine, and polyene. Antimicrob Agents Chemother 54(8):3505–3508. doi:10.1128/AAC.00227-10

Nett JE, Sanchez H, Cain MT, Andes DR (2010b) Genetic basis of Candida biofilm resistance due to drug-sequestering matrix glucan. J Infect Dis 202(1):171–175. doi:10.1086/651200

Nicodemo AC, Paez JI (2007) Antimicrobial therapy for Stenotrophomonas maltophilia infections. Eur J Clin Microbiol Infect Dis 26(4):229–237. doi:10.1007/s10096-007-0279-3

Nobile CJ, Fox EP, Nett JE, Sorrells TR, Mitrovich QM, Hernday AD, Tuch BB, Andes DR, Johnson AD (2012) A recently evolved transcriptional network controls biofilm development in Candida albicans. Cell 148(1–2):126–138. doi:10.1016/j.cell.2011.10.048

Nseir S, Di Pompeo C, Brisson H, Dewavrin F, Tissier S, Diarra M, Boulo M, Durocher A (2006) Intensive care unit-acquired Stenotrophomonas maltophilia: incidence, risk factors, and outcome. Crit Care 10(5):R143. doi:10.1186/cc5063

O'Gara JP, Humphreys H (2001) Staphylococcus epidermidis biofilms: importance and implications. J Med Microbiol 50(7):582–587

O'Riordan M, Moors MA, Portnoy DA (2003) Listeria intracellular growth and virulence require host-derived lipoic acid. Science 302(5644):462–464. doi:10.1126/science.1088170

Obert C, Sublett J, Kaushal D, Hinojosa E, Barton T, Tuomanen EI, Orihuela CJ (2006) Identification of a candidate Streptococcus pneumoniae core genome and regions of diversity correlated with invasive pneumococcal disease. Infect Immun 74(8):4766–4777. doi:10.1128/IAI.00316-06

Obritsch MD, Fish DN, MacLaren R, Jung R (2005) Nosocomial infections due to multidrug-resistant Pseudomonas aeruginosa: epidemiology and treatment options. Pharmacotherapy 25 (10):1353–1364. doi:10.1592/phco.2005.25.10.1353

Osanai A, Li SJ, Asano K, Sashinami H, Hu DL, Nakane A (2013) Fibronectin binding protein, FbpA, is an adhesin responsible for pathogenesis of Listeria monocytogenes infection. Microbiol Immunol. doi:10.1111/j.1348-0421.12030

Otto M (2009) Staphylococcus epidermidis—the 'accidental' pathogen. Nat Rev Microbiol 7 (8):555–567. doi:10.1038/nrmicro2182

Otto M (2012) Molecular basis of Staphylococcus epidermidis infections. Semin Immunopathol 34(2):201–214. doi:10.1007/s00281-011-0296-2

Palella FJ Jr, Baker RK, Moorman AC, Chmiel JS, Wood KC, Brooks JT, Holmberg SD, HIV Outpatient Study Investigators (2006) Mortality in the highly active antiretroviral therapy era: changing causes of death and disease in the HIV outpatient study. J Acquir Immune Defic Syndr 43(1):27–34. doi:10.1097/01.qai.0000233310.90484.16

Pamer EG (2004) Immune responses to Listeria monocytogenes. Nat Rev Immunol 4 (10):812–823. doi:10.1038/nri1461

Pan Y, Breidt F Jr, Kathariou S (2006) Resistance of Listeria monocytogenes biofilms to sanitizing agents in a simulated food processing environment. Appl Environ Microbiol 72 (12):7711–7717. doi:10.1128/AEM.01065-06

Pancotto L, De Angelis G, Bizzarri E, Barocchi MA, Del Giudice G, Moschioni M, Ruggiero P (2013) Expression of the Streptococcus pneumoniae pilus-1 undergoes on and off switching during colonization in mice. Sci Rep 3:2040. doi:10.1038/srep02040

Papaioannides D, Boniatsi L, Korantzopoulos P, Sinapidis D, Giotis C (2006) Acute septic arthritis due to Streptococcus sanguis. Med Princ Pract 15(1):77–79. doi:10.1159/000089391

Pappas PG, Rex JH, Lee J, Hamill RJ, Larsen RA, Powderly W, Kauffman CA, Hyslop N, Mangino JE, Chapman S, Horowitz HW, Edwards JE, Dismukes WE, NIAID Mycoses Study Group (2003) A prospective observational study of candidemia: epidemiology, therapy, and influences on mortality in hospitalized adult and pediatric patients. Clin Infect Dis 37 (5):634–643. doi:10.1086/376906

Park BJ, Wannemuehler KA, Marston BJ, Govender N, Pappas PG, Chiller TM (2009) Estimation of the current global burden of cryptococcal meningitis among persons living with HIV/AIDS. AIDS 23(4):525–530. doi:10.1097/QAD.0b013e328322ffac

Paterson Y, Guirnalda PD, Wood LM (2010) Listeria and Salmonella bacterial vectors of tumor-associated antigens for cancer immunotherapy. Semin Immunol 22(3):183–189. doi:10.1016/j.smim.2010.02.002

Paton JC, Ferrante A (1983) Inhibition of human polymorphonuclear leukocyte respiratory burst, bactericidal activity, and migration by pneumolysin. Infect Immun 41(3):1212–1216

Pedersen SS, Kharazmi A, Espersen F, Hoiby N (1990) Pseudomonas aeruginosa alginate in cystic fibrosis sputum and the inflammatory response. Infect Immun 58(10):3363–3368

Peeters A, Waer M, Michielsen P, Verbist L, Carton H (1989) Listeria monocytogenes meningitis. Clin Neurol Neurosurg 91(1):29–36

Pendleton JN, Gorman SP, Gilmore BF (2013) Clinical relevance of the ESKAPE pathogens. Expert Rev Anti Infect Ther 11(3):297–308. doi:10.1586/eri.13.12

Perfect JR, Casadevall A (2002) Cryptococcosis. Infect Dis Clin North Am 16(4):837–874, v–vi

Pfaller MA, Diekema DJ (2007) Epidemiology of invasive candidiasis: a persistent public health problem. Clin Microbiol Rev 20(1):133–163. doi:10.1128/CMR.00029-06

Pfaller MA, Diekema DJ, Procop GW, Rinaldi MG (2007) Multicenter comparison of the VITEK 2 antifungal susceptibility test with the CLSI broth microdilution reference method for testing amphotericin B, flucytosine, and voriconazole against Candida spp. J Clin Microbiol 45 (11):3522–3528. doi:10.1128/JCM.00403-07

Pfaller M, Neofytos D, Diekema D, Azie N, Meier-Kriesche HU, Quan SP, Horn D (2012) Epidemiology and outcomes of candidemia in 3648 patients: data from the Prospective Antifungal Therapy (PATH Alliance(R)) registry, 2004-2008. Diagn Microbiol Infect Dis 74 (4):323–331. doi:10.1016/j.diagmicrobio.2012.10.003

Pietrocola G, Schubert A, Visai L, Torti M, Fitzgerald JR, Foster TJ, Reinscheid DJ, Speziale P (2005) FbsA, a fibrinogen-binding protein from Streptococcus agalactiae, mediates platelet aggregation. Blood 105(3):1052–1059. doi:10.1182/blood-2004-06-2149

Planet PJ, Larussa SJ, Dana A, Smith H, Xu A, Ryan C, Uhlemann AC, Boundy S, Goldberg J, Narechania A, Kulkarni R, Ratner AJ, Geoghegan JA, Kolokotronis SO, Prince A (2013) Emergence of the epidemic methicillin-resistant Staphylococcus aureus strain USA300 coincides with horizontal transfer of the arginine catabolic mobile element and speG-mediated adaptations for survival on skin. MBio 4(6). doi:10.1128/mBio.00889-13

Podschun R, Ullmann U (1998) Klebsiella spp. as nosocomial pathogens: epidemiology, taxonomy, typing methods, and pathogenicity factors. Clin Microbiol Rev 11(4):589–603

Pompilio A, Crocetta V, Confalone P, Nicoletti M, Petrucca A, Guarnieri S, Fiscarelli E, Savini V, Piccolomini R, Di Bonaventura G (2010) Adhesion to and biofilm formation on IB3-1 bronchial cells by Stenotrophomonas maltophilia isolates from cystic fibrosis patients. BMC Microbiol 10:102. doi:10.1186/1471-2180-10-102

Popovich KJ, Weinstein RA, Hota B (2008) Are community-associated methicillin-resistant Staphylococcus aureus (MRSA) strains replacing traditional nosocomial MRSA strains? Clin Infect Dis 46(6):787–794. doi:10.1086/528716

Prasad R, Kapoor K (2005) Multidrug resistance in yeast Candida. Int Rev Cytol 242:215–248. doi:10.1016/S0074-7696(04)42005-1

Puhan MA, Van Natta ML, Palella FJ, Addessi A, Meinert C, Ocular Complications of AIDS Research Group (2010) Excess mortality in patients with AIDS in the era of highly active antiretroviral therapy: temporal changes and risk factors. Clin Infect Dis 51(8):947–956. doi:10.1086/656415

Queenan AM, Bush K (2007) Carbapenemases: the versatile beta-lactamases. Clin Microbiol Rev 20(3):440–458. doi:10.1128/CMR.00001-07, table of contents

Ramage G, Vande Walle K, Wickes BL, Lopez-Ribot JL (2001) Standardized method for in vitro antifungal susceptibility testing of Candida albicans biofilms. Antimicrob Agents Chemother 45(9):2475–2479

Ramage G, Saville SP, Thomas DP, Lopez-Ribot JL (2005) Candida biofilms: an update. Eukaryot Cell 4(4):633–638. doi:10.1128/EC.4.4.633-638.2005

Rea MC, Alemayehu D, Ross RP, Hill C (2013) Gut solutions to a gut problem: bacteriocins, probiotics and bacteriophage for control of Clostridium difficile infection. J Med Microbiol 62 (Pt 9):1369–1378. doi:10.1099/jmm.0.058933-0

Ribeiro MA, Miranda AE, Gambale W, Paula CR (2004) Prevalence and exoenzyme secretion by Candida albicans isolates from oral and vaginal mucosas of HIV-infected women. Mycopathologia 157(3):255–261

Richardson M, Lass-Florl C (2008) Changing epidemiology of systemic fungal infections. Clin Microbiol Infect 14(Suppl 4):5–24. doi:10.1111/j.1469-0691.2008.01978.x

Robbins N, Uppuluri P, Nett J, Rajendran R, Ramage G, Lopez-Ribot JL, Andes D, Cowen LE (2011) Hsp90 governs dispersion and drug resistance of fungal biofilms. PLoS Pathog 7(9): e1002257. doi:10.1371/journal.ppat.1002257

Robinson BW, Venaille TJ, Mendis AH, McAleer R (1990) Allergens as proteases: an Aspergillus fumigatus proteinase directly induces human epithelial cell detachment. J Allergy Clin Immunol 86(5):726–731

Roche FM, Downer R, Keane F, Speziale P, Park PW, Foster TJ (2004) The N-terminal A domain of fibronectin-binding proteins A and B promotes adhesion of Staphylococcus aureus to elastin. J Biol Chem 279(37):38433–38440. doi:10.1074/jbc.M402122200

Rogers KL, Fey PD, Rupp ME (2009) Coagulase-negative staphylococcal infections. Infect Dis Clin North Am 23(1):73–98. doi:10.1016/j.idc.2008.10.001

Rotstein C, Evans G, Born A, Grossman R, Light RB, Magder S, McTaggart B, Weiss K, Zhanel GG (2008) Clinical practice guidelines for hospital-acquired pneumonia and ventilator-associated pneumonia in adults. Can J Infect Dis Med Microbiol 19(1):19–53

Rupnik M, Wilcox MH, Gerding DN (2009) Clostridium difficile infection: new developments in epidemiology and pathogenesis. Nat Rev Microbiol 7(7):526–536. doi:10.1038/nrmicro2164

Saa Ibusquiza P, Herrera JJ, Cabo ML (2011) Resistance to benzalkonium chloride, peracetic acid and nisin during formation of mature biofilms by Listeria monocytogenes. Food Microbiol 28 (3):418–425. doi:10.1016/j.fm.2010.09.014

Sabet C, Toledo-Arana A, Personnic N, Lecuit M, Dubrac S, Poupel O, Gouin E, Nahori MA, Cossart P, Bierne H (2008) The Listeria monocytogenes virulence factor InlJ is specifically expressed in vivo and behaves as an adhesin. Infect Immun 76(4):1368–1378. doi:10.1128/IAI.01519-07

Sabiiti W, May RC (2012) Mechanisms of infection by the human fungal pathogen Cryptococcus neoformans. Future Microbiol 7(11):1297–1313. doi:10.2217/fmb.12.102

Safdar A, Rolston KV (2007) Stenotrophomonas maltophilia: changing spectrum of a serious bacterial pathogen in patients with cancer. Clin Infect Dis 45(12):1602–1609. doi:10.1086/522998

Sardi JC, Scorzoni L, Bernardi T, Fusco-Almeida AM, Mendes Giannini MJ (2013) Candida species: current epidemiology, pathogenicity, biofilm formation, natural antifungal products and new therapeutic options. J Med Microbiol 62(Pt 1):10–24. doi:10.1099/jmm.0.045054-0

Saville SP, Lazzell AL, Monteagudo C, Lopez-Ribot JL (2003) Engineered control of cell morphology in vivo reveals distinct roles for yeast and filamentous forms of Candida albicans during infection. Eukaryot Cell 2(5):1053–1060

Scallan E, Hoekstra RM, Angulo FJ, Tauxe RV, Widdowson MA, Roy SL, Jones JL, Griffin PM (2011) Foodborne illness acquired in the United States—major pathogens. Emerg Infect Dis 17 (1):7–15. doi:10.3201/eid1701.091101p1

Scherr TD, Heim CE, Morrison JM, Kielian T (2014) Hiding in plain sight: interplay between staphylococcal biofilms and host immunity. Front Immunol 5:37. doi:10.3389/fimmu.2014.00037

Schild L, Heyken A, de Groot PW, Hiller E, Mock M, de Koster C, Horn U, Rupp S, Hube B (2011) Proteolytic cleavage of covalently linked cell wall proteins by Candida albicans Sap9 and Sap10. Eukaryot Cell 10(1):98–109. doi:10.1128/EC.00210-10

Sebeny PJ, Riddle MS, Petersen K (2008) Acinetobacter baumannii skin and soft-tissue infection associated with war trauma. Clin Infect Dis 47(4):444–449. doi:10.1086/590568

Sepkowitz KA (2002) Opportunistic infections in patients with and patients without Acquired Immunodeficiency Syndrome. Clin Infect Dis 34(8):1098–1107. doi:10.1086/339548

Severo CB, Xavier MO, Gazzoni AF, Severo LC (2009) Cryptococcosis in children. Paediatr Respir Rev 10(4):166–171. doi:10.1016/j.prrv.2009.06.009

Shaw L, Golonka E, Potempa J, Foster SJ (2004) The role and regulation of the extracellular proteases of Staphylococcus aureus. Microbiology 150(Pt 1):217–228

Shivshankar P, Sanchez C, Rose LF, Orihuela CJ (2009) The Streptococcus pneumoniae adhesin PsrP binds to Keratin 10 on lung cells. Mol Microbiol 73(4):663–679. doi:10.1111/j.1365-2958.2009.06796.x

Shon AS, Bajwa RP, Russo TA (2013) Hypervirulent (hypermucoviscous) Klebsiella pneumoniae: a new and dangerous breed. Virulence 4(2):107–118. doi:10.4161/viru.22718

Simoons-Smit AM, Verweij-van Vught AM, MacLaren DM (1986) The role of K antigens as virulence factors in Klebsiella. J Med Microbiol 21(2):133–137

Simpson JA, Smith SE, Dean RT (1989) Scavenging by alginate of free radicals released by macrophages. Free Radic Biol Med 6(4):347–353

Singh N, Alexander BD, Lortholary O, Dromer F, Gupta KL, John GT, del Busto R, Klintmalm GB, Somani J, Lyon GM, Pursell K, Stosor V, Munoz P, Limaye AP, Kalil AC, Pruett TL, Garcia-Diaz J, Humar A, Houston S, House AA, Wray D, Orloff S, Dowdy LA, Fisher RA, Heitman J, Wagener MM, Husain S (2008) Pulmonary cryptococcosis in solid organ transplant recipients: clinical relevance of serum cryptococcal antigen. Clin Infect Dis 46(2):e12–e18. doi:10.1086/524738

Singh N, Forrest G, AST Infectious Diseases Community of Practice (2009) Cryptococcosis in solid organ transplant recipients. Am J Transplant 9(Suppl 4):S192–S198. doi:10.1111/j.1600-6143.2009.02911.x

Smith JJ, Travis SM, Greenberg EP, Welsh MJ (1996) Cystic fibrosis airway epithelia fail to kill bacteria because of abnormal airway surface fluid. Cell 85(2):229–236

Snitkin ES, Zelazny AM, Thomas PJ, Stock F, NISC Comparative Sequencing Program Group, Henderson DK, Palmore TN, Segre JA (2012) Tracking a hospital outbreak of carbapenem-resistant Klebsiella pneumoniae with whole-genome sequencing. Sci Transl Med 4 (148):148ra116. doi:10.1126/scitranslmed.3004129

Sobel JD (1997) Vaginitis. N Engl J Med 337(26):1896–1903. doi:10.1056/NEJM199712253372607

Spellerberg B, Rozdzinski E, Martin S, Weber-Heynemann J, Schnitzler N, Lutticken R, Podbielski A (1999) Lmb, a protein with similarities to the LraI adhesin family, mediates attachment of Streptococcus agalactiae to human laminin. Infect Immun 67(2):871–878
Stabler RA, Dawson LF, Phua LT, Wren BW (2008) Comparative analysis of BI/NAP1/027 hypervirulent strains reveals novel toxin B-encoding gene (tcdB) sequences. J Med Microbiol 57(Pt 6):771–775. doi:10.1099/jmm.0.47743-0
Stewart PS (2002) Mechanisms of antibiotic resistance in bacterial biofilms. Int J Med Microbiol 292(2):107–113. doi:10.1078/1438-4221-00196
Stingu CS, Eschrich K, Rodloff AC, Schaumann R, Jentsch H (2008) Periodontitis is associated with a loss of colonization by Streptococcus sanguinis. J Med Microbiol 57(Pt 4):495–499. doi:10.1099/jmm.0.47649-0
Stinson MW, Safulko K, Levine MJ (1991) Adherence of Porphyromonas (Bacteroides) gingivalis to Streptococcus sanguis in vitro. Infect Immun 59(1):102–108
Struve C, Bojer M, Krogfelt KA (2008) Characterization of Klebsiella pneumoniae type 1 fimbriae by detection of phase variation during colonization and infection and impact on virulence. Infect Immun 76(9):4055–4065. doi:10.1128/IAI.00494-08
Stubbs S, Rupnik M, Gibert M, Brazier J, Duerden B, Popoff M (2000) Production of actin-specific ADP-ribosyltransferase (binary toxin) by strains of Clostridium difficile. FEMS Microbiol Lett 186(2):307–312
Sudbery PE (2011) Growth of Candida albicans hyphae. Nat Rev Microbiol 9(10):737–748. doi:10.1038/nrmicro2636
Sun HY, Wagener MM, Singh N (2009) Cryptococcosis in solid-organ, hematopoietic stem cell, and tissue transplant recipients: evidence-based evolving trends. Clin Infect Dis 48 (11):1566–1576. doi:10.1086/598936
Sun Y, Karmakar M, Taylor PR, Rietsch A, Pearlman E (2012) ExoS and ExoT ADP ribosyltransferase activities mediate Pseudomonas aeruginosa keratitis by promoting neutrophil apoptosis and bacterial survival. J Immunol 188(4):1884–1895. doi:10.4049/jimmunol.1102148
Surewaard BG, de Haas CJ, Vervoort F, Rigby KM, DeLeo FR, Otto M, van Strijp JA, Nijland R (2013) Staphylococcal alpha-phenol soluble modulins contribute to neutrophil lysis after phagocytosis. Cell Microbiol 15(8):1427–1437. doi:10.1111/cmi.12130
Swiatecka-Urban A, Moreau-Marquis S, Maceachran DP, Connolly JP, Stanton CR, Su JR, Barnaby R, O'Toole GA, Stanton BA (2006) Pseudomonas aeruginosa inhibits endocytic recycling of CFTR in polarized human airway epithelial cells. Am J Physiol Cell Physiol 290(3):C862–C872. doi:10.1152/ajpcell.00108.2005
Tablan OC, Anderson LJ, Besser R, Bridges C, Hajjeh R, CDC, Healthcare Infection Control Practices Advisory Committee (2004) Guidelines for preventing health-care—associated pneumonia, 2003: recommendations of CDC and the Healthcare Infection Control Practices Advisory Committee. MMWR Recomm Rep 53(RR-3):1–36
Talan DA, Krishnadasan A, Gorwitz RJ, Fosheim GE, Limbago B, Albrecht V, Moran GJ, EMERGEncy ID Net Study Group (2011) Comparison of Staphylococcus aureus from skin and soft-tissue infections in US emergency department patients, 2004 and 2008. Clin Infect Dis 53(2):144–149. doi:10.1093/cid/cir308
Tan CK, Liaw SJ, Yu CJ, Teng LJ, Hsueh PR (2008) Extensively drug-resistant Stenotrophomonas maltophilia in a tertiary care hospital in Taiwan: microbiologic characteristics, clinical features, and outcomes. Diagn Microbiol Infect Dis 60(2):205–210. doi:10.1016/j.diagmicrobio.2007.09.007
Tasaka S, Tokuda H (2012) Pneumocystis jirovecii pneumonia in non-HIV-infected patients in the era of novel immunosuppressive therapies. J Infect Chemother 18(6):793–806. doi:10.1007/s10156-012-0453-0
Taslim H (2009) Clostridium difficile infection in the elderly. Acta Med Indones 41(3):148–151
Tattevin P, Schwartz BS, Graber CJ, Volinski J, Bhukhen A, Bhukhen A, Mai TT, Vo NH, Dang DN, Phan TH (2012) Concurrent epidemics of skin and soft tissue infection and bloodstream

infection due to community-associated methicillin-resistant Staphylococcus aureus. Clin Infect Dis 55(6):781–788

Thammavongsa V, Missiakas DM, Schneewind O (2013) Staphylococcus aureus degrades neutrophil extracellular traps to promote immune cell death. Science 342(6160):863–866. doi:10.1126/science.1242255

Thelestam M, Chaves-Olarte E (2000) Cytotoxic effects of the Clostridium difficile toxins. Curr Top Microbiol Immunol 250:85–96

Theriot CM, Koenigsknecht MJ, Carlson PE Jr, Hatton GE, Nelson AM, Li B, Huffnagle GB, Li JZ, Young VB (2014) Antibiotic-induced shifts in the mouse gut microbiome and metabolome increase susceptibility to Clostridium difficile infection. Nat Commun 5:3114. doi:10.1038/ncomms4114

Thomas CF Jr, Limper AH (2007) Current insights into the biology and pathogenesis of Pneumocystis pneumonia. Nat Rev Microbiol 5(4):298–308. doi:10.1038/nrmicro1621

Thurlow LR, Joshi GS, Clark JR, Spontak JS, Neely CJ, Maile R, Richardson AR (2013) Functional modularity of the arginine catabolic mobile element contributes to the success of USA300 methicillin-resistant Staphylococcus aureus. Cell Host Microbe 13(1):100–107. doi:10.1016/j.chom.2012.11.012

Thwaites GE, Gant V (2011) Are bloodstream leukocytes Trojan Horses for the metastasis of Staphylococcus aureus? Nat Rev Microbiol 9(3):215–222

Tilney LG, Portnoy DA (1989) Actin filaments and the growth, movement, and spread of the intracellular bacterial parasite, Listeria monocytogenes. J Cell Biol 109(4 Pt 1):1597–1608

Tomee JF, Wierenga AT, Hiemstra PS, Kauffman HK (1997) Proteases from Aspergillus fumigatus induce release of proinflammatory cytokines and cell detachment in airway epithelial cell lines. J Infect Dis 176(1):300–303

Tong MJ (1972) Septic complications of war wounds. JAMA 219(8):1044–1047

Tronchin G, Bouchara JP, Annaix V, Robert R, Senet JM (1991) Fungal cell adhesion molecules in Candida albicans. Eur J Epidemiol 7(1):23–33

Tucker SC, Casadevall A (2002) Replication of Cryptococcus neoformans in macrophages is accompanied by phagosomal permeabilization and accumulation of vesicles containing polysaccharide in the cytoplasm. Proc Natl Acad Sci U S A 99(5):3165–3170. doi:10.1073/pnas.052702799

Uckay I, Pittet D, Vaudaux P, Sax H, Lew D, Waldvogel F (2009) Foreign body infections due to Staphylococcus epidermidis. Ann Med 41(2):109–119. doi:10.1080/07853890802337045

UNAIDS (2012) Global summary of the AIDS epidemic

UNAIDS (2013) UNAIDS report on the global AIDS epidemic 2013

van der Linden JW, Snelders E, Kampinga GA, Rijnders BJ, Mattsson E, Debets-Ossenkopp YJ, Kuijper EJ, Van Tiel FH, Melchers WJ, Verweij PE (2011) Clinical implications of azole resistance in Aspergillus fumigatus, The Netherlands, 2007-2009. Emerg Infect Dis 17(10):1846–1854. doi:10.3201/eid1710.110226

Van de Velde S, Nguyen HA, Van Bambeke F, Tulkens PM, Grellet J, Dubois V, Quentin C, Saux M-C (2008) Contrasting effects of human THP-1 cell differentiation on levofloxacin and moxifloxacin intracellular accumulation and activity against Staphylococcus aureus and Listeria monocytogenes. J Antimicrob Chemother 62(3):518–521

Van Mellaert L, Shahrooei M, Hofmans D, Eldere JV (2012) Immunoprophylaxis and immunotherapy of Staphylococcus epidermidis infections: challenges and prospects. Expert Rev Vaccines 11(3):319–334. doi:10.1586/erv.11.190

Velagapudi R, Hsueh YP, Geunes-Boyer S, Wright JR, Heitman J (2009) Spores as infectious propagules of Cryptococcus neoformans. Infect Immun 77(10):4345–4355. doi:10.1128/IAI.00542-09

Veldkamp KE, van Strijp JA (2009) Innate immune evasion by staphylococci. Adv Exp Med Biol 666:19–31

Vincent JL, Rello J, Marshall J, Silva E, Anzueto A, Martin CD, Moreno R, Lipman J, Gomersall C, Sakr Y, Reinhart K, EPIC II Group of Investigators (2009) International study of the prevalence and outcomes of infection in intensive care units. JAMA 302(21):2323–2329. doi:10.1001/jama.2009.1754

Visvesvara GS, Moura H, Schuster FL (2007) Pathogenic and opportunistic free-living amoebae: Acanthamoeba spp., Balamuthia mandrillaris, Naegleria fowleri, and Sappinia diploidea. FEMS Immunol Med Microbiol 50(1):1–26. doi:10.1111/j.1574-695X.2007.00232.x

von Eiff C, Becker K, Machka K, Stammer H, Peters G (2001) Nasal carriage as a source of Staphylococcus aureus bacteremia. Study Group. N Engl J Med 344(1):11–16. doi:10.1056/NEJM200101043440102

Walker AS, Eyre DW, Wyllie DH, Dingle KE, Harding RM, O'Connor L, Griffiths D, Vaughan A, Finney J, Wilcox MH, Crook DW, Peto TE (2012) Characterisation of Clostridium difficile hospital ward-based transmission using extensive epidemiological data and molecular typing. PLoS Med 9(2):e1001172. doi:10.1371/journal.pmed.1001172

Walsh EJ, O'Brien LM, Liang X, Hook M, Foster TJ (2004) Clumping factor B, a fibrinogen-binding MSCRAMM (microbial surface components recognizing adhesive matrix molecules) adhesin of Staphylococcus aureus, also binds to the tail region of type I cytokeratin 10. J Biol Chem 279(49):50691–50699. doi:10.1074/jbc.M408713200

Wampler JL, Kim KP, Jaradat Z, Bhunia AK (2004) Heat shock protein 60 acts as a receptor for the Listeria adhesion protein in Caco-2 cells. Infect Immun 72(2):931–936

Wann ER, Gurusiddappa S, Hook M (2000) The fibronectin-binding MSCRAMM FnbpA of Staphylococcus aureus is a bifunctional protein that also binds to fibrinogen. J Biol Chem 275(18):13863–13871

Wargo MJ, Gross MJ, Rajamani S, Allard JL, Lundblad LK, Allen GB, Vasil ML, Leclair LW, Hogan DA (2011) Hemolytic phospholipase C inhibition protects lung function during Pseudomonas aeruginosa infection. Am J Respir Crit Care Med 184(3):345–354. doi:10.1164/rccm.201103-0374OC

Warny M, Pepin J, Fang A, Killgore G, Thompson A, Brazier J, Frost E, McDonald LC (2005) Toxin production by an emerging strain of Clostridium difficile associated with outbreaks of severe disease in North America and Europe. Lancet 366(9491):1079–1084. doi:10.1016/S0140-6736(05)67420-X

Warren CA, Guerrant RL (2011) Pathogenic C difficile is here (and everywhere) to stay. Lancet 377(9759):8–9. doi:10.1016/S0140-6736(10)61885-5

Wasylnka JA, Simmer MI, Moore MM (2001) Differences in sialic acid density in pathogenic and non-pathogenic Aspergillus species. Microbiology 147(Pt 4):869–877

Waterman SR, Hacham M, Hu G, Zhu X, Park YD, Shin S, Panepinto J, Valyi-Nagy T, Beam C, Husain S, Singh N, Williamson PR (2007) Role of a CUF1/CTR4 copper regulatory axis in the virulence of Cryptococcus neoformans. J Clin Invest 117(3):794–802. doi:10.1172/JCI30006

Waters V, Atenafu EG, Salazar JG, Lu A, Yau Y, Matukas L, Tullis E, Ratjen F (2012) Chronic Stenotrophomonas maltophilia infection and exacerbation outcomes in cystic fibrosis. J Cyst Fibros 11(1):8–13. doi:10.1016/j.jcf.2011.07.008

Watnick P, Kolter R (2000) Biofilm, city of microbes. J Bacteriol 182(10):2675–2679

Watson DA, Musher DM (1999) A brief history of the pneumococcus in biomedical research. Semin Respir Infect 14(3):198–208

Watt JP, Wolfson LJ, O'Brien KL, Henkle E, Deloria-Knoll M, McCall N, Lee E, Levine OS, Hajjeh R, Mulholland K, Cherian T, Hib and Pneumococcal Global Burden of Disease Study Team (2009) Burden of disease caused by Haemophilus influenzae type b in children younger than 5 years: global estimates. Lancet 374(9693):903–911. doi:10.1016/S0140-6736(09)61203-4

Webb EL, Ekii AO, Pala P (2012) Epidemiology and immunology of helminth-HIV interactions. Curr Opin HIV AIDS 7(3):245–253. doi:10.1097/COH.0b013e32835210cd

Weidenmaier C, Kokai-Kun JF, Kulauzovic E, Kohler T, Thumm G, Stoll H, Gotz F, Peschel A (2008) Differential roles of sortase-anchored surface proteins and wall teichoic acid in Staphylococcus aureus nasal colonization. Int J Med Microbiol 298(5–6):505–513. doi:10.1016/j.ijmm.2007.11.006

Weigel LM, Clewell DB, Gill SR, Clark NC, McDougal LK, Flannagan SE, Kolonay JF, Shetty J, Killgore GE, Tenover FC (2003) Genetic analysis of a high-level vancomycin-resistant isolate of Staphylococcus aureus. Science 302(5650):1569–1571. doi:10.1126/science.1090956

Welch MD, Rosenblatt J, Skoble J, Portnoy DA, Mitchison TJ (1998) Interaction of human Arp2/3 complex and the Listeria monocytogenes ActA protein in actin filament nucleation. Science 281(5373):105–108

Wertheim HF, Walsh E, Choudhurry R, Melles DC, Boelens HA, Miajlovic H, Verbrugh HA, Foster T, van Belkum A (2008) Key role for clumping factor B in Staphylococcus aureus nasal colonization of humans. PLoS Med 5(1):e17. doi:10.1371/journal.pmed.0050017

Wescombe PA, Heng NC, Burton JP, Chilcott CN, Tagg JR (2009) Streptococcal bacteriocins and the case for Streptococcus salivarius as model oral probiotics. Future Microbiol 4(7):819–835. doi:10.2217/fmb.09.61

Williams JC, Lucas BJ, Knee C, Renzetti M, Donahue J (1992) Acute lung injury induced by Pseudomonas aeruginosa elastase in hamsters. Exp Lung Res 18(1):155–171

Xayarath B, Marquis H, Port GC, Freitag NE (2009) Listeria monocytogenes CtaP is a multifunctional cysteine transport-associated protein required for bacterial pathogenesis. Mol Microbiol 74(4):956–973. doi:10.1111/j.1365-2958.2009.06910.x

Yong D, Toleman MA, Giske CG, Cho HS, Sundman K, Lee K, Walsh TR (2009) Characterization of a new metallo-beta-lactamase gene, bla(NDM-1), and a novel erythromycin esterase gene carried on a unique genetic structure in Klebsiella pneumoniae sequence type 14 from India. Antimicrob Agents Chemother 53(12):5046–5054. doi:10.1128/AAC.00774-09

Young VB, Schmidt TM (2004) Antibiotic-associated diarrhea accompanied by large-scale alterations in the composition of the fecal microbiota. J Clin Microbiol 42(3):1203–1206

Zaragoza O, Chrisman CJ, Castelli MV, Frases S, Cuenca-Estrella M, Rodriguez-Tudela JL, Casadevall A (2008) Capsule enlargement in Cryptococcus neoformans confers resistance to oxidative stress suggesting a mechanism for intracellular survival. Cell Microbiol 10 (10):2043–2057. doi:10.1111/j.1462-5822.2008.01186.x

Zgair AK, Chhibber S (2011) Adhesion of Stenotrophomonas maltophilia to mouse tracheal mucus is mediated through flagella. J Med Microbiol 60(Pt 7):1032–1037. doi:10.1099/jmm.0.026377-0

Zhang L, Li XZ, Poole K (2000) Multiple antibiotic resistance in Stenotrophomonas maltophilia: involvement of a multidrug efflux system. Antimicrob Agents Chemother 44(2):287–293

Zhang Q, Choo S, Finn A (2002) Immune responses to novel pneumococcal proteins pneumolysin, PspA, PsaA, and CbpA in adenoidal B cells from children. Infect Immun 70(10):5363–5369

Zhao WH, Hu ZQ (2010) Beta-lactamases identified in clinical isolates of Pseudomonas aeruginosa. Crit Rev Microbiol 36(3):245–258. doi:10.3109/1040841X.2010.481763

MIX
Papier aus verantwortungsvollen Quellen
Paper from responsible sources
FSC® C105338

If you have any concerns about our products,
you can contact us on
ProductSafety@springernature.com

In case Publisher is established outside the EU,
the EU authorized representative is:
Springer Nature Customer Service Center GmbH
Europaplatz 3, 69115 Heidelberg, Germany

Printed by Libri Plureos GmbH
in Hamburg, Germany